U0190551

中 国 手 工 纸 文 库

Library of Chinese Handmade Paper

中国手工纸文库

Library of Chinese Handmade Paper

汤书昆
总主编

浙江卷 · 下卷

Zhejiang Ⅲ

汤书昆
主 编

中国科学技术大学出版社
University of Science and Technology of China Press

图书在版编目（CIP）数据

中国手工纸文库.浙江卷.下卷/汤书昆主编.
—合肥：中国科学技术大学出版社，2021.5
国家出版基金项目
“十三五”国家重点出版物出版规划项目
ISBN 978-7-312-04833-3

Ⅰ.中…　Ⅱ.汤…　Ⅲ.手工纸—介绍—浙江
Ⅳ.TS766

中国版本图书馆CIP数据核字（2020）第205428号

中国手工纸文库

浙江卷·下卷

项目负责　伍传平　项赟飚
责任编辑　蒋劲柏　李芳宇　黄　柯
艺术指导　吕敬人
书籍设计　敬人书籍设计
　　　　　吕　旻+黄晓飞

出版发行　中国科学技术大学出版社
　　　　　地址　安徽省合肥市金寨路96号
　　　　　邮编　230026
印　　刷　北京雅昌艺术印刷有限公司
经　　销　全国新华书店
开　　本　880 mm×1230 mm　1/16
印　　张　29.25
字　　数　915千
版　　次　2021年5月第1版
印　　次　2021年5月第1次印刷
定　　价　1800.00元

总　序

造纸技艺是人类文明的重要成就。正是在这一伟大发明的推动下，我们的社会才得以在一个相当长的历史阶段获得比人类使用口语的表达与交流更便于传承的介质。纸为这个世界创造了五彩缤纷的文化记录，使一代代的后来者能够通过纸介质上绘制的图画与符号、书写的文字与数字，了解历史，学习历代文明积累的知识，从而担负起由传承而创新的文化使命。

中国是手工造纸的发源地。不仅人类文明中最早的造纸技艺发源自中国，而且中华大地上遍布着手工造纸的作坊。中国是全世界手工纸制作技艺提炼精纯与丰富的文明体。可以说，在使用手工技艺完成植物纤维制浆成纸的历史中，中国一直是人类造纸技艺与文化的主要精神家园。下图是中国早期造纸技艺刚刚萌芽阶段实物样本的一件遗存——西汉放马滩古纸。

西汉放马滩古纸残片
纸上绘制的是地图
1986年出土于甘肃省天水市
现藏于甘肃省博物馆

Map drawn on paper from Fangmatan Shoals in the Western Han Dynasty
Unearthed in Tianshui City, Gansu Province in 1986
Kept by Gansu Provincial Museum

Preface

Papermaking technique illuminates human culture by endowing the human race with a more traceable medium than oral tradition. Thanks to cultural heritage preserved in the form of images, symbols, words and figures on paper, human beings have accumulated knowledge of history and culture, and then undertaken the mission of culture transmission and innovation.

Handmade paper originated in China, one of the largest cultural communities enjoying advanced handmade papermaking techniques in abundance. China witnessed the earliest papermaking efforts in human history and embraced papermaking mills all over the country. In the history of handmade paper involving vegetable fiber pulping skills, China has always been the dominant centre. The picture illustrates ancient paper from Fangmatan Shoals in the Western Han Dynasty, which is one of the paper samples in the early period of papermaking techniques unearthed in China.

一

本项目的缘起

从2002年开始，我有较多的机缘前往东邻日本，在文化与学术交流考察的同时，多次在东京的书店街——神田神保町的旧书店里，发现日本学术界整理出版的传统手工制作和纸（日本纸的简称）的研究典籍，先后购得近20种，内容包括日本全国的手工造纸调查研究，县（相当于中国的省）一级的调查分析，更小地域和造纸家族的案例实证研究，以及日、中、韩等东亚国家手工造纸的比较研究等。如：每日新闻社主持编撰的《手漉和纸大鉴》五大本，日本东京每日新闻社昭和四十九年（1974年）五月出版，共印1 000套；久米康生著的《手漉和纸精髓》，日本东京讲谈社昭和五十年（1975年）九月出版，共印1 500本；菅野新一编的《白石纸》，日本东京美术出版社昭和四十年（1965年）十一月出版等。这些出版物多出自几十年前的日本昭和年间（1926~1988年），不仅图文并茂，而且几乎都附有系列的实物纸样，有些还有较为规范的手工纸性能、应用效果对比等技术分析数据。我阅后耳目一新，觉得这种出版物形态既有非常直观的阅读效果，又散发出很强的艺术气息。

1. Origin of the Study

Since 2002, I have been invited to Japan several times for cultural and academic communication. I have taken those opportunities to hunt for books on traditional Japanese handmade paper studies, mainly from old bookstores in Kanda Jinbo-cho, Tokyo. The books I bought cover about 20 different categories, typified by surveys on handmade paper at the national, provincial, or even lower levels, case studies of the papermaking families, as well as comparative studies of East Asian countries like Japan, Korea and China. The books include five volumes of *Tesukiwashi Taikan* (*A Collection of Traditional Handmade Japanese Papers*) compiled and published by Mainichi Shimbun in Tokyo in May 1974, which released 1 000 sets, *The Essence of Japanese Paper* by Kume Yasuo, which published 1 500 copies in September 1975 by Kodansha in Tokyo, Japan, *Shiraishi Paper* by Kanno Shinichi, published by Fine Arts Publishing House in Tokyo in November 1965. The books which were mostly published between 1926 and 1988 among the Showa reigning years, are delicately illustrated with pictures and series of paper samples, some even with data analysis on performance comparison. I was extremely impressed by the intuitive and aesthetic nature of the books.

我几乎立刻想起在中国看到的手工造纸技艺及相关的研究成果，在我们这个世界手工造纸的发源国，似乎尚未看到这种表达丰富且叙述格局如此完整出色的研究成果。对中国辽阔地域上的手工造纸技艺与文化遗存现状，研究界尚较少给予关注。除了若干名纸业态，如安徽省的泾县宣纸、四川省的夹江竹纸、浙江省的富阳竹纸与温州皮纸、云南省的香格里拉东巴纸和河北省的迁安桑皮纸等之外，大多数中国手工造纸的当代研究与传播基本上处于寂寂无闻的状态。

此后，我不断与国内一些从事非物质文化遗产及传统工艺研究的同仁交流，他们一致认为在当代中国工业化、城镇化大规模推进的背景下，如果不能在我们这一代人手中进行手工造纸技艺与文化的整体性记录、整理与传播，传统手工造纸这一中国文明的结晶很可能会在未来的时空中失去系统记忆，那真是一种令人难安的结局。但是，这种愿景宏大的文化工程又该如何着手？我们一时觉得难觅头绪。

《手漉和纸精髓》
附实物纸样的内文页
A page from *The Essence of Japanese Paper* with a sample

《白石纸》
随书的宣传夹页
A folder page from *Shiraishi Paper*

The books reminded me of handmade papermaking techniques and related researches in China, and I felt a great sadness that as the country of origin for handmade paper, China has failed to present such distinguished studies excelling both in presentation and research design, owing to the indifference to both papermaking technique and our cultural heritage. Most handmade papermaking mills remain unknown to academia and the media, but there are some famous paper brands, including Xuan paper in Jingxian County of Anhui Province, bamboo paper in Jiajiang County of Sichuan Province, bamboo paper in Fuyang District and bast paper in Wenzhou City of Zhejiang Province, Dongba paper in Shangri-la County of Yunnan Province, and mulberry paper in Qian'an City of Hebei Province.

Constant discussion with fellow colleagues in the field of intangible cultural heritage and traditional craft studies lead to a consensus that if we fail to record, clarify, and transmit handmade papermaking techniques in this age featured by a prevailing trend of industrialization and urbanization in China, regret at the loss will be irreparable. However, a workable research plan on such a grand cultural project eluded us.

2004年，中国科学技术大学人文与社会科学学院获准建设国家"985工程"的"科技史与科技文明哲学社会科学创新基地"，经基地学术委员会讨论，"中国手工纸研究与性能分析"作为一项建设性工作由基地立项支持，并成立了手工纸分析测试实验室和手工纸研究所。这一特别的机缘促成了我们对中国手工纸研究的正式启动。

2007年，中华人民共和国新闻出版总署的"十一五"国家重点图书出版规划项目开始申报。中国科学技术大学出版社时任社长郝诗仙此前知晓我们正在从事中国手工纸研究工作，于是建议正式形成出版中国手工纸研究系列成果的计划。在这一年中，我们经过国际国内的预调研及内部研讨设计，完成了《中国手工纸文库》的撰写框架设计，以及对中国手工造纸现存业态进行全国范围调查记录的田野工作计划，并将其作为国家"十一五"规划重点图书上报，获立项批准。于是，仿佛在不经意间，一项日后令我们常有难履使命之忧的工程便正式展开了。

2008年1月，《中国手工纸文库》项目组经过精心的准备，派出第一个田野调查组（一行7人）前往云南省的滇西北地区进行田野调查，这是计划中全中国手工造纸田野考察的第一站。按照项目设计，将会有很多批次的调查组走向全中国手工造纸现场，采集能获

In 2004, the Philosophy and Social Sciences Innovation Platform of History of Science and S&T Civilization of USTC was approved and supported by the National 985 Project. The academic committee members of the Platform all agreed to support a new project, "Studies and Performance Analysis of Chinese Handmade Paper". Thus, the Handmade Paper Analyzing and Testing Laboratory, and the Handmade Paper Institute were set up. Hence, the journey of Chinese handmade paper studies officially set off.

In 2007, the General Administration of Press and Publication of the People's Republic of China initiated the program of key books that will be funded by the National 11th Five-Year Plan. The former President of USTC Press, Mr. Hao Shixian, advocated that our handmade paper studies could take the opportunity to work on research designs. We immediately constructed a framework for a series of books, *Library of Chinese Handmade Paper*, and drew up the fieldwork plans aiming to study the current status of handmade paper all over China, through arduous pre-research and discussion. Our project was successfully approved and listed in the 11th Five-Year Plan for National Key Books, and then our promising yet difficult journey began.

The seven members of the *Library of Chinese Handmade Paper* Project embarked on our initial, well-prepared fieldwork journey to the northwest area of Yunnan

取的中国手工造纸的完整技艺与文化信息及实物标本。

2009年，国家出版基金首次评审重点支持的出版项目时，将《中国手工纸文库》列入首批国家重要出版物的资助计划，于是我们的中国手工纸研究设计方案与工作规划发育成为国家层面传统技艺与文化研究所关注及期待的对象。

此后，田野调查、技术分析与撰稿工作坚持不懈地推进，中国科学技术大学出版社新一届领导班子全面调动和组织社内骨干编辑，使《中国手工纸文库》的出版工程得以顺利进行。2017年，《中国手工纸文库》被列为“十三五”国家重点出版物出版规划项目。

二
对项目架构设计的说明

作为纸质媒介出版物的《中国手工纸文库》，将汇集文字记

调查组成员在香格里拉县白地村调查
2008年1月

Researchers visiting Baidi Village of Shangri-la County
January 2008

Province in January 2008. After that, based on our research design, many investigation groups would visit various handmade papermaking mills all over China, aiming to record and collect every possible papermaking technique, cultural information and sample.

In 2009, the National Publishing Fund announced the funded book list gaining its key support. Luckily, *Library of Chinese Handmade Paper* was included. Therefore, the Chinese handmade paper research plan we proposed was promoted to the national level, invariably attracting attention and expectation from the field of traditional crafts and culture studies.

Since then, field investigation, technical analysis and writing of the book have been unremittingly promoted, and the new leadership team of USTC Press has fully mobilized and organized the key editors of the press to guarantee the successful publishing of *Library of Chinese Handmade Paper*. In 2017, the book was listed in the 13th Five-Year Plan for the Publication of National Key Publications.

2. Description of Project Structure

Library of Chinese Handmade Paper compiles with many forms of ideography language: detailed descriptions and records, photographs, illustrations of paper fiber structure and transmittance images, data analysis, distribution of the papermaking sites, guide map

录与描述、摄影图片记录、样纸纤维形态及透光成像采集、实验分析数据表达、造纸地分布与到达图导引、实物纸样随文印证等多种表意语言形式，希望通过这种高度复合的叙述形态，多角度地描述中国手工造纸的技艺与文化活态。在中国手工造纸这一经典非物质文化遗产样式上，《中国手工纸文库》的这种表达方式尚属稀见。如果所有设想最终能够实现，其表达技艺与文化活态的语言方式或许会为中国非物质文化遗产研究界和保护界开辟一条新的途径。

项目无疑是围绕纸质媒介出版物《中国手工纸文库》这一中心目标展开的，但承担这一工作的项目团队已经意识到，由于采用复合度很强且极丰富的记录与刻画形态，当项目工程顺利完成后，必然会形成非常有价值的中国手工纸研究与保护的其他重要后续工作空间，以及相应的资源平台。我们预期，中国当代整体（计划覆盖34个省、市、自治区与特别行政区）的手工造纸业态按照上述记录与表述方式完成后，会留下与《中国手工纸文库》伴生的中国手工纸图像库、中国手工纸技术分析数据库、中国手工纸实物纸样库，以及中国手工纸的影像资源汇集等。基于这些伴生的集成资源的丰富性，并且这些资源集成均为首次，其后续的价值延展空间也不容小视。中国手工造纸传承与发展的创新拓展或许会给有志于继续关注中国手工造纸技艺与文化的同仁提供

to the papermaking sites, and paper samples, etc. Through such complicated and diverse presentation forms, we intend to display the technique and culture of handmade paper in China thoroughly and vividly. In the field of intangible cultural heritage, our way of presenting Chinese handmade paper was rather rare. If we could eventually achieve our goal, this new form of presentation may open up a brand-new perspective to research and preservation of Chinese intangible cultural heritage.

Undoubtedly, the *Library of Chinese Handmade Paper* Project developed with a focus on paper-based media. However, the team members realized that due to complicated and diverse ways of recording and displaying, there will be valuable follow-up work for further research and preservation of Chinese handmade paper and other related resource platforms after the completion of the project. We expect that when contemporary handmade papermaking industry in China, consisting of 34 provinces, cities, autonomous regions and special administrative regions as planned, is recorded and displayed in the above mentioned way, a Chinese handmade paper image library, a Chinese handmade paper technical data library, a Chinese handmade paper sample library, and a Chinese handmade paper video information collection will come into being, aside from the *Library of Chinese Handmade Paper*. Because of the richness of these byproducts, we should not overlook these possible follow-up

更多元的机遇。

毫无疑问，《中国手工纸文库》工作团队整体上都非常认同这一工作的历史价值与现实意义。这种认同给了我们持续的动力与激情，但在实际的推进中，确实有若干挑战使大家深感困惑。

三
我们的困惑和愿景

困惑一：

中国当代手工造纸的范围与边界在国家层面完全不清晰，因此无法在项目的田野工作完成前了解到中国到底有多少当代手工造纸地点，有多少种手工纸产品；同时也基本无法获知大多数省级区域手工造纸分布地点的情况与存活、存续状况。从调查组2008~2016年集中进行的中国南方地区（云南、贵州、广西、四川、广东、海南、浙江、安徽等）的田野与文献工作来看，能够提供上述信息支持的现状令人失望。这导致了项目组的田野工作规划处于“摸着石头过河”的境地，也带来了《中国手工纸文库》整体设计及分卷方案等工作的不确定性。

developments. Moving forward, the innovation and development of Chinese handmade paper may offer more opportunities to researchers who are interested in the techniques and culture of Chinese handmade papermaking.

Unquestionably, the whole team acknowledges the value and significance of the project, which has continuously supplied the team with motivation and passion. However, the presence of some problems have challenged us in implementing the project.

3. Our Confusions and Expectations

Problem One:

From the nationwide point of view, the scope of Chinese contemporary handmade papermaking sites is so obscure that it was impossible to know the extent of manufacturing sites and product types of present handmade paper before the fieldwork plan of the project was drawn up. At the same time, it is difficult to get information on the locations of handmade papermaking sites and their survival and subsisting situation at the provincial level. Based on the field work and literature of South China, including Yunnan, Guizhou, Guangxi, Sichuan, Guangdong, Hainan, Zhejiang and Anhui etc., carried out between 2008 and 2016, the ability to provide the information mentioned above is rather difficult. Accordingly, it placed the planning of the project's fieldwork into an obscure unplanned route,

困惑二：

中国正高速工业化与城镇化，手工造纸作为一种传统的手工技艺，面临着经济效益、环境保护、集成运营、技术进步、消费转移等重要产业与社会变迁的压力。调查组在已展开了九年的田野调查工作中发现，除了泾县、夹江、富阳等为数不多的手工造纸业态聚集地，多数乡土性手工造纸业态都处于生存的“孤岛”困境中。令人深感无奈的现状包括：大批造纸点在调查组到达时已经停止生产多年，有些在调查组到达时刚刚停止生产，有些在调查组补充回访时停止生产，仅一位老人或一对老纸工夫妇在造纸而无传承人……中国手工造纸的业态正陷于剧烈的演化阶段。这使得项目组的田野调查与实物采样工作处于非常紧迫且频繁的调整之中。

困惑三：

作为国家级重点出版物规划项目，《中国手工纸文库》在撰写开卷总序的时候，按照规范的说明要求，应该清楚地叙述分卷的标准与每一卷的覆盖范围，同时提供中国手工造纸业态及地点分布现

贵州省仁怀市五马镇
取缔手工造纸作坊的横幅
2009年4月

Banner of a handmade papermaking mill in Wuma Town of Renhuai City in Guizhou Province, saying "Handmade papermaking mills should be closed as encouraged by the local government"
April 2009

which also led to uncertainty in the planning of *Library of Chinese Handmade Paper* and that of each volume.

Problem Two:
China is currently under the process of rapid industrialization and urbanization. As a traditional manual technique, the industry of handmade papermaking is being confronted with pressures such as economic benefits, environmental protection, integrated operation, technological progress, consumption transfer, and many other important changes in industry and society. During nine years of field work, the project team found out that most handmade papermaking mills are on the verge of extinction, except a few gathering places of handmade paper production like Jingxian, Jiajiang, Fuyang, etc. Some handmade papermaking mills stopped production long before the team arrived or had just recently ceased production; others stopped production when the team paid a second visit to the mills. In some mills, only one old papermaker or an elderly couple were working, without any inheritor to learn their techniques... The whole picture of this industry is in great transition, which left our field work and sample collection scrambling with hasty and frequent changes.

Problem Three:
As a national key publication project, the preface of *Library of Chinese Handmade Paper* should clarify the standard and the scope of each volume according to the research plan. At the same time, general information such as the map with locations of Chinese handmade

状图等整体性信息。但由于前述的不确定性，开宗明义的工作只能等待田野调查全部完成或进行到尾声时再来弥补。当然，这样的流程一定程度上会给阅读者带来系统认知的先期缺失，以及项目组工作推进中的迷茫。尽管如此，作为拓荒性的中国手工造纸整体研究与田野调查就在这样的现状下全力推进着！

当然，我们的团队对《中国手工纸文库》的未来仍然满怀信心与憧憬，期待着通过项目组与国际国内支持群体的协同合作，尽最大努力实现尽可能完善的田野调查与分析研究，从而在我们这一代人手中为中国经典的非物质文化遗产样本——中国手工造纸技艺留下当代的全面记录与文化叙述，在中国非物质文化遗产基因库里绘制一份较为完整的当代手工纸文化记忆图谱。

汤书昆
2017年12月

papermaking industry should be provided. However, due to the uncertainty mentioned above, those tasks cannot be fulfilled, until all the field surveys have been completed or almost completed. Certainly, such a process will give rise to the obvious loss of readers' systematic comprehension and the team members' confusion during the following phases. Nevertheless, the pioneer research and field work of Chinese handmade paper have set out on the first step.

There is no doubt that, with confidence and anticipation, our team will make great efforts to perfect the field research and analysis as much as possible, counting on cooperation within the team, as well as help from domestic and international communities. It is our goal to keep a comprehensive record, a cultural narration of Chinese handmade paper craft as one sample of most classic intangible cultural heritage, to draw a comparatively complete map of contemporary handmade paper in the Chinese intangible cultural heritage gene library.

Tang Shukun
December 2017

编撰说明

1

关于类目的划分标准，《中国手工纸文库·浙江卷》（以下简称《浙江卷》）在充分考虑浙江地域当代手工造纸高度聚集于杭州市富阳区（县级区划）一地，而且手工纸在富阳区的传承品种依然相当丰富的特点后，决定将富阳区以外浙江造纸厂坊按市、县（区）地域分布来划分类目，如第五章“湖州市”；章之下的二级类目以县一级内的造纸企业或家庭纸坊为单元，形成节的类目，如“安吉县龙王村手工竹纸”。富阳区则按照调查时现存纸种来分类，即按照元书纸、祭祀竹纸、皮纸、造纸工具的方式划分第一级类目，形成“章”的类目单元，如第十章“富阳区祭祀竹纸”。章之下的二级类目仍以造纸企业或家庭纸坊为单元，形成“节”的类目，如第九章第四节“杭州富阳逸古斋元书纸有限公司”。

2

《浙江卷》成书内容丰富，篇幅较大，从适宜读者阅读和装帧牢固角度考虑，将其分为上、中、下三卷。上卷内容为概述及富阳区以外浙江现存手工造纸厂坊，按照地级市来分类，包括：第一章“浙江省手工纸概述”、第二章“衢州市”、第三章“温州市”、第四章“绍兴市”、第五章“湖州市”、第六章“宁波市”、第七章“丽水市”、第八章“杭州市”（不含富阳区）；中卷内容为富阳区的元书纸，包括第九章的14节；下卷内容为富阳区的祭祀竹纸、皮纸与造纸工具，包括第十章“富阳区祭祀竹纸”、第十一章“富阳区皮纸”、第十二章“工具”以及“附录”。

3

《浙江卷》第一章为概述，其格式与先期出版的《中国手工纸文库·云南卷》（以下简称《云南卷》）、《中国手工纸文库·贵州卷》（以下简称《贵州卷》）等类似。其余各章各节的标准撰写格式则因有手工纸业态高度密集的县级

Introduction to the Writing Norms

1. In *Library of Chinese Handmade Paper: Zhejiang*, the categorization standards are different from the past. After fully considering the characteristics of high concentration in Fuyang District (county-level) of Hangzhou City, the papermaking factories (mills) in the rest of Zhejiang Province are classified according to the regional distribution of cities and counties (districts), e.g., Chapter V "Huzhou City". Each chapter consists of sections accordingly listing different paper factories or family-based paper mills in counties. For instance, "Handmade Paper in Longwang Village of Anji County". For Fuyang District, chapters are set based on paper types, i.e., Yuanshu paper, bamboo paper for Sacrificial Purposes, bast paper, papermaking tools, e.g., Chapter X "Bamboo Paper for Sacrificial Purposes in Fuyang District". Sections in each chapter include papermaking enterprises or family-based paper mills, e.g. Chapter IX Section 4 "Hangzhou Fuyang Yiguzhai Yuanshu Paper Co., Ltd.".

2. Due to its rich content and great length, *Library of Chinese Handmade Paper: Zhejiang* is further divided into three sub-volumes (I, II, III) for convenience of the readers and bookbinding. Volume I consists of Chapter I "Introduction to Handmade Paper in Zhejiang Province" Chapter II "Quzhou City", Chapter III "Wenzhou City", Chapter IV "Shaoxing City", Chapter V "Huzhou City", Chapter VI "Ningbo City", Chapter VII "Lishui City", Chapter VIII "Hangzhou City (except Fuyang District)"; Volume II contains 14 sections of Chapter IX about Yuanshu paper in Fuyang District; Volume III is composed of three chapters, including "Chapter X Bamboo Paper for Sacrificial Purposes in Fuyang District", "Chapter XI Bast Paper in Fuyang District", Chapter XII "Tools", and "Appendices".

3. First chapter of Volume I is an introduction, which follows the format of *Library of Chinese Handmade Paper: Yunnan* and *Library of Chinese Handmade Paper: Guizhou*, which have already been released. Sections of other chapters follow two different writing norms, because of the concentrated distribution of county-level handmade papermaking practice, and this is different from two volumes that have been published.

First type of writing norm is similar to the *Library of Chinese Handmade Paper: Yunnan* and *Library of Chinese Handmade Paper: Guizhou*, namely, "Basic Information and Distribution" "The

区域存在，故与《云南卷》《贵州卷》所具有的单一标准撰写格式有所不同，分为两类标准撰写格式。

第一类与《云南卷》《贵州卷》相近，适应一个县域内手工造纸厂坊不密集、品种相对单纯的业态分布。通常的格式及大致名称为："××××纸的基础信息及分布""××××纸生产的人文地理环境""××××纸的历史与传承""××××纸的生产工艺与技术分析""××××纸的用途与销售情况""××××纸的品牌文化与习俗故事""××××纸的保护现状与发展思考"。如遇某一部分田野调查和文献资料均未能采集到信息，则按照实事求是原则略去标准撰写格式的相应部分。

第二类主要针对富阳区造纸厂坊聚集分布的特征，或者一个纸厂纸品很丰富、不适合采用第一类撰写格式时采用。通常的格式及大致名称为："××××纸（纸厂）的基础信息与生产环境""××××纸（纸厂）的历史与传承""××××纸（纸厂）的代表纸品及其用途与技术分析""××××纸（纸厂）的生产原料、工艺与设备""××××纸（纸厂）的市场经营状况""××××纸（纸厂）的品牌文化与习俗故事""××××纸（纸厂）的业态传承现状与发展思考"。

4

《浙江卷》选择作为专门一节记述的手工造纸厂坊的正常入选标准是：（1）项目组进行田野调查时仍在生产；（2）项目组田野调查时虽已不再生产，但保留着较完整的生产环境与设备，造纸技师仍能演示或讲述手工造纸完整技艺和相关知识。

考虑到浙江省历史上嵊州藤纸、绍兴鹿鸣纸、富阳桃花纸、温州皮纸曾经是非常著名的传统纸品，而当代业态萎缩特别明显，或处于几近消亡状态，或处于技艺刚刚恢复的试制初期，因此调查组在调查样本上放宽了"保留着较完整的生产环境与设备"这一标准。

5

《浙江卷》调查涉及的造纸点均参照国家地图标准绘制两幅示意图，一幅为造纸点在浙江省和所属县（区）的地理分布位置图，另一幅为由该县（区）县城前往造纸点的路线图，但在具体出图时，部分节会将两图合一呈现。在标示地名时，均统一标示出县城、乡镇两级，乡镇下一级则直接标注造纸点所在村，而不再做行政村、自然村、村民组之区别。示意图上的行政区划名称及编制规则均依

Cultural Geographic Environment" "History and Inheritance" "Papermaking Technique and Technical Analysis" "Uses and Sales" "Folk Customs and Culture" "Preservation and Development". Omission is also acceptable if our fieldwork efforts and literature review fail to collect certain information. This writing norm applies to the handmade papermaking practice in the area where mills and factories are not dense, and the paper produced is of single variety.

A second writing norm is applied to Fuyang District, which harbors abundant paper factories or mills, or where one factory produces diverse paper types. In this chapter, sections are usually named as: "Basic Information and Production Environment" "History and Inheritance" "Representative Paper, Its Uses and Technical Analysis" "Raw Materials, Papermaking Techniques and Tools" "Marketing Status" "Brand Culture and Traditional Stories" "Reflection on Current Status and Future Development".

4. The handmade papermaking factories (mills) included in each section of the volume conforms to the following standards: firstly, it was still under production when the research group did their fieldwork. Secondly, the papermaking tools and major sites were well preserved, and the handmade papermakers were still able to demonstrate the papermaking techniques and relevant knowledge of handmade paper, in case of ceased production.

Because Teng paper in Shengzhou City, Luming paper in Shaoxing City, Taohua paper in Fuyang District and Bast paper in Wenzhou City, are historically renowned traditional paper, their practice shrank greatly or even lingering on extinction in current days, or now in trial production to recover the papermaking practice. Thus, the research team decided to omit the requirement of comparatively complete preservation of production environment and equipment.

5. For each handmade papermaking site, we draw two standard illustrations, i.e. distribution map and roadmap from county centre to the papermaking sites (in some sections, two figures are combined). We do not distinguish the administrative village, natural village or villagers' group, and we provide county name, town name and village name of each site based on standards released by Sinomaps Press.

6. For each type of representative paper investigated in the paper factories (mills) with sufficient output included in the special

据中国地图出版社出版的相关地图。

6

《浙江卷》原则上对每一个所调查的造纸厂坊的代表纸品，均在珍稀收藏版书中相应章节后附调查组实地采集的实物纸样。采样量足的造纸点代表纸品附全页纸样；由于各种限制因素采集量不足的，附2/3、1/2、1/4或更小规格的纸样；个别因停产或小批量试验生产等，导致未能获得纸样或采样严重不足的，则不附实物纸样。

7

《浙江卷》原则上对所有在章节中具体介绍原料与工艺的代表纸品进行技术分析，包括在书中呈现实物纸样的类型，以及个别只有极少量纸样遗存，可以满足测试要求而无法在“珍稀收藏版”中附上实物纸样的类型。

全卷对所采集纸样进行的测试参考了中国宣纸的技术测试分析标准（GB/T 18739—2008），并根据浙江地域手工纸的多样性特色做了必要的调适。实测、计算了所有满足测试分析标示足量需求，并已采样的手工纸中的元书纸类、书画纸类、皮纸类、藤纸类的定量、厚度、紧度、抗张力、抗张强度、撕裂度、湿强度、白（色）度、耐老化度下降、尘埃度、吸水性、伸缩性、纤维长度和纤维宽度共14个指标；加工纸类的定量、厚度、紧度、抗张力、抗张强度、撕裂度、色度、吸水性共8个指标；竹纸类的定量、厚度、紧度、抗张力、抗张强度、色度、纤维长度和纤维宽度共8个指标。由于所采集的浙江省各类手工纸纸样的生产标准化程度不同，因而若干纸种纸品所测数据与机制纸、宣纸的标准存在一定差距。

8

测试指标说明及使用的测试设备如下：

（1） 定量 ▸ 所测纸的定量指标是指单位面积纸的质量，通过测定试样的面积及其质量，并计算定量，以g/m²为单位。
所用仪器 ▸ 上海方瑞仪器有限公司3003电子天平。

（2） 厚度 ▸ 所测纸的厚度指标是指纸在两块测量板间受一定压力时直接测量得到的厚度。根据纸的厚薄不同，可采取多层指标测量、单层指标测量，以单层指标测量的结果表示纸的厚度，以mm为单位。

edition volume, a full page is attached. We attach a piece of paper sample (2/3, 1/2 or 1/4 of a page, or even smaller) if we do not have sufficient sample available to the corresponding section. For some sections, no sample is attached for the shortage of sample paper (e.g. the papermakers had ceased production or were in trial production).

7. All the paper samples elaborated in this volume, in terms of raw materials and papermaking techniques, were tested, including those attached to the special edition, or not attached to the volume due to scarce sample which only enough for technical analysis.

The test was based on the technical analysis standards of Chinese Xuan Paper (GB/T 18739—2008), with modifications adopted according to the specific features of the handmade paper in Zhejiang Province. All paper with sufficient samples, such as Yuanshu paper, calligraphy and painting paper, bast paper, Teng paper, were tested in terms of 14 indicators, including mass per unit area, thickness, tightness, resistance force, tensile strength, tear resistance, wet strength, whiteness, ageing resistance, dirt count, absorption of water, elasticity, fiber length and width. Processed paper was tested in terms of 8 indicators, including , mass per unit area, thickness, tightness, resistance force, tensile strength, tear resistance, whiteness, and absorption of water. Bamboo paper was tested in terms of 8 indicators, including mass per unit area, thickness, tightness, resistance force, tensile strength, whiteness, fiber length and width. Due to the various production standards involved in papermaking in Zhejiang Province, the data might vary from those standards of machine-made paper and Xuan paper.

8. Test indicators and devices:
(1) Mass per unit area: the values obtained by measuring the sample mass divided by area, with the measurement unit g/m². Electronic balance (specification: 3003) we employed is produced by Fangrui Instrument Co., Ltd., Shanghai City.
(2) Thickness: the values obtained by using two measuring boards pressing the paper. In the measuring process, single layer or multiple layers of paper were employed depending on the thickness of the paper, and the single layer measurement unit is mm. The thickness measuring instruments employed are produced by Yueming Small Testing Instrument Co., Ltd., Changchun City (specification: JX-HI) and Pinxiang Science and Technology Co., Ltd., Hangzhou City (specification: PN-PT6).
(3) Tightness: mass per unit volume, obtained by measuring the

所用仪器 ▸ 长春市月明小型试验机有限责任公司JX-HI型纸张厚度仪、杭州品享科技有限公司PN-PT6厚度测定仪。

(3) 紧度 ▸ 所测纸的紧度指标是指单位体积纸的质量，由同一试样的定量和厚度计算而得，以g/m^3为单位。

(4) 抗张力 ▸ 所测纸的抗张力指标是指在标准试验方法规定的条件下，纸断裂前所能承受的最大张力，以N为单位。

所用仪器 ▸ 杭州高新自动化仪器仪表公司DN-KZ电脑抗张力试验机、杭州品享科技有限公司PN-HT300卧式电脑拉力仪。

(5) 抗张强度 ▸ 所测试纸抗张强度指标一般用在抗张强度试验仪上所测出的抗张力除以样品宽度来表示，也称为纸的绝对抗张强度，以kN/m为单位。

《浙江卷》采用的是恒速加荷法，其原理是使用抗张强度试验仪在恒速加荷的条件下，把规定尺寸的纸样拉伸至撕裂，测其抗张力，计算出抗张强度。公式如下：

$$S=F/W$$

式中，S为试样的抗张强度（kN/m），F为试样的绝对抗张力（N），W为试样的宽度（mm）。

(6) 撕裂度 ▸ 所测纸张撕裂强度的一种量度，即在测定撕裂度的仪器上，拉开预先切开一小切口的纸达到一定长度时所需要的力，以mN为单位。

所用仪器 ▸ 长春市月明小型试验机有限责任公司ZSE-1000型纸张撕裂度测定仪、杭州品享科技有限公司PN-TT1000电脑纸张撕裂度测定仪。

(7) 湿强度 ▸ 所测纸张在水中浸润规定时间后，在润湿状态下测得的机械强度，以mN为单位。

所用仪器 ▸ 长春市月明小型试验机有限责任公司ZSE-1000型纸张撕裂度测定仪、杭州品享科技有限公司PN-TT1000电脑纸张撕裂度测定仪。

(8) 白（色）度 ▸ 白度是指被测物体的表面在可见光区域内与完全白（标准白）物体漫反射辐射能的大小的比值，用百分数（%）来表示，即白色的程度。所测纸的白度指标是指在D65光源、漫射/垂射照明观测条件下，纸对主波长475 nm蓝光的漫反射因数。

所用仪器 ▸ 杭州纸邦仪器有限公司ZB-A色度仪、杭州品享科技有限公司PN-48A白度颜色测定仪。

(9) 耐老化度下降 ▸ 指所测纸张进行高温试验的温度环境变化后的参数及性能。本测试采用105℃高温恒温放置72小时后进行测试，以百分数（%）表示。

所用仪器 ▸ 上海一实仪器设备厂3GW-100型高温老化试验箱、杭州

mass per unit area and thickness, with the measurement unit g/cm^3.

(4) Resistance force: the maximum tension that the sample paper can withstand without tearing apart, when tested by the standard experimental methods. The measurement unit is N. The tensile strength testing instrument (specification: DN-KZ) is produced by Hangzhou Gaoxin Technology Company, Hangzhou City and PN-HT300 horizontal computer tensionmeter by Pinxiang Science and Technology Co., Ltd., Hangzhou City.

(5) Tensile strength: the values obtained by measuring the sample maximum resistance force against the constant loading, then divided the maximum force by the sample width, with the measurement unit kN/m.

In this volume, constant loading method was employed to measure the maximum tension the material can withstand without tearing apart. The formula is:

$$S=F/W$$

S stands for tensile strength (kN/m), F is resistance force (N) and W represents sample width (mm).

(6) Tear resistance: a measure of how well a piece of paper can withstand the effects of tearing. It measures the strength the test specimen resists the growth of any cuts when under tension. The measurement unit is mN. Paper tear resistance testing instrument (specification: ZSE-1000), produced by Yueming Small Testing Instrument Co., Ltd., Changchun City and computerized paper tear resistance testing instrument (specification: PN-TT1000) produced by Pinxiang Science and Technology Co., Ltd.

(7) Wet strength: a measure of how well the paper can resist a force of rupture when the paper is soaked in the water for a set time. The measurement unit is mN. Paper tear resistance testing instrument (specification: ZSE-1000), produced by Yueming Small Testing Instrument Co., Ltd., Changchun City and computerized paper tear resistance testing instrument (specification: PN-TT1000) produced by Pinxiang Science and Technology Co., Ltd., Hangzhou City.

(8) Whiteness: degree of whiteness, represented by percentage(%), which is the ratio obtained by comparing the radiation diffusion value of the test object in visible region to that of the completely white (standard white) object. Whiteness test in our study employed D65 light source, with dominant wavelength 475 nm of blue light, under the circumstances of diffuse reflection or vertical reflection. The whiteness testing instrument (specification: ZB-A) is produced by Zhibang Instrument Co., Ltd., Hangzhou City and whiteness tester (specification PN-48A) produced by Pinxiang Science and Technology Co., Ltd., Hangzhou City respectively.

(9) Ageing resistance: the performance and parameters of

品亨科技有限公司YNK/GW100-C50耐老化测试箱。

（10）尘埃度 ▸ 所测纸张单位面积上尘埃涉及的黑点、黄茎和双浆团个数。测试时按照标准要求计算出每一张试样正反面每组尘埃的个数，将4张试样合并计算，然后换算成每平方米的尘埃个数，计算结果取整数，以个/m²为单位。

所用仪器 ▸ 杭州品亨科技有限公司PN-PDT尘埃度测定仪。

（11）吸水性 ▸ 所测纸张在水中能吸收水分的性质。测试时使用一条垂直悬挂的纸张试样，其下端浸入水中，测定一定时间后的纸张吸液高度，以mm为单位。

所用仪器 ▸ 四川长江造纸仪器有限责任公司J-CBY100型纸与纸板吸水性测定仪、杭州品亨科技有限公司PN-KLM纸张吸水率测定仪。

（12）伸缩性 ▸ 所测纸张由于张力、潮湿的缘故，尺寸变大、变小的倾向性。分为浸湿伸缩性和风干伸缩性，以百分数（%）表示。

所用仪器 ▸ 50 cm × 50 cm × 20 cm长方形容器。

（13）纤维长度/宽度 ▸ 所测纸的纤维长度/宽度是指从所测纸里取样，测其纸浆中纤维的自身长度/宽度，分别以mm和μm为单位。测试时，取少量纸样，用水湿润，用Herzberg试剂染色，制成显微镜试片，置于显微分析仪下采用10倍及20倍物镜进行观测，部分显微镜试片在观测过程中使用了40倍物镜，并显示相应纤维形态图各一幅。

所用仪器 ▸ 珠海华伦造纸科技有限公司XWY-VI型纤维测量仪和XWY-VII型纤维测量仪。

9

《浙江卷》对每一种调查采集的纸样均采用透光摄影的方式制作成图像，以显示透光环境下的纸样纤维纹理影像，作为实物纸样的另一种表达方式。其制作过程为：先使用透光台显示纯白影像，作为拍摄手工纸纹理透光影像的背景底，然后将纸样铺平在透光台上进行拍摄。拍摄相机为佳能5D-III。

10

《浙江卷》引述的历史与当代文献均以当页脚注形式标注。所引文献原则要求为一手文献来源，并按统一标准注释，如“陈伟权. 棠云竹纸的文明传奇[J]. 文化交流，2016（5）：50-52.”“袁代绪. 浙江省手工造纸业[M]. 北京：科学出版社，1959：30-33.”“浙江设计委员会统计部. 浙江之纸业[Z]. 1930：232-234.”“谷

the sample paper when put in high temperature. In our test, temperature is set 105 degrees centigrade, and the paper is put in the environment for 72 hours. It is measured in percentage (%). The high temperature ageing test box (specification: 3GW-100) is produced by Yishi Testing Instrument Factory in Shanghai City; Ageing test box (specification: YNK/GW100-C50) produced by Pinxiang Science and Technology Co., Ltd., Hangzhou City.
(10) Dirt count: fine particles (black dots, yellow stems, fiber knots) in the test paper. It is measured by counting fine particles in every side of four pieces of sample paper, adding up and then calculate the number (integer only) of particles every square meter. It is measured by number of particles/m². Dust tester (specification: PN-PDT) produced by Pinxiang Science and Technology Co., Ltd., Hangzhou City.
(11) Absorption of water: it measures how sample paper absorbs water by dipping the sample paper vertically in water and testing the level of water. It is measured in mm. Paper and paper board water absorption tester (specification: J-CBY100) produced by Changjiang Papermaking Instrument Co., Ltd., Sichuan Province and water absorption tester (specification: PN-KLM) produced by Pinxiang Science and Technology Co., Ltd., Hangzhou City.
(12) Elasticity: continuum mechanics of paper that deform under stress or wet. It is measured in %, consists of two types, i.e. wet elasticity and dry elasticity. Testing with a rectangle container (50 cm × 50 cm × 20 cm).
(13) Fiber length (mm) and width (μm): analyzed by dying the moist sample paper with Herzberg reagent, and the fiber pictures were taken through ten times and twenty times objective lens of the microscope (part of the samples were taken through four times objective lens). And the corresponding photo of fiber was displayed respectively. We used the fiber testing instrument (specifications: XWY-VI and XWY-VII) produced by Hualun Papermaking Technology Co., Ltd., Zhuhai City.

9. Each paper sample included in *Library of Chinese Handmade Paper: Zhejiang* was photographed against a luminous background, which vividly demonstrated the fiber veins of the samples. This is a different way to present the status of our paper sample. Each piece of paper sample was spread flat-out on the LCD monitor giving white light, and photographs were taken with Canon 5D-III camera.

10. All the quoted literature are original first-hand resources and the footnotes are used for documentation with a uniform standard. For instance, “Chen Weiquan. The Legend of Bamboo Paper in Tangyun Village[J]. Culture Exchange, 2016(5): 50-52.”

宇. 浙江地区传统造纸工艺的保护研究[D]. 上海：复旦大学，2014：5.”等。

11

《浙江卷》所引述的田野调查信息原则上要求标示出调查信息的一手来源，如“调查组于2016年8月11日第一次前往作坊现场考察，通过朱金浩介绍和实地参观了解到……”“盛建桥在访谈时表示……”等。

12

《浙江卷》所使用的摄影图片主体部分为调查组成员在实地调查时所拍摄的图片，也有项目组成员在既往田野工作中积累的图片，另有少量属撰稿过程中所采用的非项目组成员的摄影作品。由于项目组成员在完成全卷过程中形成的图片的著作权属集体著作权，且在调查过程中多位成员轮流拍摄或并行拍摄为工作常态，因而全卷对图片均不标示项目组成员作者。项目组成员既往积累的图片，以及非项目组成员拍摄的图片在图题文字或后记中特别说明，并承认其个人图片著作权。

13

考虑到《浙江卷》中文简体版的国际交流需要，编著者对全卷重要或提要性内容同步给出英文表述，以便英文读者结合照片和实物纸样领略全卷的基本语义。对于文中一些晦涩的古代文献，英文翻译采用意译的方式进行解读。英文内容包括：总序、编撰说明、目录、概述、图目、表目、术语、后记，以及所有章节的标题，全部图题、表题与实物纸样名。

“浙江省手工造纸概述”为全卷正文第一章，为保持与后续各章节体例一致，除保留章节英文标题及图表标题英文名外，全章的英文译文作为附录出现。

14

《浙江卷》的名词术语附录兼有术语表、中英文对照表和索引的三重功能。其中收集了全卷中与手工纸有关的地理名、纸品名、原料与相关植物、工艺技术和工具设备、历史文化等5类术语。各个类别的名词术语按术语的汉语拼音顺序排列。每条中文名词术语后都给以英文直译，可以作中英文对照表使用，也可以当作名词索引使用。

“Yuan Daixu. Handmade Papermaking Industry in Zhejiang Province[M]. Beijing: Science Press, 1959:30-33.” “Statistics Department of Zhejiang Design Committee. Zhejiang Paper Industry[Z]. 1930:232-234.” and “Gu Yu. Protection of Traditional Papermaking Techniques in Zhejiang Region[D]. Shanghai: Fudan University, 2014:5.”etc.

11. Sources of field investigation information were attached in this volume. For instance, “On August 11, 2016, the research team firstly visited the paper mill, where through Zhu Jinhao's introduction and field trip we got that ...” “According to Sheng Jianqiao's words in the interview ...”.

12. The majority of photographs included in the volume were taken by the researchers when they were doing fieldworks of the research. Others were taken by our researchers in even earlier fieldwork errands, or by the photographers who were not involved in our research. We do not give the names of the photographers in the book, because almost all our researchers are involved in the task and they agreed to share the intellectual property of the photos. Yet, as we have claimed in the epilogue or the caption, we officially admit the copyright of all the photographers, including those who are not our researchers.

13. For the purpose of international academic exchange, English version of some important parts is provided, so that the English readers can have a basic understanding of the volume based on the English parts together with photos and samples. For the ancient literature which is hard to understand, free translation is employed to present the basic idea. English part includes Preface, Introduction to the Writing Norms, Contents, Introduction, Figures, Tables, Terminology, Epilogue, and all the titles, figure and table captions and paper sample names.

Among them, “Introduction to Handmade Paper in Zhejiang Province” is the first chapter of the volume and its translation is appended in the appendix part, apart from the section titles and table and figure titles in the chapter.

14. Terminology is appended in *Library of Chinese Handmade Paper*: *Zhejiang*, which covers five categories of Places, Paper Types, Raw Materials and Plants, Techniques and Tools, History and Culture, relevant to our handmade paper research. All the terms are listed following the alphabetical order of the Chinese character. The Chinese and English parts in the Terminology can be used as check list and index.

目 录
Contents

第十章
富阳区祭祀竹纸

Chapter X
Bamboo Paper for Sacrificial Purposes in Fuyang District

第一节
章校平纸坊

浙江省
Zhejiang Province

杭州市
Hangzhou City

富阳区
Fuyang District

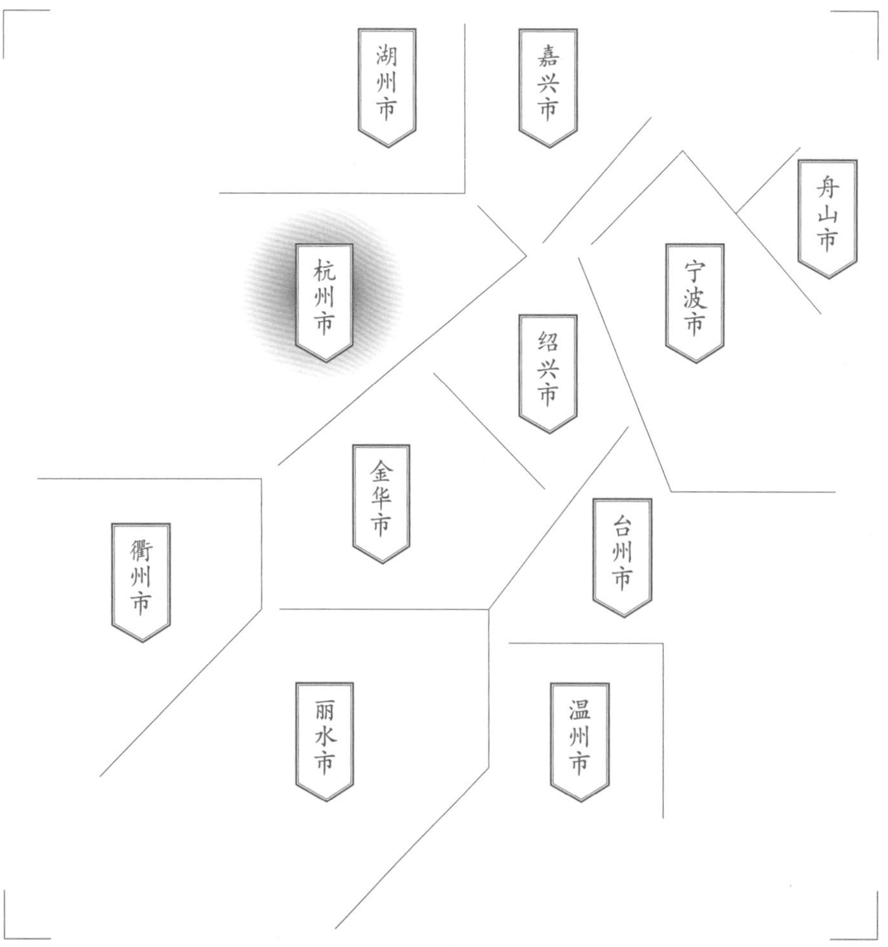

调查对象
富阳区常绿镇黄弹村
章校平纸坊
竹纸

Section 1
Zhang Xiaoping Paper Mill

Subject
Bamboo Paper of Zhang Xiaoping Paper Mill
in Huangdan Village of Changlv Town in Fuyang District

一
章校平纸坊的基础信息与生产环境

1
Basic Information and Production Environment of Zhang Xiaoping Paper Mill

章校平纸坊位于富阳区常绿镇黄弹行政村寺前自然村71号，地理坐标为：东经120° 2′ 57″，北纬29° 50′ 58″。常绿镇位于富阳区南部，地属山区，东接萧山区，南连诸暨县，西邻湖源乡，北靠上官乡、大源镇，距富阳中心城区23.5 km。常绿镇历史上曾经称为长春，其含义是全境遍布竹木林，四季翠绿如春。常绿镇的居民大部分姓章，为章姓聚居之地。

全镇山林面积38.27 km^2，其中竹林面积26 km^2，占富阳的十分之一，素有“富阳竹乡”之称。地处山区的常绿镇，毛竹资源十分丰富，历史上造毛竹纸曾经是很普遍的乡土手工业。1949年前这里的常绿土纸就远销到上海、南京、苏州等长三角中心城市，声誉甚高；1949年后也一直是一项最主要的乡土经济收入。

2016年8月25日与2019年1月23日，调查组两次前往章校平纸坊进行田野调查，第一轮现场调查获得的信息是：章校平纸坊主要生产以毛竹为原料的祭拜用祭祀竹纸，纸坊内有抄纸槽和打浆池各1个，属于微型家庭作坊，没有雇佣工人，章校平负责抄纸，妻子李祥娟负责晒纸，近3年的年产量为200余件，一件2 000张纸，即400 000张以上。

⊙1

⊙2

⊙1 黄弹村外的重重竹山
Bamboo forest around Huangdan Village

⊙2 寺前村边的山溪
Mountain stream by Siqian Village

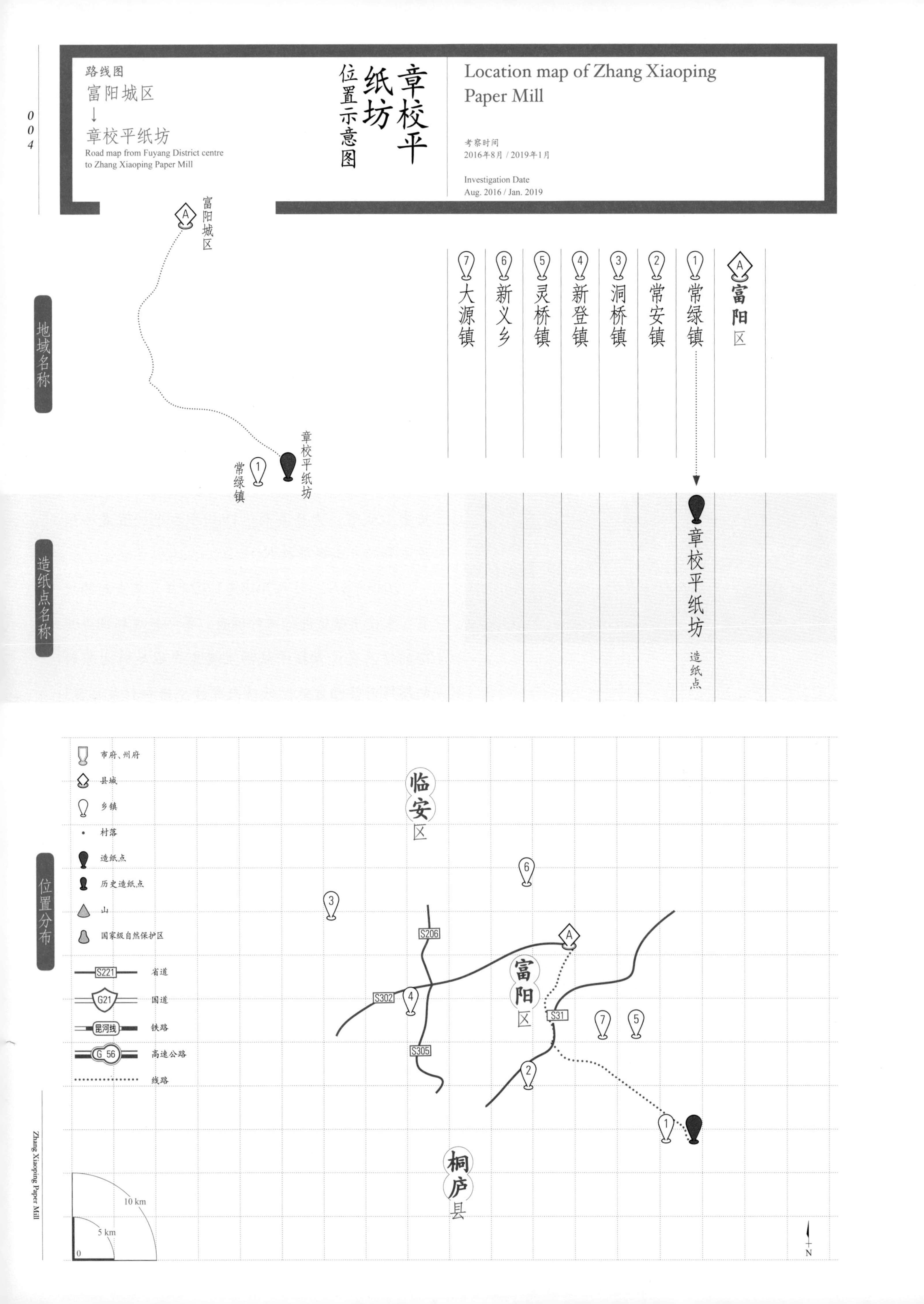

004
路线图
富阳城区
↓
章校平纸坊
Road map from Fuyang District centre to Zhang Xiaoping Paper Mill
章校平纸坊位置示意图
Location map of Zhang Xiaoping Paper Mill
考察时间
2016年8月 / 2019年1月
Investigation Date
Aug. 2016 / Jan. 2019
地域名称
造纸点名称
位置分布
A 富阳城区
1 常绿镇
章校平纸坊
A 富阳区
1 常绿镇
2 常安镇
3 洞桥镇
4 新登镇
5 灵桥镇
6 新义乡
7 大源镇
章校平纸坊
造纸点
市府、州府
县城
乡镇
村落
造纸点
历史造纸点
山
国家级自然保护区
S221 省道
G21 国道
昆河线 铁路
G 56 高速公路
线路
临安区
富阳区
桐庐县
S206
S302
S305
S31
10 km
5 km
0
N
Zhang Xiaoping Paper Mill

二
章校平纸坊的历史与传承

2
History and Inheritance of Zhang Xiaoping Paper Mill

通过访谈得知，章校平的生平经历较为坎坷。他出生于1956年，2016年调查时60岁。父亲早逝，章校平10岁不到就开始砍柴卖给生产队，小学二年级时休学一年，重新回到学校又读了一年书后，辍学到生产队当了7年的记账员，当时生产队的名称是黄弹村寺前第四生产队。17岁从生产队出来后，他又到龙门乡的坛丁林站工作了2年，在林站管辖的林区里种植杉树。章校平19岁去了竹器厂做篾工，奔波于本省的海宁、桐乡、德清等地，两年后因为篾工生意不好，大量产品被塑料品替代，无奈之下又回到家乡。

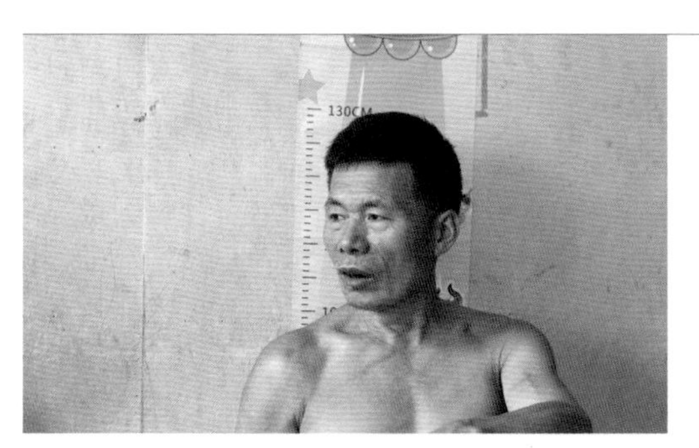

⊙1

⊙2

章校平15岁（生产队时期）开始接触手工造纸，跟随朱高良师傅。当时朱师傅已经70多岁，章校平先学习晒纸、揭纸，学了一年半后仅仅做了半年纸，之后就到林站、竹器厂工作。1983年生产队撤销，回到家乡的章校平于1984年开始自学抄纸，大概在1987年曾经给别的作坊做纸。到1988年，章校平看到别家作坊生意很好后决心自己做纸，其白天抄纸，晚上晒纸，一个人一天可抄2 000张纸，纸的规格为48 cm×48 cm。那时候章校平做的纸为祭祀黄纸（因其中加有姜黄，故叫黄纸，原纸又称小白纸）。

章校平纸坊2013年之前均使用石灰浸泡与发酵，2013年开始使用强碱类的烧碱蒸煮竹料，直至2017年因为环保问题而停止生产。章校平纸坊一直只有他和妻子两个人劳作，章校平负责抄纸，妻子李祥娟负责晒纸。

李祥娟出生于1959年，于1984年跟着章校平学会了晒纸，不会抄纸和做原料。

据章校平回忆，过去在生产队做纸时有2口槽，所以生产队分10个人为两组，一组中有1个捞纸的、2个晒纸的、1个烧料的、1个揭纸的。生产队时期主要生产元书纸，纸张呈白色的元书纸在做的过程中要求很高，一要用嫩竹；二来竹子砍下来后直接放进有帆布盖住的池子里，尽量避免

⊙1 正在讲解的章校平
Zhang Xiaoping explaining the papermaking tips

⊙2 正在揭纸的李祥娟
Li Xiangjuan peeling the paper down

接触阳光，为的是使造出来的纸更白；三在锅里蒸煮时要把石灰水沤得浓一点，把水烧到90 ℃左右，竹子放进去煮2天，配比为一小捆大约5 kg的竹料放0.75 kg的石灰，然后用碱水漂清，这样做出来的纸就会呈白色。

这些元书纸统一卖给供销社，被分为一级、二级和三级，价格上也相差较大。章校平给调查员打了个比方，一级为200元/捆（一捆2 000张），二级为150～180元/捆，三级为120元/捆。做出来的纸主要用于书法创作。如果纸张呈黄色则只能用于制作报纸（当时的报纸还是手抄报纸，不用于印刷），价格一捆只要100元多一点。

三 章校平纸坊的代表纸品及其用途与技术分析

3 Representative Paper and Its Uses and Technical Analysis of Zhang Xiaoping Paper Mill

（一）章校平纸坊代表纸品及其用途

调查员于2016年8月25日入厂调查得知，章校平纸坊基本上只生产毛竹原料祭祀竹纸（“迷信纸”）一种纸，章校平本人也管这种纸叫“小元书纸”。纸张规格为38 cm × 42 cm，选取嫩毛竹为主料进行生产，大多销往江苏省的无锡市，在无锡又被叫作传统纸，用作民间祭拜烧纸。

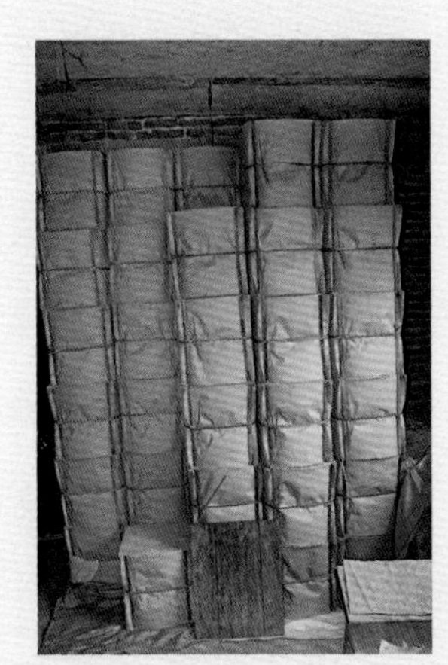

⊙1

⊙1 章校平纸坊存放的『小元书』
Piles of small-sized Yuanshu paper in Zhang Xiaoping Paper Mill

（二）章校平纸坊代表纸品性能分析

测试小组对采样自章校平纸坊的祭祀竹纸所做的性能分析，主要包括定量、厚度、紧度、抗张力、抗张强度、白度、纤维长度和纤维宽度等。按相应要求，每一指标都重复测量若干次后求平均值，其中定量抽取5个样本进行测试，厚度抽取10个样本进行测试，抗张力抽取20个样本进行测试，白度抽取10个样本进行测试，纤维长度测试了200根纤维，纤维宽度测试了300根纤维。对章校平纸坊祭祀竹纸进行测试分析所得到的相关性能参数如表10.1所示，表中列出了各参数的最大值、最小值及测量若干次所得到的平均值或者计算结果。

表10.1　章校平纸坊祭祀竹纸相关性能参数
Table 10.1　Performance parameters of bamboo paper for sacrificial purposes in Zhang Xiaoping Paper Mill

指标		单位	最大值	最小值	平均值	结果
定量		g/m^2				47.7
厚度		mm	0.151	0.101	0.136	0.136
紧度		g/cm^3				0.351
抗张力	纵向	mN	7.5	7.1	7.3	7.3
	横向	mN	7.1	4.3	5.7	5.7
抗张强度		kN/m				0.433
白度		%	24.4	24.2	24.3	24.3
纤维	长度	mm	1.7	0.1	0.5	0.5
	宽度	μm	35.7	0.4	9.0	9.0

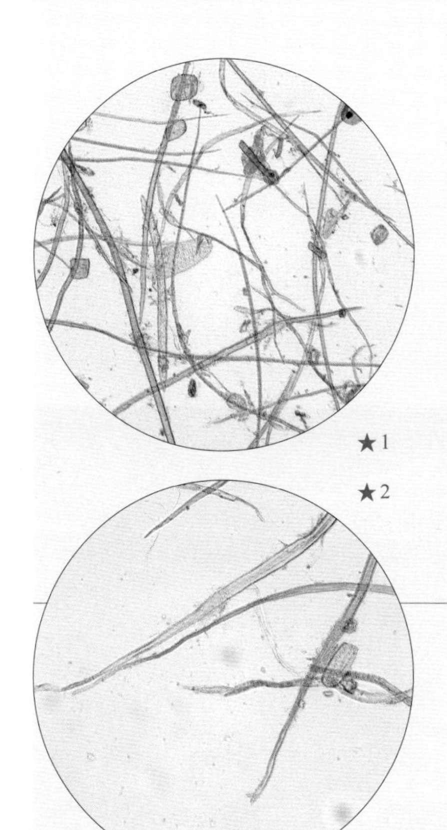

由表10.1可知，所测章校平纸坊祭祀竹纸的平均定量为47.7 g/m^2。章校平纸坊祭祀竹纸最厚约是最薄的1.495倍，经计算，其相对标准偏差为0.122。通过计算可知，章校平纸坊祭祀竹纸紧度为0.351 g/cm^3。抗张强度为0.433 kN/m。

所测章校平纸坊祭祀竹纸平均白度为24.3%。白度最大值是最小值的1.008倍，相对标准偏差为0.004。

章校平纸坊祭祀竹纸在10倍和20倍物镜下观测的纤维形态分别如图★1、图★2所示。所测章校平纸坊祭祀竹纸纤维长度：最长1.7 mm，最短0.1 mm，平均长度为0.5 mm；纤维宽度：最宽35.7 μm，最窄0.4 μm，平均宽度为9.0 μm。

★1
章校平纸坊祭祀竹纸纤维形态图（10×）
Fibers of bamboo paper for sacrificial purposes in Zhang Xiaoping Paper Mill (10× objective)

★2
章校平纸坊祭祀竹纸纤维形态图（20×）
Fibers of bamboo paper for sacrificial purposes in Zhang Xiaoping Paper Mill (20× objective)

四 章校平纸坊的生产原料、工艺与设备

4 Raw Materials, Papermaking Techniques and Tools of Zhang Xiaoping Paper Mill

（一）章校平纸坊祭祀竹纸的生产原料

1. 主料：毛竹

章校平介绍，他造祭祀竹纸通常要砍嫩毛竹，从底部开始砍，再把枝叶削掉，10根一捆运下山。

2. 辅料：水

章校平纸坊所用的水为山间流淌下的溪水，经调查组实地测试，该溪水pH为5.5～6.0，偏酸性。

⊙1

⊙2

⊙3

⊙1 寺前村旁山上的竹林 Bamboo forest up the hill near Siqian Village

⊙2 山间的溪水 Mountain stream

⊙3 水源pH测试 Testing water source pH

（二）章校平纸坊祭祀竹纸的生产工艺流程

据章校平介绍，综合调查组于2016年8月25日的现场观察，归纳其祭祀竹纸的生产工艺流程为：

壹	贰	叁	肆	伍	陆	柒	捌	玖	拾	拾壹	拾贰	拾叁
砍竹	断青	拷白	晒干	断料	蒸煮	落塘	榨水	打浆	抄纸	压榨	晒纸	成品包装

壹 砍竹

1

小满前3天开山，从小满到芒种的近半个月期间，选取嫩毛竹从底部开始砍，把枝叶削掉后10根一捆拖下山。

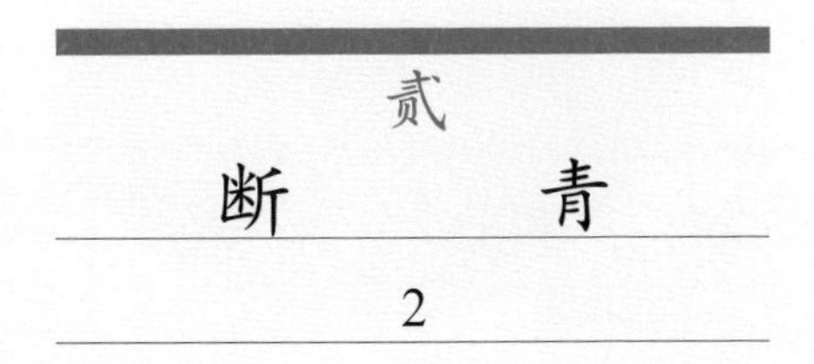

贰 断青

2

毛竹拖下山后，把毛竹切成2 m长的竹段。

⊙4

叁 拷白

3

拿着竹段一头使劲向地上敲打，然后换一头敲打。顺着裂缝将竹段全部破开，直到竹段能平摊在石头上。这时候竹段就成了竹坯。

肆 晒干

4

把竹坯铺在太阳底下晒干，晒到看不出毛竹的青色为止，一般要晒10余天。晒干后就可以进行蒸煮了，没来得及蒸煮的竹坯用尼龙布盖好，防止竹料淋雨受潮。

⊙4 运竹下山的工具 Tools for carrying the bamboo downhill

伍 断料

5 ⊙5

将竹坯用切割机切成35～40 cm长的竹段，用塑料绳将5～6段竹段捆成4～5 kg的小捆，等待蒸煮。

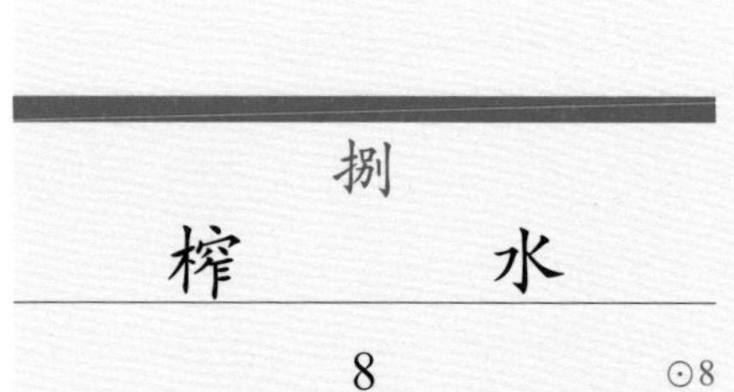

⊙5

陆 蒸煮

6 ⊙6

将捆好的竹料放入锅中，加入烧碱和水开始蒸煮。一锅放200多捆竹料，加500 kg烧碱，烧2天后再焖5～6天。排不出去的烧碱水用水缸接着，下一次蒸煮时放进锅里，再补充适量烧碱，如此循环利用。

⊙6

柒 落塘

7 ⊙7

蒸煮好之后，把竹料直接放进清水中浸4～5天，中间不用换水，让碱液慢慢沉淀。

⊙7

捌 榨水

8 ⊙8

拿出竹料，运送到压榨的地方榨干水分。

⊙8

玖 打浆

9 ⊙9

先用石磨将竹料磨成粉，一般要磨1个小时。然后用泵将石磨磨好的竹粉加水搅拌10分钟。不磨也可以打浆，但打浆的时间会相应变长。

⊙9

⊙5 切割机 Cutting machine

⊙6 蒸煮锅 Steaming and boiling utensil

⊙7 落塘池子 Soaking pool

⊙8 废弃的压榨机 Abandoned pressing machine

⊙9 打浆池 Beating pool

拾 抄纸

10　⊙10～⊙16

抄纸前将和单槽棍从自己身前按顺时针方向向外椭圆状推开，此时一定要匀速，等到和成纸槽中心成旋涡状即可。然后拿着纸帘倾斜20°左右由上到下插入槽内，再缓慢向身前方向提上来，当纸帘出水面时纸帘朝前倾斜，将多余的纸浆匀出。最后将纸帘从帘架上抬起，把抄好的湿纸放在旁边的纸架上。这样一张湿纸就抄完了。据章校平介绍，他一天大概可以捞2 000张，至少需要工作8～9个小时。

⊙10

⊙11

⊙12

⊙13

⊙14

⊙15

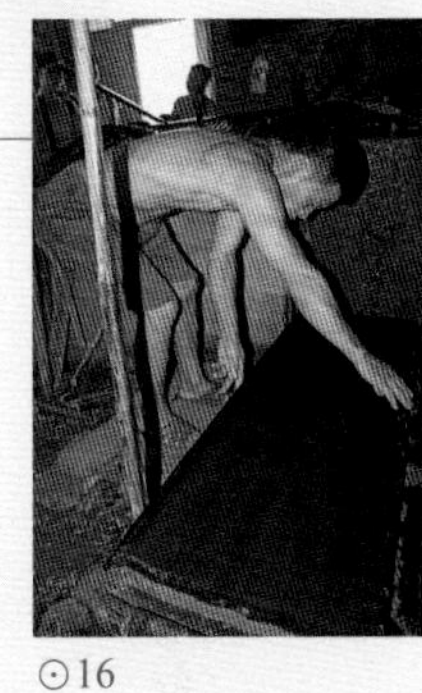

⊙16

拾壹 压榨

11　⊙17

捞完纸后，用千斤顶压榨湿纸帖。压榨时力度由小变大，动作要缓慢，否则会压坏纸张。压榨到湿纸不再出水，即结束。然后用尼龙线切割2下，即切割成3张纸。

⊙17

⊙10/16 抄纸与放帘工序关键动作 Major procedures of papermaking and turning the papermaking screen upside down on the board

⊙17 废弃的千斤顶 Abandoned lifting jack

拾贰
晒　　纸

12　⊙18～⊙22

压榨过后晒纸。用鹅榔头在压干的纸帖四边划一下，让纸松散开；然后捏住纸的右上角捻一下，这样右上角的纸就翘起来了；再用嘴巴吹一下，粘在一起的纸的一角就分开了；10张一揭，摊放在干净的地面上，上面压上竹竿，利用阳光自然晒干。

⊙18

⊙19

⊙20

⊙21

⊙22

⊙18/21 晒纸工序关键动作 Major procedures of drying the paper

⊙22 平铺晒纸 Drying the paper on a flat ground

拾叁 成品包装

13 ⊙23

将晒好的纸整理打包成捆。

⊙23

（三）章校平纸坊祭祀竹纸的主要制作工具

壹 撕纸板

1

揭纸时用来放纸的工具。实测章校平纸坊所用的撕纸板尺寸为：长47 cm，宽40 cm。

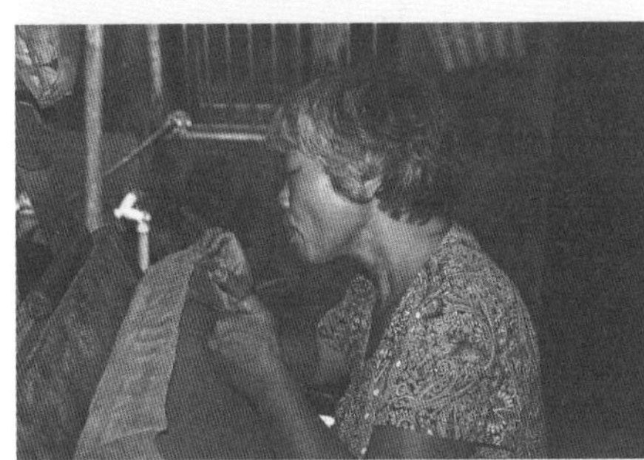
⊙24

贰 鹅榔头

2

牵纸前打松纸帖的工具，杉木制。实测章校平纸坊所用的鹅榔头尺寸为：长18 cm，直径2 cm。

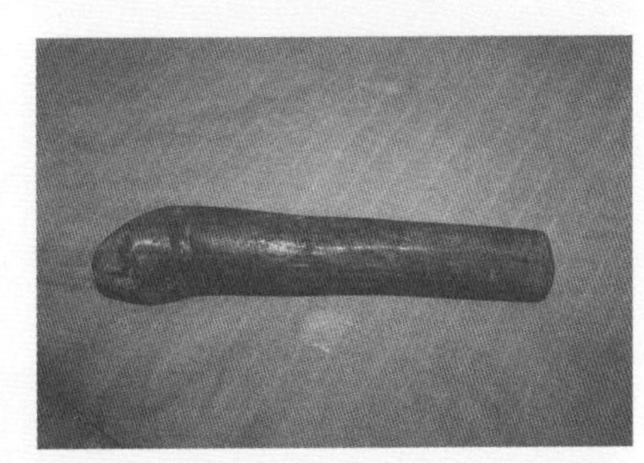
⊙25

叁 和单槽棍

3

抄纸前打匀槽中浆料的工具。实测章校平纸坊所用的和单槽棍尺寸为：长175 cm；槽头长20 cm，宽20 cm。

⊙26

⊙27

⊙23 包装好的成品纸 Packaged papers

⊙24 放在撕纸板上的纸帖 Paper pile putting on the peeling board

⊙25 鹅榔头 Goose hammer

⊙26/27 和单槽棍 Stirring stick

肆
纸　帘
4

用于抄取浆料形成湿纸膜的工具，苦竹丝编织而成，表面光滑平整，帘纹细而密集。据章校平介绍，其纸坊所用的纸帘都是从大源镇购买的，一张380元，一般可以用2年。实测章校平纸坊所用的纸帘尺寸为：长130 cm，宽50 cm。

⊙28

伍
帘　架
5

支撑纸帘的托架，硬木与细竹棍制作。据章校平介绍，其纸坊所用的帘架是请附近村里木匠做的，做一个要250元，一般可以用3年。实测章校平纸坊所用的帘架尺寸为：长143 cm，宽60 cm。

⊙29

陆
打浆池
6

用来打浆的池子。实测章校平纸坊所用的打浆池深1.5 m。

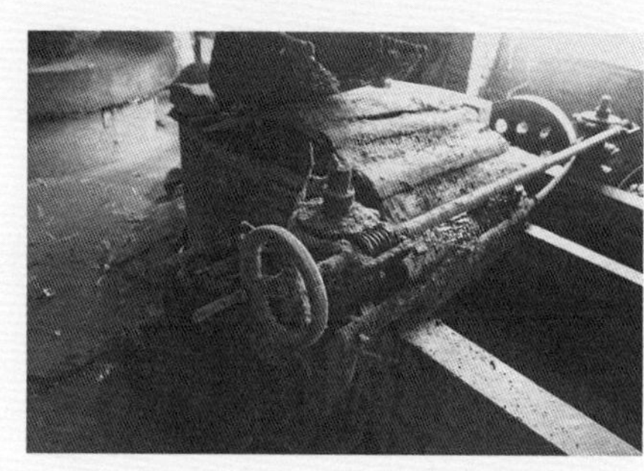
⊙30

柒
切割机
7

断料时用于切割竹段的工具。以前用竹刀，后改为切割机，更为省力。实测章校平纸坊所用的切割机尺寸为：长44 cm，宽25.5 cm；齿轮直径38 cm。

⊙31

捌
千斤顶
8

捞完纸后用于压榨湿纸。实测章校平纸坊所用的千斤顶尺寸为：长19 cm，宽15.5 cm，高29 cm；杆长104 cm。

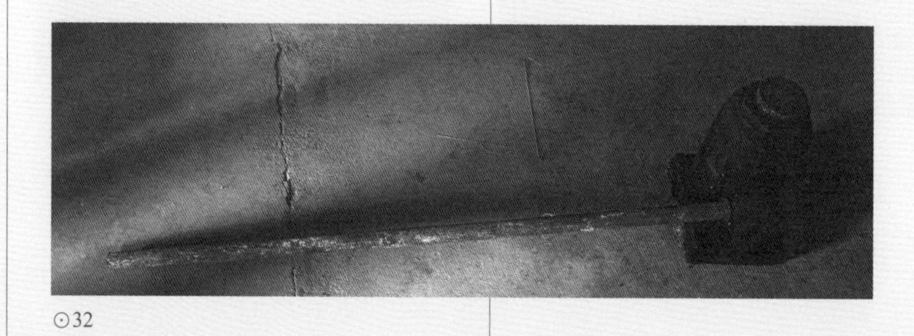
⊙32

⊙28 纸帘 Papermaking screen
⊙29 帘架 Frame for supporting the papermaking screen
⊙30 打浆机 Beating machine
⊙31 切割机 Cutting machine
⊙32 千斤顶 Lifting jack

五
章校平纸坊的市场经营状况

5
Marketing Status of Zhang Xiaoping Paper Mill

据章校平介绍，其纸坊品种单一，生产的“小元书”纸张规格为38 cm×42 cm；渠道也单一，基本上都销往无锡一个地方的市场，用作传统祭拜烧纸。2015～2016年销量为200余件，一件2 000张纸；每件售价230元，按200件算，年毛收入46 000元，若按220件算，则有50 600元的毛收入。除去成本，利润已所剩不多，加上章校平夫妻二人年纪已大，无力再做纸，故章校平于2017年关掉了纸坊。

⊙33

六
常绿镇黄弹村造纸相关传说与习俗

6
Related Legends and Stories of Huangdan Village in Changlv Town

（一）钟塔石凉亭的故事

据章校平介绍，常绿镇山上有一个凉亭，建于1928年，现在叫钟塔石凉亭。过去山路崎岖交通不便，竹子在山上，作坊在山的另一边，村民要做纸只能自己背竹子，每天从清晨4点到晚上5点，来回3趟，总共要走10多公里路，一天要背65～70 kg的毛竹，非常辛苦。上官乡周村的盛定灿那时也有自己的纸坊，时常需要往返于钟塔和上官之间，为方便自己也为方便其他过往百姓，便独立出资建了这个凉亭，专门供来往背竹子的村民休息纳凉。该凉亭与一般的凉亭不同，外形与北方的窑洞更为相似，结构相对简单，但厚厚的石墙使亭内的温度更低，夏天纳凉更为有效。

⊙33
钟塔石凉亭现场
Zhongtashi Pavilion

（二）习俗故事

1.祭拜山神保平安

每年的小满前后，黄弹村进入开山砍竹时期，因为天热，山上虫蛇较多，十分危险，砍竹人在上山之前都会准备三炷香，点蜡烛以及烧元宝来祭拜山神，祈求山公、山母能够保佑其上山砍竹的平安。章校平回忆，他每次上山砍竹之前也都会祭拜山神，因此其在山上从来没碰到过危险，每次砍竹都能顺顺利利。

2. 造纸过程中的忌讳

据章校平介绍，生产队时期有一种迷信的说法：做元书纸的时候，女人不能走到烧料的皮镬上去，不然烧出来的料质量会变差。另外生产队做元书纸的水也不能用来洗手，不管是男是女，该水有专人负责看守，抓到就要罚款，据传用洗过手的水造出的纸，时间一久会腐烂，造不出好纸。

⊙1

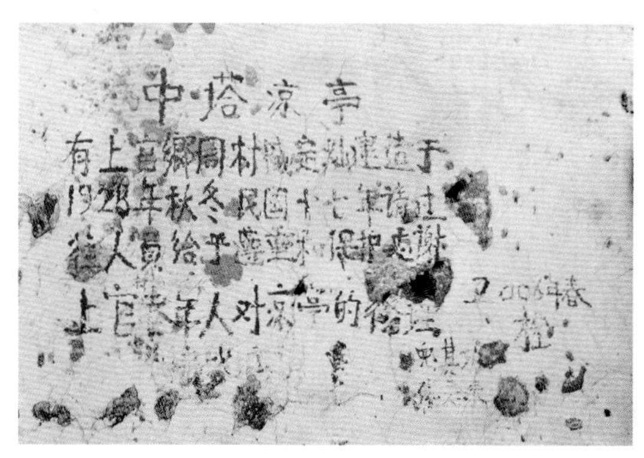

⊙2

⊙1 凉亭正式碑文
Formal inscription of the pavilion

⊙2 凉亭墙壁上题写的建亭及请过客爱护的文字
The inscription on the wall of the pavilion: introduction to the building process and calling for passersby's protection

七
黄弹村手工纸坊的业态传承现状与发展思考

7
Current Status and Development of Handmade Paper Mill in Huangdan Village

观察章校平纸坊，只有他与妻子李祥娟一对60岁的老夫妻在孤独地做纸，儿子与媳妇都没有选择传承祖业，而是选择从事与手工造纸无关的行业，调查时两人都在一所小学当老师。类似章校平纸坊的这种情况黄弹村还有数例，基本上都面临技艺传承完全断档的局面。另外当地环保要求越来越严格，对于手工造纸原料加工过程中所产生的废水管控越来越严，污水不得随意排放，纸坊已经不能再正常运营。在多种因素的影响下，黄弹村造纸作坊暂时只能无可奈何地关闭。

⊙3

⊙3
章校平、李祥娟老夫妻与孙子合影
The couple of Zhang Xiaoping and Li Xiangjuan and their grandson

章校平纸坊

019

祭祀竹纸

Bamboo Ppaper for Sacrificial Purposes
of Zhang Xiaoping Paper Mill

纯毛竹祭祀纸透光摄影图
A photo of pure *Phyllostachys edulis* paper for sacrificial purposes seen through the light

第二节
蒋位法作坊

浙江省
Zhejiang Province

杭州市
Hangzhou City

富阳区
Fuyang District

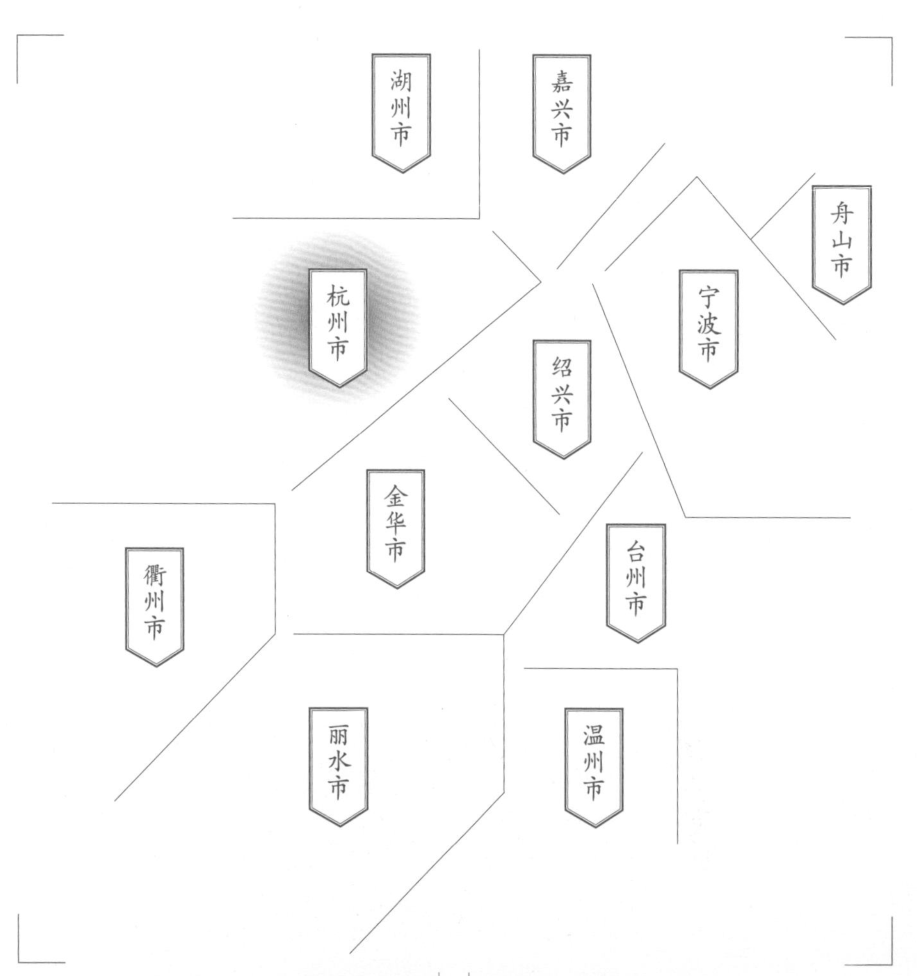

调查对象

富阳区大源镇三岭村
蒋位法作坊
竹纸

Section 2
Jiang Weifa Paper Mill

Subject

Bamboo Paper of Jiang Weifa Paper Mill
in Sanling Village of Dayuan Town in Fuyang District

一
蒋位法作坊的基础信息与生产环境

1
Basic Information and Production Environment of Jiang Weifa Paper Mill

⊙1

⊙2

⊙3

蒋位法作坊是一家专门做毛边纸的手工纸作坊，位于富阳区大源镇三岭行政村三支自然村21号，所在处当地人称为三岭村中方坊。三支村地理坐标为：东经120° 4′ 31″，北纬29° 55′ 27″。作坊负责人为蒋位法。调查组于2016年8月21日前往作坊现场考察时，通过蒋位法的描述了解到的基础信息是：作坊共有员工6人（包括蒋位法及其妻子），纸槽2帘，占地面积约100 m²，专一生产用于祭祀的毛边纸。

三岭村位于富春江南岸，距富阳市区18 km，开车行程约25分钟。村庄四周环山，房屋依山而建，沿狭长的丘陵地带分布，自然资源丰富，毛竹成片，环境优美。2007年行政村规划调整，原来的三支、岭下两个村合并调整为三岭村。现三岭村区域面积7.9 km²，境内道路全长5.5 km，路面硬化率达100%。2016年全村有农户554户，常住人口近1 800人。村中有6家以造纸为主的小型工厂，蒋位法家便是其中之一。

⊙1 蒋位法家的门头 Address plate of Jiang Weifa's house

⊙2 作坊门口的小溪 Stream by the mill

⊙3 蒋位法家后面山上的竹林 Bamboo forest in the back hill of Jiang Weifa's house

蒋位法作坊位置示意图

Location map of Jiang Weifa Paper Mill

考察时间
2016年8月 / 2019年1月

Investigation Date
Aug. 2016 / Jan. 2019

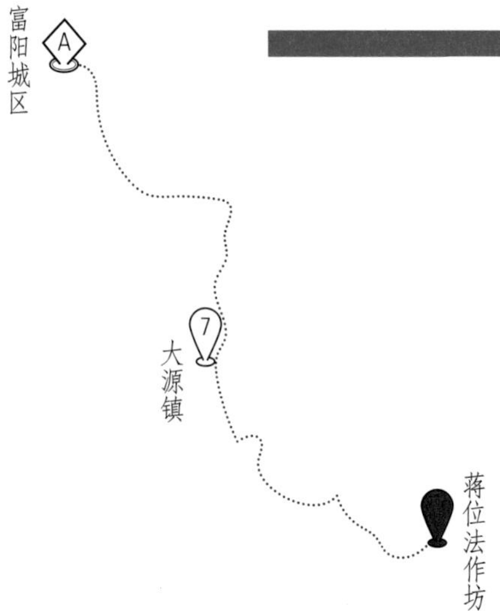

路线图
富阳城区
↓
蒋位法作坊
Road map from Fuyang District centre to Jiang Weifa Paper Mill

富阳城区
大源镇
蒋位法作坊

地域名称

A 富阳区
1 湖源乡
2 常安镇
3 洞桥镇
4 新登镇
5 灵桥镇
6 新义乡
7 大源镇

造纸点名称

蒋位法作坊 造纸点

位置分布

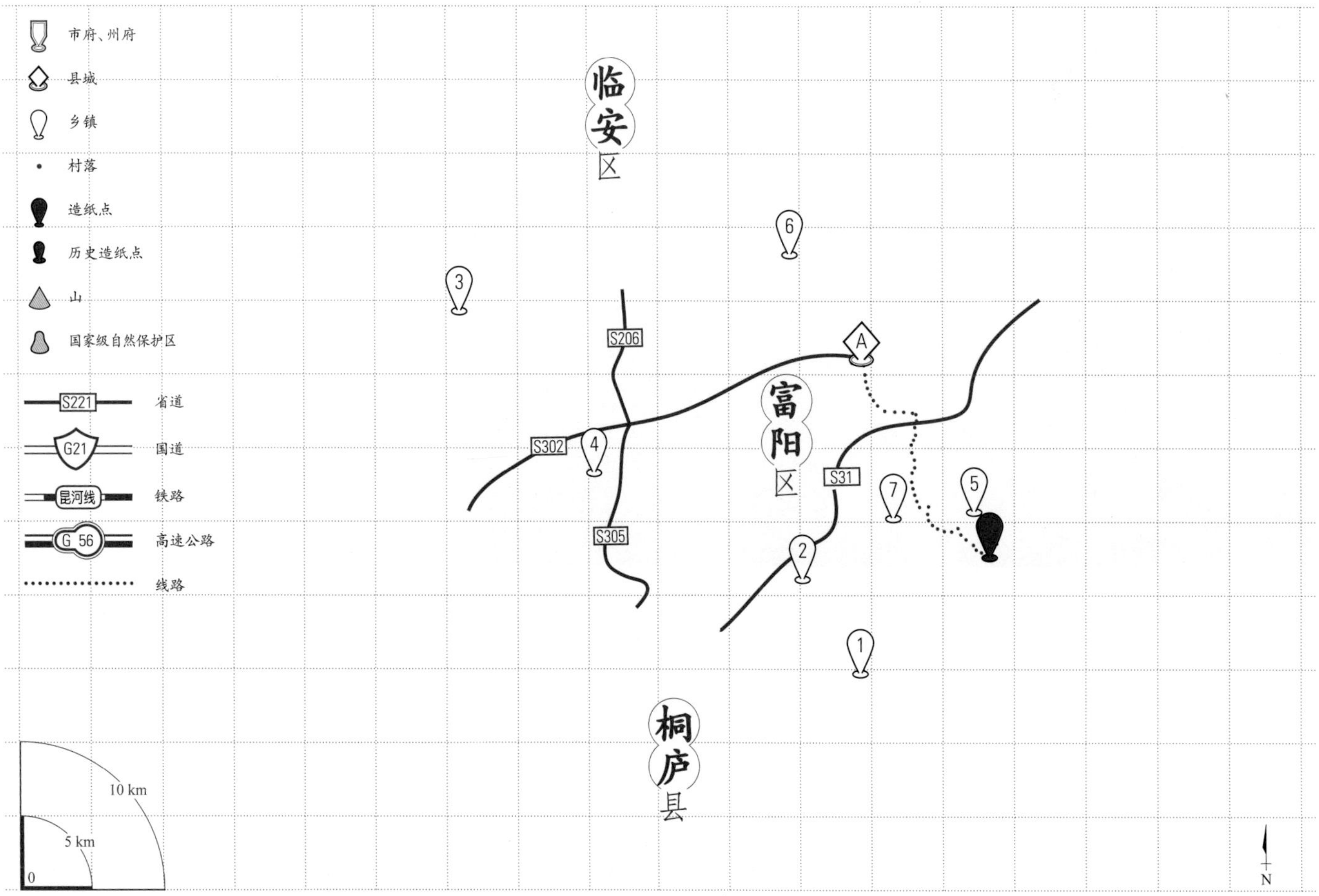

市府、州府
县城
乡镇
村落
造纸点
历史造纸点
山
国家级自然保护区
S221 省道
G21 国道
昆河线 铁路
G56 高速公路
线路

临安区
富阳区
桐庐县
S206
S302
S305
S31
10 km
5 km
0
N

二
蒋位法作坊的历史与传承

2
History and Inheritance of Jiang Weifa Paper Mill

蒋位法，富阳区大源镇三岭行政村三支自然村人，1945年生，2016年调查时已71岁，为手工纸作坊负责人，20岁时开始跟随父亲学习造纸。蒋位法的父亲名叫蒋水林，1923年出生，20多岁时就开始造纸，从事造纸行当直到2014年去世。蒋水林在世时，家中作坊生产元书纸、“老报纸”（蒋位法的叫法，指当时用来印报纸的纸，调查组在湖源乡大竹元纸坊曾经看到过当年李家村造的上海《申报》印刷用的竹纸旧纸）、四六屏等多种纸品。1994年蒋位法开始造六屏纸。2001～2003年间蒋位法曾短期前往陕西省做卷帘门生意。六屏纸一般用作祭祀焚烧纸，因当地文化用纸造纸户与产量都较少，2004年蒋位法开始改造低端文化纸。祭祀纸和低端文化纸的原料基本相同，配比都为老竹料和棉花占比三分之一，竹料与废纸边占比三分之二。两者的不同在于文化纸所用的纸边是从浙江萧山购买的高档竹料纸边，且文化纸不需像祭祀纸一样呈亮黄色，在造纸过程中不需要加入化学染料姜黄粉。

⊙1

蒋位法表示，自己不清楚家中有多少代人从事造纸，但是已知的是爷爷辈就在造纸了。蒋位法开始学造纸时学习的是捞纸，为生产队手工纸作坊工作。二十七八岁开始，自己家中开槽造纸，蒋位法也是负责捞纸，直到2012年67岁时，年纪已经比较大的蒋位法才不再从事捞纸，只负责原料的加工。

⊙2

蒋位法的妻子章菊芳，1949年出生，调查时67岁，杭州市临安区人。20岁时因蒋位法去临安削竹而结识，随后嫁入蒋位法家，在纸坊中负责晒纸。

⊙1 正在讲述造纸生涯的蒋位法 Jiang Weifa relating his papermaking experiences

⊙2 蒋位法的妻子章菊芳 Jiang Weifa's wife Zhang Jufang

蒋位法家中有1个哥哥、2个弟弟、1个妹妹。大哥蒋位昌，2016年时75岁，由于年轻时当兵而远离家乡未传习造纸技艺，当兵回来后一直在富阳从事车辆驾驶工作。四弟蒋位根，1960年出生，30岁时曾跟随蒋位法学过造纸的抄纸技艺，后转行在富阳做公交车司机。五弟蒋位洪目前在做纸，曾经也当过司机，2016年时从哥哥蒋位法处学会了抄纸。三妹蒋位娇年轻时学过造纸，掌握晒纸和揭纸技巧，自出嫁到富阳城里后就不再造纸。因此，兄弟姐妹中只有蒋位法与蒋位洪目前仍在造纸。

蒋位法有3个孩子，大女儿蒋彩珍出嫁以后务农；小女儿蒋红珍在服装厂上班；儿子蒋靖钢曾跟蒋位法学习过抄纸和晒纸技艺，但由于依靠造纸无法支持家庭经济支出，2015年转行在大源镇开面馆。2018年回访时，蒋位法表示将来儿子年纪大了可能会回来继承纸坊，目前他的3个儿女都没有从事造纸祖业。

表10.2 蒋位法传承谱系
Table 10.2 Jiang Weifa's family genealogy of papermaking inheritors

传承代数	姓名	性别	与蒋位法关系	基本情况
第一代	蒋水林	男	父亲	生于1923年，卒于2014年，大源镇三岭行政村人
第二代	蒋位法	男	—	生于1945年，大源镇三岭行政村人，纸坊负责人，20岁左右从父亲学习造纸
	蒋位洪	男	弟弟	生于1963年，大源镇三岭行政村人，2016年开始从哥哥蒋位法学习造纸，目前独立经营一家纸坊
	章芳菊	女	妻子	生于1949年，杭州市临安区人，自小和师傅习得晒纸技艺
第三代	蒋靖钢	男	儿子	生于1971年，大源镇三岭行政村人，20岁左右从父亲处习得造纸技艺

三
蒋位法作坊的代表纸品及其用途与技术分析

3
Representative Paper and Its Uses and Technical Analysis of Jiang Weifa Paper Mill

（一）蒋位法作坊代表纸品及其用途

调查组2016年8月入村调查得知：蒋位法作坊只生产一种纸——毛边纸，尺寸为48 cm×48 cm，主要用途是做祭祀用的“金元宝”。这种纸最初是从萧山传到三岭村的，现在村里还在造纸的造纸户基本都在造这种纸。

（二）蒋位法作坊代表纸品性能分析

测试小组对采样自蒋位法作坊的毛边纸所做的性能分析，主要包括定量、厚度、紧度、抗张力、抗张强度、白度、纤维长度和纤维宽度等。按相应要求，每一指标都重复测量若干次后求平均值，其中定量抽取5个样本进行测试，厚度抽取10个样本进行测试，抗张力抽取20个样本进行测试，白度抽取10个样本进行测试，纤维长度测试了200根纤维，纤维宽度测试了300根纤维。对蒋位法作坊毛边纸进行测试分析所得到的相关性能参数如表10.3所示，表中列出了各参数的最大值、最小值及测量若干次所得到的平均值或者计算结果。

⊙1
蒋位法作坊造的毛边纸
Deckle-edged paper produced in Jiang Weifa Paper Mill

表10.3　蒋位法作坊毛边纸相关性能参数
Table 10.3　Performance parameters of deckle-edged paper in Jiang Weifa Paper Mill

指标		单位	最大值	最小值	平均值	结果
定量		g/m^2				45.8
厚度		mm	0.194	0.171	0.184	0.184
紧度		g/cm^3				0.249
抗张力	纵向	mN	5.4	2.6	4.4	4.4
	横向	mN	3.3	1.8	2.4	2.4
抗张强度		kN/m				0.227
白度		%	19.0	18.4	18.8	18.8
纤维	长度	mm	2.5	0.3	1.0	1.0
	宽度	μm	49.6	0.7	10.5	10.5

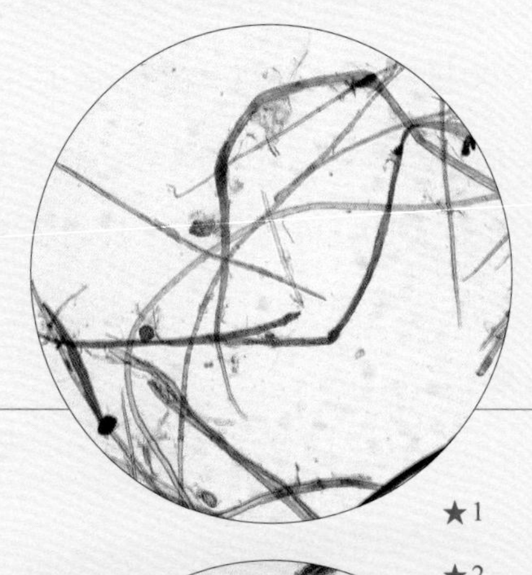

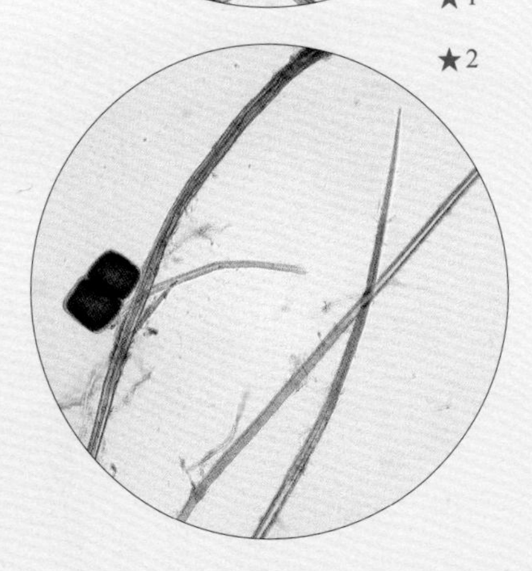

由表10.3可知，所测蒋位法作坊毛边纸的平均定量为45.8 g/m^2。蒋位法作坊毛边纸最厚约是最薄的1.134倍，经计算，其相对标准偏差为0.046。通过计算可知，蒋位法作坊毛边纸紧度为0.249 g/cm^3。抗张强度为0.227 kN/m。

所测蒋位法作坊毛边纸平均白度为18.8%。白度最大值是最小值的1.032倍，相对标准偏差为0.014。

蒋位法作坊毛边纸在10倍和20倍物镜下观测的纤维形态分别如图★1、图★2所示。所测蒋位法作坊毛边纸纤维长度：最长2.5 mm，最短0.3 mm，平均长度为1.0 mm；纤维宽度：最宽49.6 μm，最窄0.7 μm，平均宽度为10.5 μm。

★1
蒋位法作坊毛边纸纤维形态图（10×）
Fibers of deckle-edged paper in Jiang Weifa Paper Mill (10× objective)

★2
蒋位法作坊毛边纸纤维形态图（20×）
Fibers of deckle-edged paper in Jiang Weifa Paper Mill (20× objective)

四
蒋位法作坊毛边纸的生产原料、工艺与设备

4
Raw Materials, Papermaking Techniques and Tools for Deckle-edged Paper in Jiang Weifa Paper Mill

蒋位法在访谈中介绍：作坊一直沿用传统手工捞纸工艺，但为了节省成本，2008年开始不再像过去制作元书纸那样砍嫩竹做原料，而是购买加工好的竹浆板，回来碾磨加工后直接使用。

（一）蒋位法作坊毛边纸的生产原料

1. 主料：竹浆板、棉花、废纸边

蒋位法介绍，自家作坊生产的毛边纸有3种原料：棉花、老竹料和废纸边。棉花是从绍兴购买的玩具厂的旧棉花，近年来价格在1.4～2.6元/kg范围浮动；老竹料是从富阳购买的老竹浆板料，竹料有2种，纸坊使用的老竹料1.2元/kg；废纸边价格是2.4元/kg。1994年生产祭祀纸时纸边价格约是0.2元/kg，那时造纸原料竹料用得多，纸边用得少。具体各项原料的价格在不同时期有浮动，难以有一个具体明确的数字。废纸边的收购地蒋位法没有透露。所有原料都是供应商送货上门。具体的原料配比为棉花占三分之一，废纸边与竹料占三分之二。

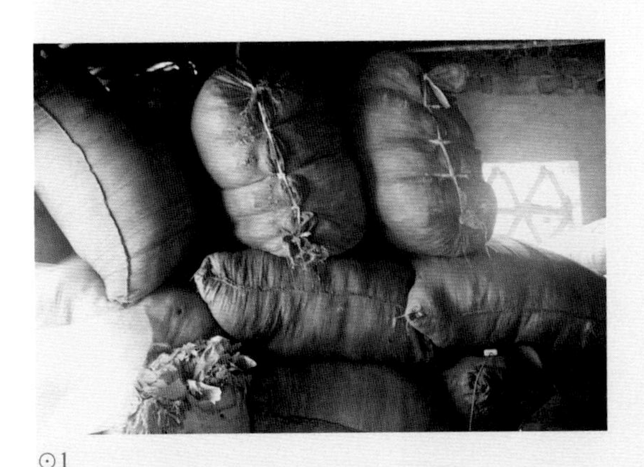
⊙1

2. 辅料：水

蒋位法作坊位于山边，山泉水常年不断流经作坊，便就地取材，使用山泉水来造纸，既保证了水源的充足，又保证了造纸用水的质量。经调查组成员现场取样检测，水的pH为5.5～6.0，偏酸性。

⊙2

⊙1 作坊中堆放的竹浆板原料 Bamboo pulp materials in the mill

⊙2 蒋位法介绍采购来的原料 Jiang Weifa introducing the purchased raw materials

（二）蒋位法作坊毛边纸的制作工艺流程

通过2016年8月21日对蒋位法作坊毛边纸生产工艺进行的实地调查和访谈，归纳纸坊毛边纸的主要制作工艺流程如下：

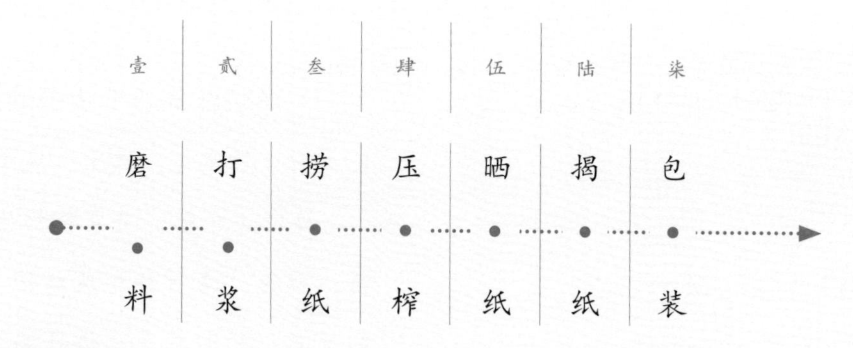

壹 磨料

1 ⊙1⊙2

制作毛边纸的原料有3种，其中废纸边和老竹料需要用石碾碾磨成粉状并充分混合。蒋位法介绍其碾磨工艺要领：废纸边和老竹料分开操作。废纸边上洒水，纸边潮湿后磨20分钟即可；竹料需要磨到没有筋呈细碎状，碾磨的时间约为40分钟。棉花不用磨，在打浆环节直接加入。

据调查组成员现场观察，蒋位法在磨老竹料时，先将石磨调成快速模式，磨约10分钟后，再降低石磨速度碾磨30分钟，快速速度约为慢速的3倍。废纸边单独碾磨，碾磨时适当洒水。蒋位法每天上午磨料，磨料时间约为4个小时。

⊙1

⊙2

⊙1 蒋位法在石碾旁给废纸边洒水
Jiang Weifa watering the waste paper materials by the stone roller

⊙2 调查组成员在石碾旁访谈蒋位法
Researchers interviewing Jiang Weifa by the stone roller

贰 打浆

2 ⊙3⊙4

将碾磨好的废纸边与老竹料放入打浆池，加水打浆，加水的过程中加入棉花，使各种原料充分混合，形成适合制作毛边纸的浆料。

⊙3

⊙4

叁 捞纸

3 ⊙5⊙6

将打好的浆料放入纸槽，用竹耙搅拌均匀后即可进行捞纸。

捞纸时，捞纸工站于纸槽的一侧，两手分别握住帘床左、右两端，将帘床缓缓斜浸入浆液中，待纸帘完全进入浆液中后，缓缓抬起帘床，抬起时靠近捞纸工的一边高度稍高于另一对边，使浆液在纸帘上均匀分布。将纸帘上多余的浆液推出，一层薄薄的湿纸便形成了。

松手后，帘床自动吊于槽中水面上方约2 cm处，捞纸工一手捏住纸帘靠近身体一边的中间位置并抬起，一手捏住纸帘相对应的另一边中间位置，将湿纸面朝下，缓缓逐步将湿纸放置于纸架上，待纸帘与纸架完全贴合，将靠近身体一端的纸帘边往下轻轻按压，再迅速揭起，放置于帘床上继续捞纸。如此重复操作，纸架上便会形成逐渐变高的湿纸堆。

捞纸工每捞一定量的纸后，因为槽中的纸浆成分变少，就会往纸槽中添加纸浆，再用竹耙进行搅拌。具体的量和浓度由捞纸工自己把握。2016年8月调查时，作坊中请了2名捞纸工，2名晒纸工，2名负责打浆压榨的工人，均为本地人。正常情况下2帘槽生产，每帘槽由3名工人负责。其中一名捞纸工张永根从45岁开始就跟随蒋位法学捞纸，已在蒋位法作坊工作了七八年。捞纸工每天早上5点开始工作，捞完4 000张纸就结束一天的工作。

⊙5

⊙6

⊙3 蒋位法介绍打浆过程
Jiang Weifa introducing the beating procedures

⊙4 捞纸工用竹耙搅拌纸浆
A papermaker using a bamboo rake to stir the pulp

⊙5 捞纸工在捞纸
A papermaker making the paper

⊙6 纸帘上形成湿纸膜
Wet paper forming on the papermaking screen

肆 压榨

4 ⊙7

在捞纸工完成一天的工作量以后，蒋位法便用千斤顶对湿纸帖进行压榨，约需压榨1个小时。

⊙7

伍 晒纸

5 ⊙8～⊙10

蒋位法作坊的晒纸方法与别的作坊不同，不使用焙壁晒纸，而是自然晾干。在蒋位法家的院中，有用毛竹搭起来的架子，经过压榨的毛边纸以每5张为1个单位，放在竹竿上晾晒，每根竹竿上可晒3个单位的毛边纸。通常晴天晒2～3天即可。在调查组访谈时，突遇下雨，蒋位法的女儿蒋彩珍便立即拿塑料膜将正在晒的纸盖住，防止纸被淋湿。

⊙8

⊙7 压榨机与千斤顶 Pressing machine and lifting jack

⊙8 竹竿晒纸 Drying the paper on bamboo poles

⊙9

⊙10

陆
揭纸
6 ⊙11

晒干的纸收回后，由揭纸工一张张揭开，并叠好堆放。据蒋位法介绍，1个揭纸工1天约可以揭2件纸，每件4 000张，工资为80元/天。

⊙11

柒
包装
7 ⊙12～⊙14

毛边纸为自然晾干，在叠成堆时不是很平整，造成纸堆较高，需要用千斤顶将其压实。压实后用包装绳将毛边纸按件捆绑，捆绑好后即可等待出售。

⊙12

⊙13

⊙14

⊙9 蒋彩珍用塑料膜盖住正在晒的纸 Jiang Caizhen covering the paper with plastic film

⊙10 调查组成员帮忙盖纸 A researcher helping Jiang Caizhen covering the paper

⊙11 揭纸工在一张张揭纸 A worker peeling the paper one by one

⊙12 蒋位法用千斤顶压已晒干的纸堆 Jiang Weifa using a lifting jack to press the dried paper

⊙13 蒋位法用包装绳捆绑毛边纸 Jiang Weifa using rope to tie up the deckle-edged paper

⊙14 蒋位法家中堆放的毛边纸 Piled deckle-edged paper in Jiang Weifa's house

（三）蒋位法作坊毛边纸的主要制作工具

壹 石碾 1

用来碾磨原料的电动工具，由石磨和磨盘组成。蒋位法表示，使用石碾磨料是20世纪80年代从大源镇学过来的。蒋位法使用的石碾是村里公用的，2008年以后因为很多作坊停产，石碾就只有蒋位法在使用了。实测蒋位法作坊使用的石碾尺寸为：磨料石直径96 cm，厚44 cm；底座直径267 cm，高54.5 cm。

⊙1

贰 打浆机 2

用来打浆的机器，可以使浆料充分混合。蒋位法作坊使用的打浆机是请村里匠人帮忙打造的，制作成本约为4 000元。实测蒋位法作坊使用的打浆机尺寸为：长252 cm，宽141 cm，高73 cm；池内侧宽度为57 cm。

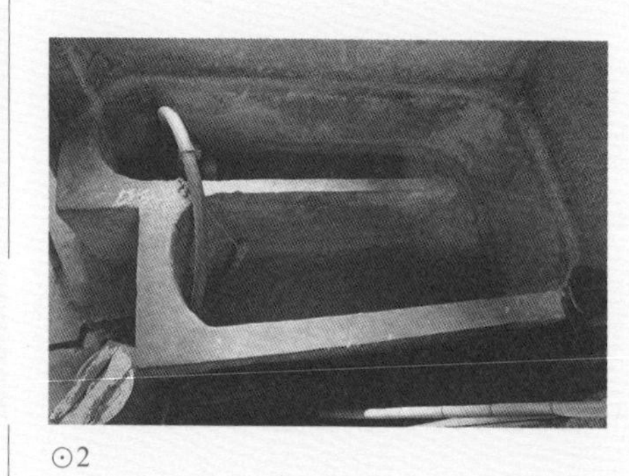

⊙2

叁 纸槽 3

用于盛放纸浆的长方形水泥制容器。捞纸时捞纸工站于其一侧。实测蒋位法作坊纸槽尺寸为：长197 cm，宽180 cm，高110 cm。纸槽所在的捞纸房是蒋位法从2010年开始租用的，一年租金约3 000元。

⊙3

肆 纸帘 4

捞纸时的重要成纸工具，用于形成湿纸膜和过滤掉多余的水分。据蒋位法介绍，纸帘是从大源镇大源帘厂购买的，购买时价格为350元，一张纸帘能用1年时间。2019年1月回访时了解到，现在纸帘的售价为500元，同时纸帘的尺寸有所改变，以前的纸帘一次只可捞2张纸，现在单次可捞3张48 cm×48 cm的毛边纸。实测蒋位法作坊纸帘尺寸为：长156 cm，宽53 cm。

⊙4

⊙1 石碾 Stone roller

⊙2 打浆机 Beating machine

⊙3 纸槽 Papermaking trough

⊙4 纸帘 Papermaking screen

伍 帘架 5

用于放置纸帘的长方形框架，捞纸时捞纸工手握其左、右两短边中点。帘架是蒋位法请村里的木工做的，花了约200元，一张帘架可用2～3年。调查时实测蒋位法作坊帘架尺寸为：长164 cm，高60 cm。

⊙5

陆 竹耙 6

用于搅拌纸浆的竹制工具，耙头为橡胶材料。调查时实测蒋位法作坊竹耙尺寸为：柄长172 cm，耙头橡胶片尺寸为26 cm × 20 cm。

⊙6

柒 千斤顶 7

用于压干湿纸块和压实干纸堆的工具。调查时实测蒋位法作坊千斤顶尺寸为：长20 cm，宽16 cm，高28 cm。

⊙7

⊙5 帘架 Frame for supporting the papermaking screen

⊙6 竹耙 Bamboo rake

⊙7 千斤顶 Lifting jack

五
蒋位法作坊的市场经营状况

5
Marketing Status of Jiang Weifa Paper Mill

据蒋位法介绍，作坊每年大约生产200件纸，一件4 000张，正常情况下1件纸能卖600元。但是从2016年下半年开始，毛边纸的价格下跌为每件500元。分析其中原因可能是毛边纸市场竞争加大，同行压低价格所致。如果计算不包括原材料、工钱、场地租金等的年造纸毛收入，按照600元/件有120 000元左右，按照500元/件则降为100 000元。2019年1月回访时了解到，近两年纸坊的纸品产量较之前有所增加，平均每年可生产400～450件纸，销售价格在580～650元，但扣除原材料和各项成本，1件的净利润只有100多元。

尽管利润下降，毛边纸的销路还是稳定的。蒋位法作坊的毛边纸由来自安徽与杭州的经销商收购，再销往外地。每次收购都是先打货款再发货，没有拖欠货款的情况。

⊙1

⊙1
蒋位法家中存放的低端文化纸
Piled low-end culture paper in Jiang Weifa's house

六
蒋位法作坊的
相关文化事象

6
Related Culture of
Jiang Weifa Paper Mill

（一）纸坊的第一位经销商

1994年之前，纸坊生产的祭祀火烧纸一直是由来自萧山的纸商购买。在做出六屏纸之后，萧山的客户也将六屏纸作为火烧纸购买。之后蒋位法在外地从事了2年的卷帘门生意。2004年成功将祭祀纸改进为文化纸后，第一个文化纸客户是来自大源镇朱家门的纸商朱金豪。当年朱金豪正在富阳地区寻找性价比高的文化纸品，在途经三岭村蒋位法作坊时，看到蒋位法的纸色泽光洁、质感光滑，便以180元/件的价格购买了1件六屏纸。自此以后，他一直对蒋位法作坊的纸品给以很高的评价，也为蒋位法介绍其他的客户，渐渐来纸坊购买纸品的人多了，客源日趋稳定。蒋位法表示，两人相互来往已有15年了，至今两家过年时期还相互拜访。

（二）“京都状元富阳纸，十件元书考进士”

富阳地区一直以来就有“造纸之乡”的美称，在富阳各地都流传着“京都状元富阳纸，十件元书考进士”的说法（1件元书纸为4 000张），说的就是写完十件富阳元书纸才有高中进士的可能。最上品的竹纸称作“元书”，传说宋真宗时期，每逢元旦庙祭都会使用上品竹纸书写祭文，因此才有“元书纸”之名。蒋位法表示，上品富阳元书纸的韧性好似丝绸，润墨性极佳，还具有不易被虫蛀、不易腐坏、不易变色等特点。古时有朝代将其作为科举试卷用纸，因而有众多考生选择其作为练习用纸。但蒋位法虽然介入了文化纸，也在宣传元书纸，作坊却并未生产过真正意义上的元书纸。

⊙2

⊙2 蒋位法家门前关于元书纸的宣传标语
Publicity posters of Yuanshu paper on the gate of Jiang Weifa's house

七
蒋位法作坊的业态传承现状与发展思考

7
Current Status and Development of Jiang Weifa Paper Mill

2019年1月回访时，调查组了解到，目前蒋位法作坊的生产存在两方面的困扰：

（一）人员结构老化，传承出现断层

据蒋位法介绍，纸坊兴盛时期有6名工人，现在作坊中请的4名工人，除了一名捞纸工50多岁，其他的工人都已经60多岁了。工人主要负责捞纸、晒纸和揭纸，不同工作的工资也有区别：捞纸工150元/天，晒纸工100元/天，揭纸工90元/天。每个工人每天工作11个小时，每天生产8 000多张纸。随着年龄的增长，工人们快干不动这样的辛苦活了，自己和妻子年纪也大了，行动能力大不如前，经常会感到力不从心。蒋位法表示：村里原先造毛边纸的作坊多数都关了，为了补贴家用和留够养老金，他才维持作坊的生产直到现在。女儿们都出嫁了难以接手作坊，蒋位法希望等儿子年纪大了，有一天能回来继承纸坊。目前村子里大多数人都是这样的状态，会造纸的人在年龄大了以后再回来。20世纪60年代，村子里有44户人家，有34户在造纸，那时有国家的供销社，定时定点收购，造纸不用担心销售。现在村子里只有5户人家还在造纸，客源也不稳定。

（二）造纸原材料紧缺

蒋位法作坊所用的竹料采购自富阳，是已加工过的老竹料，买来即可直接投入生产。但2018年因环保问题，富阳的工厂不再生产这种老竹料，目前纸坊所用的竹料还是2018年初采购的。若选择未加工的嫩竹，从砍竹到运输等

⊙1

⊙1 蒋位法 Jiang Weifa

环节的成本都很高，且在加工竹料的过程中也会面临诸多环境污染问题。回访时蒋位法表示库存原料已用完，后续造纸可能会考虑不再使用这种竹料，但用什么原料来替代，如何改进这一问题蒋位法还在探索。

毛边纸

Deckle-edged Paper
of Jiang Weifa Paper Mill

毛边纸（老竹+棉花+纸边）透光摄影图
A photo of deckle-edged paper (old bamboo + cotton + paper edges) seen through the light

第三节

李财荣纸坊

浙江省
Zhejiang Province

杭州市
Hangzhou City

富阳区
Fuyang District

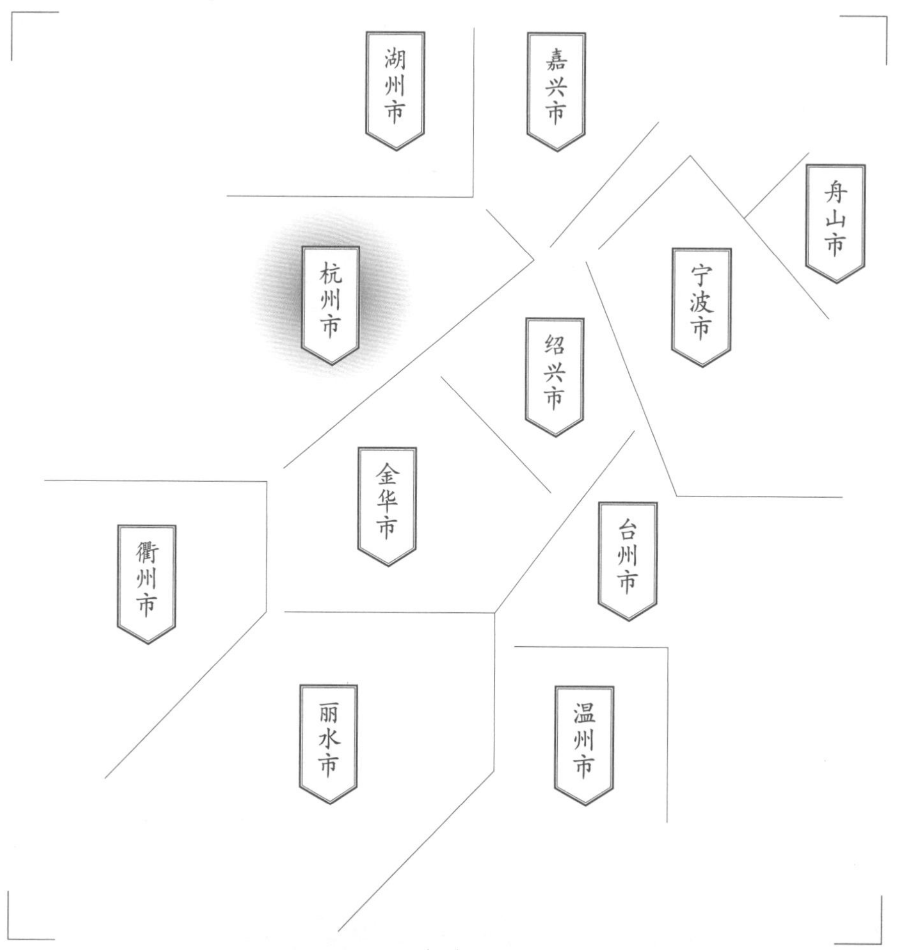

调查对象

富阳区灵桥镇新华村
李财荣纸坊
竹纸

Section 3
Li Cairong Paper Mill

Subject

Bamboo Paper of Li Cairong Paper Mill
in Xinhua Village of Lingqiao Town in Fuyang District

一
李财荣纸坊的基础信息与生产环境

1
Basic Information and Production Environment of Li Cairong Paper Mill

新华村位于富阳区灵桥镇镇域范围的中心，地理坐标为：东经120° 3′ 22″，北纬29° 58′ 3″。北临杭千高速，由原新华、倪家滩、月台3个自然村合并而成，2011年被列为富阳市中心村。2015年全村有农户588户，人口1 963人，地域面积为5.88 km²，山林面积达566.7万m²，耕地面积33万m²。在新华村，从事手工造纸业的村户主要来自原倪家滩和月台村两个村子，李财荣纸坊就是坐落在原月台村的家庭纸坊之一。

2016年8月23日，调查组入村调查得到的基础信息是：李财荣纸坊由夫妻二人合作造纸，纸品单一，为老毛竹浆料添加废纸和棉花的低端祭祀竹纸。实行工作3天半休息1天的作业模式，从原料加工到抄纸、晒纸，再到包装、售卖，整个流程都是夫妻二人完成。2016年时一年可以做400～500件纸（1件2 000张），每件售价150元，扣除成本50元/件，平均年收入为4万～5万元人民币。销售渠道为上门收购，会有专人不定期上门收购已经做好的纸，纸品大多销往富阳本地及相邻的萧山和临安县。

⊙1

⊙2

⊙1 月台村村口的村名石 Village name inscription on a stone at the entrance of Yuetai Village

⊙2 月台村入口的公路 Highway at the entrance of Yuetai Village

李财荣纸坊位置示意图

Location map of Li Cairong Paper Mill

路线图
富阳城区
↓
李财荣纸坊
Road map from Fuyang District centre to Li Cairong Paper Mill

考察时间
2016年8月 / 2019年1月

Investigation Date
Aug. 2016 / Jan. 2019

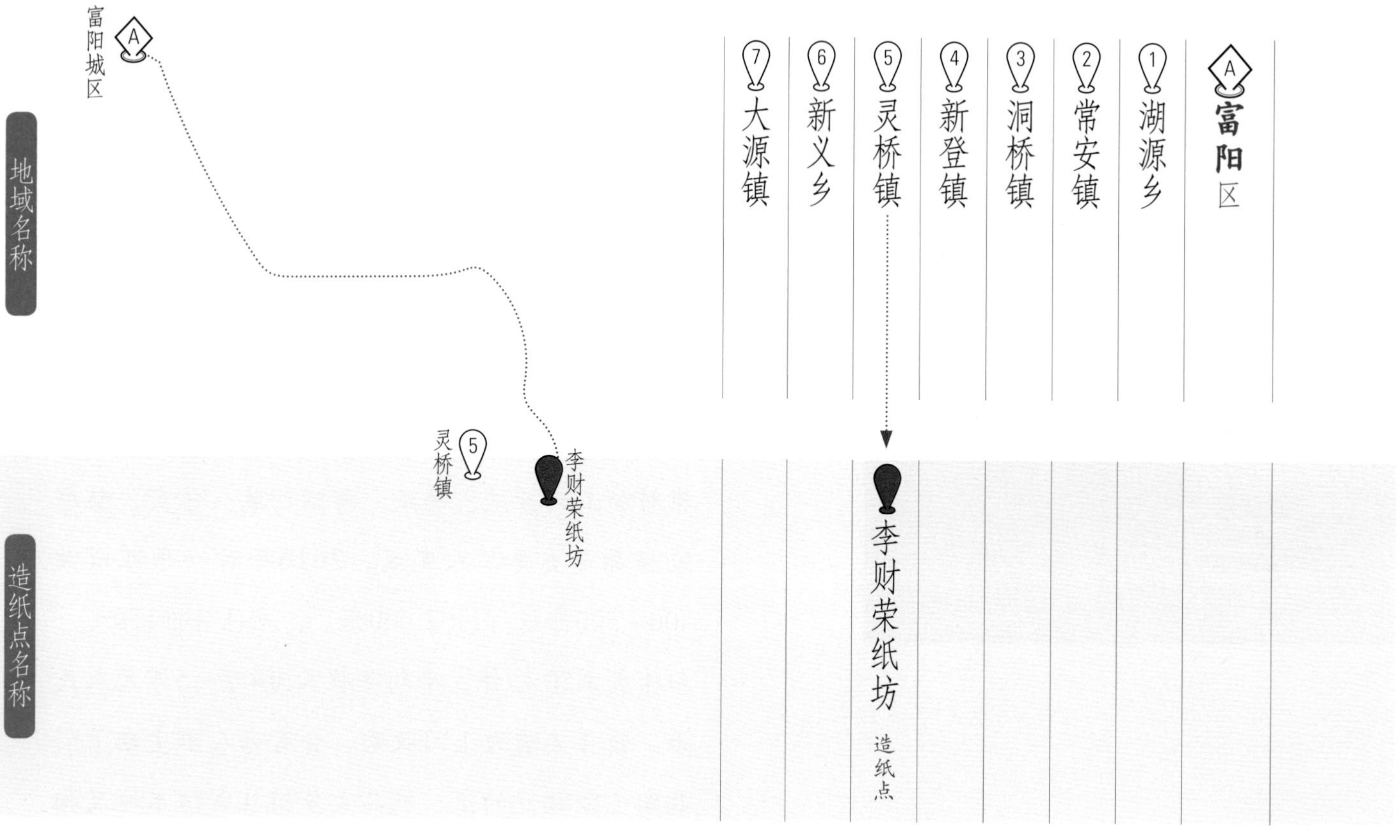

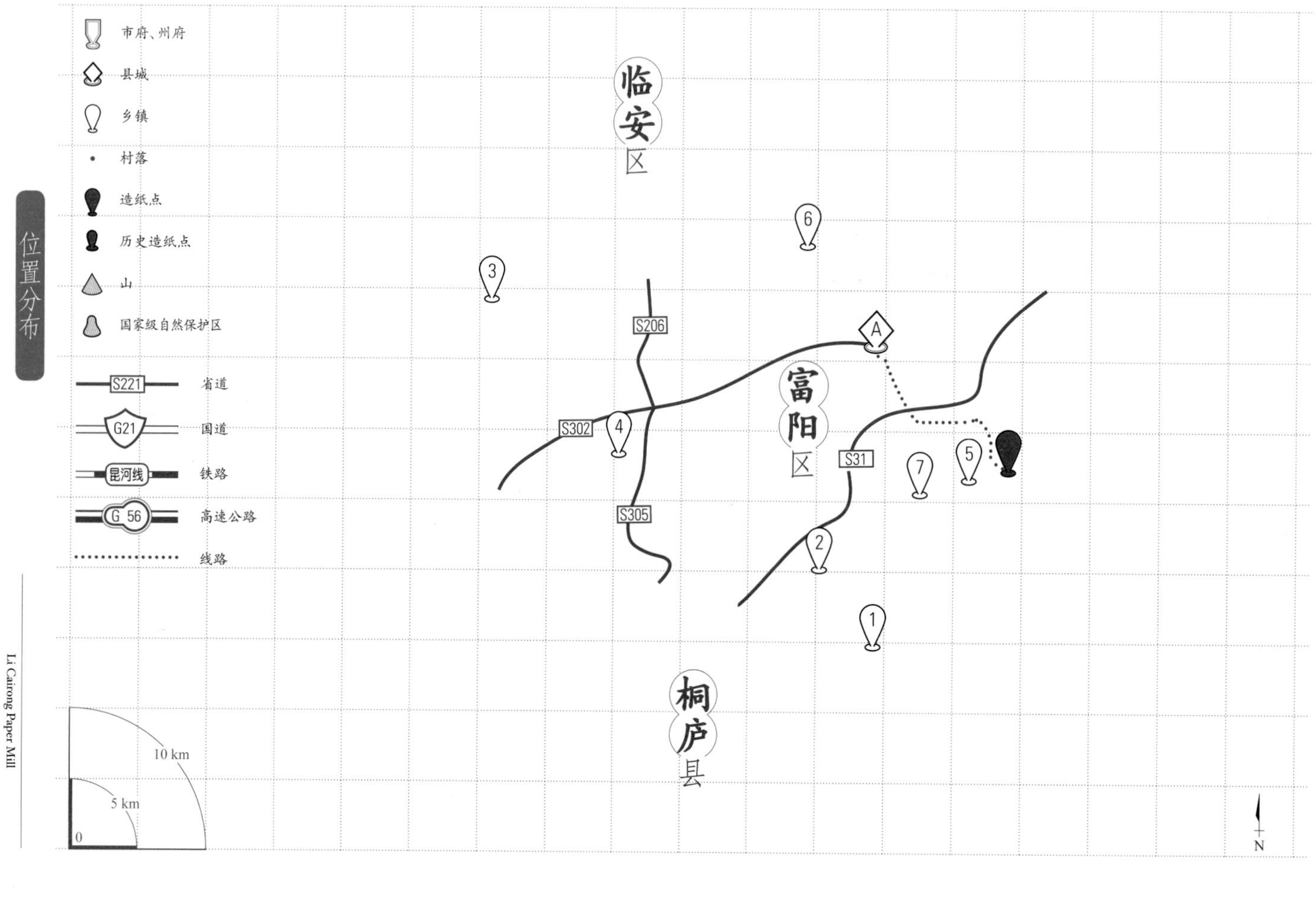

地域名称

A 富阳城区
5 灵桥镇
李财荣纸坊

A 富阳区
1 湖源乡
2 常安镇
3 洞桥镇
4 新登镇
5 灵桥镇
6 新义乡
7 大源镇

造纸点名称

李财荣纸坊 造纸点

位置分布

市府、州府
县城
乡镇
村落
造纸点
历史造纸点
山
国家级自然保护区
S221 省道
G21 国道
昆河线 铁路
G56 高速公路
线路

临安区
富阳区
桐庐县
S206
S302
S305
S31

0
5 km
10 km
N

二
李财荣纸坊的历史与传承

2
History and Inheritance of Li Cairong Paper Mill

⊙1

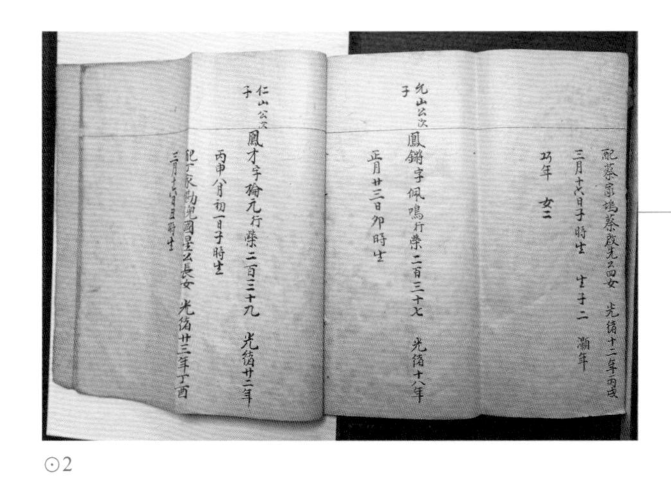

⊙2

据李财荣回忆，李家祖辈均从事手工造纸工作。调查组综合2016年8月23日、2019年1月25日两次入村访谈，家谱资料及李财荣的叙述，整理出李财荣家族造纸的历史及传承谱系如下：

李家祖辈从第一代鹿翔公至李财荣所在的第十七代*，主要男性劳动力均曾以手工造纸业为生。

李财荣的天祖父李元鋐（第十二代）生于清乾隆五十年（1785年），会做纸，李氏家庙即由其建造。李财荣高祖父李云炳（第十三代，生于嘉庆二十五年（1820年）、伯高祖父李云璨（第十三代，出生年月不详）均会做纸。李财荣的曾祖父李允山（生于1870年），叔曾祖父李富山（出生年月不详）、李仁山（出生年月不详）会做纸。李财荣的祖父李凤锵（1892—1973年），先在自己家纸坊做纸，1949年后在生产队做纸，最擅长捞纸，一直做到75岁左右；伯祖父李凤槐（出生年月不详），会做纸。李财荣的父亲李全年（1927—2012年），十七八岁开始跟其爷爷学做纸，擅长捞纸，一直做到82岁，刚开始在生产队做纸，分产到户后在自己家的纸坊做纸；伯父李开年1922年出生，会做纸。

李财荣的大哥1956年出生，晒纸技艺最好，16岁左右跟其爷爷学习造纸，调查时仍在自己家的纸坊做纸；二哥李松浩1961年出生，技艺传自父亲，高中毕业19岁左右开始学习造纸，最擅长捞纸，但已10余年不做纸，现在杭州一家电器公司上班；李财荣1965年出生，16岁开始学习捞纸，17岁时便开始自己尝试捞纸，技艺传承自父亲，所有造纸流程都会，现主要在其家庭作坊中负责捞纸。

根据李财荣的介绍，因为做手工纸不赚钱且特别辛苦，村里的年轻一代很少有人愿意从事手工造纸业，大多出去上学或者打工，所从事的工作也都与手工纸关系不大。李家第十八代即其儿

* 根据李财荣提供的家谱资料，其家族第十一代至第一代祖先分别为：士明，维正，伯儁，荣春，澄廿四，本十六，端二，仲端，士谟，芳蕙，鹿翔公。

⊙1 李氏家庙 Family Temple of the Lis

⊙2 李氏家谱图 Family genealogy of the Lis

子、侄子、侄女均不会造纸：李财荣的独子李立邦1993年出生，大学毕业后在宁波一家外资航空物流公司工作；李财荣大哥家的独子李力华，二哥家的女儿李柳青、李龙英等大学毕业后均在杭州的房地产、医院、制药厂等单位工作，未接触过手工造纸行业，李财荣认为他们也不会传承父辈的手工造纸技艺。

李财荣的妻子张利华1968年出生，为邻村倪家滩人（现属新华村），嫁入李家已27年（截至2019年）。据张利华介绍，其娘家祖辈也以造纸为生：曾祖父张正福，年龄记不清，会造纸；爷爷张关彐（“彐”为“雪”的简化字），年龄已记不清，会做纸；父亲张树顺1943年生，会做纸。

张利华家兄弟姊妹四人，其是家中长女，还有一个妹妹和两个弟弟，她与妹妹和其中一个弟弟会做纸，最小的弟弟没有接触过造纸。张利华会捞纸、晒纸，妹妹张笑来（1971年生）会晒纸，技艺均传承自父亲。1982年不再集体造纸，资源分包到户后，两姐妹一边念书，一边跟着父亲学做纸。妹妹张笑来23岁左右嫁至萧山后不再做纸。弟弟张荣飞1973年出生，十七八岁跟父亲学捞纸，2014～2015年因造纸效益不好外出做卷帘门生意，目前已不再做纸。

李财荣回忆，父亲、祖父那一辈主要做元书纸，1949年后在生产队做纸时，实行多种经营，开始大量做祭祀时使用的黄纸，元书纸只在新原料上市时才做，两种纸品主要销往江苏、山东等地。由于元书纸的收益没有黄纸高，1982年左右资源分包到户后他们的纸坊不再做元书纸，仅做祭祀黄纸。

20世纪50年代至80年代初生产队集体造纸时期是统购统销，做好的纸由村供销社销往江苏等地，所以李财荣记不清楚当时纸品的销售价格。不过据他转述的老一代人的回忆：新中国成立前当地村民均是用扁担将做好的纸挑到杭州售卖，每次最多挑4捆纸，一捆4 000张，大约50 kg。由于路途较远，外出卖纸来回至少需一个礼拜（7天）左右的时间。李财荣表示记得家中老人说，20世纪六七十年代他们家每年有4个劳动力（母亲从事农业生产，父亲及他的2个哥哥从事以造纸为主的副业生产），每人工作一天可得12个工分，每年年底根据生产队当年总收入的不同拿不同的分红，总的来说除日常开销外，一年能存下300～400元，“够养活自己，还能有富余”。

李财荣当时在新华村第二生产队，所属的集

⊙1

⊙2

⊙1
李财荣全家福（老照片翻拍）
A photo of Li Cairong' family members (retake an old photo)

⊙2
张利华（左）
Zhang Lihua (left)

体车间在3个生产队共有18口槽，平均每个生产队拥有6口槽。李财荣在生产队时主要做祭祀用的“黄纸”，原料为当地产的毛竹，再加水泥纸袋，配比大约是4：1，不加染料，纸张大小约为六尺（47 cm × 60 cm）。1982年分山到户后至20世纪90年代，原料加工中开始加入染料。从5年前（2014年左右）开始，“黄纸”的尺寸开始缩小，现在的纸张大小为40 cm × 47 cm。按照李财荣的说法，尺寸缩小是因为效益不好，缩小尺寸可以节约成本。大概从2006年开始，整个村子逐渐不再自己做原料，主要通过电话从富阳北部地区订购原料。

⊙3

⊙4

⊙5

⊙3
李财荣父母（老照片翻拍）
Li Cairong's parents (retake an old photo)

⊙4
李财荣（左）兄弟三人
Li Cairong (left) and his two brothers

⊙5
李财荣夫妇向调查组介绍家族造纸信息
Li Cairong and his wife introducing family papermaking information to a researcher

表10.4 李财荣、张利华家庭纸坊传承谱系
Table 10.4 Genealogy of papermaking inheritors in Li Cairong and Zhang Lihua's family paper mills

	传承代数	姓名	性别	与李财荣关系	基本情况
李财荣传承谱系*	第十二代	李元铉	男	天祖父	生于乾隆五十年（1785年），会做纸，李氏家庙由其建造
	第十三代	李云炳	男	高祖父	生于嘉庆二十五年（1820年），会做纸
		李云璨	男	伯高祖父	生卒年月不详，会做纸
	第十四代	李允山	男	曾祖父	生于1870年，会做纸
		李富山	男	叔曾祖父	生卒年月不详，会做纸
		李仁山	男	叔曾祖父	生卒年月不详，会做纸
	第十五代	李凤锵	男	祖父	生于1892年，卒于1973年，先在自己家纸坊做纸，1949年后在生产队做纸，最擅长捞纸，一直做到75岁左右
		李凤槐	男	伯祖父	生卒年月不详，会做纸
	第十六代	李全年	男	父亲	生于1927年，卒于2012年，十七八岁开始跟其爷爷学做纸，最擅长捞纸，一直做到82岁，刚开始在生产队做纸，分产到户后在自己家的纸坊做纸
		李开年	男	伯父	生于1922年，会做纸
	第十七代	李良松	男	大哥	生于1956年，晒纸技艺最好，16岁左右跟其爷爷学习，目前仍在自己家的纸坊做纸
		李松浩	男	二哥	生于1961年，技艺传自父亲，高中毕业19岁左右开始学习，最擅长捞纸，已经10余年不做纸，现在杭州一家电器公司上班
		李财荣	男	—	生于1965年，16岁开始学捞纸，技艺传承自父亲，所有流程都会，现主要负责捞纸
	第十八代	李立邦	男	儿子	李财荣独子，生于1993年，大学毕业后在宁波中外运敦豪公司工作，不会做纸
		李力华	男	侄子	大哥李良松独子，不会做纸
		李柳青	女	侄女	二哥李松浩的女儿，不会做纸
		李龙英	女	侄女	二哥李松浩的女儿，不会做纸

	传承代数	姓名	性别	与张利华关系	基本情况
张利华传承谱系	第一代	张正福	男	曾祖父	生卒年月不详，会做纸
	第二代	张关彐（同“雪”）	男	祖父	生卒年月不详，会做纸
	第三代	张树顺	男	父亲	生于1943年，倪家滩（现属新华村）人，会做纸
	第四代	张利华	女	—	生于1968年，倪家滩（现属新华村）人，会捞纸、晒纸，技艺均传承自父亲
		张笑来	女	妹妹	生于1971年，1982年左右分产到户后开始学习晒纸，23岁左右嫁到萧山后不再从事造纸行业
		张荣飞	男	弟弟	生于1973年，十七八岁时跟父亲学习造纸，2014～2015年因造纸效益差不再做纸去外地做卷帘门生意

*：根据李财荣的回忆及其家谱，其家族中第十一代至第一代祖先分别为：士明，维正，伯信，荣春，澄廿四，本十六，端二，仲端，士谟，芳蕙，鹿翔公，均从事造纸。由于年代较久，不纳入详细谱系，本谱系图从第十二代开始。

三 李财荣纸坊的代表纸品及其用途与技术分析

3 Representative Paper and Its Uses and Technical Analysis of Li Cairong Paper Mill

（一）李财荣纸坊代表纸品及其用途

李财荣纸坊的纸品很单一，全部都是祭祀竹纸，以前叫“黄纸”。纸的原料是老毛竹掺棉花和水泥袋，配比是60%老毛竹+20%水泥袋+20%棉花，造出的纸销往富阳、萧山和临安等地，主要用于祭祀祖先和逝去的亲人，还可折成元宝状焚烧，但新华村的造纸人只负责造原纸，不负责折成元宝。

（二）李财荣纸坊祭祀黄纸性能分析

测试小组对采样自李财荣纸坊的祭祀黄纸所做的性能分析，主要包括定量、厚度、紧度、抗张力、抗张强度、白度、纤维长度和纤维宽度等。按相应要求，每一指标都重复测量若干次后求平均值，其中定量抽取5个样本进行测试，厚度抽取10个样本进行测试，抗张力抽取20个样本进行测试，白度抽取10个样本进行测试，纤维长度测试了200根纤维，纤维宽度测试了300根纤维。对李财荣纸坊祭祀黄纸进行测试分析所得到的相关性能参数如表10.5所示，表中列出了各参数的最大值、最小值及测量若干次所得到的平均值或者计算结果。

⊙1

⊙1 李财荣家中存放的整捆祭祀黄纸 Bundles of yellow paper for sacrificial purposes in Li Cairong's house

表10.5 李财荣纸坊祭祀黄纸相关性能参数
Table 10.5 Performance parameters of yellow paper for sacrificial purposes in Li Cairong Paper Mill

指标		单位	最大值	最小值	平均值	结果
定量		g/m²				42.5
厚度		mm	0.230	0.185	0.204	0.204
紧度		g/cm³				0.208
抗张力	纵向	mN	4.4	3.3	3.8	3.8
	横向	mN	4.5	3.3	3.5	3.5
抗张强度		kN/m				0.240
白度		%	61.9	61.0	61.4	61.4
纤维	长度	mm	2.5	0.1	0.7	0.7
	宽度	μm	32.0	0.7	9.3	9.3

性能分析

由表10.5可知，所测李财荣纸坊祭祀黄纸的平均定量为42.5 g/m^2。李财荣纸坊祭祀黄纸最厚约是最薄的1.243倍，经计算，其相对标准偏差为0.128。通过计算可知，李财荣纸坊祭祀黄纸紧度为0.208 g/cm^3。抗张强度为0.240 kN/m。

所测李财荣纸坊祭祀黄纸平均白度为61.4%。白度最大值是最小值的1.064倍，相对标准偏差为0.039。

李财荣纸坊祭祀黄纸在10倍和20倍物镜下观测的纤维形态分别如图★1、图★2所示。所测李财荣纸坊祭祀黄纸纤维长度：最长1.5 mm，最短0.1 mm，平均长度为0.7 mm；纤维宽度：最宽32.0 μm，最窄0.7 μm，平均宽度为9.3 μm。

★1
李财荣纸坊祭祀黄纸纤维形态图（10×）
Fibers of yellow paper for sacrificial purposes in Li Cairong Paper Mill (10× objective)

★2
李财荣纸坊祭祀黄纸纤维形态图（20×）
Fibers of yellow paper for sacrificial purposes in Li Cairong Paper Mill (20× objective)

四 李财荣纸坊祭祀黄纸的生产原料、工艺与设备

4 Raw Materials, Papermaking Techniques and Tools for Yellow Paper for Sacrificial Purposes in Li Cairong Paper Mill

（一）李财荣纸坊祭祀黄纸的生产原料

1. 主料一：老毛竹

李财荣纸坊使用的竹原料是从市面上购入的老毛竹浆料，2016年8月的零售规格是25 kg/袋，1元/kg，每袋售价25元人民币。据李财荣回忆，2006年之前，当地造纸户自己上山砍竹做料，后因工人数量少、工资成本高、上山砍竹太辛苦和危险、本地竹子较少等原因，2006年之后开始从富阳北部买经过化学处理并粉碎好的老竹浆料。李财荣纸坊每年需购买约1万kg竹料。

2. 主料二：棉花

李财荣纸坊使用的棉花是废弃棉花，多为从绍兴、萧山等地纺织厂购入的废弃边角料，2009年即开始使用，购入的价格为2.4元/kg。

3. 主料三：水泥袋

李财荣纸坊使用的水泥袋从本地建筑工地购买，单个售价0.6元。

⊙1

⊙2

⊙3

⊙4

⊙1/2 李财荣纸坊购买的老竹浆料 Old bamboo pulp materials purchased in Li Cairong Paper Mill

⊙3 废弃棉花 Abandoned cotton

⊙4 水泥袋 Cement bags

4. 辅料一：水

水的好坏一定程度上决定着纸的好坏，虽然李财荣造的是低端黄纸，但水源质量依然对纸品质量有影响。调查组成员现场测试李财荣纸坊使用的清水pH为5.5～6.0，偏酸性。

5. 辅料二：嫩黄染料

李财荣纸坊使用的染料叫嫩黄，是一种化工原料，购自天津，1983年左右售价为11元/kg，现在售价为2.5元/kg。李财荣回忆，虽然其使用的嫩黄一直购自天津，但近年来（2009年开始）随着价格的下降，质量也下降不少，现在需加入0.5 kg嫩黄才能达到当年2调羹量的染色效果。

⊙1

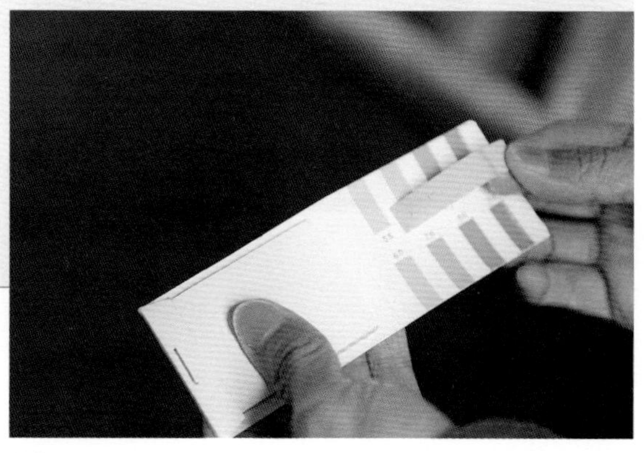

⊙2

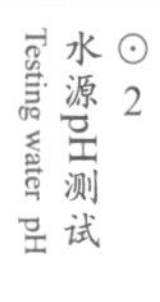

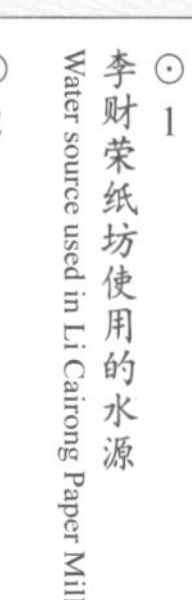

⊙1
李财荣纸坊使用的水源
Water source used in Li Cairong Paper Mill

⊙2
水源pH测试
Testing water pH

（二）李财荣纸坊祭祀黄纸的生产工艺流程

据李财荣介绍，其纸坊生产祭祀黄纸的主要工艺流程为：

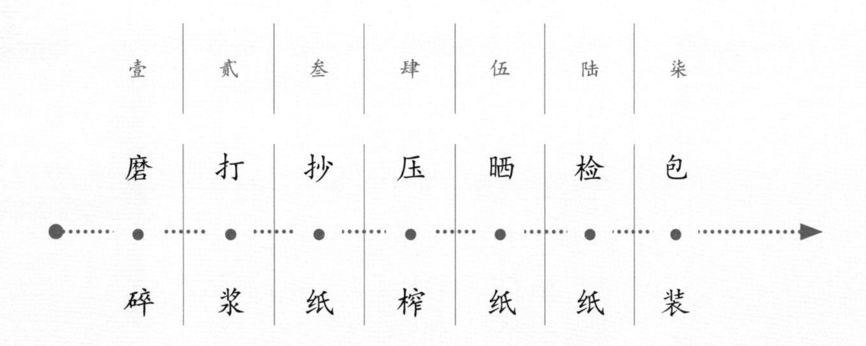

壹 磨碎

1 ⊙3

购买的老竹浆料和水泥袋，需先用磨浆机磨碎。棉花不用磨碎，可直接进行打浆。

⊙3

贰 打浆

2 ⊙4

先在打浆池中加入清水，将一定量的棉花放入打浆池中搅拌5分钟左右，再放入磨碎的老毛竹和水泥袋原料一起打浆，一般打15～20分钟，再用磨浆机匀速搅拌好待用。

⊙4

⊙5

叁 抄纸

3 ⊙5

抄纸前首先将和单槽棍从自己身前向外顺时针椭圆状推开，此时一定要匀速，等到和成纸槽中心成旋涡状即可。然后抄纸工拿着纸帘，从上到下倾斜20°左右下到槽内，再缓慢向身前方向提上来，出水面时纸帘朝前倾斜，将多余的纸浆匀出。最后将纸帘从帘架上抬起，把抄好的湿纸放在旁边的纸架上，这样一张湿纸就完成了。纸帖是倾斜放的，这样可以让水流到一边。

调查组现场询问李财荣得知，李财荣每天需要工作8个小时，一般一天可以捞6 000张（1 500帘，每帘可分为4张纸）40 cm×47 cm的祭祀黄纸。

⊙3 石磨碾料 Grinding the materials with a stone roller

⊙4 打浆池 Beating pool

⊙5 抄纸 Papermaking

肆 压榨

4 ⊙6

李财荣介绍，其纸坊每天要压榨2次湿纸帖，上午、下午各1次。压榨时使用千斤顶，力度从小到大缓慢进行，直到手挤不出水时，压榨完成。

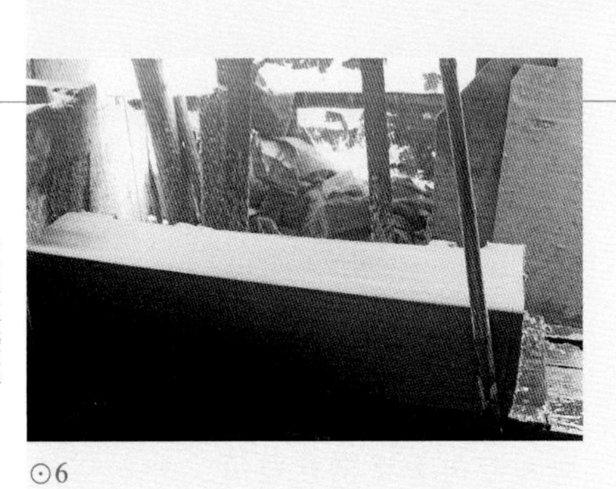

⊙6

伍 晒纸

5 ⊙7～⊙10

首先用鹅榔头在压干的纸帖四边划一下，让纸松散开。然后捏住纸的右上角捻一下，这样右上角的纸就翘起来了。再用嘴巴吹一下，粘在一起的纸就分开了。最后，用手沿着纸的右上角将纸帖中的纸揭下来，铺在向阳的地面上或者用杆子架起来晒干。张利华访谈中说，她每天从事晒纸工作至少15个小时，早上6点左右开始晒，大概每天晒4件（8 000张），当天捞的纸当天晒。天气好时晒1天即可，冬天需晒2天左右。

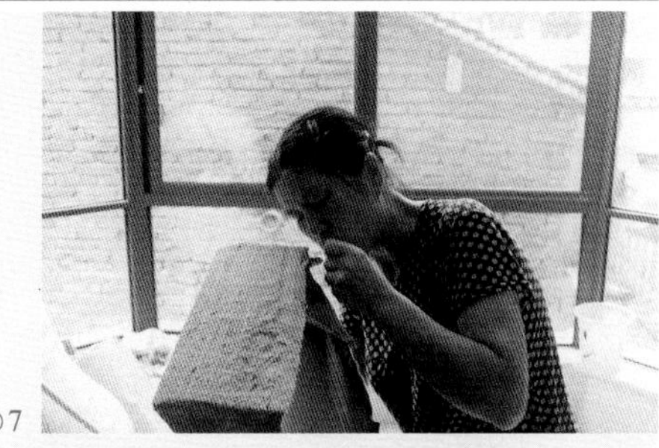

⊙7

⊙8

⊙9

⊙10

⊙6 待压榨的纸帖 Paper pile to be pressed

⊙7/8 揭纸 Peeling the paper down

⊙9/10 晒纸 Drying the Paper

陆 检纸

6 ⊙11

对晒好的纸进行检验，挑选出合格的纸，不合格的纸回笼打浆。由于规格是确定的，李财荣家不需要剪纸边。

⊙11

柒 包装

7 ⊙12⊙13

晒好、检好的纸放在一个干燥通风的房间，按照客户要求或者按照2 000张1捆打包。

⊙12

⊙13

⊙11 检纸 Checking the paper

⊙12 整理 Sorting the paper

⊙13 包装 Packaging the paper

（三）李财荣纸坊祭祀黄纸的主要制作工具

壹 打浆槽 1

用来盛浆打浆的设施。实测李财荣纸坊所用的打浆槽尺寸为：长80 cm，宽70 cm，高73.5 cm。张利华介绍，该村仅有1个打浆槽，由15户共用，每户使用时会做好一个星期使用的浆料。打浆机从大源镇购买，张利华称价格已记不清。

⊙1

贰 手工抄纸槽 2

调查时为水泥浇筑，捞纸时使用。实测李财荣纸坊的抄纸槽尺寸为：长222 cm，宽188 cm，高98.5 cm。

⊙2

叁 石　磨 3

用于磨碎竹料、水泥袋等原料。李财荣纸坊使用的石磨为新华村现存的15户造纸户集资购买，每人出资1.5万元。每次使用可磨1个星期的量。实测李财荣纸坊所用的石磨尺寸为：底部圆盘直径271 cm，高68 cm；石磨直径113 cm，厚43 cm。

⊙3

肆 泡料池 4

李财荣纸坊使用的泡料池（共有7口）为集体（7户）共有，每家使用1口泡料池，本村剩余8户造纸户的泡料池在本村别处。

⊙4

⊙1 打浆槽 Beating trough
⊙2 手工抄纸槽 Papermaking trough
⊙3 石磨 Stone roller
⊙4 泡料池 Soaking pool

伍

滑　板

5

在石磨上磨料需使用滑板翻料。实测李财荣纸坊使用的滑板尺寸为：总长141.5 cm；板长25 cm，宽15 cm；柄长127 cm。

⊙5

陆

压榨机

6

榨湿纸帖时使用。实测李财荣纸坊使用的压榨机尺寸为：臂长86 cm，高139 cm，底长85.5 cm，底宽81.5 cm。

⊙6

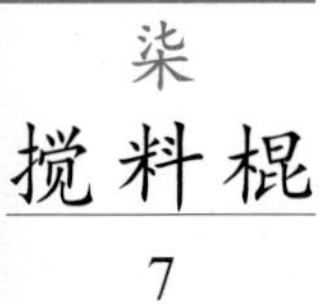

柒

搅料棍

7

⊙7

打浆时使用，用于搅拌浆料。实测李财荣纸坊使用的搅料棍尺寸为：长166 cm，板长27 cm，板宽27 cm。

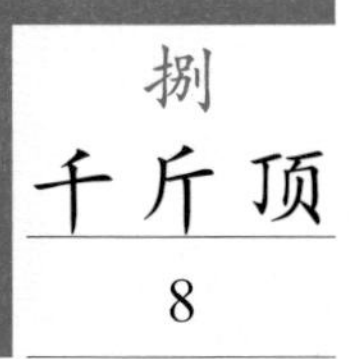

捌

千斤顶

8

榨纸时使用。实测李财荣纸坊使用的千斤顶尺寸为：高18 cm，底宽15 cm，底长17 cm。

⊙8

玖

纸　帘

9

抄纸时使用，苦竹丝编织而成，表面光滑平整，帘纹细而密集。实测李财荣纸坊所用的纸帘尺寸为：长144 cm，宽47 cm。纸帘从灵桥镇光明村购买，2015年时价格为500元/张，一般可以用1年左右。

⊙9

⊙5 滑板 Stick for stirring the paper pulp

⊙6 压榨机 Pressing machine

⊙7 搅料棍 Stirring stick

⊙8 千斤顶 Lifting jack

⊙9 纸帘 Papermaking screen

拾 帘架 10

支承纸帘的架子，硬木制作。实测李财荣纸坊所用的帘架尺寸为：长137 cm，宽38 cm。帘架是李财荣自己请木工加工的，一般可以用2年左右，帘子和帘床共花费350元左右。

拾壹 帘滤 11

帘滤是纸槽中滤浆的帘子，使用时将其竖放在纸槽中间，用于分隔浆料，使纸槽中的纸浆不至于太浓。竹子所编，专门请人加工，价格为120元/张。实测李财荣纸坊所用的帘滤尺寸为：长192 cm，宽100 cm。

拾贰 鹅榔头 12

牵纸前打松纸帖的工具，杉木制作。实测李财荣纸坊所用的鹅榔头尺寸为：长15.5 cm，直径2 cm。

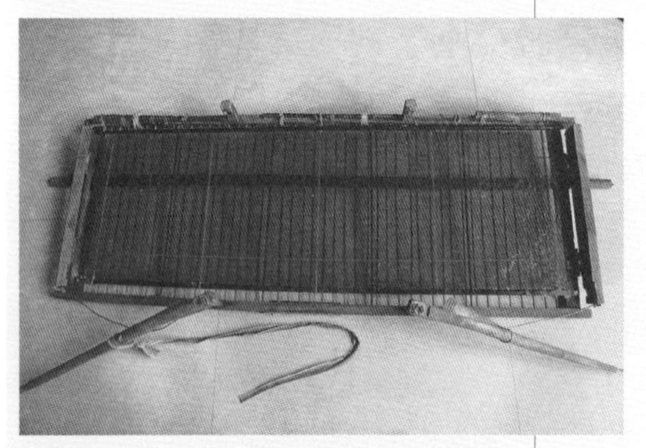

⊙10

⊙11

⊙12

⊙10 帘架 Frame for supporting the papermaking screen

⊙11 帘滤 Screen filter

⊙12 鹅榔头 Goose hammer

五
李财荣纸坊的市场经营状况

5
Marketing Status of Li Cairong Paper Mill

综合2016年8月、2019年1月调查组两次入纸坊调查了解到的信息，李财荣纸坊市场经营情况如下：

据李财荣介绍，其造纸坊2016年时年产400～500件祭祀黄纸（一件2 000张），每件售价150元，年销售额为60 000～75 000元。生产出的祭祀黄纸一般销往富阳本地，以及紧邻的杭州市萧山、临安区。

2019年1月回访时李财荣表示：由于现在当地做黄纸的人少了，人工工资增高，2018～2019年每件售价增至180元。按照年产800件纸的数量计算，销售额14万元左右，扣去每件100元的成本，纯利润保守估计为5万～6万元。

据李财荣、张利华提供的信息，15年前（2004年左右）当地还有80余户人家做纸，2019年1月25日调查组回访时该村只有15户人家做纸。当地造纸户从不主动出去卖纸，会有人上门收纸，但李财荣并不清楚收购商的个人信息。收购商收纸后，销往富阳、萧山、临安等地。

⊙1

⊙1
李财荣家生产的祭祀黄纸
Yellow paper for sacrificial purposes in Li Cairong Paper Mill

六 李财荣纸坊的文化与习俗故事

6 Culture and Stories of Li Cairong Paper Mill

（一）生产队时期的回忆：捞纸能手奖励“小红旗”

结合李财荣、张利华的回忆，以前村里做纸都是男性，女性因为体力小很少从事捞纸工作，生产队解体分槽到户以后，女性开始学习晒纸。

李财荣回忆，生产队一般5个人一组，1个捞纸，2个晒纸，2个做料，李财荣主要负责捞纸，捞纸分到的钱比其他工序高10%左右。为提高积极性，生产队设立奖励机制，按月计算，一口槽捞纸超出100件有奖励，多劳多得。具体来说，捞得多的人，会被奖励一个红旗挂在自己使用的纸槽边。另外，同样份量的原料做的件数多，说明你节约了原料，也会有相应的红旗奖励。根据李财荣的回忆，当时他与另外两个小伙子负责捞纸，因年纪小动作快，经常会得到红旗奖励。

⊙2

（二）习俗趣事

1.“贱骨金口”——罗隐秀才的故事

2019年1月25日调查组访谈李财荣时得知，当地广泛流传着罗隐秀才（又称龙游秀才）的故事。为了能详细生动地介绍罗隐秀才的故事，李财荣专门请当地另一位擅长讲故事的造纸户李善强（1954年生，目前仍在造纸）为调查组讲述。调查组结合李善强的口述整理如下：

⊙3

据称罗隐小时候便有皇帝命，读书时从李家庙门前走过，庙里的菩萨会向他低头弯腰作揖。罗隐将此事告诉母亲，母亲不相信，拿了三个鸡蛋放到菩萨腿上，如果菩萨真的低头作揖，鸡蛋会掉下来。果然，罗隐再经过时，菩萨腿上放着的鸡蛋掉落，母亲才相信儿子是皇帝命。

有一次母亲做饭时在灶台前絮絮叨叨，将此前受的苦以及别人对他们的不好一件件说出

⊙2 正在捞纸的李财荣 Li Cairong making the paper

⊙3 为调查组讲故事的造纸户李善强 Papermaker Li Shanqiang telling stories to the researchers

来，一边说话，一边用火钳打灶沿，“谁不借我们米，我们饿了好几天，一遭；谁不肯借钱，一遭；谁欺负了他们，又一遭”。母亲说的话惊动了灶王爷（灶神），灶王爷错将“一遭，一遭，又一遭”听成“一刀，一刀，又一刀”，以为她要大开杀戒，便向玉皇大帝告状，称如果让罗隐当皇帝，可能会滥杀无辜，百姓要遭难。玉皇大帝听了灶王爷的话后深信不疑，派人给罗隐换骨，将其皇帝命换成乞丐命。天兵天将下凡为罗隐换骨时，罗隐躲在母亲的围裙下，嘴巴紧紧咬着围裙布。后虽然被换骨，由皇帝命变成乞丐命，但咬着围裙布的嘴巴没有被换掉，仍然是一张金口，便是“贱骨金口”。

传闻罗隐的“金口”说话很灵验。有一次罗隐路途中被一辆水车挡了路，便向踩水车的村民恳请让路，村民不耐烦，让他从水车底下穿过去。罗隐随口说“人往车底走，水往沙里走”。结果河流里的水真的没了，水车自然也抽不出水来。

富阳大源镇一带山脉毛竹众多，从山脚长到山上，传闻便是由于罗隐秀才开了“金口”。据称，罗隐到大源时口渴向当地村民讨茶喝，村民大方地送了他一杯茶。罗隐很感动，看到他们正在种毛竹，便告诉他们，“你们只要在下面种种，竹子会自己长到山顶上的”。而到富阳小源（今新华村一带）时，当地村民吝于提供茶水饭菜，罗隐便“金口”一开，预言小源一带的毛竹“种死也长不到山顶”。调查组现场观察时确实发现，新华村当地的毛竹只长到半山腰，山顶没有毛竹。

2. 逢初一烧纸钱，祭祖烧“真经”

调查组了解到，李财荣家有“逢初一烧纸钱，祭祖烧‘真经’”的习俗。李财荣家每月初一烧的纸钱（又称元宝）由自家生产的黄纸所做，黄纸被叠成元宝形状，每套元宝中放置12个小元宝，元宝上用红颜料写或印上“太平经”“关公经”“灶司经”等字样。

张利华介绍，每家可根据需要制作和选择不同的“经”，他们家最喜欢做“太平经”“关公经”“灶司经”。不同的“经”有着不同的寓意，如“太平经”寓意平平安安，“关公经”寓意身体健康，“灶司经”则为祭祀灶神。

每逢初一，直接在家中用从当地购买的不锈钢炉子（以前新房没盖好时直接在家里的水泥地上）烧纸钱，每次烧3套（1套12个）。

祭祀祖先时，使用的是“真经”，也用自家的黄纸制作。除春节、清明等节日外，当地也会在七月半“鬼节”、冬至等时间烧纸缅怀先人。

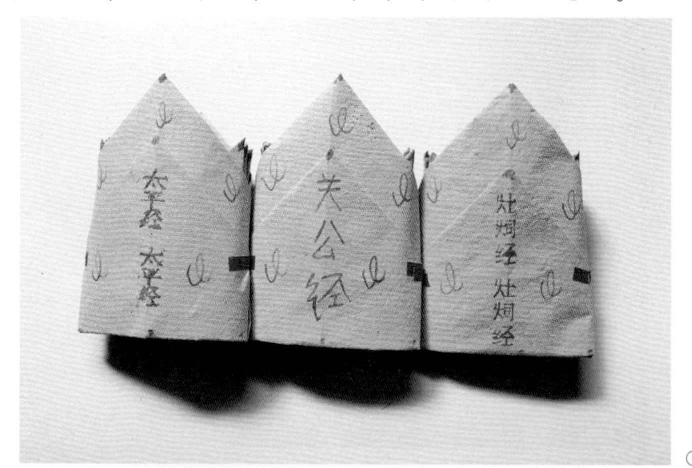

⊙1

⊙2

⊙1 纸钱（元宝）
Joss paper (paper ingot)

⊙2 烧纸时使用的不锈钢炉子
Stainless steel stove used for burning joss paper

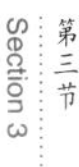

⊙3

⊙4

⊙3
新华村的毛竹多长至山腰
Phyllostachys edulis forest on a hill by Xinhua Village

⊙4
祭祀祖先时使用的『真经』
"True Scriptures" used in sacrificial ceremonies

3. 祭山公山母

造纸户李善强介绍，当地旧时每年砍竹时，曾有祭山公山母（土地公公、土地娘娘）的传统。祭山公山母时，会摆上猪头、鸡、黄酒、蜡烛、黄纸做的元宝（“太平经”“土地经”）等物，向山公山母鞠躬祭拜，请求山公山母保佑砍竹时平安顺利，待蜡烛和香烧好后便可离开。

在李善强的叙述里，祭山公山母有一些比较神秘的色彩。据他所说，山公山母会变成狗来吃祭品。他还听说，某次祭祀中一条狗突然窜出，叼走了祭祀用的肉，被村民打死，可能是因为犯了“众怒”，本来很平安的山脉，当年砍竹时蛇多且不太平安。

⊙1

⊙1
扛纸上山的张利华
Zhang Lihua carring the paper up the hill

七
李财荣纸坊的业态传承现状与发展思考

7
Current Status and Development of Li Cairong Paper Mill

（一）“纸二代”无人造纸，新华村手工造纸面临传承难题

访谈中李财荣表示，李家虽然祖辈都做纸，其父亲和爷爷都是捞纸（元书纸）工人，但由于手工造纸辛苦且赚不到钱，像其儿子、侄女一样的“纸二代”们大学毕业后均在外地工作，村里其他年轻人也选择去外地上学、工作，即便同辈人所从事的工作也大多和手工纸脱离了关系。随着老一代人的去世或渐渐失去劳动能力，当地造纸户正逐年减少，祭祀黄纸这种低端纸的制作技艺面临后继无人的尴尬境地。

李财荣妻子张利华娘家所在的倪家滩（现亦属新华村），也曾是世世代代做纸的村落，现在已无人进行手工纸生产，当地造纸户多外出务工或做卷帘门生意。未来新华村是否会面临同样的境地，是值得思考的问题。

（二）销售过于依赖收购商

调查组调研时了解到，除生产队时统购统销外，当地造纸户从不去外地推销纸品，而是等收购商上门收购。虽然近年来由于人工成本增加，销售价格小幅上涨，但这种过于依赖收购商的单一销售模式可能会面临收购商压价等风险，不利于当地手工造纸业的长远发展。

祭祀黄纸

Yellow Paper for Sacrificial Purposes
of Li Cairong Paper Mill

祭祀黄纸（老竹+棉花+废纸袋）透光摄影图
A photo of yellow paper for sacrificial purposes(old bamboo + cotton + waste paperbag) seen through the light

第四节

李申言金钱纸作坊

浙江省
Zhejiang Province

杭州市
Hangzhou City

富阳区
Fuyang District

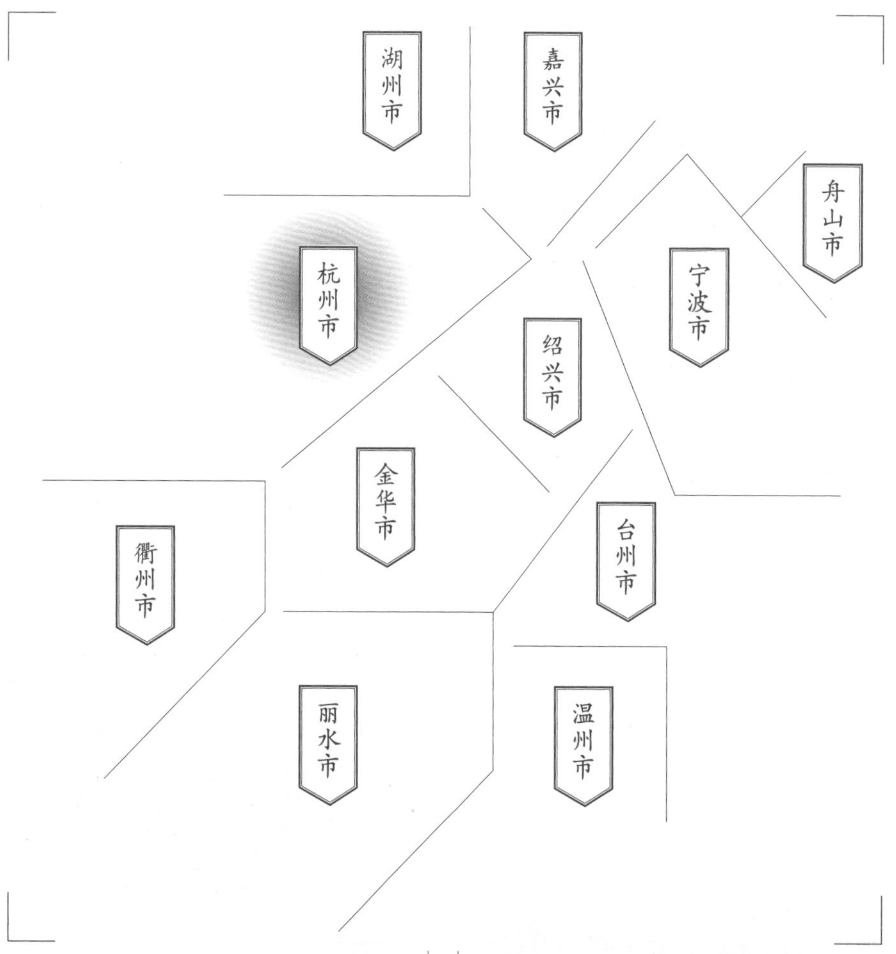

调查对象

富阳区常安镇大田村
李申言金钱纸作坊
竹纸

Section 4
Li Shenyan Joss Paper Mill

Subject

Bamboo Paper of Li Shenyan Joss Paper Mill
in Datian Village of Chang'an Town in Fuyang District

一
李申言金钱纸作坊的基础信息与生产环境

1
Basic Information and Production Environment of Li Shenyan Joss Paper Mill

李申言金钱纸作坊是一家专门生产苦竹原料金钱纸的家庭纸坊，位于富阳区常安镇大田行政村32号，大田村地理坐标为：东经119° 54′ 35″，北纬29° 58′ 32″。作坊负责人为李申言，纸坊以主人名字命名。

2016年8月17日、2016年11月3日以及2019年1月27日调查组三次前往纸坊现场考察，通过李申言的描述了解到的基础信息为：作坊为纯粹家庭式，除了在腌料时请本村的人帮忙外，其他全部工序都由李申言夫妻二人完成；作坊内有纸槽1帘，只生产祭祀时使用的金钱纸这种专门用来打制纸钱的黄色苦竹纸。

大田村为常安镇管辖的16个行政村之一，位于常安镇中心地带，2008年由原来的大田村与六石村合并而成，村域面积7 km²，有6个自然村、16个村民小组。2016年入村调查时村中有七八户以造纸为主业的纸坊，李申言金钱纸作坊便是其中一户。

⊙1

⊙2

⊙3

⊙4

⊙1 李申言家周边的村巷 Alleys near Li Shenyan's house

⊙2/3 李申言捞纸房外景 External view of Li Shenyan's papermaking workshop

⊙4 李申言捞纸房内景 Internal view of Li Shenyan's papermaking workshop

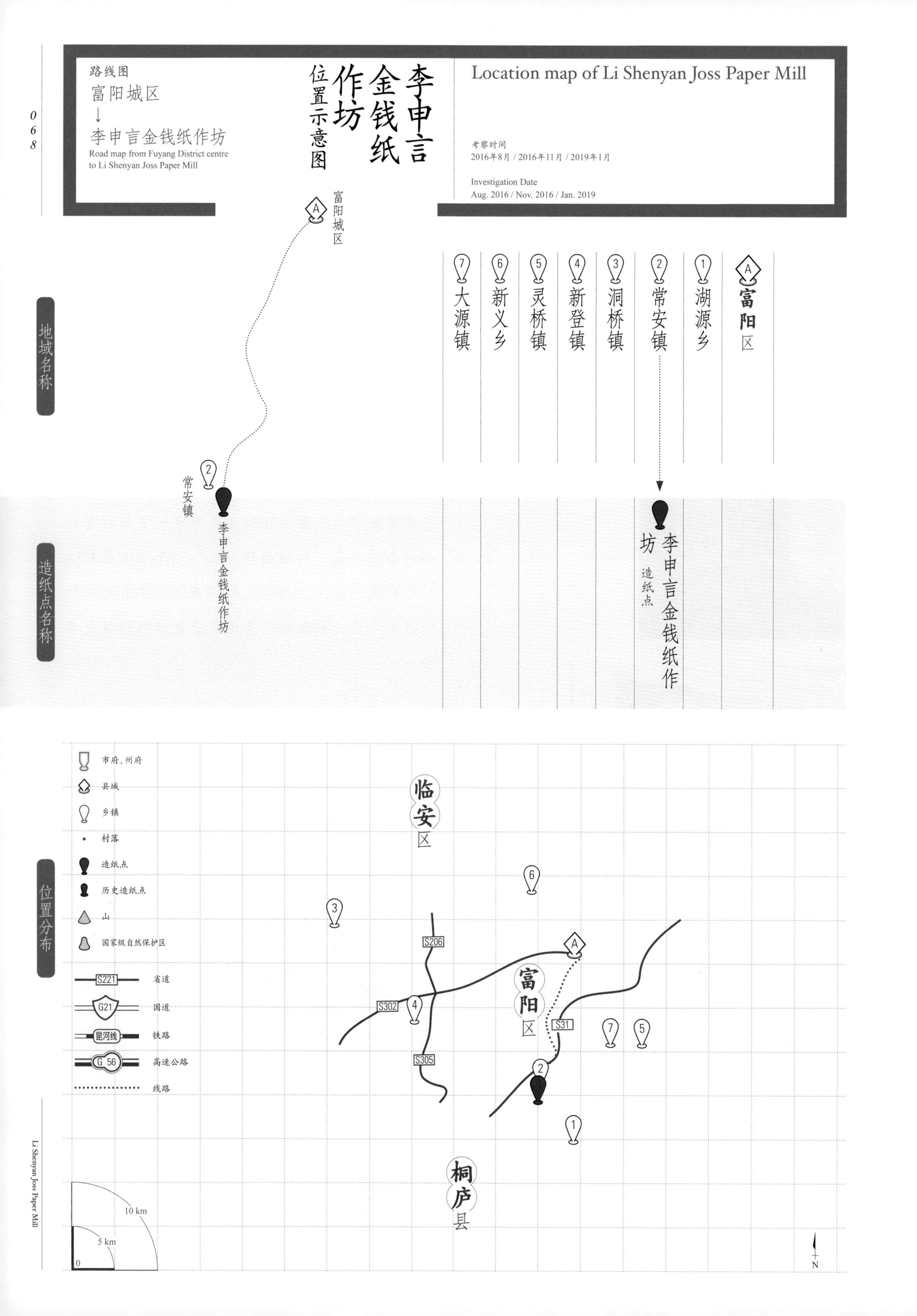
路线图
富阳城区
↓
李申言金钱纸作坊
Road map from Fuyang District centre to Li Shenyan Joss Paper Mill
李申言金钱纸作坊位置示意图
Location map of Li Shenyan Joss Paper Mill
考察时间
2016年8月 / 2016年11月 / 2019年1月
Investigation Date
Aug. 2016 / Nov. 2016 / Jan. 2019
地域名称
造纸点名称
位置分布
A 富阳城区
2 常安镇
李申言金钱纸作坊
A 富阳区
1 湖源乡
2 常安镇
3 洞桥镇
4 新登镇
5 灵桥镇
6 新义乡
7 大源镇
李申言金钱纸作坊 造纸点
市府、州府
县城
乡镇
村落
造纸点
历史造纸点
山
国家级自然保护区
S221 省道
G21 国道
昆河线 铁路
G 56 高速公路
线路
临安区
富阳区
桐庐县
S206
S302
S305
S31
10 km
5 km
0
N

二
李申言金钱纸作坊的历史与传承

2
History and Inheritance of Li Shenyan Joss Paper Mill

李申言，富阳区常安镇大田村人，1951年生，初中学历，2016年调查时65岁。据李申言自述：父亲李增焕1953年就已去世（当时李申言才18个月大），李申言1970年约19岁时开始跟随叔叔李增洪（1978年左右去世）学习造纸。能够造纸后，刚开始的时候是在生产队中负责捞纸，后来到20世纪80年代初生产队解散了，便在自家开槽造纸，一直独立造纸直到现在，已经接近40年了。李申言曾在1984年前后当了3年村干部，但在当村干部期间也一直没有放弃在家中作坊造纸。

在李申言的记忆里，自己家中造纸已经有很多代了，但具体有多少代说不清楚，只是家里的老人们说过，在其高祖时造纸达到鼎盛时期，按照20年一代来推算，至少也有150年以上，也就是清代的咸丰、同治时期。佐证是那时候只有造纸大户才能有印章，而自己家中便有“东升印制”的印章，相当于当时的“老字号”或“名纸号”。但是因为时间久远，李申言自己并没有见过印有印章的纸，也没有见过传说中的印章，只是从祖辈那里听说过有这回事。李申言的妻子李香妹，1952年生，2016年时64岁，也是大田村人。李香妹从小接触造纸，自述从7岁开始就在家中帮忙拔碓。嫁入李申言家以后，一直帮忙晒纸。李申言有1个儿子和1个女儿，儿子初中学历，2016年调查时40岁，会造纸技艺但是不愿意留在村里造纸，于是选择了去外地当油漆工；女儿大学学历，2016年调查时37岁，不会造纸，在药店当药剂师。李申言家庭纸坊常年由老夫妻二人负责全流程运作。

另外一个值得关注的传统是，由于大田村没有竹子资源，据说自祖辈造纸以来，一直是从外地购买苦竹来造纸，这一传统一直延续至今，主要购于桐庐、新登、临安、余杭等地。但是为什么会形成这一情况村里造纸人也不明白。

⊙1

⊙2

⊙1 访谈中的李申言
Interviewing Li Shenyan

⊙2 李申言夫妻二人与调查组成员（右一）
Li Shenyan and his wife with a researcher (first one from the right)

表10.6 李申言家族造纸传承谱系
Table 10.6 Li Shenyan's family genealogy of papermaking inheritors

传承代数	姓名	性别	与李申言关系	基本情况
第一代	李元梁	男	祖父	1970年去世，造纸的各道工序技艺熟练
	汪阿兰	女	祖母	1970年去世，会晒纸
第二代	李增焕	男	父亲	1923年出生，1953年去世，造纸的各道工序技艺熟练
	李增洪	男	叔叔	生卒年月不详，造纸的各道工序技艺熟练
第三代	李申言	男	—	1951年出生，造纸的各道工序技艺熟练
第四代	李杰	男	儿子	1977年出生，会造纸的多道工序

三
李申言金钱纸作坊的代表纸品及其用途与技术分析

3
Representative Paper and Its Uses and Technical Analysis of Li Shenyan Joss Paper Mill

（一）李申言金钱纸作坊代表纸品及其用途

据2016年入村调查的信息：李申言作坊生产的纸有3种尺寸规格：90 cm × 26 cm，100 cm × 23 cm以及88 cm × 17 cm，3种纸的原料完全一样，都是苦竹，只是在尺寸和捞纸工序上有细微差别。用途也一样，都是用来制作祭祀用的纸钱。不过李申言表示，在30多年前供销社收购纸的时期，村里造的88 cm × 17 cm的竹纸曾被一分为四用作卫生纸，现在当地称为四六屏纸。李申言造的100 cm × 23 cm的纸当地称为长帘纸，卖至江苏苏州后被称为金钱纸。90 cm × 26 cm的纸也叫长帘纸。

（二）李申言金钱纸作坊金钱纸性能分析

测试小组对采样自李申言纸坊的金钱纸所做的性能分析，主要包括定量、厚度、紧度、抗张力、抗张强度、白度、纤维长度和纤维宽度等。按相应要求，每一指标都重复测量若干次后求平均值，其中定量抽取5个样本进行测试，厚度抽取10个样本进行测试，抗张力抽取20个样本进行测试，白度抽取10个样本进行测试，纤维长度测试了200根纤维，纤维宽度测试了300根纤维。对李申言纸坊金钱纸进行测试分析所得到的相关性能参数如表10.7所示，表中列出了各参数的最大值、最小值及测量若干次所得到的平均值或者计算结果。

表10.7　李申言金钱纸作坊金钱纸相关性能参数
Table 10.7　Performance parameters of joss paper in Li Shenyan Joss Paper Mill

指标		单位	最大值	最小值	平均值	结果
定量		g/m^2				58.9
厚度		mm	0.218	0.198	0.212	0.212
紧度		g/cm^3				0.278
抗张力	纵向	mN	15.9	14.3	15.5	15.5
	横向	mN	8.1	7.4	7.9	7.9
抗张强度		kN/m				0.780
白度		%	42.0	38.7	40.4	40.4
纤维	长度	mm	1.4	0.1	0.5	0.5
	宽度	μm	54.6	0.7	9.9	9.9

由表10.7可知，所测李申言纸坊金钱纸的平均定量为58.9 g/m^2。李申言纸坊金钱纸最厚约是最薄的1.101倍，经计算，其相对标准偏差为0.018。通过计算可知，李申言纸坊金钱纸紧度为0.278 g/cm^3。抗张强度为0.780 kN/m。

所测李申言纸坊金钱纸平均白度为40.4%。白度最大值是最小值的1.085倍，相对标准偏差为0.030。

李申言纸坊金钱纸在10倍和20倍物镜下观测的纤维形态分别如图★1、图★2所示。所测李申言纸坊金钱纸纤维长度：最长1.4 mm，最短0.1 mm，平均长度为0.5 mm；纤维宽度：最宽54.6 μm，最窄0.7 μm，平均宽度为9.9 μm。

★1
李申言纸坊金钱纸纤维形态图（10×）
Fibers of joss paper in Li Shenyan Joss Paper Mill (10× objective)

★2
李申言纸坊金钱纸纤维形态图（20×）
Fibers of joss paper in Li Shenyan Joss Paper Mill (20× objective)

四
李申言金钱纸作坊金钱纸的生产原料、工艺与设备

4
Raw Materials, Papermaking Techniques and Tools for Joss Paper in Li Shenyan Joss Paper Mill

（一）金钱纸的生产原料

1. 主料：苦竹

李申言纸坊生产的金钱纸原料是苦竹。大田村只有良田，没有竹子生长，在农民无法靠种田自给自足的年代，不知是什么缘由或缘分，大田村的农户想到了购买苦竹造纸。于是，用苦竹造纸成了大田村的传统。据李申言介绍，造金钱纸用的苦竹必须是生长数年的老苦竹，因为老苦竹的胀性好，容易发胀，出浆率高，嫩苦竹和早笋竹都不行，造出来的纸会很脆。李申言纸坊用的苦竹是从邻近的余杭区购买的，每年下半年的11月份，经销商砍好苦竹以后送上门，2015年的购买价格是 68元/100 kg。李申言2015年购买了50 000 kg苦竹。

2. 辅料：水

大田村水源丰富，李申言直接就近取水，引附近的山泉水入作坊，作为制作浆料和捞纸的重要辅料。调查组成员现场取样检测，山泉水的pH为5.5左右，偏酸性。

⊙1

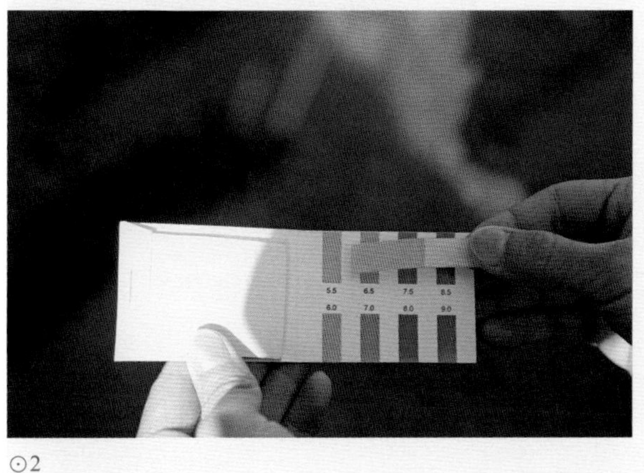

⊙2

⊙1 苦竹原料 Raw material of *Pleioblastus amarus*

⊙2 水源pH测试 Testing the water source pH

（二）金钱纸的生产工艺流程

根据李申言对工艺的描述，综合调查组2016年8月17日、2016年11月3日和2019年1月27日在纸坊的实地工序考察，归纳李申言纸坊金钱纸的生产工艺流程为：

壹	贰	叁	肆	伍	陆	柒	捌	玖	拾	拾壹	拾贰	拾叁
砍断	敲碎	浸泡	腌料	堆蓬	清洗	磨料	踩料	捞纸	压榨	跌打拆松	晒纸	打捆

壹 砍断

1 ⊙1

对买回来的苦竹进行整理，粗的苦竹和细的苦竹要分开，分开后对其分别进行砍断，将长4～5 m的苦竹砍成大约1.8 m一段。据李申言介绍，过去砍竹是人工用刀砍，现在全部都是用机器砍断了。

贰 敲碎

2 ⊙2

砍断苦竹以后，用机器将其敲碎。四五根一起敲，2个人敲竹的话，一天可敲约1 500 kg，50 000 kg苦竹需要1个多月才能敲完。更早以前大田村采用木柱压的方式，通过木柱的重量来压碎竹子。敲碎的苦竹在自然环境中放置3个月左右，使其经历日晒雨淋，目的是使竹料变脆，更易磨料。

叁 浸泡

3 ⊙3

将敲碎淋晒过的苦竹打捆，每30～35 kg为1捆，放入流动的溪水中浸泡约1个月时间。浸泡好以后，将苦竹捞出来沥干。

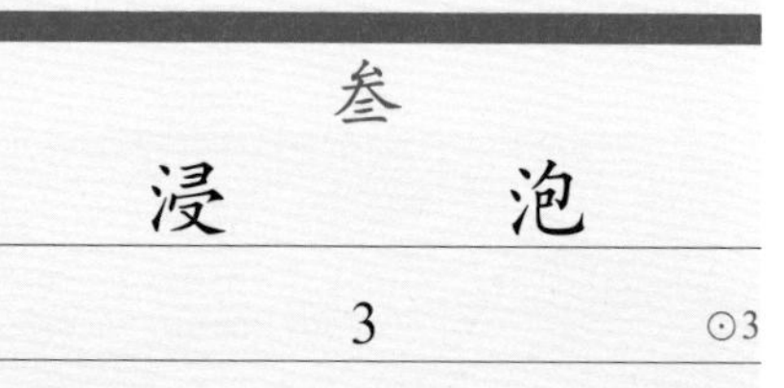

⊙2

⊙1

⊙3

⊙1 砍竹用的竹刀 Knife for cutting the bamboo

⊙2 早前用于敲竹的木柱 Wooden pillar for beating the bamboo

⊙3 李申言家附近的溪水 Stream near Li Shenyan's house

肆 腌料

4 ⊙4⊙5

腌料是整个原料加工过程中最为关键的一步，也是最消耗人力的一步。加工步骤：第一步，先将石灰倒入石灰池中，并用灰耙搅拌，直至石灰水呈均匀浓稠状态。据李申言描述，每腌30～35 kg料需要10～11 kg石灰，一口石灰池要放超过1 500 kg石灰进去，所以石灰水浓度很高。石灰只能多不能少，如果石灰水浓度不够高的话，苦竹料腌得不充分，就很容易发酸以至于烂掉。搅拌石灰的灰耙质量通常超过50 kg，需要7～8个工人合力才能将其抬起工作。第二步，将沥干的苦竹放入石灰池中，用灰耙继续搅拌，大约1捆料腌1分钟即可，捞起来后在石灰池旁堆放。如此重复，一天可腌110～120捆苦竹。

⊙4

伍 堆蓬

5 ⊙6

将腌好的料逐层堆起来，在料的上方盖上稻草，再将石灰池里面的渣捞出来放到稻草上面盖住稻草，堆放至少6个月时间。据李申言介绍，必须要让堆蓬的竹料经历夏季的高温才能充分发酵，因此堆放的时间越长越好。

⊙5

⊙6

陆 清洗

6 ⊙7

将堆蓬后的竹料放入溪水中浸泡清洗，大约需要清洗1天1夜，目的是洗净竹料中的石灰。

⊙7

⊙4 石灰池 Lime pool

⊙5 捞出腌好的料 Picking out the soaked materials

⊙6 堆蓬 Piled materials

⊙7 清洗 Cleaning the materials

柒 磨料

7

⊙8

清洗干净的竹料就可以放置于石碾上进行磨料了。李申言介绍，每2捆料磨一次，大约30分钟即可磨碎。石碾以前村里有很多，而现在只有3户人家（李国良、李雪余、李根才）有。李申言此前一般去李国良家磨料，需要自己出电费。

捌 踩料

8

⊙9

将经过石碾碾磨的细末料运至浆池中，再兑水用脚踩，大约踩30分钟即可。

⊙8

⊙9

⊙8 磨料 Grinding the materials

⊙9 正在踩料的李申言 Li Shenyan stamping the materials

玖
捞　　纸
9

⊙10～⊙14

将踩好的料运送至纸槽中，加水，并用打槽棍搅拌。搅拌均匀后即可捞纸。捞纸时，手握帘床左右两端，垂直放入水中并迅速端平抬起，抖去纸帘上的水，留下厚厚的一层湿纸膜。再捏住纸帘一条长边的中点，拿起后将纸帘反扣在纸架上，发出“提－提－呱”的声音，一张湿纸便形成了。

李申言强调，在捞纸的时候一定要将纸帘上的水抖干净，抖水的时候会发出“锵锵锵锵”的声音。回水越多的纸就越平整，拉力也会大一些，纸不容易破；而回水少的纸则相反。在捞纸过程中，浆料会下沉，所以每隔一段时间，就要用槽耙将槽底部的浆料抄起来，并搅拌使浆料均匀。

⊙10

⊙11

⊙12

⊙13

⊙14

⊙10 打槽 Stirring the materials

⊙11 正在打槽的李申言 Li Shenyan beating the materials

⊙12 捞纸 Papermaking

⊙13 抖水 Shaking the papermaking screen

⊙14 湿纸块 Wet paper pile

拾
压　榨

10　⊙15

将捞好的湿纸块用千斤顶压榨，压到半干不出水为止。

拾壹
跌　打　拆　松

11　⊙16～⊙18

将压干的纸进行跌打，使其变松，并将纸弯曲拆松，以便于将纸一张张分开。

⊙16

⊙15

⊙17

拾贰
晒　纸

12　⊙19～⊙20

晒纸时，厚的纸每3张一起晒，薄的纸每7～8张一起晒，天晴的情况下晒1天就可以了。

⊙18

⊙19

⊙20

⊙15 压榨 Pressing the paper

⊙16/17 李申言示范跌打 Li Shenyan demonstrating tumbling procedures

⊙18 拆松 Loosening the paper

⊙19 晒纸 Drying the paper

⊙20 正在晒纸的李申言 Li Shenyan drying the paper

拾叁 打捆

13 ⊙21

将晒好的纸理平并对齐放置好，用夹竹（上、下各两根）将纸张定形后再用包装绳进行打捆。尺寸90 cm×26 cm的纸每14 kg为1捆，尺寸100 cm×23 cm的纸每15 kg为1捆，尺寸88 cm×17 cm的纸每1 000张为1捆。捆好后即可等待出售。

⊙21

（三）金钱纸的主要制作工具

壹 灰耙

1

腌料时用于搅拌石灰和翻竹料的工具，由耙头和耙杆组成。实测李申言作坊使用的灰耙耙头尺寸为：85 cm×25 cm×12 cm，由于耙杆已经断了，无法测得实际尺寸，据李申言描述，耙杆有6～7 m长。

贰 石灰池

2

用来盛放石灰水和腌料的池子。实测李申言作坊使用的石灰池尺寸为：长290 cm，宽155 cm，高100 cm。（2019年1月回访时李申言已不再造纸，石灰池已被填。）

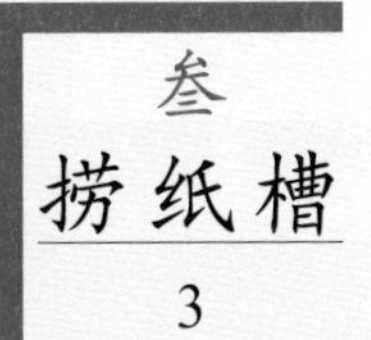

叁 捞纸槽

3

用于放置纸浆的长方体容器，捞纸时工人站于其侧。实测李申言作坊使用的捞纸槽尺寸为：长172 cm，宽172 cm，高76 cm。建造成本约1 000元。

⊙22

⊙23

⊙21 李申言示范打捆 Li Shenyan showing binding procedure

⊙22 灰耙耙头 The grey rake head

⊙23 捞纸槽 Papermaking trough

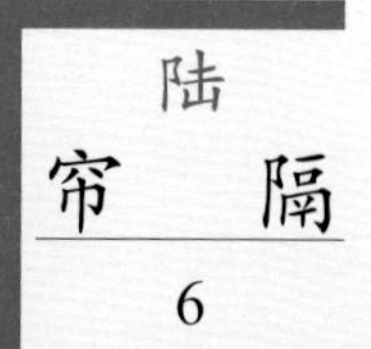

肆 纸帘 4

由细竹丝编织而成的捞纸工具，纸浆在纸帘上形成湿纸膜。实测李申言作坊使用的两种纸帘尺寸分别为：长92 cm，宽17 cm；长104 cm，宽23 cm。2015～2016年纸帘价格为150～160元/张，购于富阳的大源镇。

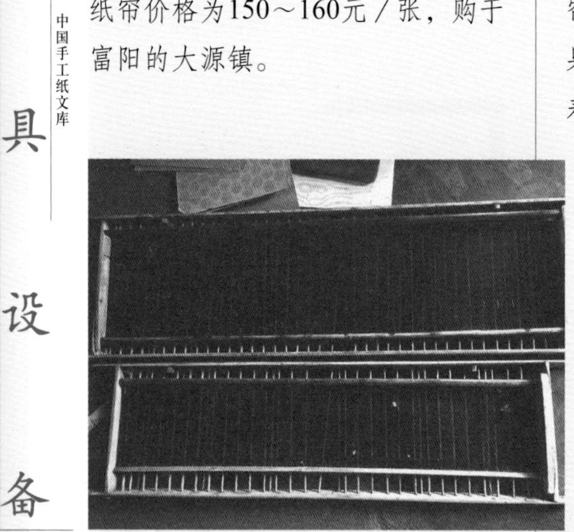

⊙24

伍 帘床 5

捞纸时用于放置纸帘的长方形木质框架。李申言作坊使用的两种帘床尺寸分别为：长93 cm，宽27 cm；长105 cm，宽32 cm。两种帘床都是李申言自己用从上官乡买来的废帘改装而成，成本就几十块钱，如果认识人的话，则不需要钱。帘床寿命大概在1年左右。

陆 帘隔 6

捞纸时放置于纸帘上方、用于压住纸帘的长方体木块，不同尺寸的纸帘和帘床使用不同大小的帘隔。李申言作坊内两种帘隔的尺寸分别为：23.5 cm × 3 cm × 3 cm和28 cm × 3 cm × 3 cm。

⊙25

柒 和单槽棍 7

用于搅拌捞纸槽中的浆料，使捞纸的过程中浆料更加均匀。实测李申言作坊使用的和单槽棍尺寸为：总长159 cm；板长27.5 cm，宽12 cm。

⊙26

捌 切纸刀 8

用于成品纸的裁切，使纸张大小一致。实测李申言作坊使用的切纸刀尺寸为：长42 cm，宽29 cm；柄长35.5 cm。

⊙27

⊙24 纸帘和帘床 Papermaking screen and its supporting frame

⊙25 帘隔 Screen blocks

⊙26 和单槽棍 Stirring stick

⊙27 遗留下来的老切纸刀 The remained old paper knife

五
李申言金钱纸作坊的市场经营状况

5
Marketing Status of Li Shenyan Joss Paper Mill

2016年11月调查时据李申言的介绍，作坊内生产的金钱纸都是销往江苏省苏州市东山镇的，不过一直是销售原纸，至于打制纸钱就不是造纸人的事了。2015年李申言生产的3种不同规格纸的销售价格分别为：90 cm×26 cm的每捆110元（每捆1 000张），100 cm×23 cm的每捆210元，88 cm×17 cm的每捆250元（此款纸薄，质量好，会做的人也不多，因此价格相对较高）。2015年，李申言作坊销售了各类纸约200捆，销售额约4.5万元，李申言估计其中成本大约为1.5万元。

2019年回访时李申言说其作坊2017年就已停止生产，目前还剩有一些原料，打算2020年再做1～2个月，原料做完后，以后都不再造纸了。一方面因为其年纪大了，另一方面纸的价格越来越低，据李申言估计如果2019年造一年纸只能挣1万多元，另外会腌料的人也越来越少，且岁数都越来越大，大多在50～60岁，无法担起需要重体力的腌料工作。

⊙1

⊙1
金钱纸
Joss paper

六
李申言金钱纸作坊文化习俗与造纸故事

6
Cultural Custom and Papermaking Stories of Li Shenyan Joss Paper Mill

⊙2

⊙2 堆蓬 Piled materials

（一）造纸习俗

1. 腌料俗语中的工艺要领

据李申言回忆，关于腌料，有一句老话叫“热要火里去，冷要水里去”。大致的意思是：腌料要选择夏天，使竹料通过高温发酵，更容易熟化，而洗料和捞纸要选择冬天寒冷的时候，这样纸的紧密度好。腌料和洗料之间相差的半年时间则用来堆蓬。

2. “仙人指路”话“额头”

调查组成员考察造纸工具的时候，意外发现在帘床上有一小缺口，便询问李申言这一缺口的用处。李申言解释道，这一缺口叫作额头，捞纸的时候，额头的位置会多出一点纸浆，使纸在这一位置突出。每张湿纸被放置到纸架上的时候都要沿着额头的位置对齐，如此便可形成整齐的湿纸块。因此，额头相当于“仙人”，为捞纸工指明了放置湿纸的正确位置，即为“仙人指路”。

3. 女人不能去石灰池旁

大田村有一传统，腌料的时候女人不能去石灰池旁边，造纸人传下来的说法是：一旦违背了这一传统，纸就做不好。大致的原因有两点：（1）男人们在腌料的时候，任务很重，每个人都要竭尽全力、齐心协力，才能完成工作。石灰池旁多一个人就会干扰他们干活，特别是女人去了会分散男人的注意力。（2）石灰池旁边温度很高，劳动强度也很大，女人承受不了这种劳动强度，去了石灰池旁也不能帮忙干活。

（二）纸的故事

1. “李百万”与大田村

李申言介绍，现在大田村所有的纸全部销往江苏苏州，但是在若干年之前，村里的纸是销往湖州的。大约在民国前期，有个湖州的有钱商人，专门做纸的生意，在大田村开了一间门店专门收购金钱纸。所有村民的纸都经过这个商人的手被运到湖州，再从湖州通过水运运到江苏，解决了村民们纸的销路问题。因为这个专门在村里做纸生意的商人姓李，村民们又都觉得他好像非常有钱，所以私下里都喊他“李百万”。

⊙1

⊙2

⊙1
额头
Paper marker

⊙2
仙人指路
Sign for piling the wet paper, metaphorically called "the immortal is showing the way"

七 李申言金钱纸作坊的业态传承现状

7 Current Status of Li Shenyan Joss Paper Mill

2016年调查时据李申言介绍，自己和妻子年纪都大了，出去打工没人要，才继续在家造纸。每年真正捞纸的时间只有2～3个月，造纸的数量也不多，仅够维持家中日常开销。除了造纸外，李申言还在村中开了个小百货商店，赚点小钱补贴家用。由于儿女都不愿从事造纸行业，等到自己和妻子干不动了，祖祖辈辈传下来的造纸事业也就只能放弃了。

2019年1月回访时，李申言纸坊实际上已经停产，只是还有原料没有用完，设施也还齐全，李申言觉得不能浪费了，准备把原料用完就不再造纸了。

⊙3

⊙3 李申言和妻子、孙子 Li Shenyan, his wife and grandson

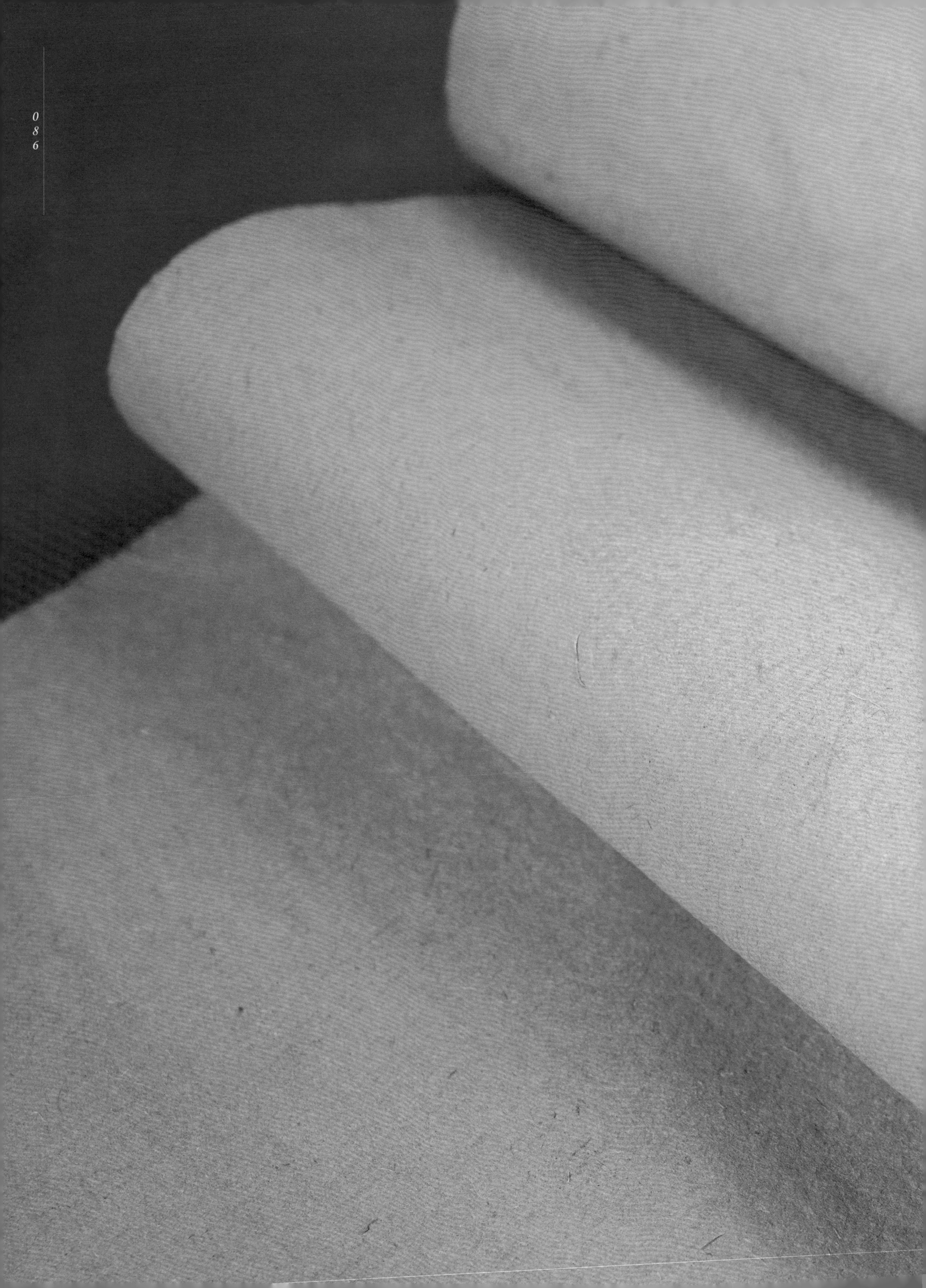

金钱纸

Joss Paper
of Li Shenyan Joss Paper Mill

纯苦竹金钱纸透光摄影图

A photo of pure *Pleioblastus amarus* joss paper seen through the light

第五节

李雪余屏纸作坊

浙江省
Zhejiang Province

杭州市
Hangzhou City

富阳区
Fuyang District

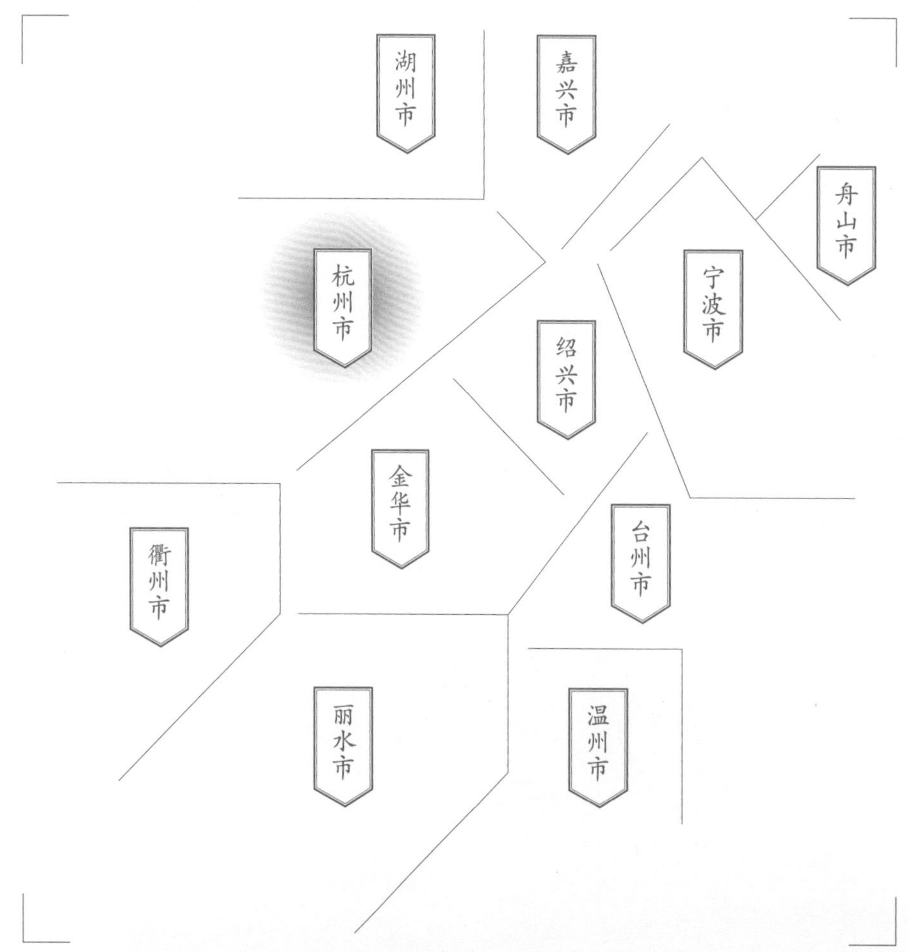

调查对象
富阳区常安镇大田村
李雪余屏纸作坊
竹纸

Section 5
Li Xueyu Ping Paper Mill

Subject
Bamboo Paper of Li Xueyu Ping Paper Mill
in Datian Village of Chang'an Town in Fuyang District

一
李雪余屏纸作坊的基础信息与生产环境

1
Basic Information and Production Environment of Li Xueyu Ping Paper Mill

李雪余屏纸作坊是一家专门用苦竹造屏纸的家庭式手工纸作坊，位于富阳区常安镇大田行政村105号，地理坐标为：东经119° 53′ 49″，北纬29° 52′ 54″。作坊负责人为李雪余，作坊直接用负责人名字命名。

调查组于2016年8月17日第一次前往作坊现场考察，通过李雪余的介绍，了解到的基础信息为：作坊为家庭式造纸坊，平时都是夫妻二人在造纸，只有腌料工序因需要的劳动力多，必须外请10个工人临时来帮忙，每年只需请1次，其余工序都由李雪余夫妻二人完成。纸坊有1帘纸槽，只生产用于祭祀的苦竹纸，李雪余称其为屏纸（浙江省温州市瓯海区泽雅镇一带将以水竹为主要原材料生产的用于祭祀的竹纸称为屏纸）。然而在2019年1月25日调查组回访调研时，李雪余作坊已停止生产了。

常安镇位于富阳区南部，距区政府所在地21 km。全境地域总面积63.31 km^2，辖行政村16个、村民小组264个。常安镇大田村作为16个行政村之一，位于常安镇的中心地带，是集镇所在地，面向永安山，有六石溪穿村而过，背山面水。今日的大田村系2008年在行政村规模调整时，由原来的大田村与六石村合并而成，村域面积7 km^2，辖6个自然村、16个村民小组。2015年全村有住户627户，有7～8户会偶尔造纸，李雪余是村里少有的常年以造纸为生的农户。

⊙1

⊙2

⊙1
李雪余屏纸作坊外的竹料堆放场
Yard storing bamboo materials outside Li Xueyu Ping Paper Mill

⊙2
流经李雪余家门后的小溪
Stream running through the backside of Li Xueyu's house

李雪余屏纸作坊位置示意图

Location map of Li Xueyu Ping Paper Mill

路线图
富阳城区
↓
李雪余屏纸作坊
Road map from Fuyang District centre to Li Xueyu Ping Paper Mill

考察时间
2016年8月 / 2019年1月

Investigation Date
Aug. 2016 / Jan. 2019

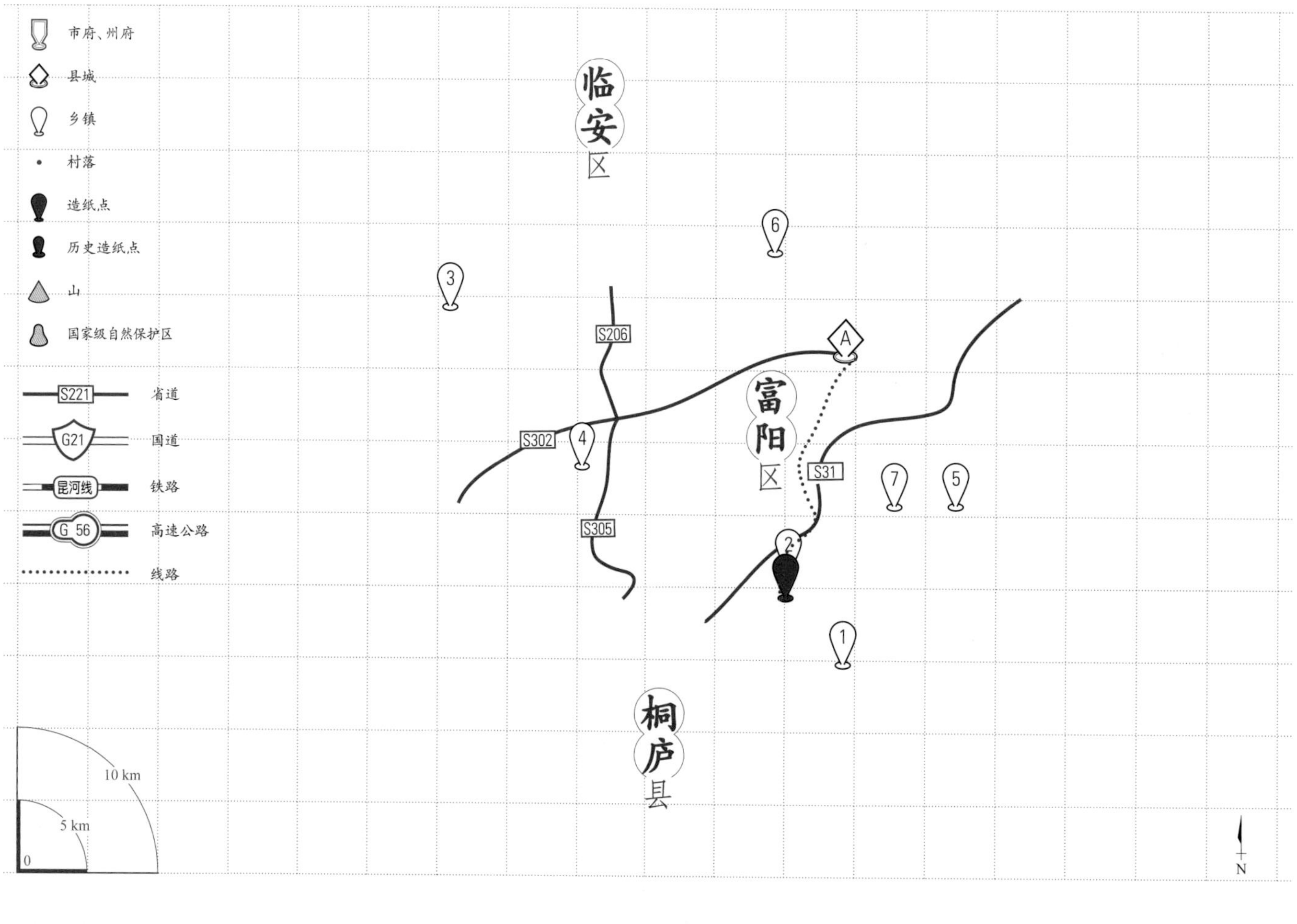

二
李雪余屏纸作坊的历史与传承

2
History and Inheritance of Li Xueyu Ping Paper Mill

李雪余，1954年出生于常安镇大田村，2016年调查时62岁，为该作坊的负责人。据李雪余自述，他12岁便开始跟随父亲学习造纸。父亲名叫李高木，1915年前后出生（调查时早已去世），父亲从年幼便开始接触造纸，之后从事造纸工作直到60岁左右。母亲倪道仙，1945年左右从常安镇沧洲村嫁到大田村，婚后负责晒纸。

关于家庭造纸的起源和传承谱系，李雪余表示，自己家中世世代代造纸，已知的至少有10多代，只是不同时期所造的纸种类不同。比如，在供销社收购纸时期（20世纪60年代至80年代初），李雪余父亲造的是四六屏的纸，尺寸为16 cm×100 cm；而调查时李雪余作坊中生产的屏纸尺寸为26 cm×96 cm。据李雪余回忆，他造屏纸至少已经有20年时间了，在造屏纸之前，李雪余做过一段时间的纸筋，以小苦竹为原料来制作，通常用作火烧纸。但20世纪70年代集体化的时候严禁开展“迷信”活动，因而火烧纸的用量小，大部分纸被拌入石灰中，作为粉刷墙的一种建筑材料，可以避免墙体开裂。

李雪余在家中排行老二，共有兄弟姐妹4人，哥哥早年当兵，弟弟和妹妹都没有从事造纸业。李雪余有2个儿子，调查时大儿子38岁，在铝制板厂开车，小儿子34岁，在空调厂装空调，都不从事造纸。因此，李雪余所说的世世代代传承下来的造纸祖业仅由李雪余夫妻两位老人在艰难地维持着。

李雪余的妻子李玉娣是富阳区小源镇人，2019年调查时66岁，自26岁嫁过来以后一直帮助李雪余晒纸。据李玉娣介绍，其祖上也会造纸，父亲李叶根已去世十余年，在世时可熟练操作造纸的各种工序。母亲徐金仙出生于1927年，2018年去世，生前也会造纸的多项工序。李玉娣从小跟父亲学造纸，在未嫁给李雪余之前，在家中负责抄纸，母亲负责揭纸，而父亲负责其余工序。

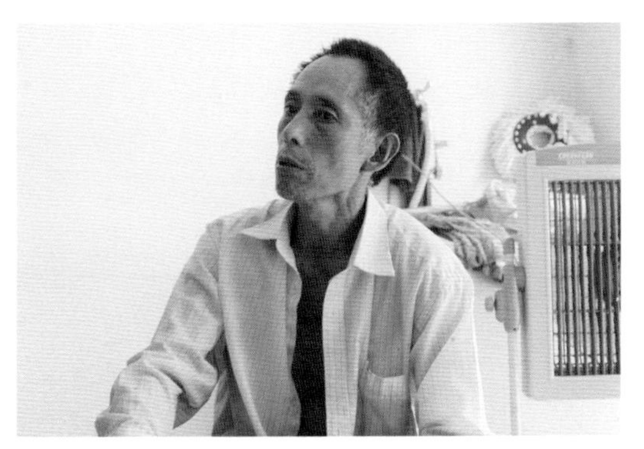
⊙1

⊙2

⊙1 访谈中非常投入的李雪余 Li Xueyu involved in the interview

⊙2 李雪余夫妻合影 Li Xueyu and his wife

李玉娣的祖父与祖母也都熟悉造纸的各项工序。李玉娣在家中排行老三，兄弟姐妹共7人，有1个哥哥、1个姐姐、2个弟弟、2个妹妹，但6人都没有从事造纸。

表10.8 李雪余屏纸作坊李雪余一系传承谱系
Table 10.8 Li Xueyu's family genealogy of papermaking in heritors in Li Xueyu Ping Paper Mill

传承代数	姓名	性别	与李雪余关系	基本情况
第一代	不详	男	祖父	出生年月不详，熟练掌握造纸的各道工序
	不详	女	祖母	出生年月不详
第二代	李高木	男	父亲	1915年前后出生，从小接触造纸，熟练掌握造纸的各道工序
	倪道仙	女	母亲	出生年月不详，1945年前后嫁给李高木后负责晒纸
第三代	李雪余	男	—	1954年出生，12岁开始学习抄纸，熟练掌握造纸的各道工序

表10.9 李雪余屏纸作坊李玉娣一系传承谱系
Table 10.9 Li Yudi's family genealogy of papermaking inheritors in Li Xueyu Ping Paper Mill

传承代数	姓名	性别	与李玉娣关系	基本情况
第一代	不详	男	祖父	出生年月不详，熟练掌握造纸的各道工序
	不详	女	祖母	出生年月不详，熟练掌握造纸的各道工序
第二代	李叶根	男	父亲	出生年月不详，熟练掌握造纸的各道工序
	徐金仙	女	母亲	1927年出生，2018年去世，生前在家中负责揭纸
第三代	李玉娣	女	—	1953年出生，从小随父学习造纸，熟练掌握造纸的各道工序

三 李雪余屏纸作坊的代表纸品及其用途与技术分析

3 Representative Paper and Its Uses and Technical Analysis of Li Xueyu Ping Paper Mill

（一）李雪余屏纸作坊代表纸品及其用途

2016年8月17日入大田村调查得知：李雪余纸坊只生产一种纸——苦竹屏纸，尺寸为26 cm×96 cm，主要用来做祭祀用的纸钱。在卫生纸没有规范地工业化量产之前，屏纸还曾经一度被用作卫生纸，用作卫生纸的时候需要将纸一切为四，尺寸为26 cm×24 cm。此外，大田村过去所造纸还曾被用作卡片纸、包装纸等中端的文化生活用纸。据李雪余介绍，大田村里现存的几家造纸户，造的也都是这种只用于祭祀纸钱的祭祀竹纸，没有其他的纸品，只是各家在尺寸和厚度上稍有不同而已。

（二）李雪余屏纸作坊代表纸品性能分析

测试小组对采样自李雪余纸坊的屏纸所做的性能分析，主要包括定量、厚度、紧度、抗张力、抗张强度、白度、纤维长度和纤维宽度等。按相应要求，每一指标都重复测量若干次后求平均值，其中定量抽取了5个样本进行测试，厚度抽取10个样本进行测试，抗张力抽取20个样本进行测试，白度抽取10个样本进行测试，纤维长度测试了200根纤维，纤维宽度测试了300根纤维。对李雪余纸坊屏纸进行测试分析所得到的相关性能参数如表10.10所示。表中列出了各参数的最大值、最小值及测量若干次所得到的平均值或者计算结果。

表10.10　李雪余屏纸作坊屏纸相关性能参数
Table 10.10　Performance parameters of Ping paper in Li Xueyu Ping Paper Mill

指标单位			最大值	最小值	平均值	结果
定量		g/m²				224.3
厚度		mm	0.786	0.617	0.725	0.725
紧度		g/cm³				0.309
抗张力	纵向	mN	31.3	25.2	27.6	27.6
	横向	mN	24.6	19.3	22.0	22.0
抗张强度		kN/m				1.653
白度		%	38.9	33.8	35.6	35.6
纤维	长度	mm	1.3	0.1	0.6	0.6
	宽度	μm	61.4	3.5	8.4	8.4

性能分析

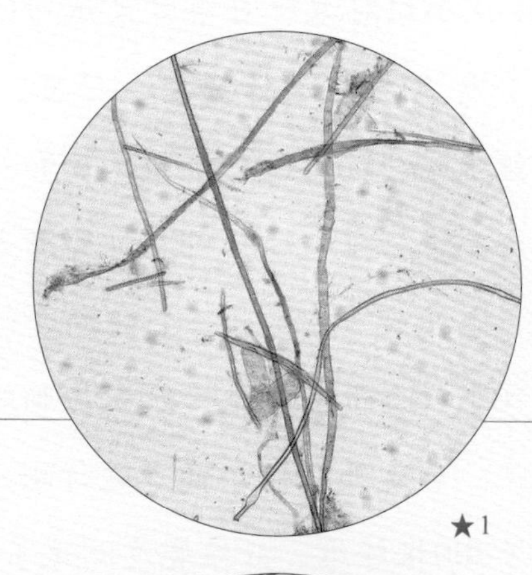

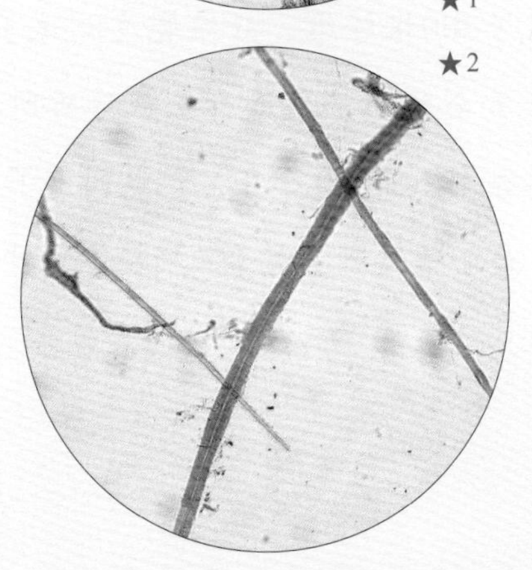

由表10.10可知，所测李雪余纸坊屏纸的平均定量为224.3 g/m²。李雪余纸坊屏纸最厚约是最薄的1.274倍，经计算，其相对标准偏差为0.087。通过计算可知，李雪余纸坊屏纸紧度为0.309 g/cm³，抗张强度为1.653 kN/m。

所测李雪余纸坊屏纸平均白度为35.6%。白度最大值是最小值的1.151倍，相对标准偏差为0.048。

李雪余纸坊屏纸在10倍和20倍物镜下观测的纤维形态分别如图★1、图★2所示。所测李雪余屏纸纤维长度：最长1.3 mm，最短0.1 mm，平均长度为0.6 mm；纤维宽度：最宽61.4 μm，最窄3.5 μm，平均宽度为8.4 μm。

★1 李雪余纸坊屏纸纤维形态图（10×）
Fibers of Ping paper in Li Xueyu Paper Mill (10× objective)

★2 李雪余纸坊屏纸纤维形态图（20×）
Fibers of Ping paper in Li Xueyu Paper Mill (20× objective)

四
李雪余屏纸作坊的生产原料、工艺与设备

4
Raw Materials, Papermaking Techniques and Tools of Li Xueyu Ping Paper Mill

⊙1

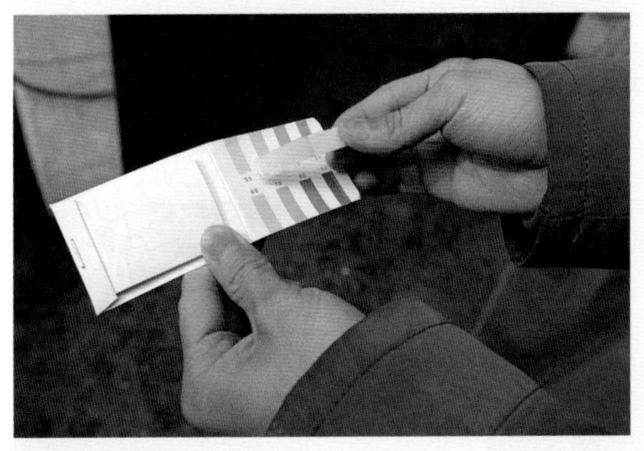
⊙2

⊙1 做原料剩余的一捆苦竹
A bundle of *Pleioblastus amarus* as raw material

⊙2 水源pH测试
Testing the water source pH

（一）李雪余屏纸作坊的生产原料

1. 主料：苦竹

李雪余屏纸作坊生产的屏纸原料是苦竹。据李雪余说，这是因为大田村没有毛竹，无法像大源镇大同村、湖源乡新二村和新三村那样就地取材使用毛竹造纸。世世代代传下来的习俗是购买外地的苦竹造纸。李雪余在访谈时向调查组成员介绍，制作屏纸必须使用老苦竹，村里似乎没有人尝试过更换材料，而且大田村现存的作坊也都全部购买苦竹作原料，未改变传统苦竹纸制造工艺的任何环节。因此，屏纸从外观上看保留着传统工艺苦竹纸的特色与质感。

据李雪余透露，作坊使用的苦竹是从浙江省余杭、桐庐诸地买的，2015年的购买价格是70元/100 kg。通常每年11月份买苦竹，因为秋冬的苦竹比较老，适宜造屏纸。以前作坊屏纸产量高的时候，每年约需购买50 000 kg苦竹，而现在一年只需要购买20 000～25 000 kg。

2. 辅料：水

大田村背山面水，水源丰富，而且造屏纸对水的要求也没有造元书纸高，不论是山泉水还是地下水，都可以用来造屏纸。李雪余纸坊使用的水是从附近引来的山泉水，经调查组成员现场取样检测，山泉水的pH为5.5～6.0，偏酸性。

（二）李雪余屏纸作坊的生产工艺流程

根据李雪余的工艺描述，综合调查组2016年8月17日在纸坊的实地工序调查，总结李雪余纸坊屏纸的生产工艺流程为：

壹	贰	叁	肆	伍	陆	柒	捌	玖	拾	拾壹	拾贰
砍断	敲碎	浸泡	腌料	堆蓬	清洗	磨料	踩料	捞纸	压榨	晒纸	打捆

壹 砍断

1 ◎1

买回来的苦竹长4～5 m，为了使其便于加工，需要用机器将其砍断成2 m左右的竹段。据李雪余介绍，过去是人工用刀砍竹，现在用的是机器，可以节省不少人力，也可以提高效率。（调查组成员到访期间不是砍竹的季节，在作坊内没有看到砍竹机器。）

贰 敲碎

2 ◎2

砍断苦竹以后，用机器将其敲碎，4～5根一起敲，越碎越好。

叁 浸泡

3 ◎3

将敲碎的苦竹打捆，每35 kg为1捆，再放入流动的水中浸泡约20天。浸泡好了以后，将苦竹捞出来沥干。

◎2

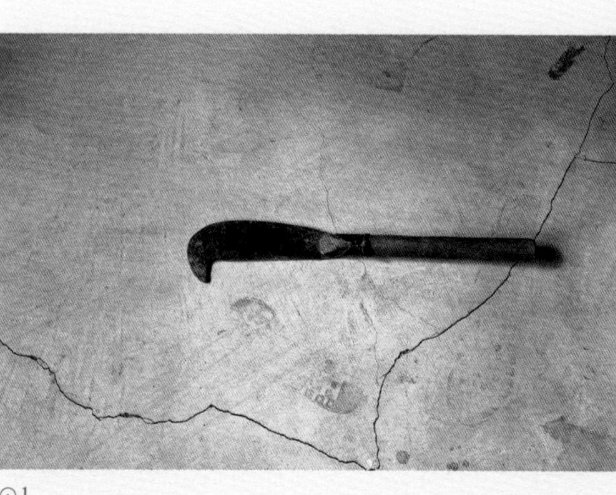

◎1

◎3

◎1 砍竹刀 Knife for cutting the bamboo

◎2 敲碎苦竹的机器 Machine for knocking *Pleioblastus amarus* to Pieces

◎3 浸泡苦竹的小溪 Stream for soaking *Pleioblastus amarus*

肆

腌　料

4　⊙4

腌料是苦竹造屏纸原料加工过程中最为关键的一步，也是最消耗人力的一步。先将石灰倒入石灰池中，并用灰耙搅拌，直至石灰水呈均匀浓稠状态。据李雪余描述，每腌70 kg料需要20 kg石灰，所以石灰水浓度很高，像浓粥一样稠。搅拌石灰的灰耙有50 kg重，需要10个工人合力才能将其抬起工作。然后将沥干的苦竹放入石灰池中，用灰耙继续搅拌，1捆料大约腌1分钟即可。腌好后捞起，堆放在石灰池旁。如此重复，1天可腌20 000 kg苦竹，其间需耗费6 000 kg石灰。腌料时请的都是本村的工人，工人一天约工作10个小时，2015年时可得到300～400元。

⊙4

伍

堆　蓬

5　⊙5

将腌好的料逐层堆起来，堆好后的高度约为12层苦竹的直径。在料的上方盖上稻草，再将石灰池里面的渣捞出来放到稻草上面盖住稻草。至少堆放4个月，最好是堆放1年。

⊙5

陆

清　洗

6　⊙6

将堆蓬后的竹料放入溪水中浸泡清洗，大约需要清洗1天1夜，目的是洗净竹料中的石灰。

⊙6

柒

磨　料

7　⊙7

清洗干净的竹料就可以放置于石碾上进行磨料了。据李雪余介绍，每2捆料磨1次，大约40分钟即可磨碎。堆蓬4个月的竹料需要碾磨久一点，超过4个月的竹料碾磨时间则可短一些。

⊙7

捌

踩　料

8　⊙8

制作屏纸特有的一道工序就是踩料。将经过石碾碾磨的细末料运至浆池中，兑水后再用脚踩。据李雪余介绍，踩料的时长不确定，只要能使浆料均匀即可，一般情况下他自己要踩10分钟。

⊙8

⊙4 石灰池 Lime pool

⊙5 堆蓬 Piled materials

⊙6 在溪水中清洗 Cleaning the materials in the stream

⊙7 磨料的石磨 Grinding the materials with a stone roller

⊙8 踩料浆池 Trough for stamping the materials

玖 捞纸

9 ⊙9⊙10

将踩好的料运送至纸槽中，加水，并用打槽棍搅拌。搅拌均匀后即可捞纸。捞纸时，手握帘床左右两端，垂直放入水中并迅速端平抬起，抖去纸帘上的水，留下厚厚的一层湿纸膜。再捏住纸帘一条长边的中点，拿起后将纸帘反扣在纸架上，一张湿纸便形成了。李雪余介绍，在捞纸过程中，浆料会下沉，所以每隔一段时间，就要用槽耙将槽底部的浆料抄起来，并搅拌使浆料均匀。具体的间隔时间由李雪余自己把控。

⊙9

⊙10

拾 压榨

10 ⊙11

将捞好的湿纸块用千斤顶压榨，大约压榨1个小时即可。

⊙11

拾壹 晒纸

11

对压干的纸进行跌打，使其变松，易于一张张分开。然后每3张一揭，揭下来后放在平地上铺着晒。阳光好的话，一天就可以晒干了。

拾贰 打捆

12 ⊙12

将晒好的纸理平并对齐放置好，用包装绳进行打捆，1捆200～220张纸。捆好后即可等待出售。

⊙12

⊙9 打槽 Stirring the materials
⊙10 捞纸 Papermaking
⊙11 待压榨的湿纸块 A wet paper pile to be pressed
⊙12 打捆 Binding the paper

（三）李雪余屏纸作坊的主要制作工具

壹
灰　耙
1

腌料时用于搅拌石灰和翻竹料的工具。由耙头和耙杆组成，实测李雪余纸坊耙头尺寸为：长82 cm，宽28 cm，高11 cm；耙杆长7～8 m。

⊙13

贰
浆　池
2

用来放碾磨好的浆料的长方体容器，工人在浆池中踩料。实测李雪余纸坊的浆池尺寸为：长190 cm，宽148 cm，高36 cm。

⊙14

叁
捞纸槽
3

用于放置纸浆的长方体容器，捞纸时工人站于其侧。实测李雪余纸坊的捞纸槽尺寸为：长205 cm，宽156 cm，高94 cm。

⊙15

肆
打槽棍
4

用于搅拌纸槽中浆料的棍子。实测李雪余纸坊使用的打槽棍尺寸为：长112 cm，直径2 cm。

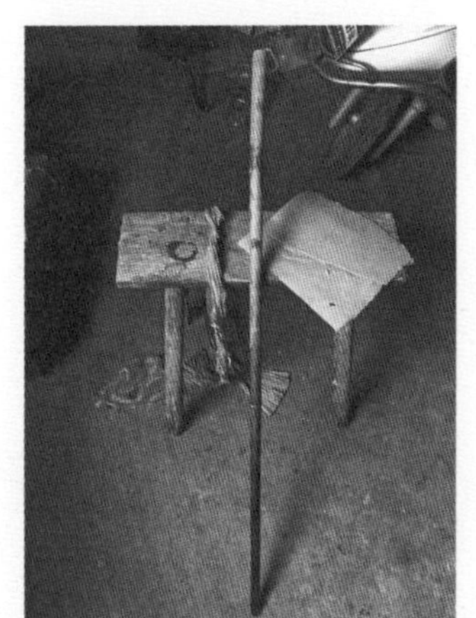

⊙16

伍
槽　耙
5

用于搅拌纸槽中浆料的耙子。实测李雪余纸坊使用的槽耙杆尺寸为：长140 cm；耙头长20 cm，宽10 cm。

⊙17

陆
纸　帘
6

由细竹丝编织而成的捞纸工具，纸浆在纸帘上形成湿纸膜。实测李雪余纸坊使用的纸帘尺寸为：长92 cm，宽29 cm。

⊙18

⊙13 灰耙耙头 Grey rake head

⊙14 浆池兼踩料池 Pulp pool which can stamp the materials

⊙15 捞纸槽 Papermaking trough

⊙16 打槽棍 Stirring stick

⊙17 槽耙 Stirring rake

⊙18 遗留下来的废纸帘 Abandoned former papermaking screen

柒 帘床 7

捞纸时用于放置纸帘的长方形木质框架。实测李雪余纸坊使用的帘床尺寸为：长97 cm，宽36 cm。

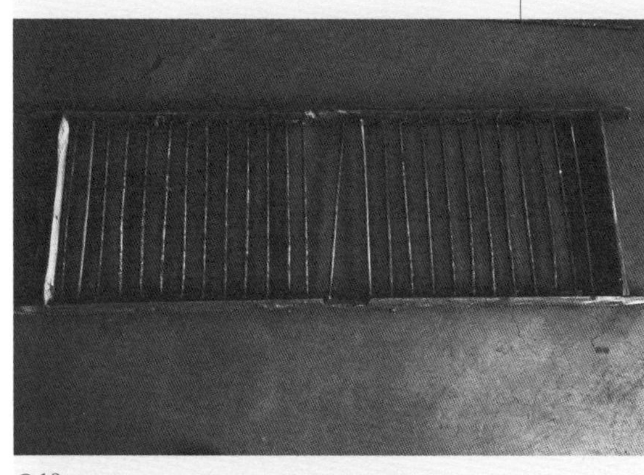

⊙19

捌 千斤顶 8

用于压干捞出的湿纸块。实测李雪余纸坊使用的千斤顶尺寸为：长15 cm，宽13.5 cm，高24.5 cm。

⊙20

⊙19 遗留下来的废帘床 Abandoned former frame for supporting the papermaking screen

⊙20 千斤顶 Lifting jack

五
李雪余屏纸作坊的市场经营状况

5
Marketing Status of Li Xueyu Ping Paper Mill

2016年调查时据李雪余介绍，目前作坊内生产的所有屏纸全部销往江苏省苏州市东山镇，用来制作祭祀用的纸钱以及供养虾和螃蟹户用作烧纸，当地淡水养殖户信奉烧纸越多财气越旺的说法。销售的时候按市斤（1市斤=500 g）计，每28市斤纸最贵可卖120元，即120元/14 kg。据李雪余粗略统计，平均一年可以卖600捆纸，每捆约200张，共约12万张纸，按照不同年份的价格波动，销售额为5万～8万元，其中成本大约2万元。由于所有屏纸都是由当地的李洪军和住在东山镇的李东华两个经销商上门收购，所以不需要担心销路，也不需要自己去找市场和客户。但2019年1月回访时了解到，2017年两家经销商合并后开始压价，将屏纸价格压至85元/14 kg，这也是导致李雪余不再继续造屏纸的一个重要原因。

⊙21

六
李雪余屏纸作坊的造纸习俗

6
Papermaking Custom of Li Xueyu Ping Paper Mill

（一）屏纸优于机械纸

李雪余家生产的纸烧出来的灰是白色的，而机械纸烧出来的灰是黑色的。当地有“烧出来的灰一定要是白灰，代表着吉祥”的说法。并且在剪纸钱的过程中，当地生产的手工纸材质上更为松散，易于剪切，而机械纸更为紧实，不利于加工成纸钱。将加工完的纸钱串成串，价格会比原来的纸贵很多，例如原本一张纸是0.5元，加工后得到的纸钱能卖到1.5元。

⊙21 李雪余手中的整张屏纸
A complete Ping paper hold by Li Xueyu

（二）烧纸祭祖

大田村当地一般在七月半（中元节、祭祖节）和清明节上山去祖坟烧纸，烧的纸无需再加工，而是直接烧整张纸。在冬至，当地人也会在家门口烧纸，以怀念去世的长辈。

七
李雪余屏纸作坊的业态传承现状

7
Current Status of Li Xueyu Ping Paper Mill

常安镇大田村的造纸历史悠久，在鼎盛时期，全村有700多人参与造纸，共有200多帘槽。但是，随着外出打工队伍的不断壮大，造纸户越来越少，愿意去学习造纸的年轻人更是寥寥无几。截至2016年8月入村调查时，大田村仅剩下8～9帘槽投入生产，且从事造纸工作的人年纪基本都在60～70岁，老龄化趋势十分严重。

李雪余家除了他本人和妻子，没有人会造纸，也没有人愿意去学习造纸。2016年8月17日访谈时李雪余说道："以前造纸是为了生存，但现在拿养老金，每个月有稳定的2 000～3 000元收入。虽然每年3万～6万元的利润还不错，但太辛苦，像我们老夫妻体力跟不上，也确实没太大必要再造纸了。说不定明年就不做了。"

2019年1月25日第二轮调查回访时，李雪余作坊已经暂时停止生产了。

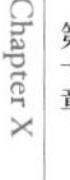

屏纸

李雪余屏纸作坊

Ping Paper of Li Xueyu Ping Paper Mill

纯苦竹屏纸透光摄影图

A photo of pure *Pleioblastus amarus* Ping paper seen through the light

第六节

姜明生纸坊

浙江省
Zhejiang Province

杭州市
Hangzhou City

富阳区
Fuyang District

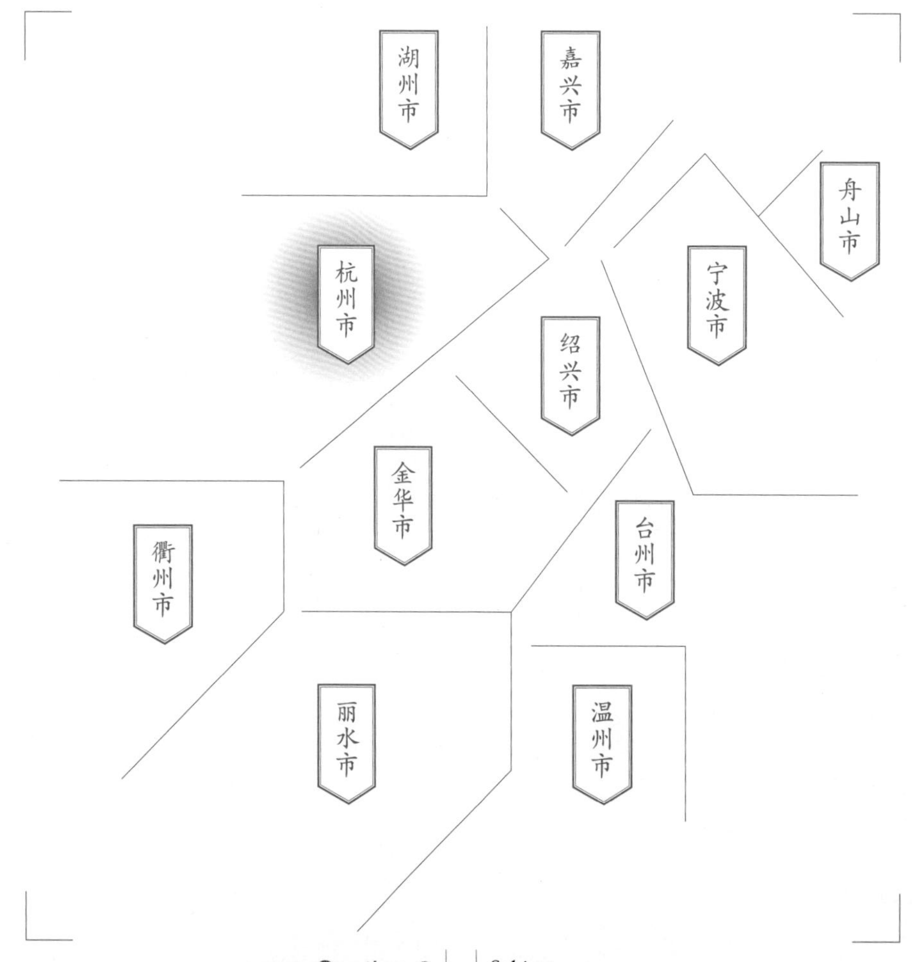

调查对象

富阳区灵桥镇山基村
姜明生纸坊
竹纸

Section 6
Jiang Mingsheng Paper Mill

Subject

Bamboo Paper of Jiang Mingsheng Paper Mill
in Shanji Village of Lingqiao Town in Fuyang District

一 姜明生纸坊的基础信息与生产环境

1 Basic Information and Production Environment of Jiang Mingsheng Paper Mill

姜明生纸坊位于富阳区灵桥镇山基村境内。山基村位于富春江南岸灵桥镇东南部的山谷中，距离灵桥镇中心约5 km。山基村呈条带状分布，山基溪从山上流下，穿村而过，村与溪共名，村民多依溪而居。纸坊所在地的地理坐标为：东经120° 2′ 38″，北纬29° 59′ 46″。

2016年9月28日，调查组成员前往姜明生纸坊，确认纸坊的具体地址是山基村20号。根据姜明生的叙述了解到的基础信息为：纸坊浸料的料塘是自己建造的，而浆料、煮料需要的设施场地为村民共用，最后独立抄制成纸。2016年纸坊有员工4名（包括姜明生夫妇）、纸槽1帘，每天能做3捆祭祀纸，1捆2 000张，即6 000张祭祀纸。每年除过年期间的休息外，基本上做满一整年。2019年1月调查组回访时，姜明生纸坊与2016年初次调查的情况相差无几，姜明生一天照旧可做6 000张祭祀纸，做到年末方才暂停纸坊事务，转而去忙活家务事好迎接新年。

山基村是一个历史上几乎家家户户造纸的村落，一方面的原因是周边纯净的溪水和山上茂密的毛竹为竹纸生产提供了便利条件和丰富资源；另一方面的原因与耕种用地的缺乏有关，使得村民不得不更多地靠乡土手工业为生。在这两方面因素影响下，山基村具有十分悠久的造纸传统，历史上其产品质量在富阳乃至全国亦属上乘。例如，1915年，山基村所产的“姜芹波忠记”昌山纸（毛竹元书纸品种之一），曾获国家农商部嘉奖及巴拿马万国商品博览会二等奖，这是山基村造纸辉煌历史的见证。

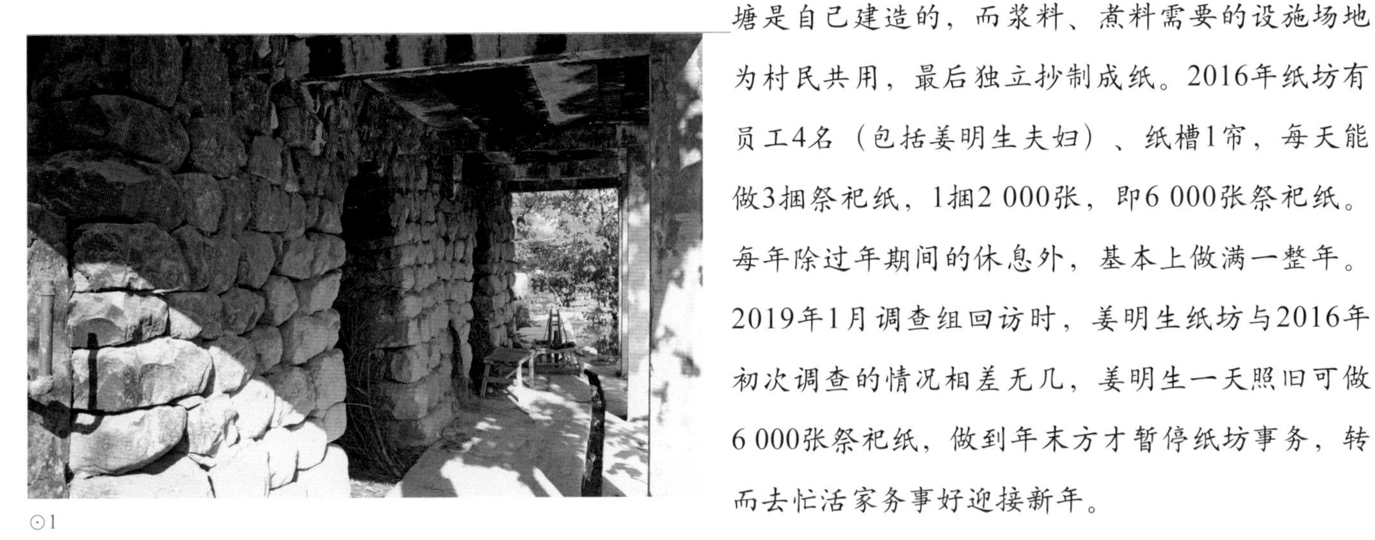

⊙1

⊙2

⊙1 姜明生纸坊中的石砌民居 Stone residence in Jiang Mingsheng Paper Mill

⊙2 山中小村——山基村 Shanji Village surrounded by mountains

路线图
富阳城区
↓
姜明生纸坊
Road map from Fuyang District centre to Jiang Mingsheng Paper Mill

姜明生纸坊位置示意图

Location map of Jiang Mingsheng Paper Mill

考察时间
2016年9月 / 2019年1月

Investigation Date
Sep. 2016 / Jan. 2019

地域名称

富阳区 湖源乡 常安镇 洞桥镇 新登镇 灵桥镇 新义乡 大源镇

富阳城区

姜明生纸坊

灵桥镇

造纸点名称

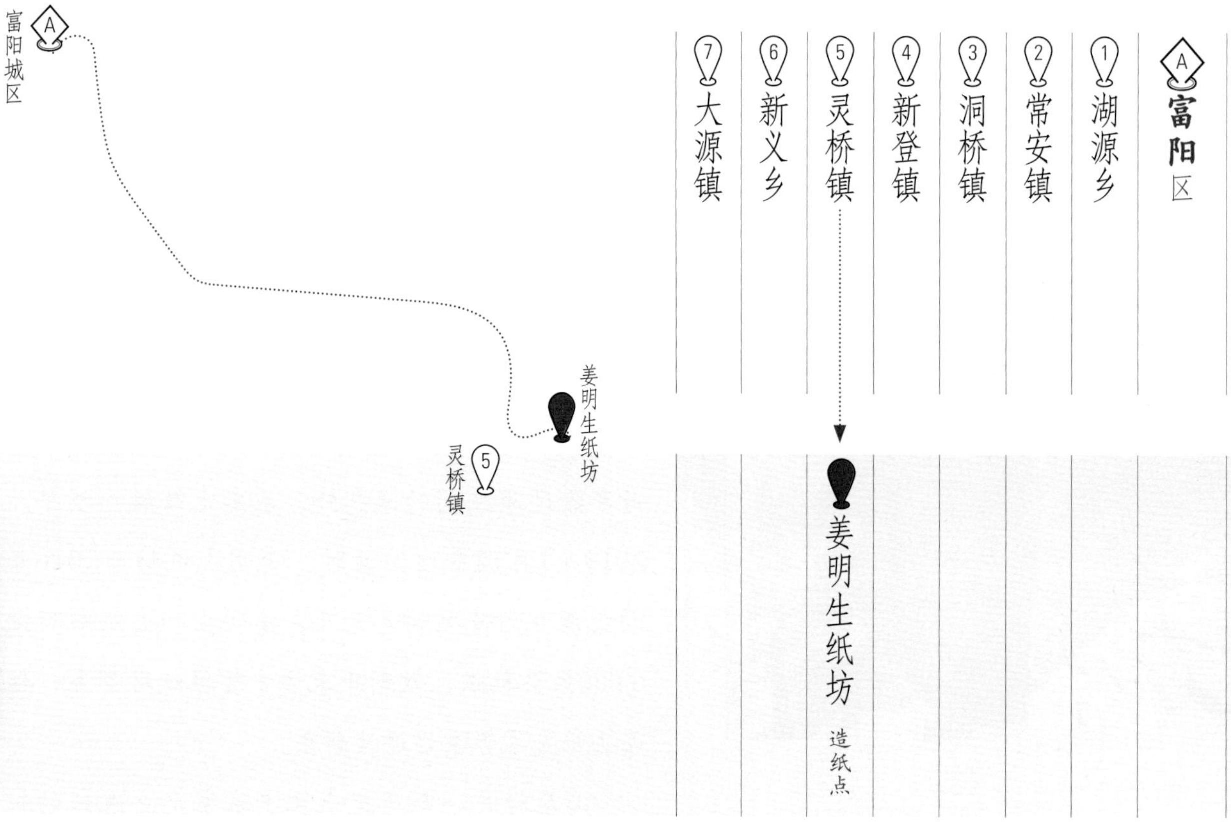

姜明生纸坊
造纸点

位置分布

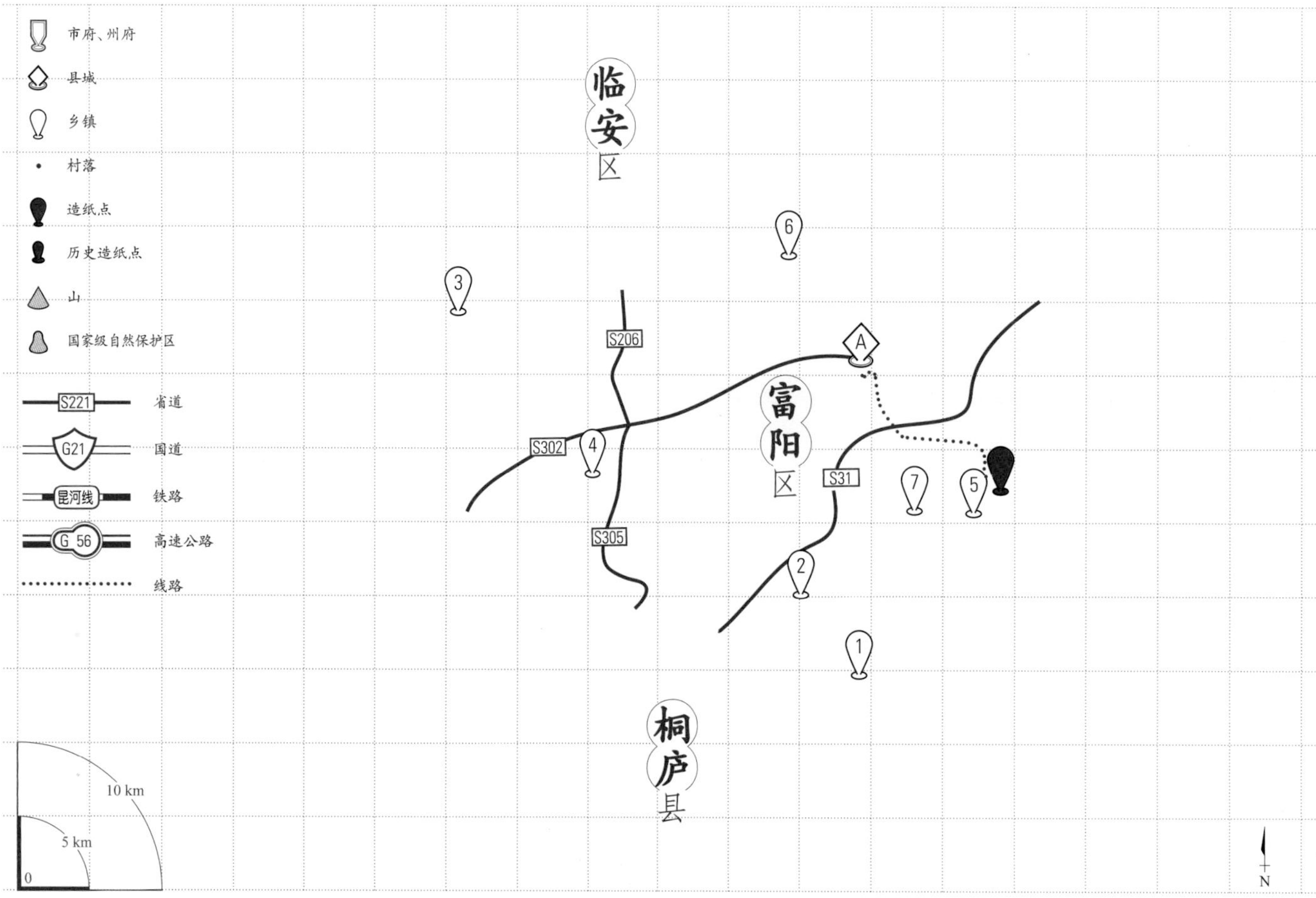

市府、州府
县城
乡镇
村落
造纸点
历史造纸点
山
国家级自然保护区
S221 省道
G21 国道
昆河线 铁路
G 56 高速公路
线路

临安区
富阳区
桐庐县
S206
S302
S305
S31
10 km
5 km
0
N

二
姜明生纸坊的历史和传承

2
History and Inheritance of Jiang Mingsheng Paper Mill

富阳当地民间流传的故事说到，早在东汉时期山基村就已使用桑皮做纸，但具体的信息来源已完全不可考。王羲之写兰亭集序时，浙江的造纸业已开始发育，有当时用桑皮、溪藤造纸的记述，至迟在魏晋时期，浙江确实就已经是纸产区了。到了宋代，富阳生产的竹纸“元书纸”被选为御用上品。

山基村在历史上以生产元书纸为主，但到了现代则以造祭祀纸和厕纸为主，20世纪90年代后则只造祭祀纸。姜姓在山基村算是大姓，调查时还可见到旧日建造的姜氏宗祠。姜明生手工造纸坊乃是由祖辈传下来的，但到底传了多少代，访谈中姜明生和其他村民也无法说清楚，只知道姜明生的爷爷和父亲一辈子都是以做纸为生的。姜明生本人参与砍料、洗料、捞纸等工序，其妻子负责捆扎白坯等工作，另外雇有两名员工负责日常的浸坯洗料。纸坊近乎全年不休，每年一般只有过年期间才休息，传统是过年前三天放假，到正月十六开工。

⊙1

⊙1 姜氏宗祠 The Ancestral Hall of the Jiangs

调查组2019年1月第二次走访姜明生纸坊，得知其家族造纸技艺传承情况为：

姜明生的爷爷姜进法，生于19世纪90年代的清朝末期，于改革开放初期去世，终年80余岁。姜进法的造纸技术是祖辈相传的，最拿手的造纸技艺是晒纸和揭纸。

姜明生的父亲姜荣清，1924年生，1981年去世，终年57岁，父亲是跟着爷爷学的晒纸和揭纸技艺，一直以造纸为生。二伯姜华清，1935年生，2017年去世，终年82岁，晒纸、捞纸的技艺姜华清也都会。姜明生感叹最会造纸的是三伯姜建清，其1938年生，调查时81岁，生产队时期在生产队里专门做元书纸，不过改革开放后他就不做纸了，转而专心于农业生产。姜明生小伯姜浩清，1943年生，2013年去世，终年70岁，生前专门从事拌料工作，一直帮别人帮工拌料到50多岁。

⊙1

⊙1
捞纸中的姜明生
Jiang Mingsheng doing the papermaking work

三
姜明生纸坊的代表纸品及其用途与技术分析

3
Representative Paper and Its Uses and Technical Analysis of Jiang Mingsheng Paper Mill

（一）姜明生纸坊代表纸品及其用途

2016年9月28日调查时得知，姜明生纸坊只生产祭祀纸，品种和规格都很单一。据姜明生介绍，以前全部采用传统工艺与设备造纸，近年来随着人力成本的提高和现代机械技术的发展，对传统造纸过程中使用的部分工具设备进行了升级改造，如在舂料和制浆环节改用电动打浆机，又如原先是上山砍伐当地的毛竹自己制浆，现在因人工成本提高，改为从绍兴的夏履镇等地购买制成品竹料。姜明生纸坊生产的祭祀纸销售时均按捆卖，不零卖，1捆2 000张，售价200元左右。纸张尺寸为33 cm × 40 cm。

⊙2

（二）姜明生纸坊代表纸品性能分析

测试小组对采样自姜明生纸坊祭祀纸所做的性能分析，主要包括定量、厚度、紧度、抗张力、抗张强度、白度、纤维长度和纤维宽度等。按相应要求，每一指标都重复测量若干次后求平均值，其中定量抽取5个样本进行测试，厚度抽取10个样本进行测试，抗张力抽取20个样本进行测试，白度抽取10个样本进行测试，纤维长度测试了200根纤维，纤维宽度测试了300根纤维。对姜明生纸坊祭祀纸进行测试分析所得到的相关性能参数如表10.11所示，表中列出了各参数的最大值、最小值及测量若干次所得到的平均值或者计算结果。

⊙2 姜明生纸坊主要纸品——祭祀纸 Joss paper, the representative paper of Jiang Mingsheng Paper Mill

表10.11　姜明生纸坊祭祀纸相关性能参数
Table 10.11　Performance parameters of joss paper in Jiang Mingsheng Paper Mill

指标		单位	最大值	最小值	平均值	结果
定量		g/m²				35.3
厚度		mm	0.183	0.131	0.151	0.151
紧度		g/cm³				0.234
抗张力	纵向	mN	3.8	3.4	3.6	3.6
	横向	mN	3.5	3.2	3.4	3.4
抗张强度		kN/m				0.233
白度		%	40.0	37.7	38.8	38.8
纤维	长度	mm	1.9	0.1	0.7	0.7
	宽度	μm	32.5	0.9	13.0	13.0

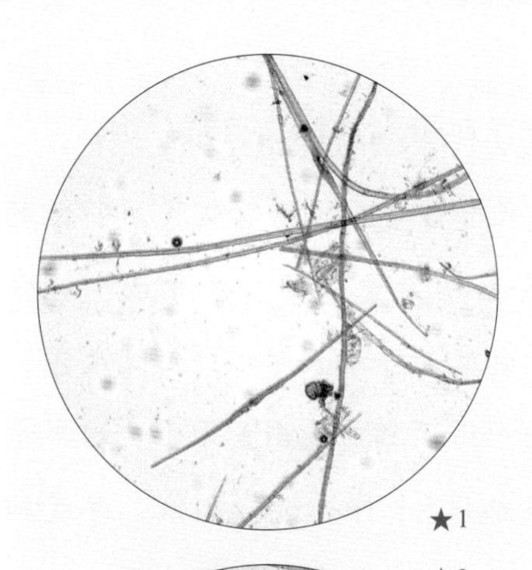

★1

★2

由表10.11可知，所测姜明生纸坊祭祀纸的平均定量为35.3 g/m^2。姜明生纸坊祭祀纸最厚约是最薄的1.397倍，经计算，其相对标准偏差为0.026。通过计算可知，姜明生纸坊祭祀纸紧度为0.234 g/cm^3，抗张强度为0.233 kN/m。

所测姜明生纸坊祭祀纸平均白度为38.8%。白度最大值是最小值的1.061倍，相对标准偏差为0.009。

姜明生纸坊祭祀纸在10倍和20倍物镜下观测的纤维形态分别如图★1、图★2所示。所测姜明生纸坊祭祀纸纤维长度：最长1.9 mm，最短0.1 mm，平均长度为0.7 mm；纤维宽度：最宽32.5 μm，最窄0.9 μm，平均宽度为13.0 μm。

★1 姜明生纸坊祭祀纸纤维形态图（10×）
Fibers of joss paper in Jiang Mingsheng Paper Mill (10× objective)

★2 姜明生纸坊祭祀纸纤维形态图（20×）
Fibers of joss paper in Jiang Mingsheng Paper Mill (20× objective)

四
姜明生纸坊祭祀纸的生产原料、工艺与设备

4
Raw Materials, Papermaking Techniques and Tool for Joss Paper in Jiang Mingsheng Paper Mill

（一）姜明生纸坊祭祀纸的生产原料

1. 主料：毛竹

富阳中高端竹纸的原料均为嫩毛竹，一般在小满前后上山砍伐。山基村依山傍水，竹林茂盛，毛竹资源较为丰富。

调查中姜明生介绍，他们生产的祭祀纸从传统工艺上来说，原料最好选用嫩竹料，但最终选取的毛竹不会特别嫩，因为涉及得浆率与加工成本问题，如果只用嫩毛竹做低端的祭祀纸，成本会抬高，而现在的用户对祭祀纸的品质并不太敏感，不会接受品质优化带来的涨价。山基村的造纸户包括姜明生纸坊在内过去砍竹后还要将竹子去青皮，要求是比较高的。去皮后的竹料颜色越红意味着竹料越老，越黄代表着越嫩，红色的部分纤维会固化。

⊙1

由于山基村曾经是家家户户造竹纸，本村山上的毛竹产量不够，加上山高路险，年轻人留在村里的少，年纪大的人上山砍竹既辛苦又危险，因此21世纪以来山基村纸坊用的毛竹基本上都是从萧山、余姚、绍兴和诸暨等地购买的。姜明生告诉调查组，他们在外地购买的都是当年生的嫩毛竹，一般是成车地购进毛竹，5 kg的新鲜嫩竹可沥1 kg干竹，2016年干料2.4元/kg，嫩毛竹0.7元/kg。2015、2016年，姜明生纸坊每年购进30 000～35 000 kg嫩毛竹。

⊙2

2. 辅料一：纸边

调查时，姜明生纸坊生产祭祀纸会在毛竹主料的基础上添加一些废纸边和水泥纸袋以降低成本。约有一半以上的纸浆会掺入纸边等别的辅料。据姜明生介绍，纸边的原材料是水竹。纸坊

⊙1 附近生长毛竹的高山
Phyllostachys edulis forest grown on the high mountain

⊙2 姜明生纸坊所用的废纸边
Abandoned paper edges in Jiang Mingsheng Paper Mill

通常不到外面去购买废纸边，外地有生意人知道山基村有批量造纸户对废弃纸边有较大的需求，会根据需求不定期将纸边送到村里。2016年纸边价格为3.3元/kg。

3. 辅料二：水

姜明生纸坊制作祭祀纸需要大量的水，选择的是山基溪里流淌的溪水。山基溪的水清澈见底，含有的金属、氯盐、硫酸盐很少，水的硬度很低，所做祭祀纸寿命较长。同时，河水水温较低，使得纸胶料不易分解和变质，可以减少发酵时尿素的用量，更利于生态保护。调查组前往山基溪，实测溪水的pH约为5.5，呈酸性。

4. 辅料三：尿素

尿素，又称碳酰胺（carbamide），是一种白色晶体。因为在人尿中含有这种物质，所以得名尿素。尿素可使水质变“酸”，有助于发酵，促使竹料腐烂，加速其熟化软化的过程，使得原料毛竹最终能够成为纸浆。姜明生介绍，他们制作熟料祭祀纸时通常会在蒸锅里加入尿素，因尿素易溶于水，将它直接放入或是一层层撒放均可。制作加工生料祭祀纸时无需加入尿素，让竹料自然发酵即可。

5. 辅料四：颜料

化学提取碱性嫩黄颜料（粉）是从山东济南购入的，2015年时1袋（25 kg）价格是280元。姜明生纸坊采用的染色方法是将颜料直接放到捞纸槽里，每次用量约为喝汤用的汤匙20勺，具体根据个人经验添加。添料的时候把隔离网拿掉，将

⊙1

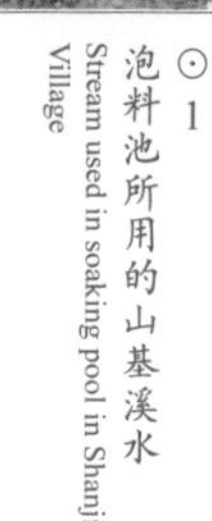

⊙1
泡料池所用的山基溪水
Stream used in soaking pool in Shanji Village

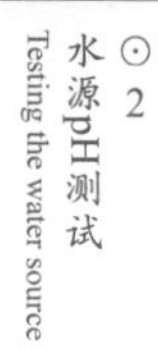

⊙2
水源pH测试
Testing the water source pH

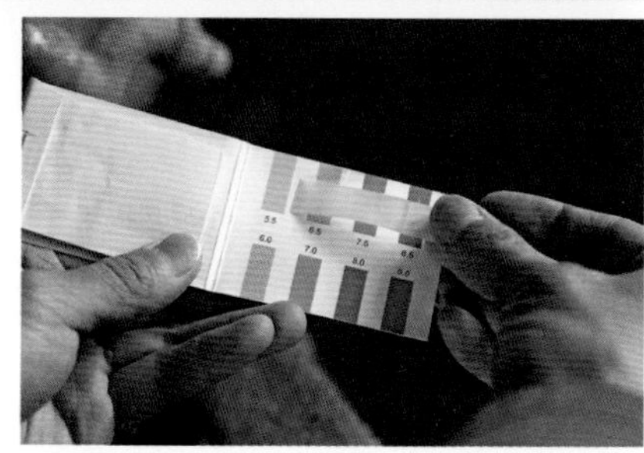

⊙2

纸浆料用泵冲均匀了，再放碱性嫩黄粉。调查人员看过姜明生被嫩黄化学染料染黄的手指，据其介绍，不做纸的话，至少需要一年才能褪去颜色。

除了手指被染黄之外，还有一些人对嫩黄化学染料会过敏，姜明生便属于这类人。姜明生说，一旦过敏发作起来，身上痒，喉咙痛，需要吃抗过敏药片才行，“夏天天气热，容易出汗，就不过敏，冬天不出汗，过敏持续过三四个月”。姜明生展示了他的手臂和抗敏药，“去医院看过，说是化工过敏”。为了造纸，捞纸师傅是不能戴手套的，只能忍着过敏的痛痒坚持捞纸。

⊙3

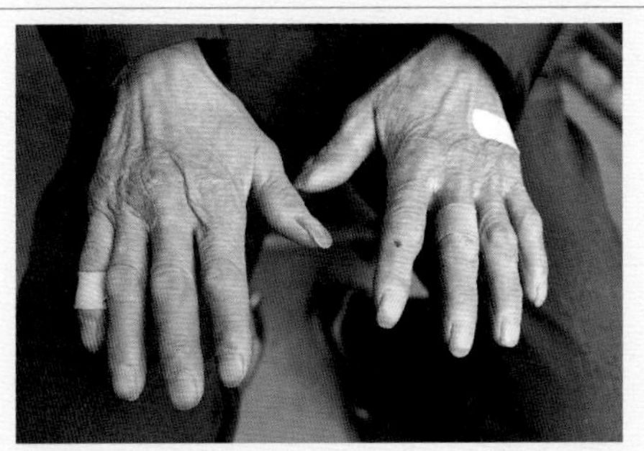
⊙4

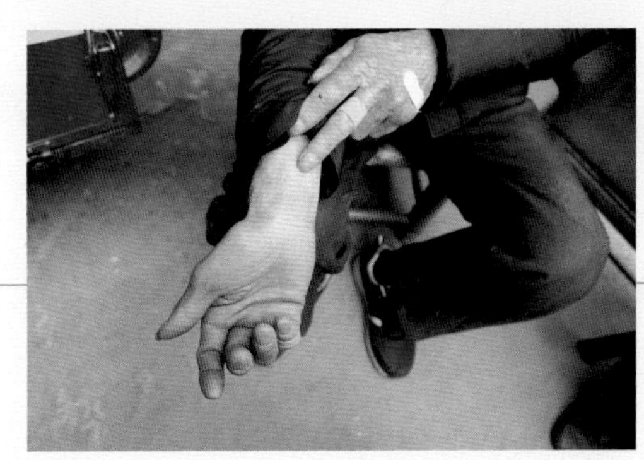
⊙5

⊙3 颜料——碱性嫩黄粉 Powder pigment-auramine

⊙4 姜明生被嫩黄化学染料染黄的双手 Jiang Mingsheng's hands dyed yellow by the chemical dyestuff

⊙5 姜明生展示过敏痕迹 Jiang Mingsheng showing his allergies

（二）姜明生纸坊祭祀纸的生产工艺流程

山基村在历史上以生产毛竹制成的书写用纸——元书纸为主，整体造纸的大小工序有数十道，经过历代不断的改进，到明清时期已基本完善。祭祀用的毛竹纸属于百姓生活用纸，相对要求没有文化用纸高。根据姜明生对纸坊祭祀纸的口述介绍，综合调查组2016年9月28日的实地调查，归纳姜明生纸坊祭祀纸的生产传统与当前工艺流程为：

壹	贰	叁	肆	伍	陆	柒	捌	玖	拾	拾壹	拾贰
砍竹	削青	拷白	落塘	断料	浸坯	浆料	煮料	翻滩漂洗	淋尿堆蓬	落塘	榨水

拾叁	拾肆	拾伍	拾陆	拾柒	拾捌	拾玖
磨料	打浆	捞纸	榨纸	牵纸晒纸	数纸检纸	成品包装

壹 砍竹

1 ⊙1

以往是在小满前后上山砍嫩毛竹，现在是购买绍兴等地的半嫩毛竹。老竹有很多纤维会固化，非纤维类的物质较多，做出来的纸杂质多，力度也不够紧，纸张也就不好。

⊙1

贰 削青

2

在削竹场用砍竹刀将嫩竹截成约2 m长的竹筒，再削去嫩竹的青皮。

叁 拷白

3

把削去青皮的嫩竹筒在石头上摔打，使其分裂开，形成白坯。

肆 落塘

4

可以做成干料之后再浸泡，也可湿毛竹直接浸泡。湿竹料拷白后，用竹篾将白坯扎成捆，放入清水塘中浸泡4～5天。

⊙1
山基村山上的毛竹
Phyllostachys edulis on mountains in Shanji Village

伍
断料

5 ⊙2⊙3

将浸泡好的白坯用切割机切割成长约40 cm的小段，扎成约15 kg重的小捆。

⊙2

⊙3

陆
浸坯

6 ⊙4

生料法通常将没有浸泡的白坯盖起来，而浸泡好的白坯则用石灰沤。工艺为：将成捆的白坯放入料塘中浸泡，至少浸泡1个月，多的话可以浸泡6个月，浸泡的白坯不需要等到发“酸”，用水冲洗干净就可以拿去发酵了。

⊙4

⊙5

⊙6

柒
浆料

7 ⊙5⊙6

把浸好的白坯在腌料场用石灰浆浆腌，然后在灰池边堆置1～2天。

⊙2 断料 Cutting the materials

⊙3 捆料 Binding the materials

⊙4 浸坯 Soaking the white bark

⊙5 浆腌 Fermenting the materials

⊙6 堆放发酵 Piled materials for fermenting

捌 煮料

8 ⊙7

将浆好的白坯在熟料坊的皮镬里面煮焖7天（煮2天焖5天），煮熟后取出，浸入料塘水中，沤1个星期后清洗，将碱水漂掉（这道工序是制作熟料时要进行的，制作生料时无此工序）。

⊙7

玖 翻滩漂洗

9 ⊙8

熟料法是将煮好的白坯在料塘中浸泡、漂洗1个星期，期间翻动冲洗5～6遍，将纤维洗净。清洗生料的过程比较复杂，需要先将白坯摔打很多次，因竹子较硬，所以需要用力摔。方法是：先把料叠放在一起，然后摔打，之后拿清水清洗，每摔打几次就用瓢舀水冲洗一下竹料，把里面的灰尘、沙子等洗掉，如此重复几遍，使每根竹子都被冲刷、碰撞几次。完成标准是白坯里面流出的水是清水就可以了，否则需要继续摔打清洗。每捆料重20 kg左右。

⊙8

拾壹 落塘

11 ⊙10

将发酵后的竹料连同原来的水，再加清水一起浸泡，一般是夏天浸泡约1个月，水色转红变黑、发青说明竹料已经熟化，可以进行下一道工序。

拾 淋尿堆蓬

10 ⊙9

将洗净的竹料重新整理捆扎后放入尿桶中用尿淋浸，促使纤维软化。捆两道之后堆叠成蓬，用青干草覆盖包裹，让其自然加温促使微生物繁殖发酵。堆叠时间以竹料是否柔软为准，一般为1～2周，10天左右的时间。之后再堆料，嫩竹料堆1个星期就够了，老竹料要堆10天以上。闷热的环境再加上淋尿后竹内富含的营养体适合微生物发酵，能较好地实现竹料熟化和纤维净化。山基村的通常做法是在料堆上面盖一些塑料雨布之类，以阻挡杂质的进入。

⊙9

⊙10

拾贰 榨水

12

将浸泡好的竹料榨干水分，即可用来制作竹纸浆料。

拾叁 磨料

13

山基村造纸工序中，磨料和打浆是同时进行的，将竹料准备好，放入打浆池，用打浆机磨碎。据姜明生的叙述，之前山基村造纸户都是用石磨进行磨料的，后来约在20世纪80年代中后期，为了省电省时，省略了这一步骤，直接打浆。

⊙7 煮料 Boiling the materials

⊙8 翻滩漂洗 Repeat cleaning the materials

⊙9 堆蓬 Piled materials

⊙10 落塘 Soaking the materials in the pool

拾肆

打　　浆

14 ⊙11

先在纸槽中放满清水，再放纸浆。将发酵好的竹料放入打浆机中，之后不断搅拌，直至其达到均匀黏稠状态。将打好的浆料通过管道运输至捞纸槽中，在纸槽中放入清水，用木耙反复搅拌。姜明生纸坊打浆时是一边加竹料一边加纸边直至搅拌均匀，一次打浆2个小时，可打18捆料，1捆约20 kg，再加纸边65 kg以上。纸边与竹浆配比为1∶7.2左右。姜明生家是用打浆机直接打好一天的量，大约可以捞3捆纸。

⊙11

拾伍

捞　　纸

15 ⊙12

姜明生纸坊每次打出来的纸浆约为150 kg，可以捞2 000帘纸。姜明生会先根据自己的生产需求和捞纸习惯对浆料进行搅拌，捞纸时会依据浆料的厚薄度多次使用抻料耙进行搅拌。之后的工序为：用竹帘在槽内纸浆中抄提，在帘上形成纸浆膜，扣在纸架上便为湿纸。添料的时候把隔离网拿掉，放个泵将纸浆冲得均匀了，再放碱性嫩黄颜料粉，一次添浆的时间约为30分钟。一般在晚上歇工前会再加一次浆，为第二天的捞纸做好准备。第二天需拿抻料耙搅拌一下纸槽里的水和纸浆，使料上浮，因为料沉在槽底太久会分布不均。

⊙12

拾陆

榨　　纸

16 ⊙13⊙14

待湿纸积攒到约1 000帘后，移至榨床，先用木板再用千斤顶榨干水分。姜明生一天压榨2次，早上4点半至9点半捞纸，9点半后压榨上午捞出来的1 000帘纸（一帘隔三纸），即3 000张纸；10点继续捞纸，下午3点再压榨1 000帘纸（一帘隔三纸）。也就是说，姜明生一天可捞、压榨2 000帘纸，即6 000张祭祀纸。

⊙13

⊙14

⊙11 打好的浆料 Pulp materials after beating

⊙12 捞纸 Papermaking

⊙13 压榨用的榨床 Pressing device used in the pressing procedure

⊙14 压榨用的千斤顶 Lifting jack used in the pressing procedure

拾柒
牵纸晒纸

17 ⊙15～⊙18

熟练的牵纸工一般用嘴蹭开或者吹开纸帖，数出6～7张纸作为一叠，逐步从湿纸帖上“牵”下来，放在旁边，等候下一步的晒纸。山基村晒纸很特别，不是用焙壁烘纸，也不是放在太阳底下晒，而是放在阴凉处自然晾干。据姜明生介绍，在生产队时期，大家造元书纸的时候是用焙壁烘纸的，之后大家都造祭祀纸的时候，就不再用焙壁了，而是放在山间平地或者庭院中的竹架上晾干。阴晾所需时间随气温变化有所不同，夏天的话一早晾出去，下午就可以收纸了，冬天的话需要阴晾三四天，一旦下雨则需20多天，而且下雨的话可能会造成祭祀纸褪色，使纸的质量下降。姜明生认为不用焙壁有三个原因：一是焙壁要烧柴火；二是很多工人已经不会在焙壁上晒纸；三是造焙壁费钱。

⊙15

⊙17

⊙16

⊙18

⊙15 牵纸 Peeling the paper down

⊙16 牵出待晒的湿纸 Wet paper to be dried

⊙17 晒纸 Drying the paper

⊙18 晒纸竹架 Bamboo frame for drying the paper

拾捌

数纸检纸

18 ⊙19

数纸工将晾干的纸按照一定的数量叠好，同时将破损的、褪色的等质量不好的纸张挑选出来。

拾玖

成品包装

19 ⊙20⊙21

用竹篾或者绳索将成品纸张扎成捆，将四面锉平、磨光，使其整齐美观。

⊙19

⊙20

⊙21

⊙19 数纸检纸 Counting and checking the paper

⊙20 包扎成捆的成品纸 Bundles of paper

⊙21 完成包装入库存放的成品竹纸 Packaged bamboo paper in the storehouse

（三）姜明生纸坊祭祀纸的主要制作工具

壹 纸帘 1

捞纸工具，用于形成湿纸膜和过滤多余的水分。由细竹丝编织而成，表面刷有黑色土漆，光滑平整。姜明生纸坊所用纸帘是在大源镇上的大源纸帘厂购买的，线是由姜明生纸坊的人自己缝的。1张帘子一般可以用7～8个月，2014～2015年价格为350元/张，2019年1月回访的时候，纸帘已经是400元一张了。实测姜明生纸坊所用纸帘尺寸为：长110 cm，宽43 cm。

贰 帘床 2

木制的长方形框架，面积稍大于纸帘，纸帘可完全嵌于框架中，作用是捞纸时承载纸浆和纸帘。帘床两侧有两根绳子垂直将其吊于纸槽内水面上方，使其不会在捞纸工松手后随意掉落。帘床是木匠做的，用的材料是杉木和毛竹，300元/张。实测姜明生纸坊所用帘床尺寸为：长112 cm，宽54 cm。

叁 打浆机 3

用来制作竹浆料的工具，调查时姜明生纸坊使用的是依靠电力打浆的打浆机。实测姜明生纸坊所用打浆机尺寸为：长546 cm，宽158 cm，高67 cm。

⊙24

⊙22

⊙23

肆 灰耙 4

腌料前用于搅拌石灰水的工具，以使石灰与水充分接触、搅拌均匀。实测姜明生纸坊所用灰耙尺寸为：柄长232 cm；耙头宽14.5 cm，长39 cm。

⊙25

⊙26

伍 两齿耙 5

腌料时用于抖、捞页料的工具，因有两齿插入页料中，可自如控制页料在石灰水中的位置。实测姜明生纸坊所用两齿耙尺寸为：大的两齿耙柄长217 cm，齿长39 cm；小的两齿耙柄长115 cm，齿长35.5 cm。

⊙22 一隔三的纸帘 Papermaking screen that can make three piece of paper simultaneously

⊙23 帘床 Frame for supporting the papermaking screen

⊙24 电动打浆机 Electric beating machine

⊙25 灰耙 Grey rake

⊙26 两齿耙 Two tine rake

陆
料　耙
6

放浆池或捞纸槽中搅拌原料所用，呈一木棍形状，其一头处镶嵌橡胶方片。实测姜明生纸坊所用料耙尺寸为：柄长175 cm；橡胶方片长25 cm，宽21 cm。

⊙27

玖
断料凳
9

断料时使用的辅助工具。将浸泡好的白坯用切割机切割成长约40 cm的小段后，将其放在断料凳的反面，这样可方便工人将其捆扎成约15 kg的小捆。实测姜明生纸坊所用断料凳尺寸为：长48 cm，宽17 cm，高32 cm。

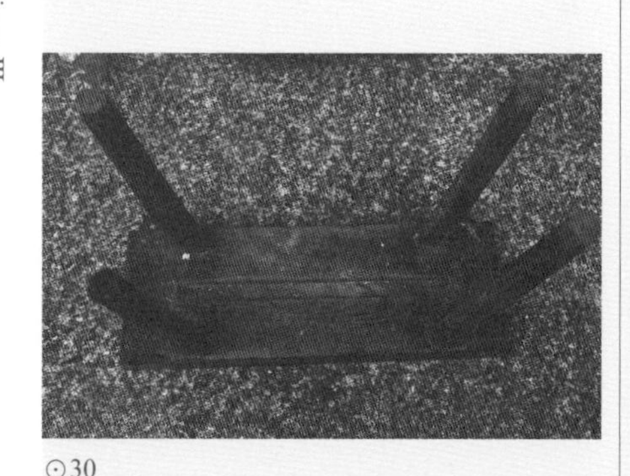
⊙30

柒
榨　床
7

用铁做成的架子，底下放一块木板，可以将捞好的湿纸帖放在木板上面，利用杠杆原理，榨干湿纸帖中的水分。实测姜明生纸坊所用榨床尺寸为：长80 cm，宽125 cm，高148 cm。

⊙28

拾
砍竹刀
10

实测砍竹刀的尺寸为：刀长43 cm，柄长15 cm。

捌
纸　槽
8

盛放纸浆的设施，长方形，捞纸工站于其侧进行工作。纸槽所在的屋子叫作造纸坊。据姜明生介绍，纸槽是40年前生产队卖给他的，那时候他一天的工资为1.3元，买这口纸槽花了近1 200元，不过是连着地皮一起买的，姜明生认为虽然贵，但很值得。实测姜明生纸坊所用纸槽尺寸为：长227 cm，宽180 cm，高79 cm。

⊙29

⊙31

⊙27 料耙 Stirring rake
⊙28 榨床 Pressing device
⊙29 纸槽 Papermaking trough
⊙30 断料凳 Bench for cutting the materials
⊙31 砍竹刀 Knife for cutting the bamboo

五
姜明生纸坊的市场经营状况

5
Marketing Status of Jiang Mingsheng Paper Mill

2016年调查时，姜明生纸坊祭祀纸年产量在1 000捆左右，1天生产3捆，一捆20刀共2 000张，即1年可生产200万张纸。2019年1月回访时，姜明生纸坊年产量有所下降，多的时候800多捆，少的时候600捆，即120万～160万张纸。纸的价格倒是没有变化，一般情况下稳定在200元/件，质量好点的230元/件，褪色的纸只能卖180元/件。姜明生纸坊每年生产的纸都能卖掉，几乎没有库存。按照姜明生的说法，除去工人工资及相关成本，一年能赚6万～7万元。造纸坊做纸不是按照订单来做，而是每天按照生产流程按部就班地生产纸张，做到家里没有地方堆放纸张为止。据姜明生介绍，这两年还是有人来收购纸张的，一般是经销商主动打电话来山基村收纸，纸被卖到海宁、余杭、绍兴、萧山、余姚和江苏等地。只要有纸他们就会开车来收，卖光就来。

现在手工造纸的人越来越少，大都受到机械祭祀纸冲击。姜明生表示，再做几年，年纪大了他就少做点，等到70岁，身体不行做不动就不做了。现在的年轻人也没有愿意做纸的了，子女晚辈均无人愿意继承。姜明生用了一个很形象的说法来形容如今的手工造纸行业："以前是2个人2口纸槽。现在是1个人可以有很多纸槽，但却找不到人从事手工造纸了。"

⊙32

⊙32
仓库中待出售的纸
Paper to be sold in the storehouse

六 山基村和姜明生纸坊的造纸习俗与故事

6 Papermaking Custom and Stories of Shanji Village and Jiang Mingsheng Paper Mill

（一）“姜芹波忠记”昌山纸

提起山基村的造纸业，不得不提“姜芹波忠记”的昌山纸。1915年，昌山纸获得国家农商部嘉奖，并被认为是最高特货，同年在巴拿马万国商品博览会上获得三等奖。山基村造纸户均以昌山纸为荣，姜明生也热情地向调查组讲述了昌山纸的故事。昌山纸是以嫩毛竹为原料造的一种文化用纸，产于灵桥镇的菖蒲坑村和山基村，各取两村村名中的一字而得名为“昌山”。据《富春姜氏宗谱》记载，姜昌忠的祖父姜元君于清嘉庆年间由梓树村迁至山基村，自此利用当地的毛竹资源开办纸业，创办“姜芹波纸号”。清末民初，传至第三代传承人姜昌忠和姜昌柳，分别开办了“姜芹波忠记”和“姜芹波生记”。姜昌忠聘请造纸能手汪志明合力造出质量上乘的“姜芹波忠记”昌山纸，闻名遐迩。据姜明生介绍，姜昌忠后人姜仁来现在仍在从事造纸业，和山基村其他纸户一样，在做祭祀用纸。

⊙1

⊙2

（二）纸坊不闭户，偷盗禁三年

山基村家家户户的作坊都是露天的，人不在纸坊的时候也不会特意去锁门，这边的门都是敞开或者掩上的。姜明生很自豪地告诉调查组，这是山基村的传统，是祖辈流传下来的规矩。如果有人偷拿了别人家作坊里的东西，一经发现，三年内不准造纸，而且在此期间后辈也不能做纸，所以这边的作坊都是露天的。多年以来大家也都很守规矩，“你不来拿我的，我也不去拿你的”。

⊙1 姜仁来纸坊外观 External view of Jiang Renlai

⊙2 敞开或虚掩上门的纸坊 Paper mill with unclosed doors

七
姜明生纸坊的业态传承现状与发展思考

7
Current Status and Development of Jiang Mingsheng Paper Mill

⊙3

⊙4

⊙5

⊙3
山基村村名碑石
Village name monument stone of Shanji Village

⊙4/5
村里仍在造纸的老人
Old man who still making paper in the village

（一）业态传承现状

在20世纪90年代之前，传统祭祀用纸的生产一直是富阳最具特色、占据支柱地位的手工业。以1912年至1935年为例，当时富阳全县五分之一的人口从事纸业生产，土纸产量占全国土纸总产量的四分之一，可见当时富阳传统造纸业之发达。然而，在现代造纸业兴起及习俗变迁的冲击下，富阳的传统土纸尤其是祭祀纸正在渐渐地退出历史舞台。众多关于土纸生产的技艺逐渐失传，大量土纸生产的工具和作坊快速消失，仅在少数村落还保留着这一古老的手工业，山基村便是其中之一。

山基村之前是靠造纸为生的村落，生产队时期有200多家造纸户，几乎是家家造纸。2016年调查时有70户左右的人家造纸，大多造纸人跟姜明生年纪差不多，都在60岁以上，全部做的是低端加纸边、强碱竹料、化工染色的祭祀用纸。2019年1月调查组回访山基村时，只有31户仍然在坚持造纸。从2016年至2019年1月，时隔不到三年就关闭了一大半造纸工坊。姜明生无奈地说："有些人老了，有些人死了，有的人转行去打工了，可不就消失了吗？"再过个5～10年，当他们这一辈年纪再大些做不动了，又没有年轻人学习继承，那么山基村的手工纸技艺和业态的消失恐难以避免。

姜明生告诉调查组，现在灵桥镇政府政策好，老年人养老有保障，山基村老人中男性到60岁、女性到50岁就可以领取养老金了，每人每个月2 000元，一对老夫妻一个月就有养老金4 000元，这个金额对于老年人来说，足够了。姜明生和村里其他一些老人还在坚持造纸并非为了能多挣多少钱，而是因为他们已经造了大半辈子纸了，放不下自己的纸坊，放不下自己的纸，就想着趁还有力气多造出一张纸，多维持一天小纸坊。

（二）发展思考

在山基村造纸园区可以看到，几乎所有料塘和泡料池上面都加盖有竹排，显得十分独特。据姜明生所述，这是2018年9月份的时候，为了建设美丽乡村项目，美化乡村环境，灵桥镇政府出资购买了毛竹让造纸户盖一下，自此山基村30多户造纸坊都给自家的泡料池戴上了“毛竹帽”。结合调查组在山基村看到的特色景点，诸如山基休闲登山道、蔡九华同志墓、直坞水库等，以及山基村优美的环境，灵桥镇政府可以参考将山基村打造成“造纸+旅游+休息”的特色小镇的思路，改进当前的祭祀纸业，鼓励年轻人学习、继承、优化山基村祖传手工造纸业，发扬“姜芹波忠记”昌山纸的名号价值，发展可结合旅游的高质量纸品和时尚文创产品。

⊙2

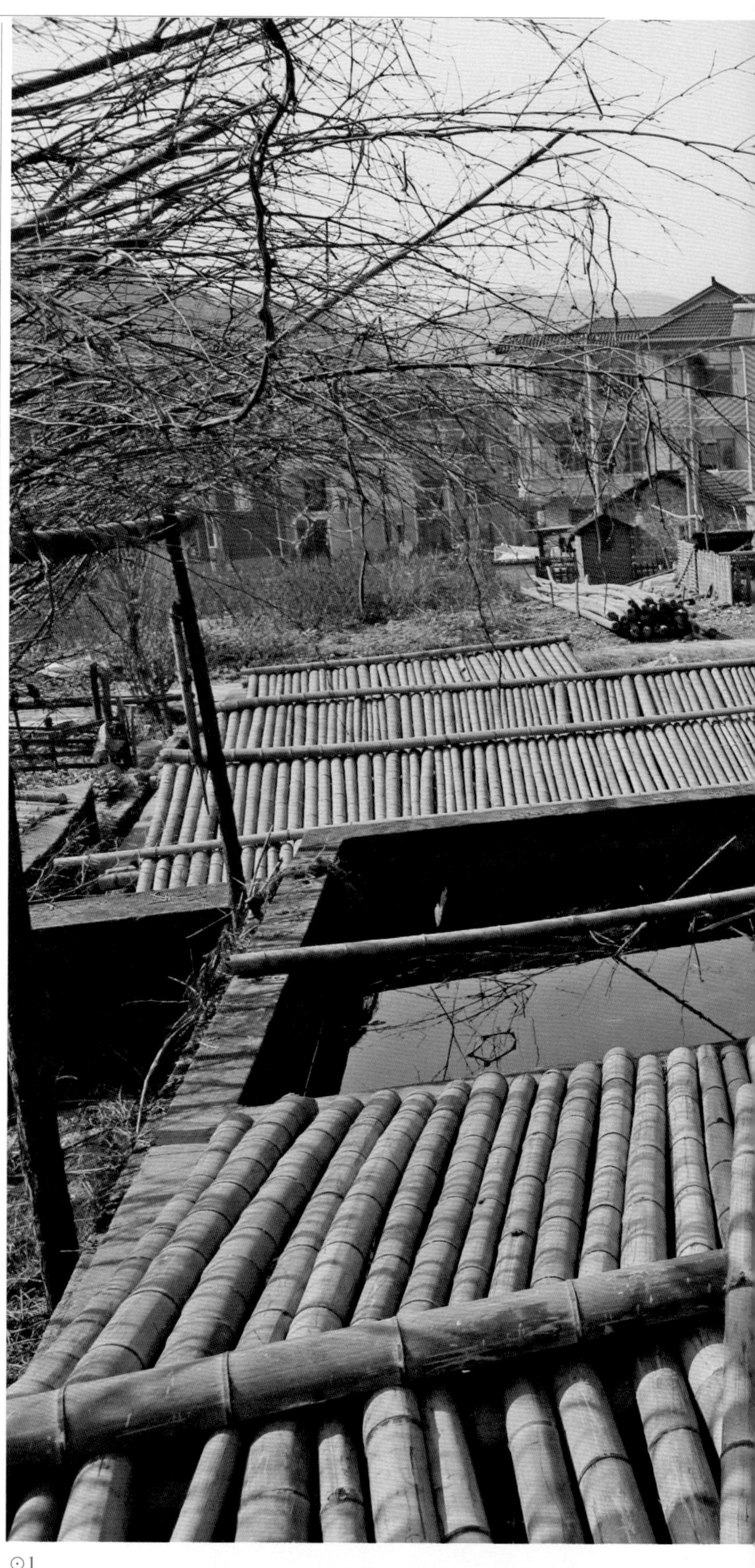

⊙1

⊙1 山基村盖着毛竹的料塘 Soaking pool covered with *Phyllostachys edulis* in Shanji Village

⊙2 山基村登山休闲步道介绍 Introduction to the hiking pathway of Shanji Village

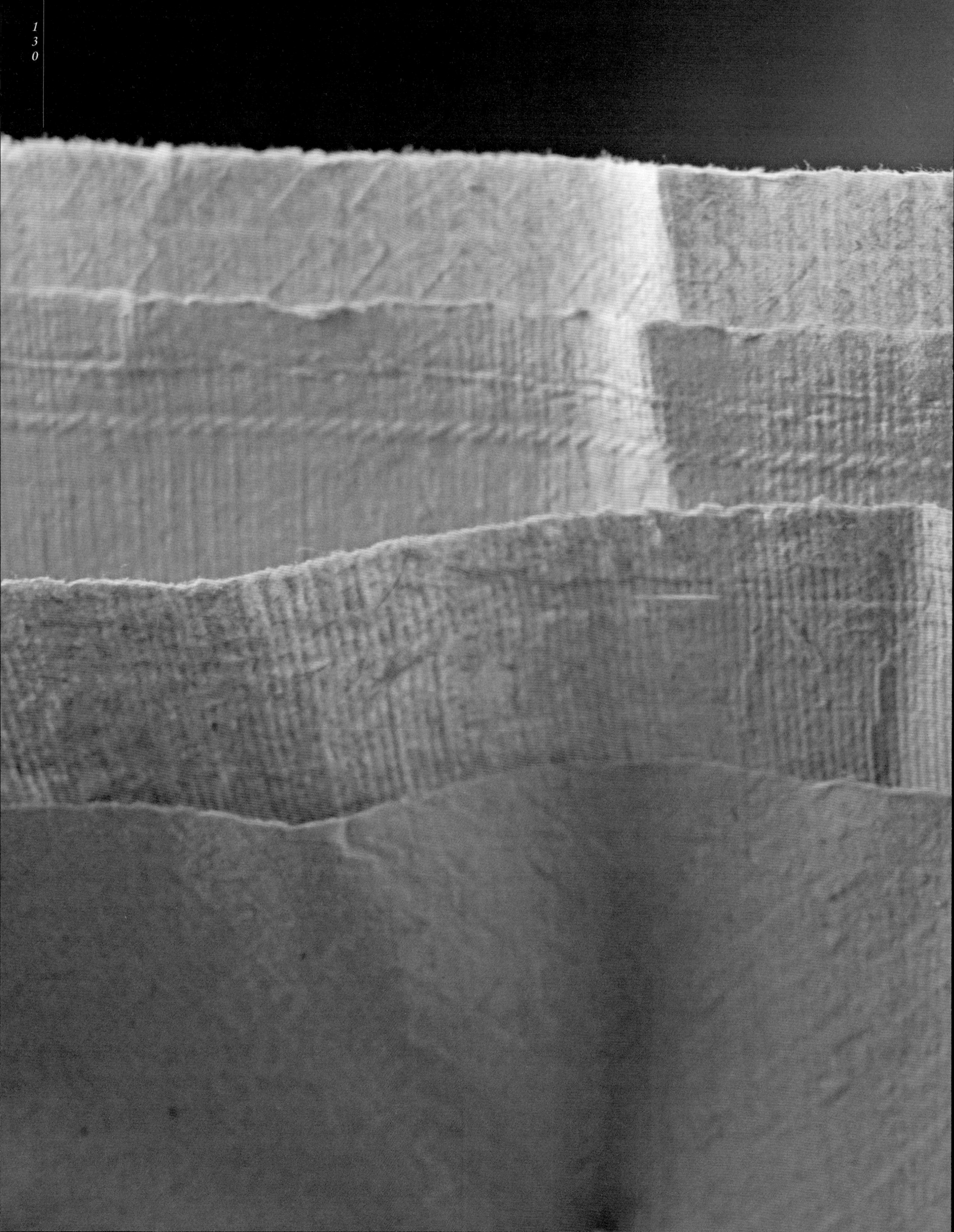

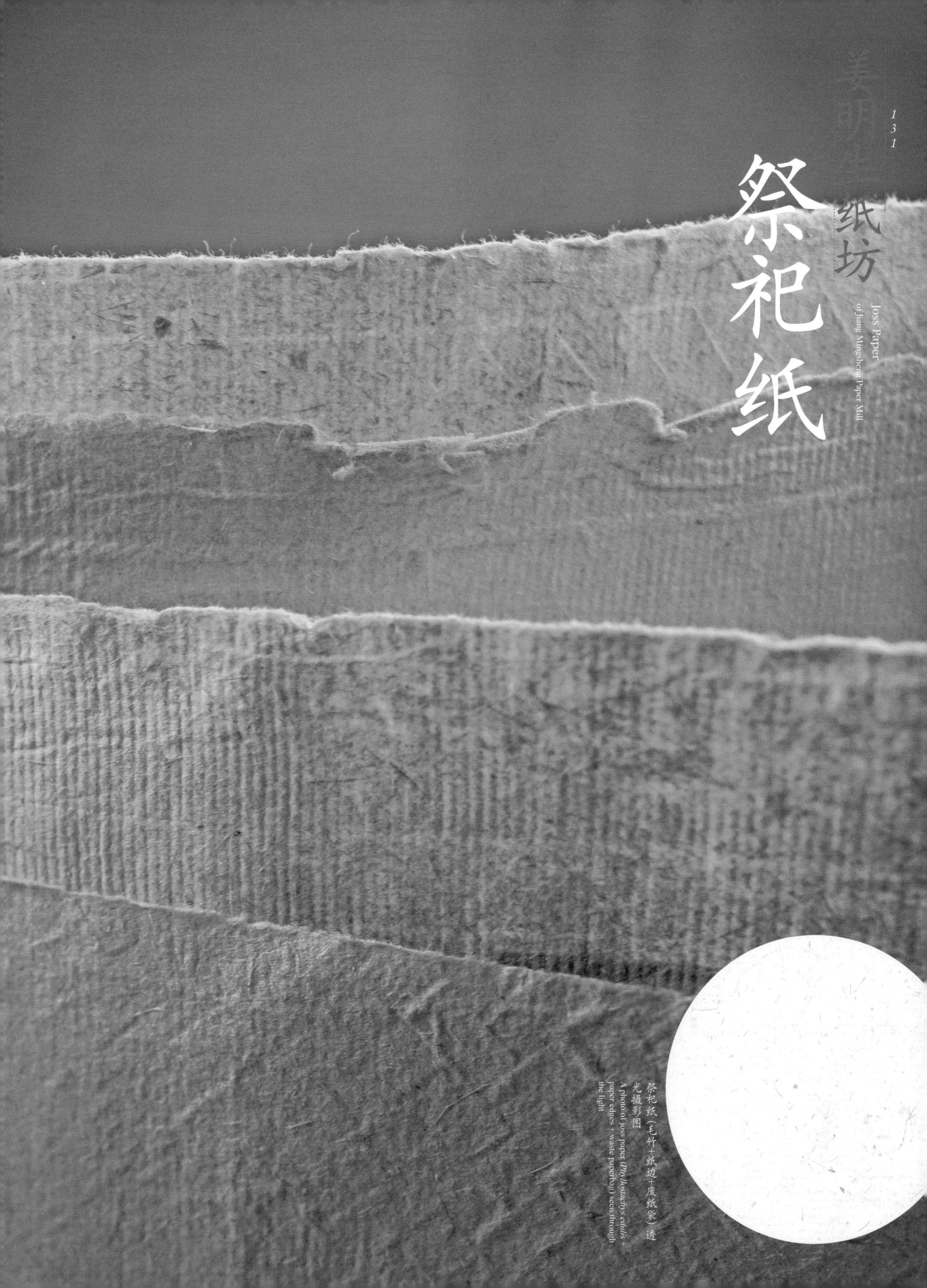

祭祀纸

Joss Paper
of Jiang Mingsheng Paper Mill

祭祀纸（毛竹+纸边+废纸袋）透光摄影图
A photo of joss paper (*Phyllostachys edulis* + paper edges + waste paper bag) seen through the light

第七节

戚吾樵纸坊

浙江省
Zhejiang Province

杭州市
Hangzhou City

富阳区
Fuyang District

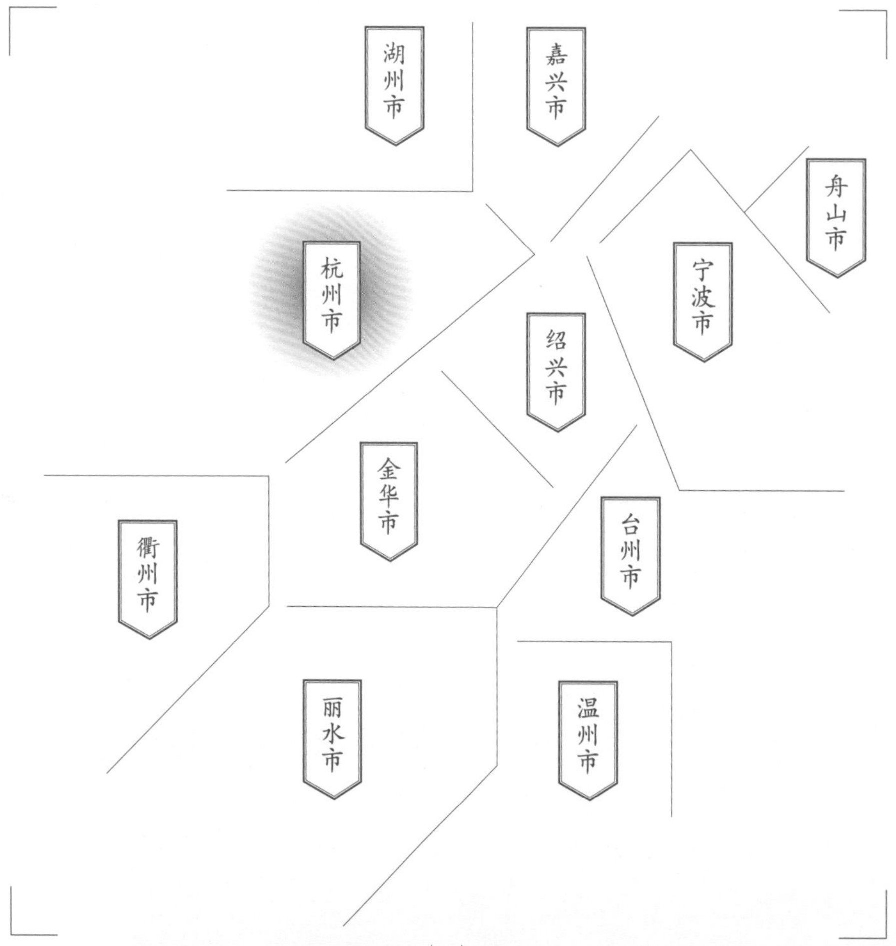

调查对象

富阳区渔山乡大葛村
戚吾樵纸坊
竹纸

Section 7
Qi Wuqiao Paper Mill

Subject

Bamboo Paper of Qi Wuqiao Paper Mill
in Dage Village of Yushan Village in Fuyang District

一
戚吾樵纸坊的基础信息与生产环境

1
Basic Information and Production Environment of Qi Wuqiao Paper Mill

戚吾樵纸坊位于富阳区渔山乡大莴行政村莴村自然村，地理坐标为：北纬30°0′34″，东经120°6′43″。大莴村位于渔山乡南端，与萧山区戴村镇相邻，森林覆盖率高，自然景色优美，五金加工业发达。

大莴村历史文化悠久而深厚，村内古寺大西庵有千年以上的历史，相传明太祖朱元璋曾避难于庵中。村旁石牛山为古代船行至钱塘江出海口时的航标，望石牛山而不迷路。今地方政府建有大西庵游客步道，沿途虎闸弄、阅台、剑门石、藏君洞、元昙庙遗址、掉头葫芦景观及传说颇有逸趣。

2007年11月，行政村规模调整，新大莴村由原旭光、莴村、勤建村三个自然村合并组成。新村落辖区面积11.75 km^2，有农户1 160户，常住人口4 010人。

2016年8月24日，调查组第一次前往戚吾樵纸坊，获得的基础生产信息为：纸坊制作嫩毛竹料祭祀焚烧纸，坊内有1口抄纸槽、1台石磨和1口打浆槽。据戚吾樵介绍，纸坊一年大概做200多件纸，1件20刀计2 000张，大多销往江苏省常熟市虞山镇。2019年回访时，戚吾樵回忆在生产队刚解散时期，纸坊生意较好就雇了两个工人，但随着纸品需求降低，纸坊已有20多年都是自己和妻子两人工作，偶有砍竹时请临时工来帮忙。目前纸坊年产量200余件，销售地以常熟为主，需要售纸时便打电话联系纸商上门收购。

⊙1

⊙2

⊙3

⊙1 大西庵遗址 Relics of Daxi Nunnery

⊙2 石牛山 Shiniu Mountain

⊙3 戚吾樵纸坊 Qi Wuqiao Paper Mill

戚吾樵纸坊位置示意图

Location map of Qi Wuqiao Paper Mill

路线图
富阳城区
↓
戚吾樵纸坊
Road map from Fuyang District centre to Qi Wuqiao Paper Mill

考察时间
2016年8月 / 2019年1月

Investigation Date
Aug. 2016 / Jan. 2019

地域名称

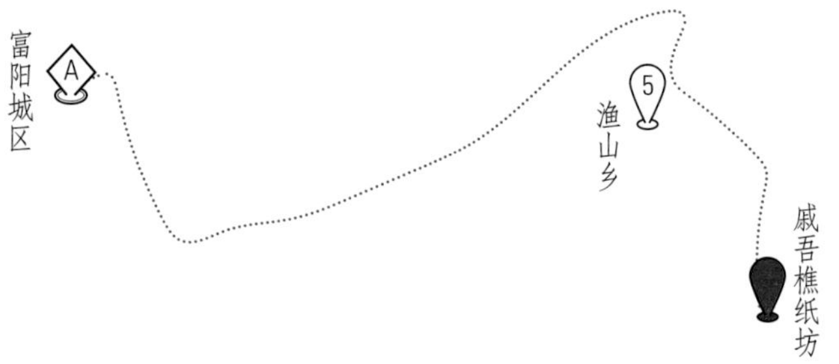

富阳城区
渔山乡
戚吾樵纸坊

7 大源镇
6 新义乡
5 渔山乡
4 新登镇
3 洞桥镇
2 常安镇
1 湖源乡
A 富阳区

造纸点名称

戚吾樵纸坊
造纸点

位置分布

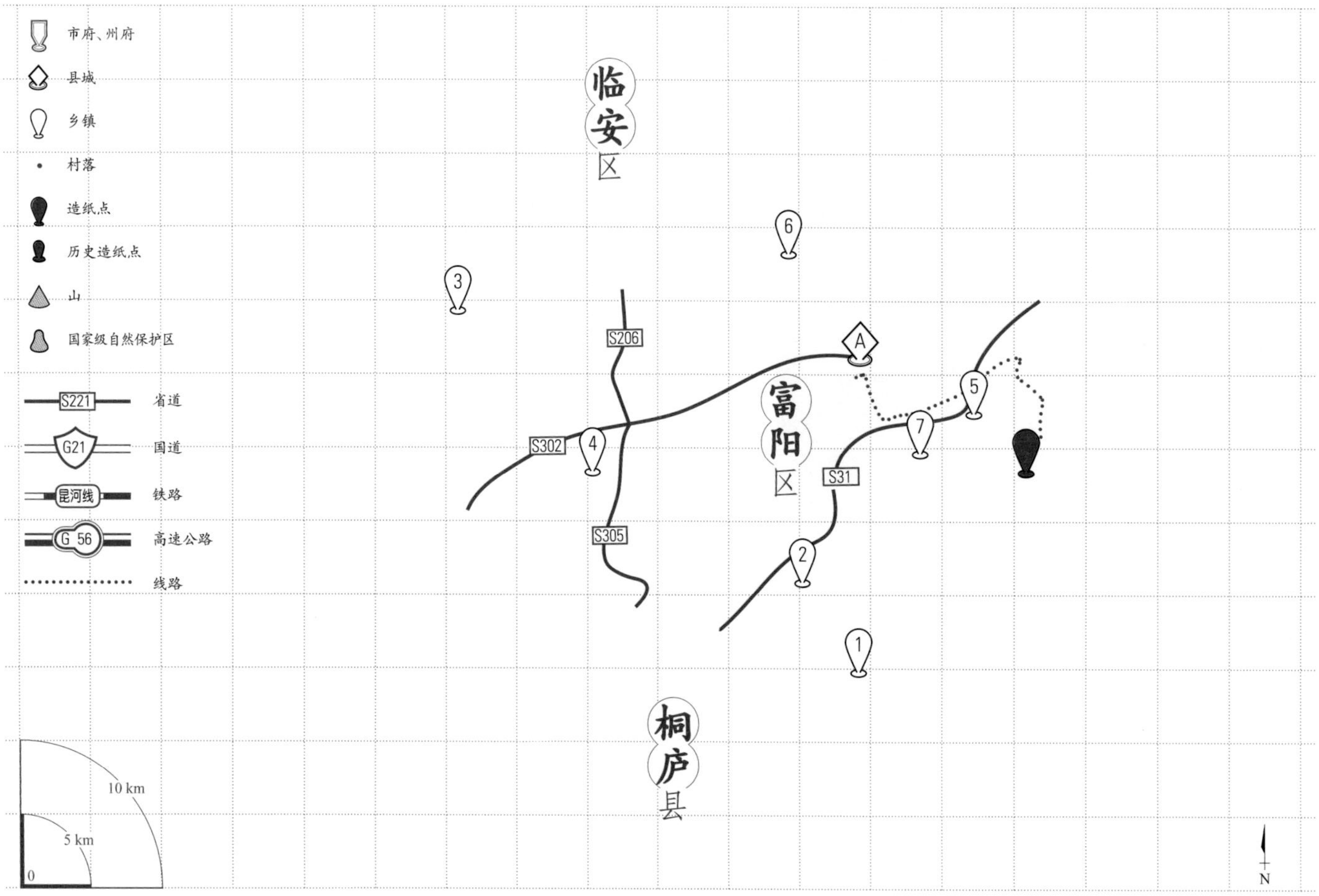

市府、州府
县城
乡镇
村落
造纸点
历史造纸点
山
国家级自然保护区
S221 省道
G21 国道
昆河线 铁路
G 56 高速公路
线路

临安区
富阳区
桐庐县
S206
S302
S305
S31
10 km
5 km
0
N

二
戚吾樵纸坊的历史与传承

2
History and Inheritance of Qi Wuqiao Paper Mill

据戚吾樵回忆，在生产队时期，蔼村共有40多口纸槽投入生产，当时有6个生产小队，1个小队配2～10口不等的纸槽，1口槽配6个工人，具体分工为：1个捞纸工、2个踩料工、2个晒纸工、1个打浆工。

戚吾樵出生于1954年，2016年调查时62岁。他13岁开始跟着伯伯戚松根在红星第二生产队学习做纸。最初学习的是踩料，顾名思义，就是用脚踩毛竹原料。戚吾樵自述，刚开始因年龄太小没有资格学抄纸技术，都是在别人做纸的时候他在旁边观看学习，后来开始自己动手试着做，逐渐掌握了造纸技艺。1982年，集体性质造纸的生产队解散以后，他便开始以私人家庭手工作坊的形式造纸。

伯伯戚松根1923年出生，1977年因病去世，自小和村里的师傅学习造纸技艺，熟悉造纸的一系列工序。父亲戚德银1930年出生，79岁时过世，土地改革时期到生产队解散前一直是村干部，1987年时曾任村电镀厂厂长。据戚吾樵介绍，父亲会造纸的全套技艺，抄纸技艺传承于爷爷戚仁钱。母亲蔼位仙，1929年出生，2019年回访时已90岁，对造纸环节中的晒纸很熟悉。爷爷戚仁钱，1965年左右去世，熟悉造纸的整套流程。

戚吾樵的妻子蔼夏芹，1956年出生，大蔼村人，在家庭纸坊里负责晒纸和包装。戚吾樵夫妻有两个儿子，大儿子戚钰铨，1975年出生，未习得造纸技艺，目前在零件加工厂工作；小儿子戚钰峰，1977年出生，15岁时随戚吾樵学习造纸技艺，18岁左右离开纸坊，在小学食堂从事厨师工作。还有一个19岁的孙女。后辈们所从事的工作都与手工纸无关。

表10.12 戚吾樵纸坊戚吾樵、葛夏芹传承谱系
Table 10.12 Qi Wuqiao and Ge Xiaqin's family genealogy of papermaking inheritors in Qi Wuqiao Paper Mill

传承代数	姓名	性别	与戚吾樵关系	基本情况
第一代	戚仁钱	男	爷爷	出生年份不详，渔山乡葛村人，熟悉造纸的一系列工序，1965年左右去世
第二代	戚德银	男	父亲	生于1930年，渔山乡葛村人，曾任葛村村干部、村电镀厂厂长，从父亲戚仁钱处学得造纸全套工序，2009年去世
	戚松根	男	伯伯	生于1923年，渔山乡葛村人，自幼与村里的师傅学习造纸技艺，1977年因病去世
	葛志坤	男	岳父	生于1930年，渔山乡葛村人，儿时从村里师傅处习得捞纸和晒纸技艺。生产队期间一直负责捞纸工作，生产队解散后转行从事篾匠工作。2005年左右去世
第三代	戚吾樵	男	—	生于1954年，渔山乡大葛村人，纸坊负责人，13岁时从戚松根处学习踩料，自学其他造纸工序
	葛夏芹	女	妻子	生于1956年，渔山乡葛村人，晒纸技艺从父亲葛志坤处习得，目前主要负责晒纸和捆扎包装工序

⊙1

⊙2

⊙1 纸坊中的戚吾樵 Qi Wuqiao in the paper mill

⊙2 葛夏芹正在揭纸 Ge Xiaqin peeling the paper down

三
戚吾樵纸坊的代表纸品及其用途与技术分析

3
Representative Paper and Its Use and Technical Analysis of Qi Wuqiao Paper Mill

（一）戚吾樵纸坊代表纸品及其用途

据调查组2016年8月24日的调查得知，戚吾樵纸坊所产纸品单一，只生产祭祀竹纸一种纸，当地也习称"迷信纸"，以毛竹为原料制作而成。调查时生产的纸张规格也单一，为38 cm×43 cm。主要销往江苏，用作祭祀祖先和逝去亲人时的焚烧纸，如折元宝、做纸钱等。但纸坊只造纸，后续加工是客户自己的事。2019年回访时，据戚吾樵介绍因原料竹末已不再生产，纸坊调整了以前70%毛竹＋30%竹末的原料配比，目前生产的纸品原料配比约为50%毛竹＋50%废纸边。

（二）戚吾樵纸坊代表纸品性能分析

测试小组对采样自戚吾樵纸坊的祭祀竹纸所做的性能分析，主要包括定量、厚度、紧度、抗张力、抗张强度、白度、纤维长度和纤维宽度等。按相应要求，每一指标都重复测量若干次后求平均值，其中定量抽取5个样本进行测试，厚度抽取10个样本进行测试，抗张力抽取20个样本进行测试，白度抽取10个样本进行测试，纤维长度测试了200根纤维，纤维宽度测试了300根纤维。对戚吾樵纸坊祭祀竹纸进行测试分析所得到的相关性能参数如表10.13所示，表中列出了各参数的最大值、最小值及测量若干次所得到的平均值或者计算结果。

⊙3

⊙3 捆扎好的祭祀竹纸 A bundle of bamboo paper for sacrificial purposes

表10.13 戚吾樵纸坊祭祀竹纸相关性能参数
Table 10.13 Performance parameters of bamboo paper for sacrificial purposes in Qi Wuqiao Paper Mill

指标		单位	最大值	最小值	平均值	结果
定量		g/m²				34.4
厚度		mm	0.217	0.123	0.152	0.152
紧度		g/cm³				0.226
抗张力	纵向	mN	5.3	4.0	4.9	4.9
	横向	mN	5.2	3.6	4.3	4.3
抗张强度		kN/m				0.307
白度		%	18.9	18.7	18.8	18.8
纤维	长度	mm	2.3	0.1	0.6	0.6
	宽度	μm	29.2	0.4	9.3	9.3

由表10.13可知，所测戚吾樵纸坊祭祀竹纸的平均定量为34.4 g/m^2。戚吾樵纸坊祭祀竹纸最厚约是最薄的1.764倍，经计算，其相对标准偏差为0.230。通过计算可知，戚吾樵纸坊祭祀竹纸紧度为0.226 g/cm^3，抗张强度为0.307 kN/m。

所测戚吾樵纸坊祭祀竹纸平均白度为18.8%。白度最大值是最小值的1.011倍，相对标准偏差为0.005。

戚吾樵纸坊祭祀竹纸在10倍和20倍物镜下观测的纤维形态分别如图★1、图★2所示。所测戚吾樵纸坊祭祀竹纸纤维长度：最长2.3 mm，最短0.1 mm，平均长度为0.6 mm；纤维宽度：最宽29.2 μm，最窄0.4 μm，平均宽度为9.3 μm。

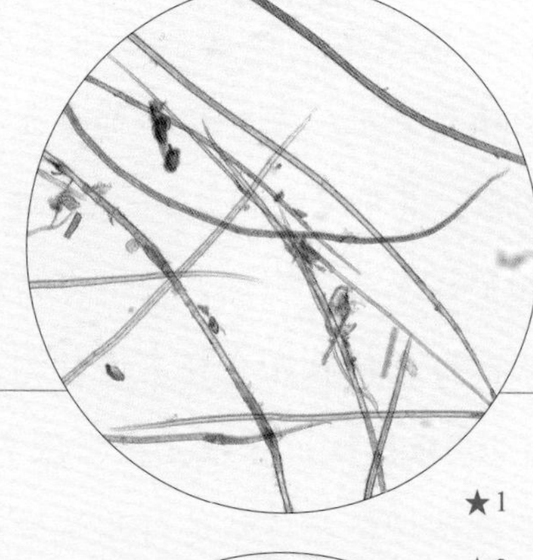

★1
戚吾樵纸坊祭祀竹纸纤维形态图（10×）
Fibers of bamboo paper for sacrificial purposes in Qi Wuqiao Paper Mill (10× objective)

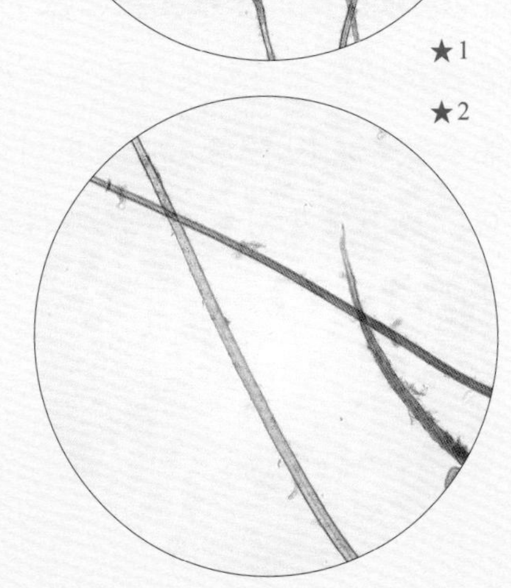

★2
戚吾樵纸坊祭祀竹纸纤维形态图（20×）
Fibers of bamboo paper for sacrificial purposes in Qi Wuqiao Paper Mill (20× objective)

四 戚吾樵纸坊祭祀竹纸的生产原料、工艺与设备

4 Raw Materials, Papermaking Techniques and Tools for Bamboo Paper for Sacrificial Purposes in Qi Wuqiao Paper Mill

（一）戚吾樵纸坊祭祀竹纸的生产原料

1. 主料一：嫩毛竹

嫩毛竹是当年农历小满前后生的毛竹。据戚吾樵介绍，做纸的原料采用当年6月的新竹，一年要砍超过10 000 kg的竹子用于造纸。戚吾樵还透露，一些竹艺加工厂的竹子碎末经烧碱处理后，也可以被纸坊用作原料，这样成本会有所降低。富阳区大青镇的竹艺加工厂一般会以0.5元/kg销售竹末，由卖家送货到纸坊。2019年回访时，戚吾樵表示因生产竹末需用到大量的烧碱，易造成环境污染，2017年大青镇竹艺加工厂就已停产竹末，目前纸坊已使用废纸边来代替竹末用于造纸。

⊙1

2. 主料二：废纸边

戚吾樵介绍，自2017年开始纸坊就从萧山购买废纸边，近几年购买价格稳定在2.8元/kg左右。每年购买量不固定，2018年购买了2 000 kg左右的废纸边。

3. 辅料：水

造纸需要大量的水，而水源的好坏在一定程度上对纸的好坏影响很大。戚吾樵纸坊选用的是当地方言称为“头把水”的源头水，水质较好。据调查组成员在现场的测试，戚吾樵纸坊使用的水pH为5.5～6.0，偏弱酸性。

⊙2

⊙3

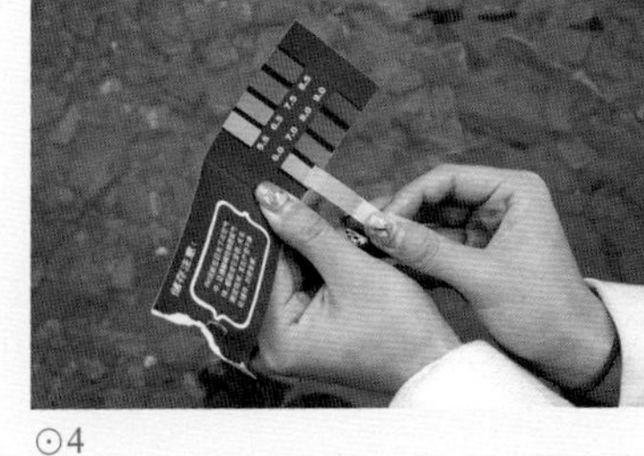

⊙4

⊙1 老竹碎末 Scrap of old bamboo

⊙2 废纸边 Abandoned paper edge

⊙3 山溪源头水 Source of the mountain stream

⊙4 水源pH测试 Testing the water source pH

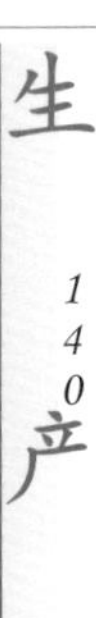

（二）戚吾樵纸坊祭祀竹纸的生产工艺流程

根据戚吾樵的介绍，以及调查组的现场观察，其手工祭祀竹纸的生产工艺流程为：

壹	贰	叁	肆	伍	陆	柒	捌	玖	拾	拾壹	拾贰
砍竹	断青	碾碎	落塘	浸坯	翻滩	淋尿	堆蓬	落塘	榨水	打浆	抄纸

拾叁	拾肆	拾伍	拾陆	拾柒
压榨	晒纸	数纸	压纸	成品捆扎

壹 砍竹

1

每年6月份的时候会上山砍嫩竹，砍7天左右。选取当年生长的竹子，从还没有发根的根部上面1 cm处开始砍，砍下来之后再运下山。

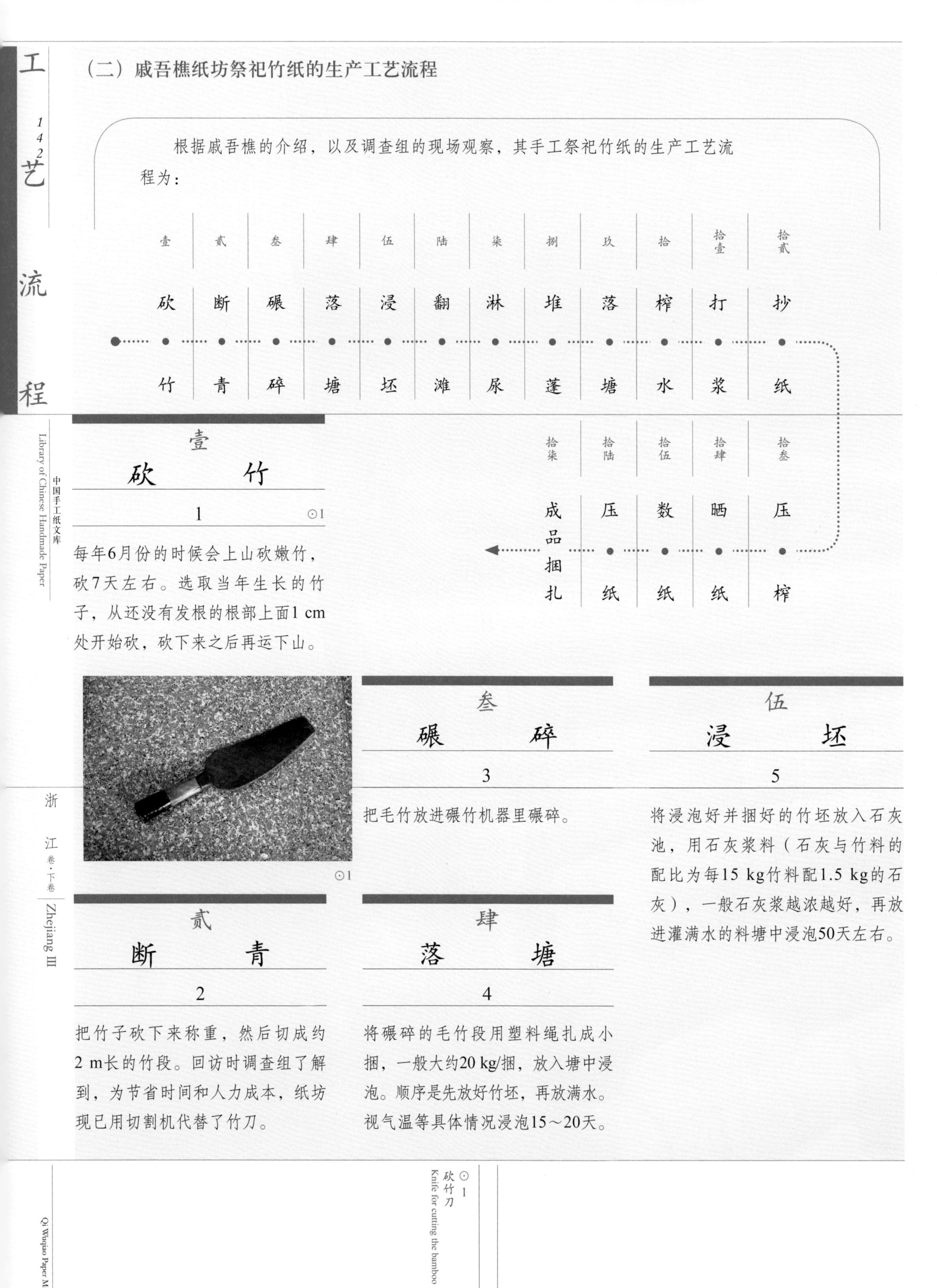

⊙1 砍竹刀 Knife for cutting the bamboo

贰 断青

2

把竹子砍下来称重，然后切成约2 m长的竹段。回访时调查组了解到，为节省时间和人力成本，纸坊现已用切割机代替了竹刀。

叁 碾碎

3

把毛竹放进碾竹机器里碾碎。

肆 落塘

4

将碾碎的毛竹段用塑料绳扎成小捆，一般大约20 kg/捆，放入塘中浸泡。顺序是先放好竹坯，再放满水。视气温等具体情况浸泡15～20天。

伍 浸坯

5

将浸泡好并捆好的竹坯放入石灰池，用石灰浆料（石灰与竹料的配比为每15 kg竹料配1.5 kg的石灰），一般石灰浆越浓越好，再放进灌满水的料塘中浸泡50天左右。

陆

翻滩

6

翻滩即是清洗，需要连续洗7次，每天洗1次，连续洗3天，然后隔1～2天洗1次。最后1次要看有没有泡沫，如果洗完竹坯出现黑水、泡沫，即洗好了。每次洗完就立即堆放在池里，一共洗半个月。这道工序的目的是把竹料上黏附的石灰残渣等洗出来。

柒

淋尿

7

淋尿相当于加药引子，尿是酸性的，竹子是碱性的，酸碱中合。把洗干净的竹料放在尿桶里，尿桶里面是尿和水的混合物，从下到上浸泡后立刻拿出来堆在一边。

捌

堆蓬

8

将堆在一边的竹料用塑料布盖起来，防止落灰，堆放发酵。如果天热，需放10天左右，天凉则需要30天左右。

玖

落塘

9 ⊙2

将发酵好的竹料竖着放在料塘里，放满清水，浸泡到水变红或变黑，长出蘑菇或菌丝就可以拿出来用了。浸泡时长一般为夏季时1个月，冬季时2～3个月。

拾

榨水

10 ⊙3

将料塘里拿出来的竹料放到压榨机上，榨干水分用于打浆。

⊙2

⊙3

⊙2 落塘 Soaking the materials in the pool

⊙3 榨干的竹料 Dried bamboo materials

拾壹
打　　浆

11 ⊙4

先用石磨将竹料磨成粉，视竹子老嫩程度磨30～40分钟即可。然后用泵将石磨磨好的竹粉加水搅拌10分钟，一般磨1次料可以做1.5天的纸。

⊙4

拾贰
抄　　纸

12 ⊙5⊙6

抄纸前首先需要将和单槽棍从自己身前向外按照顺时针方向椭圆状推开，搅拌速度要均匀，等到和纸槽中心形成旋涡即可。然后由抄纸工拿着纸帘，上下倾斜20° 左右下到槽内，再缓慢向身前方向提上来。当纸帘出水时，纸帘朝前倾斜，将多余的纸浆匀出。最后将纸帘从帘架上抬起，把抄好的湿纸放在旁边的纸架上。这样一张湿纸就制成了。纸帖是倾斜着放的，这样可以让水流到一边，不容易弄湿衣服。调查组现场访问戚吾樵时得知：戚吾樵每天早上5:30左右开始抄纸，1天需要劳作9个小时左右，捞1次可得2张纸，即一隔二的抄纸帘，一般1天捞2 800张左右。2019年回访时戚吾樵表示，因体力下降，已无法达到2016年时每天2 800张的产量，回访时一天的产量在2 500张左右。

⊙5

⊙6

⊙4 石磨磨料 Grinding the materials with a stone roller

⊙5/6 抄纸 Papermaking

拾叁 压榨

13 ⊙7

捞完一定量湿纸后，将这些湿纸放在木榨上，使用千斤顶缓慢压榨出水分。压榨时力度由小变大，动作要缓慢，否则会压坏纸张。待压榨到湿纸不再出水，压榨即结束。据戚吾樵介绍，通常当天捞的纸要在当天压榨完，他一般在下午1点左右开始压榨，压榨工作一般持续1个小时。

⊙7

拾肆 晒纸

14 ⊙8⊙9

戚吾樵纸坊不使用焙墙晒纸，而是将湿纸放在地上晒。工序为：首先，用鹅榔头将压干的纸帖沿四边划一下，让纸松散开；然后捏住纸的右上角捻一下，让右上角的纸翘起来，随后用嘴巴吹一下，粘在一起的纸角就分开了；最后，用手沿着纸的右上角将纸帖中的纸揭下来，10张一晒，放在家里阴干或者铺在太阳直射的地面上晒干。据戚吾樵介绍，如果天气太过炎热，人在太阳直射下晒纸太过辛苦，所以条件允许时，也可以放在家里阴干。

⊙8

⊙9

⊙7 压榨后的湿纸块 Wet paper pile after pressing

⊙8 分开纸张 Separating the paper

⊙9 院子里晒的纸 Drying paper in the yard

拾伍

数　纸

15　⊙10

收集晒干的纸张，放于室内数纸，每数出100张纸，便将小竹签放于纸的上方以示标记，数满2 000张后可开始压纸。

拾陆

压　纸

16　⊙11

在纸堆的上方盖一块石板，石板上堆放50 kg左右的石头。压纸的时间没有具体要求，待纸张被压平整后即可打捆包装。

拾柒

成品捆扎

17　⊙12⊙13

压好的纸张以2 000张为1捆。将1条编织绳从纸堆的下方穿过，捆扎时人站在纸堆上，双手各抓着编织绳的一端，绳子交叉打上结后，人需抓紧绳子的两端不断摇晃，利用惯性和压力让纸堆变得更紧实，最后再系上一道绳结即可完成捆扎。

⊙10

⊙12

⊙11

⊙13

⊙10 数纸 Counting the paper

⊙11 压纸 Pressing the paper

⊙12/13 捆扎纸品 Binding the paper

（三）戚吾樵纸坊祭祀竹纸的主要制作工具

壹 石 磨 1

用来磨竹料，将竹料磨成粉进而打浆，磨料时间一般在半小时到1小时，竹料被碾磨得细碎后可用于制浆。实测戚吾樵纸坊所用石磨尺寸为：底座直径243 cm，高50 cm；磨盘直径94 cm，厚39 cm。

⊙14

贰 打浆槽 2

用于打浆的设施，在其中将加水浸湿的废纸边和打磨过的竹料充分混合，制成纸浆。实测戚吾樵纸坊所用打浆槽尺寸为：长152 cm，宽100 cm，高84 cm。

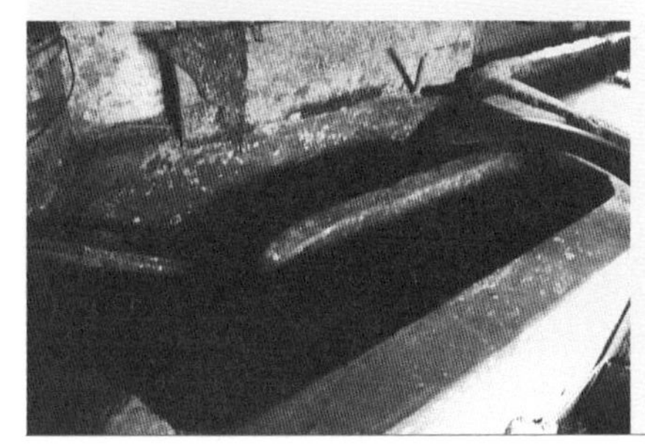

⊙15

叁 抄纸槽 3

调查时系水泥浇筑，用来抄纸的工具。实测戚吾樵纸坊所用的抄纸槽尺寸为：长206 cm，宽170 cm，高96 cm。

葛夏芹表示，纸坊为生产队解散后从其他村民处购买的，石磨、打浆槽、抄纸槽等大型工具也是当时一并购买的，纸坊及工具共计花费1 000多元，距今已使用了近40年。那时木工的工资仅为1.3元/天。

⊙16

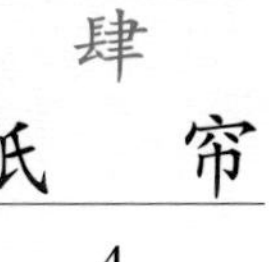

肆 纸 帘 4

用于抄纸的竹帘，苦竹丝编织而成。纸帘购自大源镇光明制帘厂，一般可使用1年，购买价格为270元/个。买来后需要手工缝三道线，分别在纸帘两边和中间缝线，以便在捞纸中将纸片一分为二。实测戚吾樵纸坊所用的纸帘尺寸为：长90 cm，宽45 cm。

⊙17

⊙14 石磨 Stone roller

⊙15 打浆槽 Beating though

⊙16 抄纸槽 Papermaking trough

⊙17 纸帘 Papermaking screen

伍 帘架 5

抄纸时支承纸帘的架子，杉木所做。据戚吾樵介绍，纸坊所用的帘架是请村里的木匠制作的，也是270元/个，一般能用2年。2019年回访时，帘架价格已上涨到300元/个，实测戚吾樵纸坊所用的帘架尺寸为：长100 cm，宽58 cm。

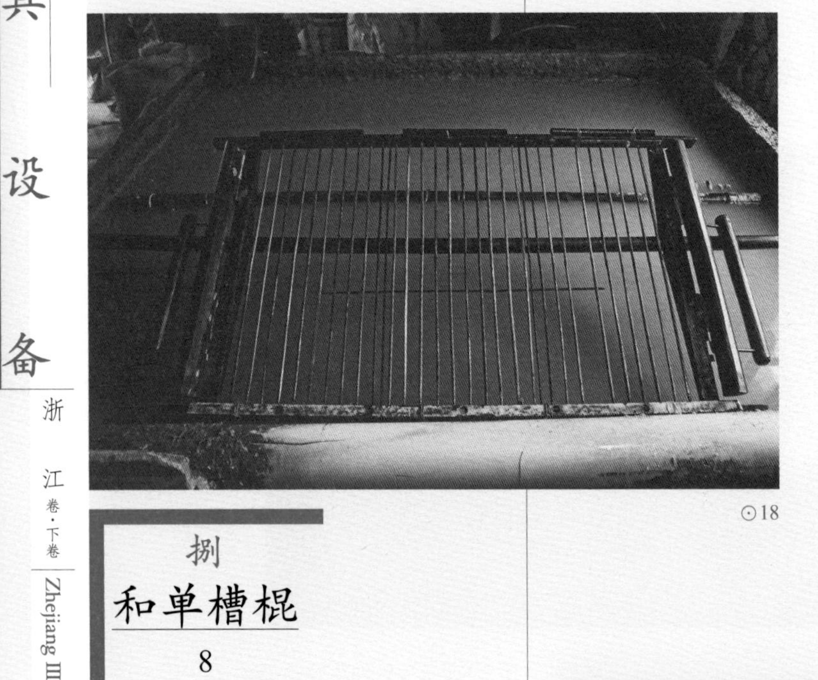

⊙18

陆 鹅榔头 6

牵纸前用于打松纸帖的工具，檀木所做。实测戚吾樵纸坊所用的鹅榔头尺寸为：长20 cm，直径3 cm。

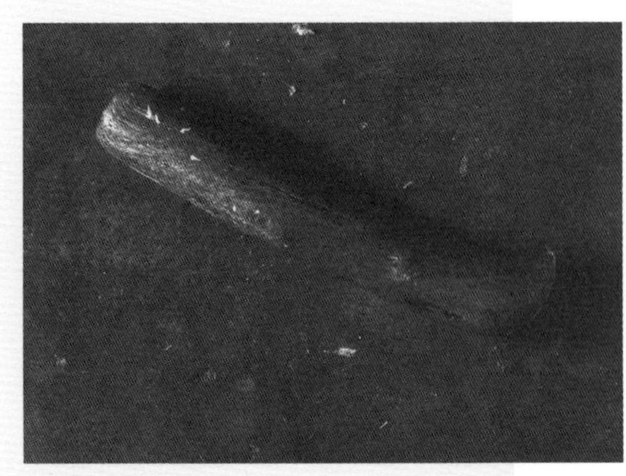

⊙19

柒 浆瓢 7

杉木制作用来舀浆的工具。实测戚吾樵纸坊所用浆瓢尺寸为：直径25 cm，高14 cm；把手长14 cm。

⊙20

捌 和单槽棍 8

竹子与松木制作，用来搅拌纸浆的工具，以使纸浆更加均匀。实测戚吾樵纸坊所用和单槽棍尺寸为：柄长190 cm；槽拨头长24 cm；槽头长23 cm，宽19 cm。

⊙21

⊙18 帘架 Frame for supporting the papermaking screen

⊙19 鹅榔头 Goose hammer

⊙20 浆瓢 Ladle for scooping the pulp

⊙21 和单槽棍 Stirring stick

五 戚吾樵纸坊的市场经营状况

5 Marketing Status of Qi Wuqiao Paper Mill

据戚吾樵介绍，其手工造纸坊的生产具有一定季节性，比如，6月份要上山砍竹子，那做纸就会比较少，春节期间也是会休息的。每年12个月做8个月的活，平均1个月做20天，年产量200余件，1件2 000张。如果按200件算，大约400 000张，如果按250件算，约500 000张。根据实地调研得知，2015年戚吾樵纸坊的净利润为5万元。生产的祭祀竹纸纸张规格为38 cm × 43 cm，主要销往江苏省常熟市的虞山镇，也有部分在杭州市的萧山区、临安区等地销售，用于烧纸或者折元宝。

纸坊目前仅有戚吾樵和妻子葛夏芹两人负责生产，仅在每年6月砍竹时以近260元/天的价格请临时工帮忙，一般砍竹时间为15～20天。2019年回访时，戚吾樵表示因年龄较大，体力大不如前，纸坊一天生产纸品近2 500张，年产量略有下降，但仍在200件左右。2019年1件纸的收购价格为220元，略高于2016年的210元/件。2018年纸坊的净利润为5万元左右。

⊙22

六 戚吾樵纸坊的文化与传承故事

6 Culture and Stories of Qi Wuqiao Paper Mill

（一）焚烧纸元宝，诵经求平安

调查组2019年回访时，在大葛村村口一处小屋内看到有多名老人在一边诵经，一边用纸叠着元宝。经询问后得知，几天前有两辆车在村口发生擦碰，每当发生像这样的安全事故，村里的老人们都会在村口的小屋里用纸叠元宝，老人们叠好元宝后，会将元宝放在嘴边念几句诸如“阿弥陀佛”之类的佛号。小屋前会点上香和红烛，寓意提醒菩萨这里有人在向他祈愿。老人们会在晚上将叠好的元宝于村口烧掉来祈求平安。

⊙22 戚吾樵家中阴干的祭祀竹纸 Paper for sacrificial purposes drying in Qi Wuqiao's house

用于折元宝的纸为长14 cm、宽6 cm左右的长条状，老人们会用剪刀在纸上剪出两排共12个小三角。问及为何在叠元宝前要这样做时，老人们表示自己也说不上来，但这是长辈们传下来的做法。做元宝的纸分手工纸、机器纸两种，机器纸居多。据老人们介绍，以前的元宝只使用手工纸，但近几年手工纸价格上涨到15元/刀，而机器纸价格仅10元/刀，且纸张破损率远低于手工纸。虽然现在两种纸都在用，但从价格方面考虑，以后会更多地使用机器纸来叠元宝。

（二）纸品光洁韧性强，纸商合作十八载

戚吾樵纸坊的纸品几乎都供给了一直合作的纸商，当纸坊纸品库存达到一定数量时，由戚吾樵联系纸商，纸商再上门来收购。回访时，据葛夏芹介绍，合作时间最长的纸商是小源镇的丁光明。两家自2000年左右就开始合作，距今已有18年了，一般做好了纸再电话联系丁光明，由他上门来收购，一次一般收购100多捆。

谈及如何认识丁光明的，葛夏芹表示最初是丁光明开车在大葛村四处收购纸品，自家的纸张不添加姜黄粉等染色剂，纸品洁白且光滑细腻，韧性较强，不易破损，丁光明大赞纸品质量极佳，就此在大葛村选择了戚吾樵纸坊达成长期合作关系。至今，戚吾樵纸坊每年产出的纸品近一大半都被丁光明收购。

⊙1

⊙2

⊙3

⊙1
叠元宝的老人
The elderly folding the paper ingot

⊙2
小屋前燃烧的香和红烛
Joss sticks and red candles burned in front of the hut

⊙3
捆扎好的纸品
A bundle of paper

七
戚吾樵纸坊的业态传承现状

7
Current Status of Qi Wuqiao Paper Mill

2016年8月调查时，据戚吾樵介绍，生产队时期的大葛村造纸业十分发达，仅葛村就有40多口纸槽投入生产，造纸工人近300人。2000年左右，村里还有近20户仍以造纸为主业。而现在在大葛村，从事手工造纸的已经没有年轻人了，目前大葛村从事这一行的最年轻的造纸人也已经有59岁了。谈到原因，戚吾樵表示，这主要是因为手工造纸既辛苦又没有多少钱赚，现在外面能赚到钱的工作很多，而且很多工作也不像造纸这样要学习技术，年轻力壮的村里人都不会留在村里造纸的。戚吾樵感慨，再过10年，可能大葛村就没人从事手工造纸业了。

2019年1月回访时，戚吾樵介绍，旭光村最后一个造纸户戚礼全已经64岁了，打算当年把纸坊中的造纸原料用完后就不再造纸；葛村除戚吾樵之外，还有葛发根、葛浩明两位60多岁的造纸户仍在造纸；勤建村仅存的两户纸坊传承人也近70岁了，未来两年内也将不再造纸。

2016年葛村还有5户人家从事手工造纸业，其中有1户把造纸当副业，并不长年造，其他4户包括戚吾樵家均以造纸为主业。2019年回访时，葛村仅有3户造纸，原先两户葛志开与葛浩田因造纸收入不高、收购商压价等原因，均放弃了造纸业，葛志开转行去渔山乡的工厂里做泥工，葛浩田转行去做了拆除违章建筑相关的工作。

造纸匠人转行、造纸技艺无人传承是手工纸技艺发展中要深思的问题，若不加以保护和转型，这个行业就会同大葛村村委会告示栏中介绍手工造纸技艺提到的那样：“已传承了千年的手工造纸业，只能随着时代的步伐走进历史的博物馆。”

⊙4

⊙5

⊙4 废弃的蒸锅 Abandoned steamer

⊙5 废弃的料塘 Abandoned soaking pool

戚吾樵纸坊

153

祭祀竹纸

Bamboo Paper for Sacrificial Purposes of Qi Wuqiao Paper Mill

祭祀纸（毛竹+废纸边）透光摄影图
A photo of joss paper (*Phyllostachys edulis* + waste paper edges) seen through the light

第八节

张根水纸坊

浙江省
Zhejiang Province

杭州市
Hangzhou City

富阳区
Fuyang District

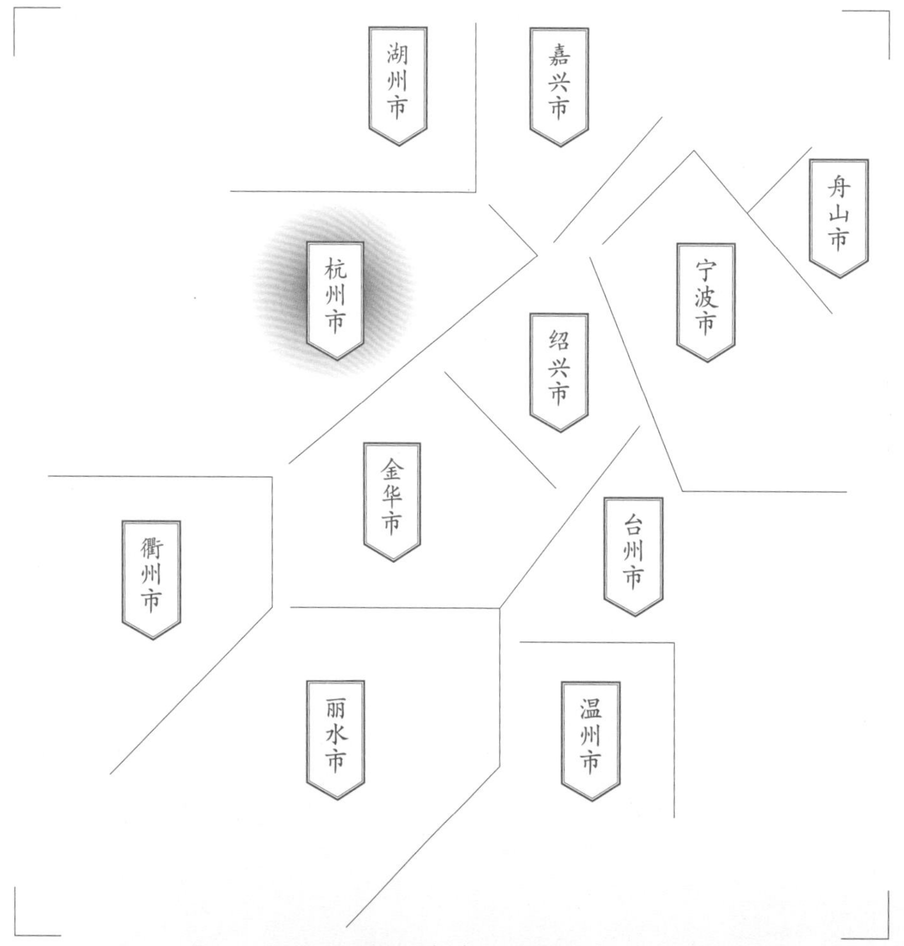

调查对象

富阳区湖源乡新三村
张根水纸坊
竹纸

Section 8

Zhang Genshui Paper Mill

Subject

Bamboo Paper of Zhang Genshui Paper Mill
in Xinsan Village of Huyuan Town in Fuyang District

一
张根水纸坊的基础信息与生产环境

1
Basic Information and Production Environment of Zhang Genshui Paper Mill

张根水纸坊坐落于素有“中国竹纸之乡”美誉的富阳区湖源乡的新三行政村，地理坐标为：东经119° 59′ 3″，北纬29° 48′ 46″。

2016年9月30日，调查组前往张根水纸坊，经此次观察确认，张根水纸坊的准确位置在湖源乡新三村颜家公变李家分线54号电线杆边上，是一个规模很小的家庭纸坊。

根据张根水的介绍，掌握到的基础生产信息为：手工造纸坊是张根水自己造的，有十五六年了，大约是在1998～1999年造的。面积很小，30 m^2左右，2016年有员工3人（包括张根水夫妇），2019年回访时，有员工4人，其中张根水捞纸，张茶花烧火，另外两名员工负责晒纸。纸坊有纸槽1帘，料塘4口（其中两口料塘为租用其他村民的）。每天做70多刀纸（1刀纸为100张），2016年约生产350天，到了2019年约生产200天。调查时，张根水手工造纸坊只生产毛竹祭祀纸，但张根水表示在1984年之前，手工纸销路广、销量好的时候也一度以做书画练习使用的元书纸为生。

⊙1

⊙2

⊙3

⊙4

⊙1
纸坊外的乡村小景
View of the village outside the paper mill

⊙2/3
张根水纸坊的准确位置标识
The exact location of Zhang Genshui Paper Mill

⊙4
纸坊附近的生态环境
Ecological environment near the paper mill

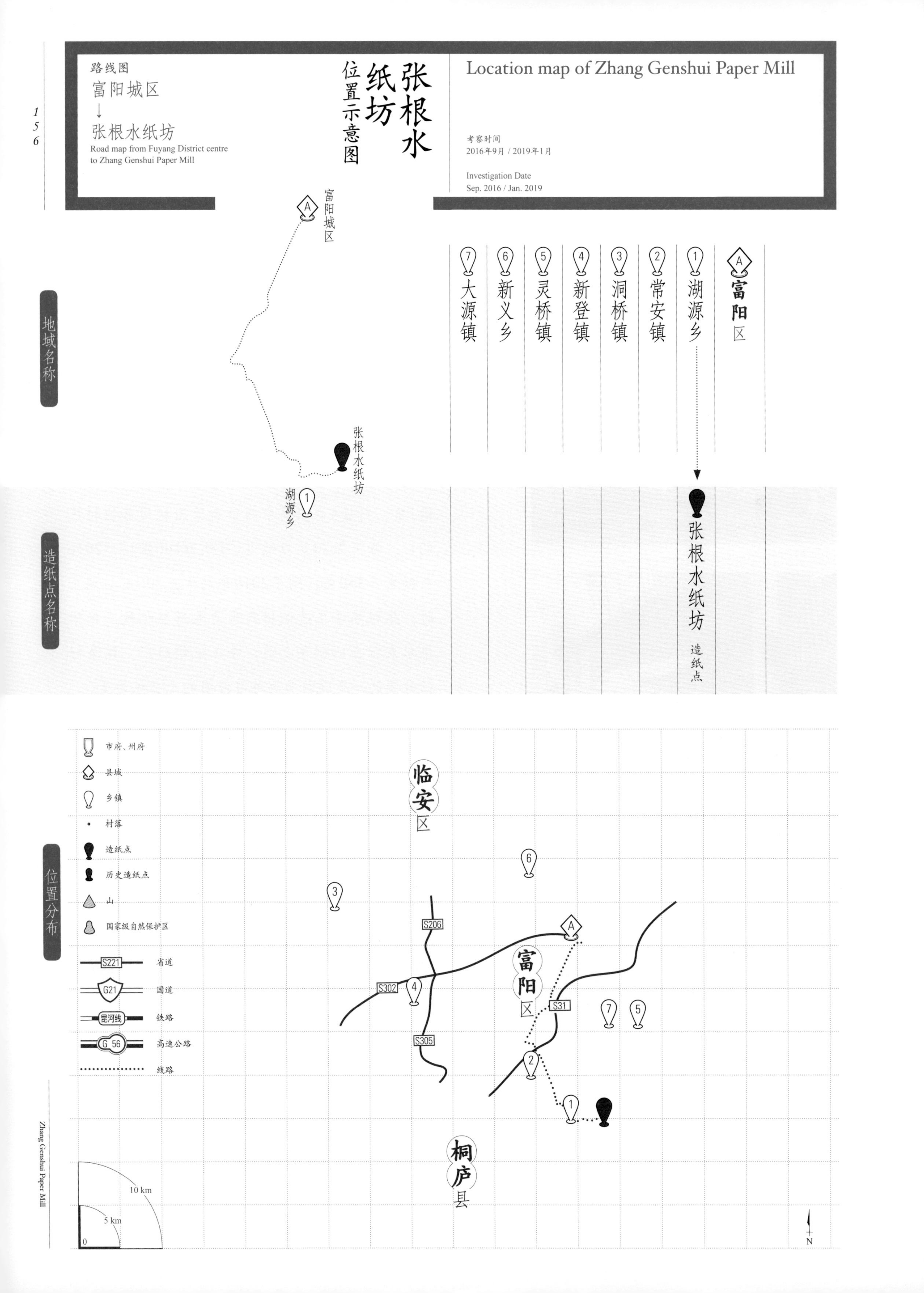

张根水纸坊位置示意图
Location map of Zhang Genshui Paper Mill
路线图
富阳城区
↓
张根水纸坊
Road map from Fuyang District centre to Zhang Genshui Paper Mill
考察时间
2016年9月 / 2019年1月
Investigation Date
Sep. 2016 / Jan. 2019
地域名称
造纸点名称
位置分布
富阳城区
张根水纸坊
湖源乡
A 富阳区
1 湖源乡
2 常安镇
3 洞桥镇
4 新登镇
5 灵桥镇
6 新义乡
7 大源镇
张根水纸坊
造纸点
市府、州府
县城
乡镇
村落
造纸点
历史造纸点
山
国家级自然保护区
S221 省道
G21 国道
昆河线 铁路
G56 高速公路
线路
临安区
富阳区
桐庐县
S206
S302
S305
S31
10 km
5 km
0
N

二
张根水纸坊的历史与传承

2
History and Inheritance of Zhang Genshui Paper Mill

2016年9月通过访谈得知，张根水纸坊衍生于20世纪七八十年代湖源乡新三村第二生产大队的造纸组，张根水在所属村落生产队解散后，便靠着自己的技艺，在妻子以及周围人的帮助下开始了个体家庭式手工纸作坊的造纸历程。张根水本人负责捞纸，妻子张茶花负责晒纸过程中焙笼的加热，另外雇有1人负责日常的晒纸工作。2019年1月回访时，纸坊雇佣2人负责日常晒纸的工作。

张根水1954年于湖源乡出生，1971年17岁时开始在新三村第二生产大队学做纸，一开始因年纪轻、力量大，在生产队造纸组里被派去参加砍伐原料、磨料、烧料等需要耗费较多力气的工作。他先后在原料制作部门做了3年，后来跟生产队的师傅学习了5年捞纸，不过张根水表示当时的师傅已经去世，无具体姓名、住址可考。20世纪80年代前期，生产队解散后，张根水便开始独立在家里做纸，约在1998～1999年自己修建了今日造纸的作坊。

访谈中张根水向调查组介绍，生产队时期，纸做得好的话，生产队会发奖状。元书纸分1～5级，每个等级的元书纸张根水都做过。当时造纸作为村里的副业，生产地位是很高的，造纸的工分都是当地最高的12个工分，1天可得1.2元钱，当年在村里算得上是收入很高的了。当年的元书纸质量做得确实好，连中央政府也会用湖源地方生产的手工纸。从早期加入生产队做纸算起，到如今的手工造纸坊，迄今张根水已有40多年的手工造纸经历了。

张茶花是张根水的妻子，1957年出生于富阳三桥乡点口村，家人是从新安江千岛湖搬迁到富阳的。张根水22岁时曾到富阳做过4个月的挖河道的工人，在那时认识了张茶花，随后两人回到新三村重新开始造纸。张茶花1979年22岁嫁给张根水后便一直跟随丈夫造纸，从不间断，迄今也已有30多年的手工造纸经历了。

⊙1

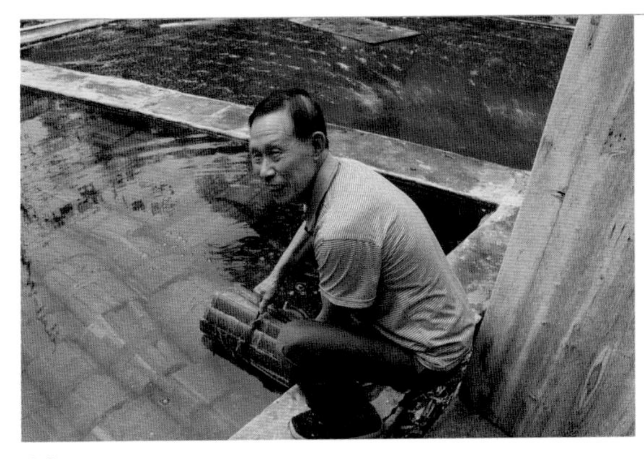
⊙2

⊙1 纸坊外观 The external view of the paper mill

⊙2 张根水在料塘边介绍原料制作工艺 Zhang Genshui introducing the raw material production procedure by the soaking pool

张根水和张茶花夫妇共育有4个孩子，3女1男，其中一个女儿因家庭当时经济情况较差，改姓送往亲戚处养大，因此实际带在身边的孩子有3个。交流中，调查组了解到，张家子女对做手工纸都不感兴趣，认为较辛苦且回报少，因而均无继承的想法。

⊙1

⊙1
张茶花在晒纸间晒纸
Zhang Chahua drying the paper in the drying room

三
张根水纸坊的代表纸品及其用途与技术分析

3
Representative Paper and Its Use and Technical Analysis of Zhang Genshui Paper Mill

（一）张根水纸坊代表纸品及其用途

2016年9月30日调查得知，张根水纸坊虽秉承富阳传统元书纸制作的工艺流程，并加上了一些改进的技艺，但纸坊近年来生产的均为祭祀用的竹纸（张根水夫妇也称其为白纸）。主要原材料为当地的毛竹，同时购买富阳当地或是外地的废纸边作为补充原料，毛竹与废纸边按照1：1的比例混合用于生产，产品规格根据客户要求定为42 cm × 46 cm。

因原料纤维较粗，故张根水纸坊的纸品较厚，符合外地客户对于祭祀纸厚度的要求。张根水介绍，2016年每件纸的销售价为660元（每件为4 800～4 900张），较之前降了约100元/件，之前几年是740～750元/件。2019年回访时，该纸品的售价为780元/件。因张家客户对纸的质量要求较高，故纸品价格也比村里其他造纸户略高点。

该祭祀纸的主要使用对象为老年人，因手工制作的祭祀纸质量较好，因而许多老人会选择拿它来做阴寿用。调查组访谈中还获知，张根水纸坊早期一度以生产可书写的元书纸为主，该种纸品造法古朴，沿袭了古元书纸的技艺，质量精良，但后期因销路变窄，销量锐减，便改成做今日的祭祀纸了。

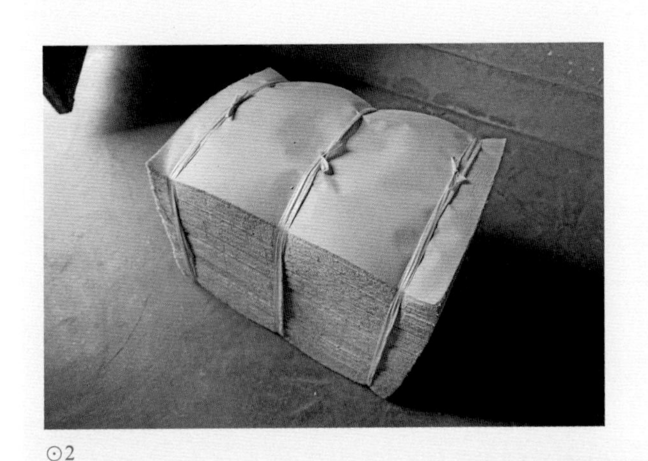

⊙2

⊙3

（二）张根水纸坊代表纸品性能分析

测试小组对采样自张根水纸坊的祭祀纸所做的性能分析，主要包括定量、厚度、紧度、抗张力、抗张强度、白度、纤维长度和纤维宽度等。按相应要求，每一指标都重复测量若干次后求平均值，其中定量抽取5个样本进行测试，厚度抽

⊙2 纸坊主要纸品——祭祀纸『白纸』
The main paper type of the paper mill - joss paper named "white paper"

⊙3 堆放的祭祀纸
Piled joss paper

取10个样本进行测试，抗张力抽取20个样本进行测试，白度抽取10个样本进行测试，纤维长度测试了200根纤维，纤维宽度测试了300根纤维。对张根水纸坊祭祀纸进行测试分析所得到的相关性能参数如表10.14所示，表中列出了各参数的最大值、最小值及测量若干次所得到的平均值或者计算结果。

表10.14　张根水纸坊祭祀纸相关性能参数
Table 10.14　Performance parameters of joss paper in Zhang Genshui Paper Mill

指标		单位	最大值	最小值	平均值	结果
定量		g/m²				35.1
厚度		mm	0.181	0.123	0.153	0.153
紧度		g/cm³				0.229
抗张力	纵向	mN	8.7	3.6	5.8	5.8
	横向	mN	4.4	3.8	4.1	4.1
抗张强度		kN/m				0.330
白度		%	19.4	17.2	18.4	18.4
纤维	长度	mm	2.5	0.2	1.0	1.0
	宽度	μm	51.1	0.7	13.8	13.8

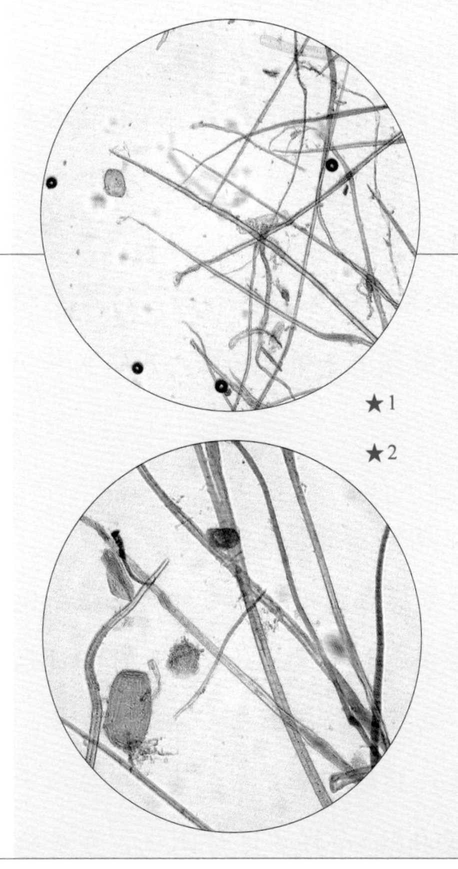

★1 张根水纸坊祭祀纸纤维形态图（10×）
Fibers of joss paper in Zhang Genshui Paper Mill (10× objective)

★2 张根水纸坊祭祀纸纤维形态图（20×）
Fibers of joss paper in Zhang Genshui Paper Mill (20× objective)

由表10.14可知，所测张根水纸坊祭祀纸的平均定量为35.1 g/m²。张根水纸坊祭祀纸最厚约是最薄的1.472倍，经计算，其相对标准偏差为0.160。通过计算可知，张根水纸坊祭祀纸紧度为0.229 g/cm³。抗张强度为0.330 kN/m。

所测张根水纸坊祭祀纸平均白度为18.4%。白度最大值是最小值的1.128倍，相对标准偏差为0.053。

张根水纸坊祭祀纸在10倍和20倍物镜下观测的纤维形态分别如图★1、图★2所示。所测张根水纸坊祭祀纸纤维长度：最长2.5 mm，最短0.2 mm，平均长度为1.0 mm；纤维宽度：最宽51.1 μm，最窄0.7 μm，平均宽度为13.8 μm。

四
张根水纸坊祭祀纸的生产原料、工艺与设备

4
Raw Materials, Papermaking Techniques and Tools for Joss Paper in Zhang Genshui Paper Mill

（一）张根水纸坊祭祀纸的生产原料

1. 主料：毛竹

据张根水介绍，他们所做的祭祀用纸原料选取的是农历芒种过后15天砍的老毛竹，这样做出来的纸会因发酵分解出来的东西少、纤维粗而相对较厚，这也是应客户需求所致。毛竹购于湖源，售价约为5元/kg，2016年购买了25 000 kg，2019年回访时购买了15 000 kg。

2. 辅料一：纸边

张根水纸坊制作祭祀用纸会在毛竹主料的基础上添加富阳当地或是外地送来的白、黄色废纸边，该纸边多为机械纸，收购价分为两个档次，质量好无太多损坏的4元/kg，磨损较多品相差的3元/kg。添加辅料时会将好的差的掺着一起放，另外黄色纸边和白色纸边放置的比例约为1∶1，张根水表示两者在质量上并无太大区分，这样配比只是习惯使然。做纸时，毛竹料与纸边按照1∶1的比例混合。

3. 辅料二：烧碱

烧碱易溶于水，可帮助发酵，促使浸泡在其中的竹料腐烂，加速其熟化软化的过程。据张根水介绍，纸坊制作祭祀用纸时通常会在蒸锅里加入烧碱，按照一层烧碱一层竹料的方式叠放在蒸锅中，然后加入水即可。

4. 辅料三：水

张根水纸坊制作祭祀用纸选择的是新三村山上流下的山涧水，山涧水清澈见底，水的硬度和水温都较低。据调查组成员在现场的测试，山涧水pH为6.0～6.5。

⊙1

⊙2

⊙3

⊙4

⊙1
纸坊料塘边的毛竹
Phyllostachys edulis by the soaking pool of the paper mill

⊙2
黄色废纸边
Abandoned yellow paper edge

⊙3
纸坊附近的山涧水
Mountain stream near the paper mill

⊙4
水源pH测试
Testing the water source pH

（二）张根水纸坊祭祀纸的生产工艺流程

根据张根水的口述，结合调查组2016年9月30日的实地调查，张根水纸坊祭祀纸的生产工艺流程可归纳为：

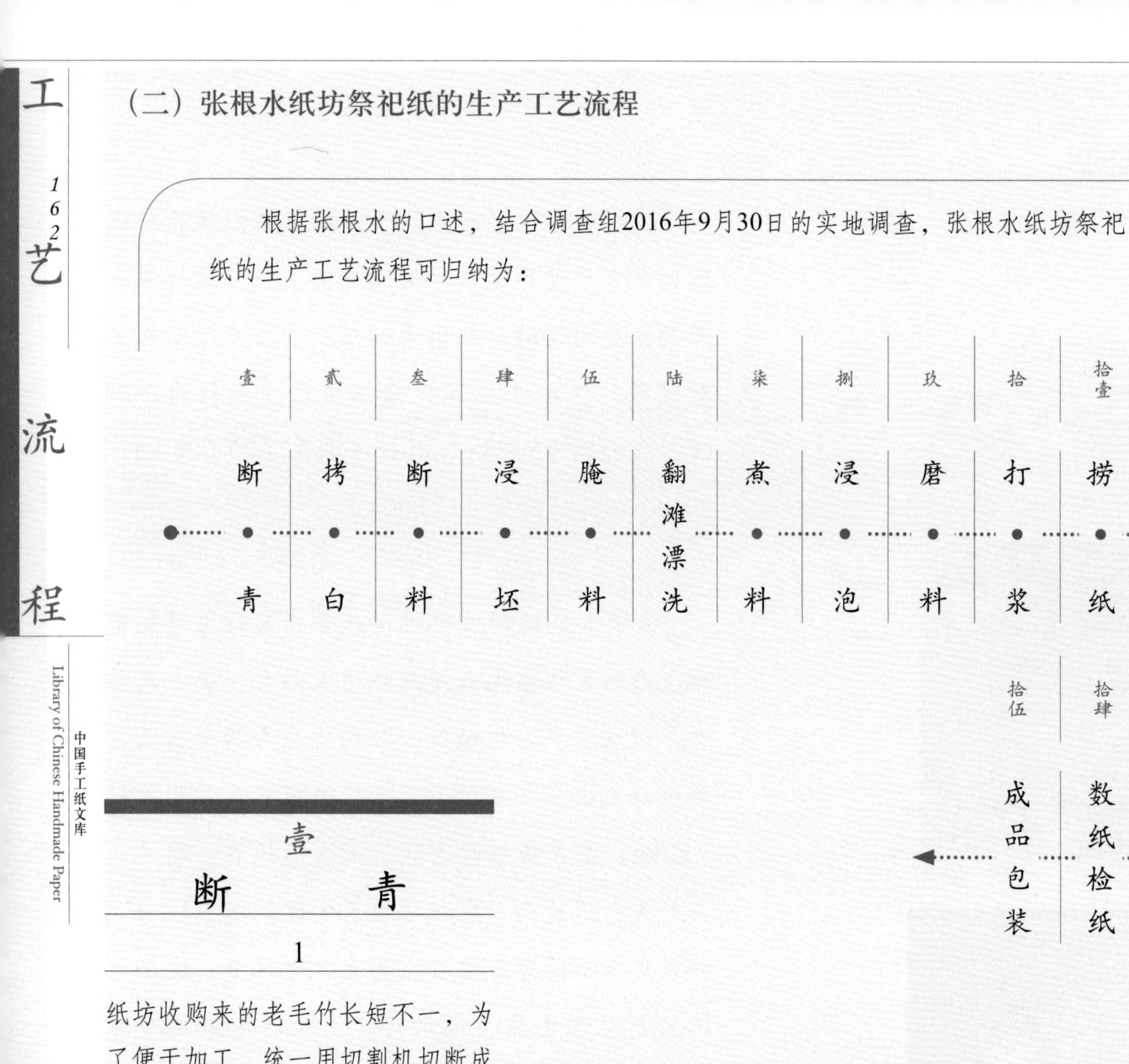

壹 断青

1

纸坊收购来的老毛竹长短不一，为了便于加工，统一用切割机切断成长度为2 m的竹段。

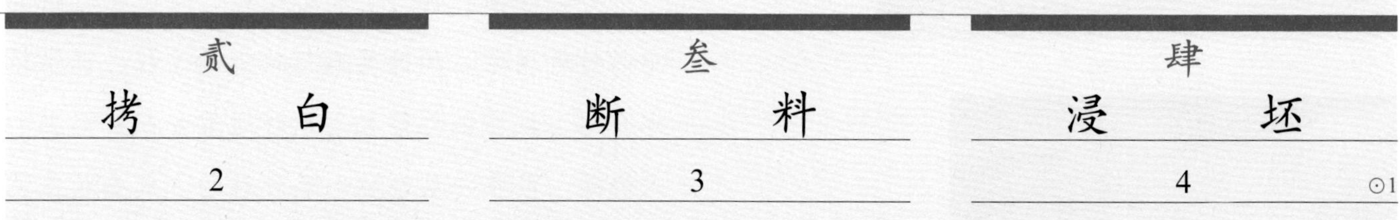

贰 拷白

2

将毛竹竹段铺在路上，用装有货物的拖拉机在竹段上反复碾压，直至毛竹被压破为止。

叁 断料

3

将碾压完成的竹段进行切割，把长度为2 m的竹段截成5段，每段长度为40 cm。用竹篾把竹段捆扎好，每12.5 kg为一捆。

肆 浸坯

4

⊙1

将用竹篾扎成小捆的竹料放入清水塘进行浸泡，让清水渗透进白坯的内部组织中，促使组织结构膨胀，将白坯自身内部汁液逐渐排出结构体外。浸泡15～30天待清水变滑发臭后，便可将白坯从水中取出。

伍
腌料

5

⊙2

将浸泡好的竹料按照1捆料2 kg石灰的比例放入腌料池中，加入水，浸泡2～3个月之后，水变红时，腌料工作就完成了。

⊙2

⊙1

⊙1
料塘中浸泡的白坯
Bamboo soaked in the soaking pool

⊙2
腌料
Fermenting the materials

陆

翻滩漂洗

6 ⊙3

将腌好的竹料放入料塘中，用清水泡15天左右。浸泡期间需要冲洗竹料4次左右，每冲洗一次，需更换塘中清水继续浸泡1天，直到把石灰完全去除干净。每口料塘大约能泡200～300页料不等，冲洗时将料放置于料塘上的木凳上。

⊙3

柒

煮料

7 ⊙4

在蒸锅内放6 000 kg的料和75 kg的烧碱，先放烧碱，再放洗干净的料，一层烧碱一层料地叠加，放好后加水，水要能将料完全淹没，水没过料2～3 cm。锅底烧火日夜蒸煮，一般蒸3天，烧1天焖2天，水温要达到100℃左右。煮好之后的竹料拿出来泡在池子里，不需要清洗。蒸锅直径约为2 m。

捌

浸泡

8 ⊙5

将蒸煮好的料放入料塘中，注入清水浸泡，使用时从料塘中取出即可。

⊙4

⊙5

⊙3 浸泡竹料 Soaking the bamboo materials

⊙4 正在煮料的蒸锅 A steamer boiling the materials

⊙5 浸泡竹料 Soaking the bamboo materials

玖 磨料

9 ⊙6

从料塘中取出料运至石碾房碾压，碾碎毛竹中的纤维，直到其成为细末。每次磨料需加少许水。由于竹子在料塘里沤烂程度不同，碾压所需的时长也不同，沤烂程度较重的竹子约30分钟可磨好，程度较轻的竹子需要60分钟左右。毛竹料被磨细之后再掺纸边，继续磨10多分钟即可。

拾 打浆

10 ⊙7

先将水放满，再放浆。将磨好的细末料放入打浆机中，之后不断进行搅拌，直至其达到均匀黏稠状态。打好的浆料通过管道运输至捞纸槽中，在纸槽中放入清水，用木耙反复搅拌。

拾壹 捞纸

11 ⊙8～⊙10

张根水纸坊由张根水本人负责捞纸，每天工作6个小时，捞纸数量约为7 000张。张根水会先根据自己的生产需求和捞纸习惯对浆料进行搅拌，捞纸过程中还会依据浆料的厚薄度多次用电动搅拌机进行搅拌。

捞纸工需要通过使用手腕的力量晃动纸帘，将纸槽内的浆液荡在纸帘上，待到纸帘上均匀分布着浆液的时候，将帘床向前倾斜，推出多余的浆液，最后仅留一层薄薄的纸浆在纸帘上。

湿纸形成后，捞纸工将手放松，帘架自动吊于水面上方约2 cm处，捞纸工拿起纸帘，将有纸浆的一面朝下，从离其最近的一边开始将纸帘缓缓放置于纸架上，待到整个纸帘完全与帘床贴合，从离其最近的一边迅速将纸帘抬起，继续捞纸。张根水每天捞纸的量用帘床边上的铁棍上的标记衡量，当湿纸堆叠到标记处时，即表明今天的任务量已完成。

捞纸的技巧之一是需要时刻观察纸浆是否均匀，抬起纸帘的时候速度要放慢，一般均为一次捞成。张根水捞纸中间会用搅拌机进行搅拌，目的是将沉在槽底的浆翻上来，叫作翻浆。具体的翻浆时间依张根水本人经验判断而定。帘床的四边有4条线，在湿纸压干后可沿着线把不整齐的纸边掰掉。

⊙7

⊙6

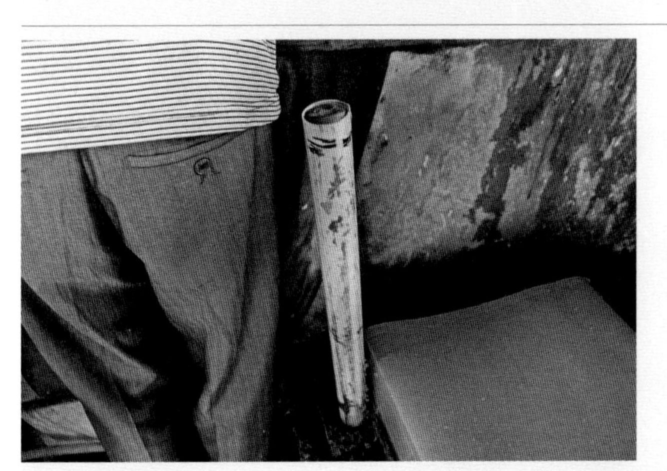

⊙8

⊙6 石磨磨料 Grinding the materials with a stone roller

⊙7 打好的浆料 Well-beaten pulp materials

⊙8 带有标记的铁棍 An iron stick with mark

⊙9

⊙10

拾贰 压 榨

12 ⊙11

每天凌晨5点左右，张根水在捞完70刀左右的纸后，用液压机（从生产队购买的一种老式压榨机，当时购买价格为1 000多元）将湿纸压干，再送去晒纸间，准备第二天由晒纸工晒干。

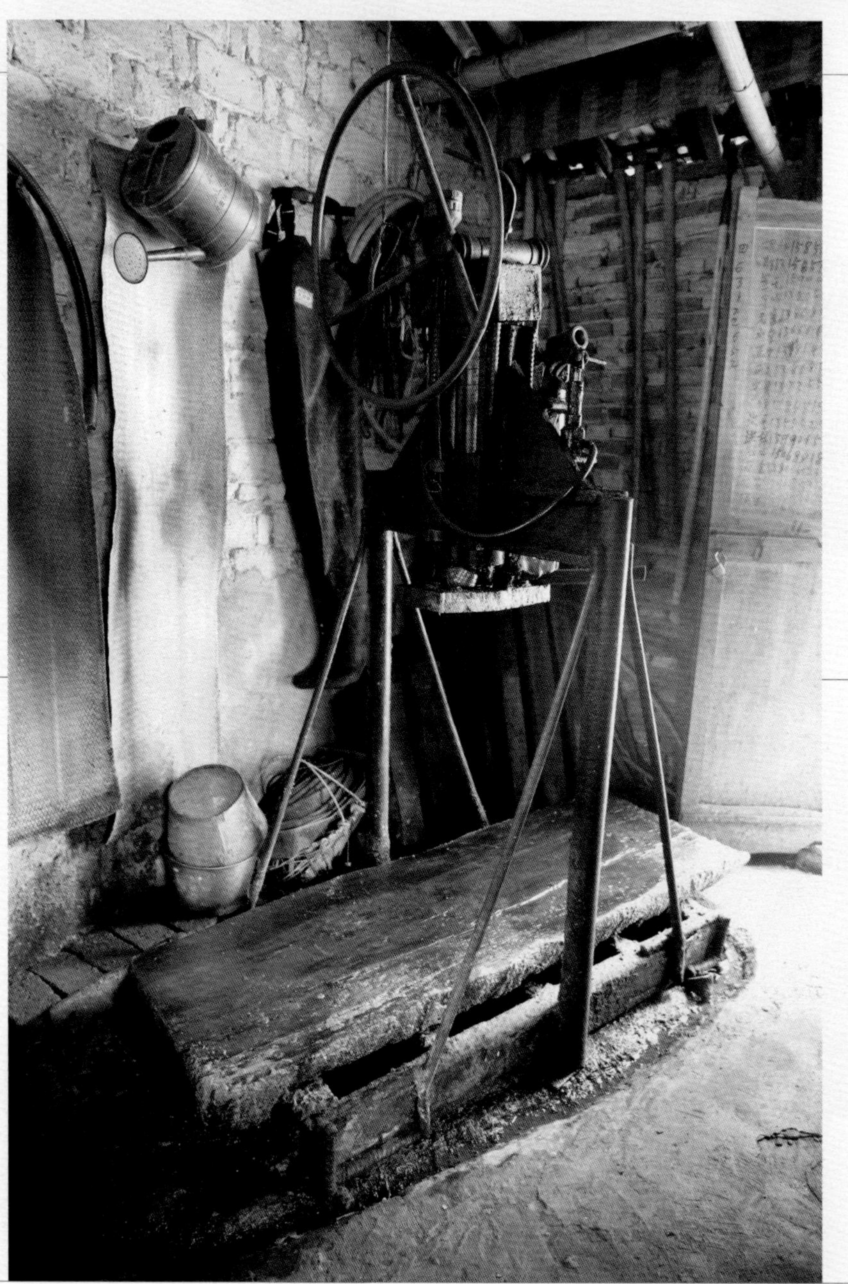
⊙11

⊙9/10 张根水正在捞纸 Zhang Genshui making the paper

⊙11 液压机 Hydraulic press

拾叁 晒纸

13 ⊙12

基本步骤是：第一步，晒纸工用鹅榔头在湿纸块上划几下，捏住纸块的右上角捻一捻，使一侧的纸角翘起；第二步，对着纸角吹一口气，用手逐张撕起，贴在刷着稀米糊的竖直的焙壁之上；第三步，用松毛刷在纸上迅速刷四五下，使湿纸与焙壁完全贴合。沿着焙壁从左往右依次晒，当晒完最右侧的纸，最左侧的纸的边角自然翘起，就到了可以顺利揭下来的时候。焙壁的温度要设置得刚刚好。在晒纸之前会在焙壁上刷一层稀米糊，之后如果纸刷到焙壁上粘不牢时就要复刷米糊。张根水纸坊通常是4张纸一起晒，据介绍，这样做是为了节约时间，纸张如果过厚，就无法在焙壁上粘牢，4张纸的厚度则刚好，也能有效晒干。

晒纸过程中若发现破碎的纸则放置于一边，可回笼打浆。据张根水的妻子张茶花介绍，目前晒纸房内除了她本人，还有2名晒纸工。张茶花负责焙笼的加热以及保温。晒纸工每天半夜3点开始晒纸，基本上要工作到下午4点才结束，每天至少工作12个小时。

拾肆 数纸检纸

14 ⊙13

把晒干的纸整理好并用木榨压平，检验后按每刀100张的规格数好。

⊙13

⊙12

⊙12 焙墙上将要晒干的纸 Paper to dried on the drying wall

⊙13 张茶花数纸、检纸 Zhang Chahua counting and checking the paper

拾伍
成品包装
15

⊙14

将数好的纸一刀刀放平整，用塑料绳捆扎好，不需要盖章（销给中间商后，他人会再加盖章），等待出售。

⊙14

（三）张根水纸坊祭祀纸的主要制作工具

壹 石磨 1

将竹料碾碎以便打浆的工具，主要由碾槽、碾砣等组成。实测张根水纸坊所用石磨的石辗直径106 cm，宽44 cm；石磨台子高60 cm，直径258 cm。张根水纸坊现在使用的石磨是租借他人的，半年付一次租金，需2 000多元。

⊙15

贰 松毛刷 2

用来将湿纸刷到焙壁上的工具。刷子很容易损坏，因此需要经常更换。实测张根水纸坊所用松毛刷尺寸为：手柄长35 cm；刷毛部分长38 cm，高3 cm。

⊙16

⊙14 等待出售的成品纸 Finished paper for sale

⊙15 租用村民的石碾 Rented stone roller from a villager

⊙16 松毛刷 Brush made of pine needle

叁
老式压榨机
3

利用木棍及上方铁块等的压力将湿纸的水分压干的机械装置。实测张根水纸坊所用老式压榨机尺寸为：底板长128 cm，宽76 cm，高175 cm；圆靶直径60 cm。

⊙17

肆
鹅榔头
4

顶部光滑的榔头，用于晒纸前划纸帖使其变松，檀木制作。实测张根水纸坊所用鹅榔头尺寸为：长25 cm，直径2 cm。

⊙18

伍
纸　帘
5

捞纸工具，用于形成湿纸和过滤多余的水分。由细竹丝编织而成，表面刷有黑色土漆，光滑平整。实测张根水纸坊所用纸帘尺寸为：长136 cm，宽51 cm。

⊙19

陆
帘　床
6

木制的长方形框架，作用是在捞纸时承载纸帘，面积稍大于纸帘，纸帘可完全嵌于框架中。帘床两侧有两根绳子垂直将其吊于纸槽内水面上方，使其不会在捞纸工松手后掉落。实测张根水纸坊所用帘床尺寸为：长137 cm，宽62 cm。

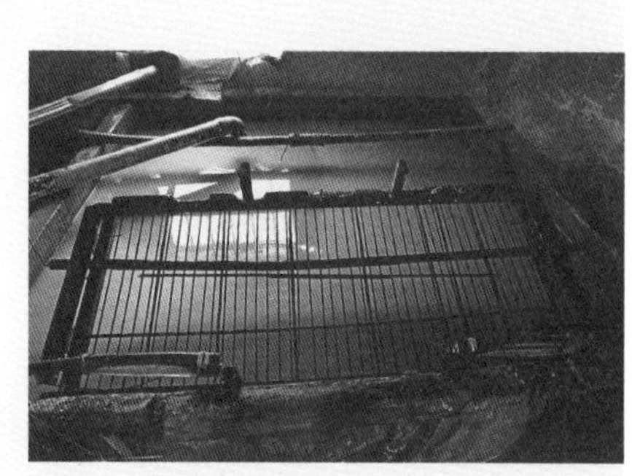
⊙20

⊙17 老式压榨机 Old-fashioned pressing machine
⊙18 鹅榔头 Goose hammer
⊙19 纸帘 Papermaking screen
⊙20 帘床 Frame for supporting the papermaking screen

柒
打浆池
7

用来制作竹浆细料，纸坊依靠电力打浆。实测张根水纸坊所用打浆池尺寸为：长187 cm，宽150 cm，高58 cm。

⊙21

捌
两齿耙
8

腌料时用于抖、捞页料的工具，因有两齿插入页料中，可自如控制页料在石灰水中的位置。实测张根水纸坊所用两齿耙尺寸为：柄长130 cm；耙头长25 cm，宽5 cm。

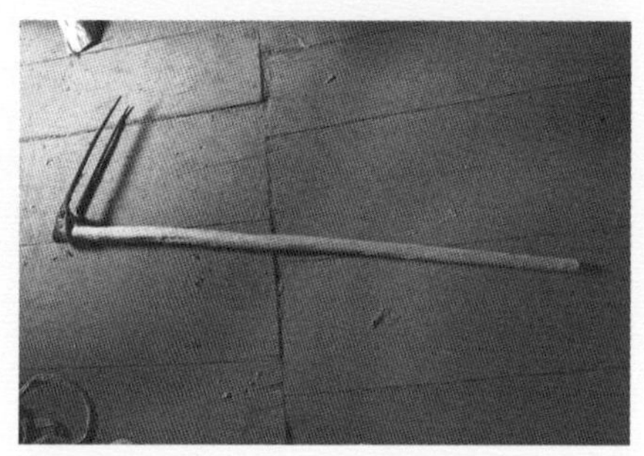

⊙22

玖
纸　槽
9

盛放纸浆的工具，长方形，捞纸工站于其一侧进行工作。纸槽所在的屋子叫作造纸坊。实测张根水纸坊所用纸槽尺寸为：长202 cm，宽196 cm，高101 cm。

⊙23

⊙24

拾
焙　壁
10

用来烘干湿纸，由两块长方形的钢板焊接而成，表面光滑，中空处流经加热的水蒸气。焙壁加热到有蒸汽冒出后两面都可以晒纸。实测张根水纸坊所用焙壁尺寸为：长264 cm，宽109 cm，高172 cm。

⊙21 打浆池 Pulp trough

⊙22 两齿耙 Two-toothed rake

⊙23 捞纸槽 Papermaking trough

⊙24 正在晒纸的焙壁 Drying wall for drying the paper

五
张根水纸坊的市场经营状况

5
Marketing Status of Zhang Genshui Paper Mill

⊙25

2016年张根水纸坊一天可做70多刀纸，每年约做350天。尺寸为42 cm×46cm的白纸售价为660元/件，一件4 800张；尺寸为37 cm × 42 cm的黄纸售价为200元/件，一件2 000张；尺寸为33 cm × 41 cm的黄纸售价为480元/件，一件2 450张。2019年纸坊一天可做70多刀纸，一年做满200天，一年纸产量约为14 000刀。尺寸为42 cm × 46 cm的白纸售价为780元/件；尺寸为37 cm × 42 cm的黄纸售价为260元/件；尺寸为33 cm × 41 cm的黄纸售价为320元/件。其中尺寸为37 cm × 42 cm的纸品销量最好。纸坊全年销售额约为20万元，净利润约为十几万元。

据张根水介绍，纸品主要销售给萧山的老板杨九斤，两人合作已有25年了，两人最初在售卖纸品的地方相识，后来达成了长期合作，杨九斤平均两周来收购一次，将收购来的纸品售卖到萧山、绍兴、上虞等地。

六
张根水纸坊的造纸习俗与文化故事

6
Papermaking Customs and Stories of Zhang Genshui Paper Mill

据张根水介绍，浙江当地有这么一种非常特别的旧习俗：人有12个属相，当有亲人去世时，当地往往会找12个不同属相的老婆婆聚集在一起捻一个佛包，即所谓的“十二烧佛”，然后一起念“心经”“太平经”“平安经”给去世的人以做悼念。张根水纸坊造的祭祀竹纸因售价较高，故富阳当地购买者较少，只有少数年纪大的人购买少许纸准备去世时做“十二烧佛”之用。

当地造纸前需要祭拜祖宗，祈求一年可以获得丰厚利润，生活平安如意。在造纸前首先需要选定开工的日子，初三、初六、初九是不宜造纸的，避开这些日子，带上酒、纸、豆腐、肉到山上祭拜祖宗，来祈求一年平安如意。

⊙25
张根水和杨九斤
Zhang Genshui and Yang Jiujin

⊙1

⊙1
张根水纸坊周边山区环境
Mountain environment around Zhang Genshui Paper Mill

七
张根水纸坊的业态传承现状

7
Current Status of Zhang Genshui Paper Mill

（一）孤独坚守、谋生养家

张根水自17岁开始在生产队参与造纸，2019年他已经65岁，一直从事造纸行业，其妻子在嫁给他之后也一直跟随他造纸。张根水本人是1～5级的元书纸都做过的造纸能手，到现在主要做的是祭祀用纸，可以说是一生都坚守在这个行业的生产第一线。目前，儿子在富阳做厨师，女儿也嫁去了富阳城里。因老俩口平时几乎都在小村里无太多花费，做手工纸挣的钱大多给了儿女。女儿结婚时夫妻俩把家里可提供的物件让女儿都带走了，后期儿子结婚时，张根水两夫妻又出钱给儿子在富阳买车买房。可以说，张根水手工造纸坊的存在造福了张根水的一家子，销售手工纸的收入既让他们的家庭基本衣食无忧，又提高了生活水平。

（二）技艺在身、无人可传

令人无奈的现状是，因手工纸技艺本身繁琐复杂、劳动强度很大、工作环境简陋艰苦、后人无传承意识等多方面因素，张根水纸坊实际上已处于无后人可传的境遇。谈及未来手工纸的发展，张根水发出一声声叹息，当询问是否愿意教授他人技艺时，张根水表示愿意无偿传艺，但无奈的是现在并没有人会来学习制作手工纸。

（三）传承渠道、不得其门

说到手工造纸的传承与发展，张根水与陪同调查的造纸师傅朱中华均表示，虽然国家鼓励发展传统文化产业，但因种种原因，切实的支持措施无法到达生产第一线的情况很普遍，如今的年轻人都觉得该项工艺繁琐复杂且收益达不到预期。因而许多跟张根水一样的造纸户虽然希望将手艺传承下去，但也因无人愿意学习而传承不得。

可以预见的是，不远的某一天，当张根水夫妇因身体状况或其他原因终止造纸时，在富阳坚持了30年的一间手工造纸坊也就消失了。

⊙2

⊙2 废弃的蒸锅 Abandoned steamer

张根水纸坊祭祀纸

Joss Paper
of Zhang Genshui Paper Mill

祭祀纸（毛竹+废纸边）透光摄影图
A photo of joss paper (*Phyllostachys edulis* + waste paper edges) seen through the light

第九节

祝南书纸坊

浙江省
Zhejiang Province

杭州市
Hangzhou City

富阳区
Fuyang District

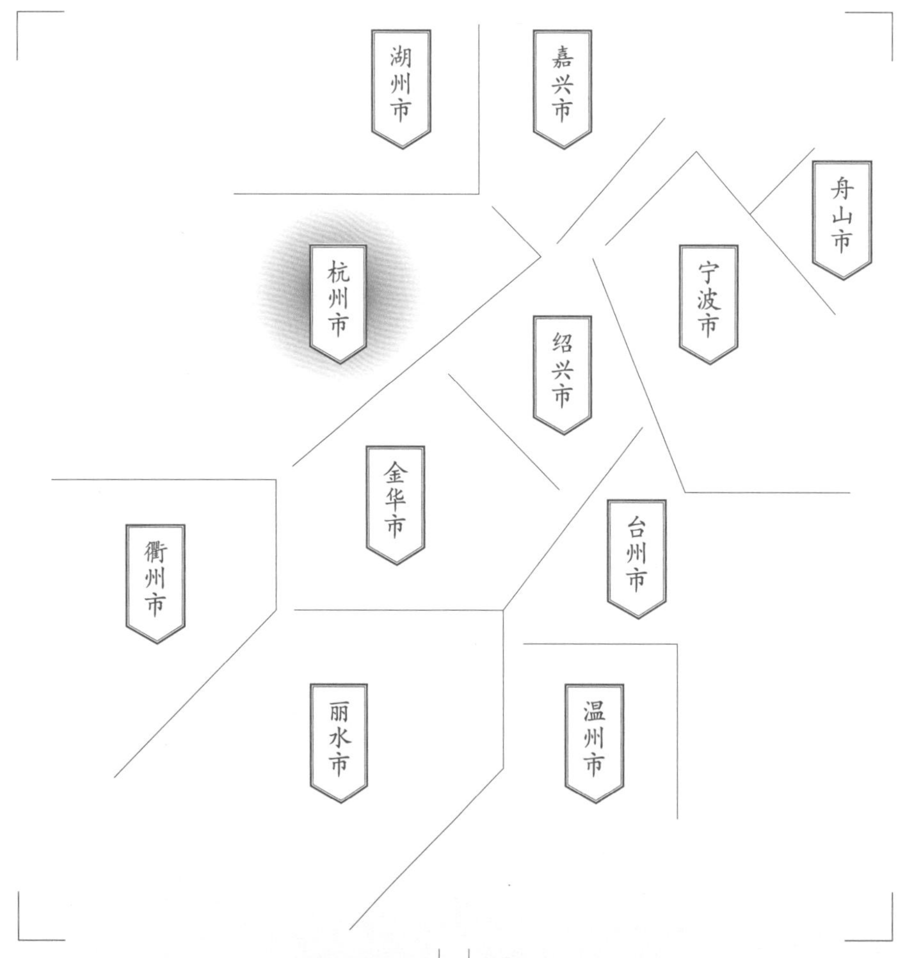

调查对象

富阳区灵桥镇山基村
祝南书纸坊
竹纸

Section 9
Zhu Nanshu Paper Mill

Subject

Bamboo Paper of Zhu Nanshu Paper Mill
in Shanji Village of Lingqiao Town in Fuyang District

一 祝南书纸坊的基础信息与生产环境

1 Basic Information and Production Environment of Zhu Nanshu Paper Mill

祝南书纸坊位于富阳区灵桥镇山基村村委会附近，地理坐标为：东经119° 38′ 28″，北纬30° 3′ 56″。山基村位于富阳区富春江南岸的灵桥镇东南部山谷中，距离灵桥镇中心约5 km。整村呈条带状分布，与村庄同名的山基溪穿村而过，村民多依溪水而居。

2016年9月28日，调查组成员前往祝南书纸坊进行田野调查。据祝南书的描述了解到的基础信息是：他们家的造纸坊是生产队解散后分户时继承下来的，祝南书捞纸所用的纸槽为清朝时期建造的古槽，是由祝家祖辈先人传下来的。调查时纸坊只有员工2人，即祝南书夫妇，没有外雇工人。正常状态是每天早上5点起床，做到下午4点，根据自己的身体情况决定工作或是休息，一年工作时间在300天左右。

祝南书纸坊生产队时期以生产用于书画的元书纸为主，自生产队解散后就只生产祭祀用纸。这一方面是由于祭祀用纸的制作工序较为简单，需要的人力成本和财力成本远远小于制作元书纸；另一方面也与传统元书纸类书画用纸的市场订单日益减少，山基村的纯粹家庭纸坊获得订单很难有关。

灵桥镇位于富阳区东部，北濒富春江，与富阳区东洲办事处隔江相望，西南与造纸名镇大源镇接壤，距杭州市区45 km，是富阳区东部一个水陆交通便利、工农业发达的经济重镇。灵桥埠自古以来就是富春江四大商埠之一，尤以手工元书纸而闻名，1916年姜芹波生产的“忠记昌山纸”曾获巴拿马万国货物博览会二等奖和国家农商部特等国货奖。

⊙1

⊙2

⊙1 祝南书纸坊 Zhu Nanshu Paper Mill

⊙2 灵桥新沙玫瑰园 Xinsha Rose Garden in Lingqiao Town

路线图
富阳城区
↓
祝南书纸坊
Road map from Fuyang District centre to Zhu Nanshu Paper Mill

祝南书纸坊位置示意图

Location map of Zhu Nanshu Paper Mill

考察时间
2016年9月 / 2019年3月

Investigation Date
Sep. 2016 / Mar. 2019

地域名称

富阳区
① 湖源乡
② 常安镇
③ 洞桥镇
④ 新登镇
⑤ 灵桥镇
⑥ 新义乡
⑦ 大源镇

造纸点名称

祝南书纸坊 造纸点

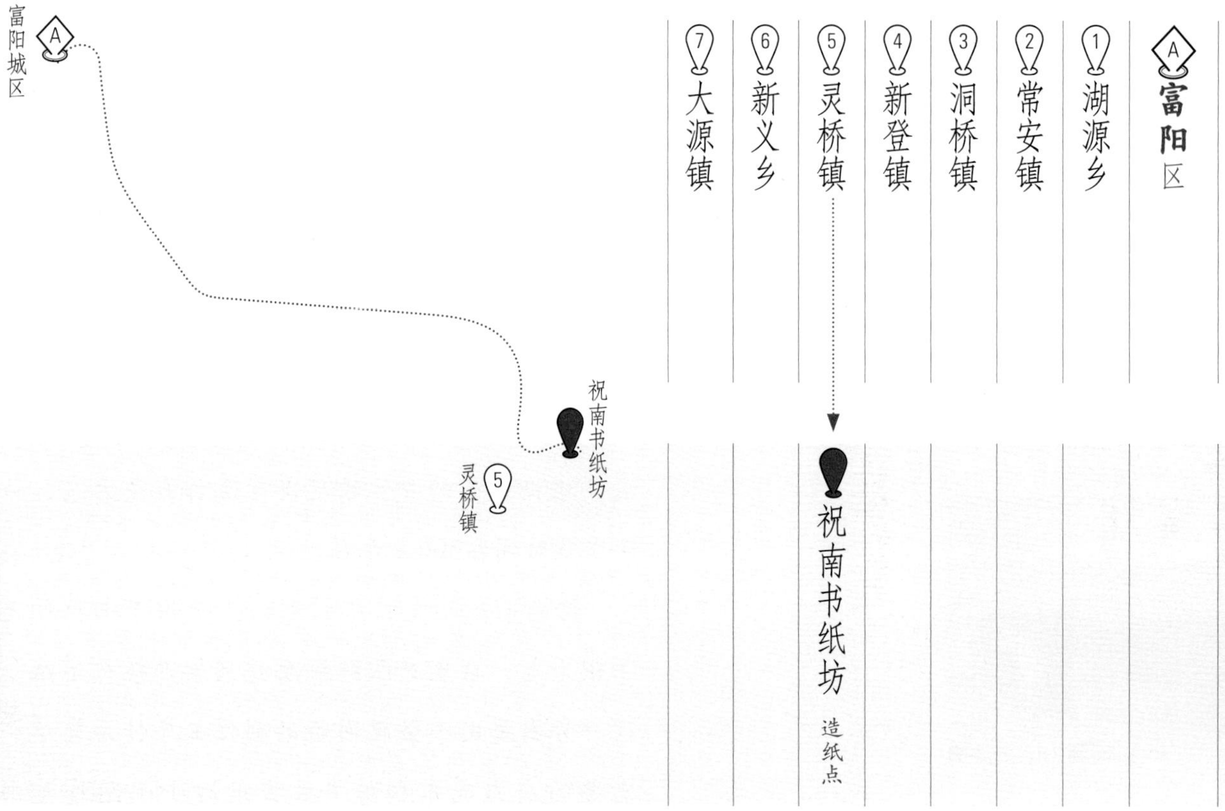

位置分布

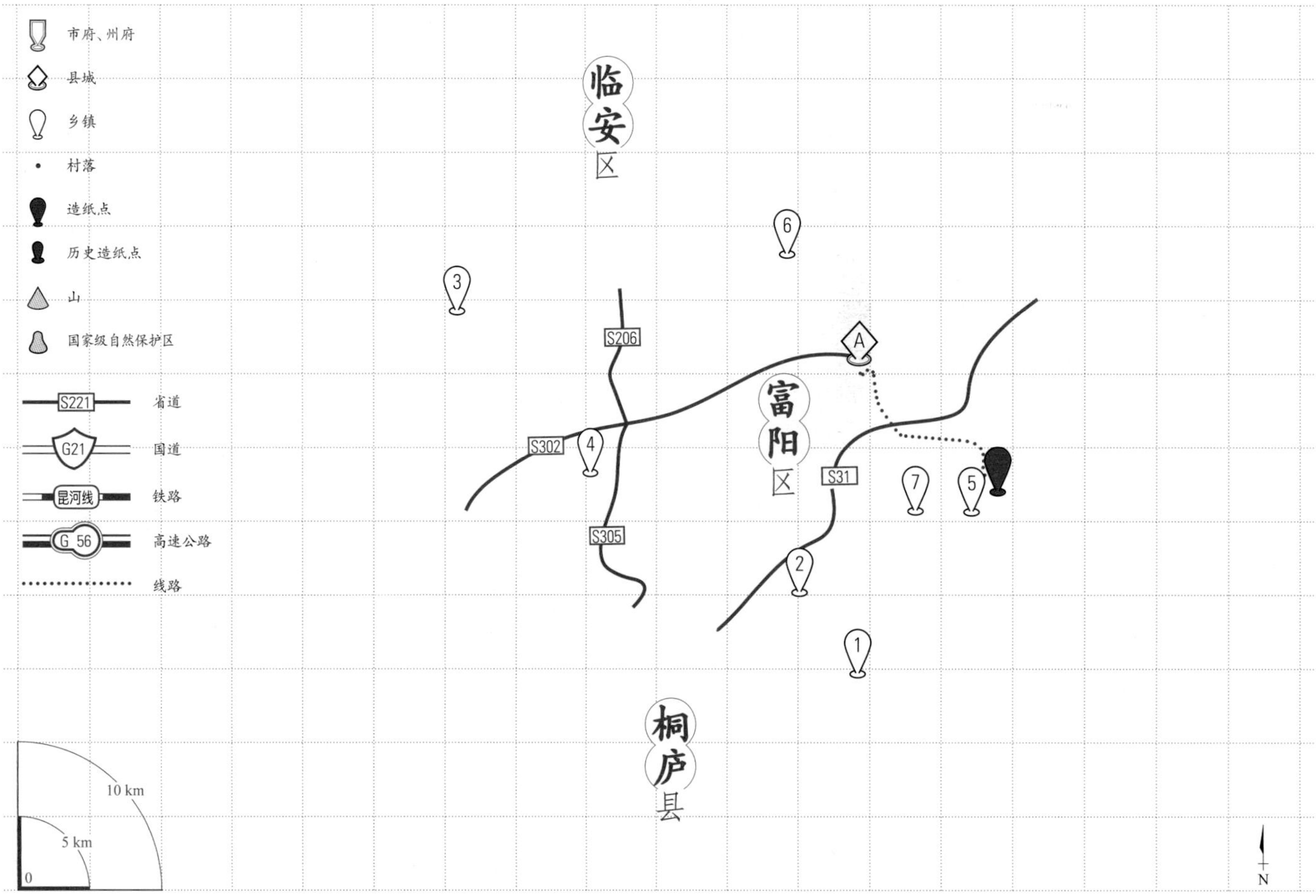

二
祝南书纸坊的历史与传承

2
History and Inheritance of Zhu Nanshu Paper Mill

2019年3月5日调查组回访时，村内一位严姓晒纸工表示，小时听长辈们谈起过山基村的造纸历史，传说皮纸生产源于东汉，竹纸生产始于东晋，但这种说法在文献中迄今仍缺少信史记载。村子里之前都是竹林，改革开放后村内还有近160户人家以造纸为主业，到20世纪晚期，因乡村道路急需拓宽，将原先部分造纸工坊拆掉用来修了路，随着村内造纸匠人年纪渐长且无后人继承，山基村手工造纸业慢慢衰落下去，年产量也越来越低。

⊙1

祝南书纸坊的建立虽然在技艺传承上有村落及家族的脉络，但直接的基础却是源自祝南书年轻时跟村里的老师傅学习捞纸。生产队解散后，祝南书凭着所学技艺，在自己妻子及其他亲戚的帮助下开启了独立建纸坊捞纸的生涯。祝南书本人负责捞纸，妻子负责晒纸。

祝南书，1970年出生于灵桥镇山基村，2016年时46岁。作为优秀的捞纸手艺传承人，一整天下来能抄3捆（每捆2 000张）左右的纸张，2016年每捆可卖出180元。祝南书15岁开始和爷爷祝志恒学习晒纸技艺，16～17岁开始和村里的老师傅姜关根学习捞纸技艺，学成后一直坚持造纸到现在。

祝南书访谈时表示，他所掌握的手工造纸技艺都是祖上传下来的。学造纸时村子里的惯例是向本村人开放，任何已掌握手艺者都可以是其师傅。祝南书的曾祖父和祖父辈都曾开办纸坊，如祝南书曾祖父祝兰和的兄弟祝家耀，1930年左右创立了祝庆山纸坊，集体化后就不再使用单人名号。因以前使用名号时造的是元书纸，之后纸坊生产的是祭祀纸，便不再使用祝庆山纸坊的名号。谈及为何以“祝庆山”为纸坊名号，祝南书表示可能是沿用某位造纸长辈的名字，但具体要追溯到哪一代无法考究。

据祝南书介绍，父亲祝钟杏1946年出生，一直务农，生产队时期只有两口纸槽，从事造纸工

⊙1 祝南书演示捞纸 Zhu Nanshu demonstrating papermaking

作的人手充足，父亲便未习得相关造纸技艺。爷爷祝志恒1911年出生，1980年左右去世，一直从事捣料工作。弟弟祝南军1973年生，由祝南书教授造纸的一系列工序，在村里的纸坊工作到30多岁，因造纸工作繁重且收入不高，后在浙江进行纸品销售的工作。妹妹祝敏今年37岁，十几岁时从祝南书处习得晒纸技艺，嫁到嘉兴后从事房地产相关工作。祝南书有一儿一女，均未习得相关造纸技艺，未来也没有继承纸坊的想法。

2019年回访时，祝南书的叔伯祝波泉表示，祝志恒有三个兄弟：大哥祝志升1905年出生，1962年去世；三弟祝志忠1916年出生，1980年左右去世；四弟祝志良20多岁时就去世了，兄弟四人在生产队时期都从事捣料、制浆等工作。造纸技艺均从祝南书曾祖父祝兰和处习得，祝兰和1900年左右去世，对造纸的各项工序都很熟练。

祝志升的儿子祝波泉1951年出生，从生产队的师傅处习得造纸的全部工序，14岁开始学习捣料，15岁开始学习捞纸，从事造纸工作40多年。祝志忠的大儿子祝波堂，1940年出生，从父亲祝志忠处习得造纸技艺，改革开放后帮其女儿照看卷帘门生意，便不再造纸；小儿子祝关根，今年73岁，因年龄渐长，自2018年开始就不再造纸。

祝南书的妻子江春芹，1969年生，会捞纸、晒纸等一系列技艺，山基村本地人，从父亲处习得造纸技艺。其妹江夏芹，1971年生，习得全部造纸技艺，结婚后未从事造纸工作。其兄江建林，1967年生，会造纸的全部技艺，近40岁时外

表10.15　祝南书造纸传承谱系
Table.10.15　Genealogy of papermaking inheritors in Zhu Nanshu’s family

传承代数	姓名	性别	与祝南书关系	基本情况
第一代	祝兰和	男	曾祖父	出生年月不详，1900年左右去世，灵桥镇山基村人，熟练掌握造纸的各项技艺
	祝家耀	男	曾叔父	生卒年不详，灵桥镇山基村人，“祝庆山纸坊”创立者，熟练掌握造纸的各项技艺
第二代	祝志升	男	大叔公	生于1905年，自小和父亲祝兰和学习造纸技艺，生产队时期负责捣料工作。卒于1962年
	祝志恒	男	爷爷	生于1911年，从父亲处习得造纸相关技艺，生产队时期负责捣料工作。卒于1980年
	祝志忠	男	二叔公	生于1916年，从父亲处习得造纸技艺，生产队时期负责捣料工作。卒于1980年
	祝志良	男	三叔公	生卒年不详，生产队时期和其他三个兄弟负责捣料工作，20岁左右去世
第三代	祝波泉	男	叔伯	祝志升之子，生于1951年，从生产队的师傅处习得造纸的全部工序，14岁开始学习捣料，15岁开始学习捞纸，从事造纸工作近50年
	祝关根	男	叔伯	祝志忠次子，生于1946年，从父亲祝志忠处习得捞纸技艺，从事捞纸工作40余年，2018年因年龄较大不再造纸
第四代	祝南书	男	—	生于1970年，纸坊负责人，自小和生产队师傅姜关根学习造纸技艺，生产队解体后至今仍独立经营纸坊
	江春芹	女	妻子	生于1969年，祝南书纸坊经营者，负责晒纸工作，其晒纸技艺习自于父亲江培友

出从事卷帘门销售工作，现已不接触造纸工作。祝南书的岳父江培友，2019年时已80余岁，以前造过元书纸，包产到户后因体力不支未再从事造纸工作。

⊙1

⊙2

⊙1
正在检验纸张的祝波泉
Zhu Boquan checking the paper

⊙2
祝南书搅拌纸槽里的纸浆
Zhu Nanshu stirring the pulp in the papermaking trough

三 祝南书纸坊的代表纸品及其用途与技术分析

3 Representative Paper and Its Uses and Technical Analysis of Zhu Nanshu Paper Mill

（一）祝南书纸坊代表纸品及其用途

调查时祝南书纸坊生产规格为33 cm×38 cm的祭祀纸，亦名火烧纸。纸品原料配比约为85%毛竹料+10%废纸边+5%棉花。火烧纸的主要用途为家中老年人去世之后祭奠之用。祝南书表示，以前在生产队时期也造过传统手工书画用途元书纸，20世纪80年代分产到户后就基本上不做书画纸了。纸坊采用的原料是从余姚购买的毛竹。由于祭祀纸的质量要求远远低于书画纸，故其工艺较书画纸而言亦有大幅简化。祝南书纸坊生产的纸张销售地主要为杭州市区、萧山区、富阳区等周边地区。每捆祭祀纸2 000张，一捆售价为180元左右。2019年回访时了解到，目前纸坊生产的祭祀纸售价有所上涨，价格为240元/捆。

⊙1

⊙1 祝南书纸坊祭祀纸 Joss paper in Zhu Nanshu Paper Mill

（二）祝南书纸坊代表纸品性能分析

测试小组对采样自祝南书纸坊的祭祀纸所做的性能分析，主要包括定量、厚度、紧度、抗张力、抗张强度、白度、纤维长度和纤维宽度等。按相应要求，每一指标都重复测量若干次后求平均值，其中定量抽取5个样本进行测试，厚度抽取10个样本进行测试，抗张力抽取20个样本进行测试，白度抽取10个样本进行测试，纤维长度测试了200根纤维，纤维宽度测试了300根纤维。对祝南书纸坊祭祀纸进行测试分析所得到的相关性能参数如表10.16所示，表中列出了各参数的最大值、最小值及测量若干次所得到的平均值或者计算结果。

表10.16　祝南书纸坊祭祀纸相关性能参数
Table 10.16　Performance parameters of joss paper in Zhu Nanshu Paper Mill

指标		单位	最大值	最小值	平均值	结果
定量		g/m^2				45.9
厚度		mm	0.194	0.171	0.184	0.184
紧度		g/cm^3				0.249
抗张力	纵向	mN	5.4	2.4	4.4	4.4
	横向	mN	3.3	1.8	2.4	2.4
抗张强度		kN/m				0.227
白度		%	19.0	18.4	18.8	18.8
纤维	长度	mm	2.5	0.3	1.0	1.0
	宽度	μm	49.6	0.7	10.5	10.5

由表10.16可知，所测祝南书纸坊祭祀纸的平均定量为45.9 g/m^2。祝南书纸坊祭祀纸最厚约是最薄的1.134倍，经计算，其相对标准偏差为0.046。通过计算可知，祝南书纸坊祭祀纸紧度为0.249 g/cm^3。抗张强度为0.227 kN/m。

所测祝南书纸坊祭祀纸平均白度为18.8%。白度最大值是最小值的1.033倍，相对标准偏差为0.014。

祝南书纸坊祭祀纸在10倍和20倍物镜下观测的纤维形态分别如图★1、图★2所示。所测祝南书纸坊祭祀纸纤维长度：最长2.5 mm，最短0.3 mm，平均长度为1.0 mm；纤维宽度：最宽49.6 μm，最窄0.7 μm，平均宽度为10.5 μm。

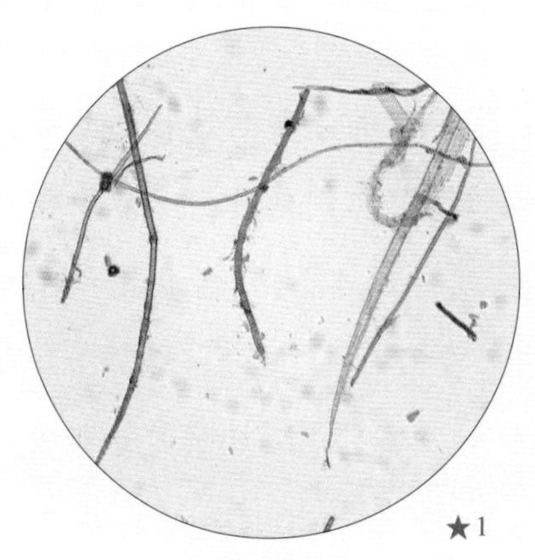

★1
祝南书纸坊祭祀纸纤维形态图（10×）
Fibers of joss paper in Zhu Nanshu Paper Mill (10× objective)

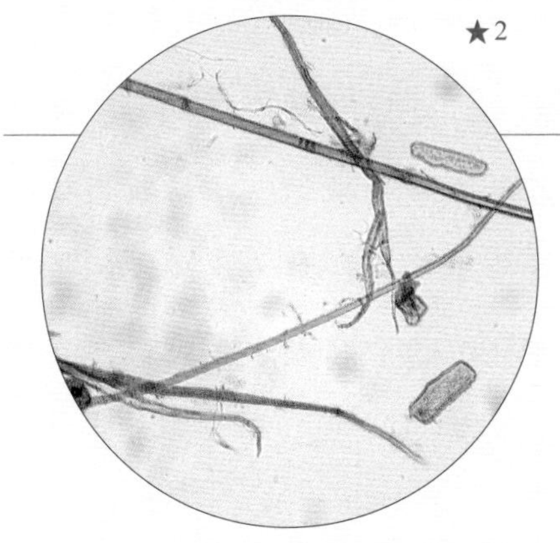

★2
祝南书纸坊祭祀纸纤维形态图（20×）
Fibers of joss paper in Zhu Nanshu Paper Mill (20× objective)

四
祝南书纸坊祭祀纸的生产原料、工艺与设备

4
Raw Materials, Papermaking Techniques and Tools for Joss Paper in Zhu Nanshu Paper Mill

（一）祝南书纸坊祭祀纸的生产原料

1. 主料一：毛竹

富阳竹纸的原料为嫩毛竹，一般在小满前后上山砍伐。从山基村的资源条件来看，山基溪流经山基村，水源丰富，山地广阔，海拔适宜，是毛竹生长的适宜之地。回访时，祝南书表示因砍竹所需人力与时间成本较高，纸坊已有七八年未自己组织工人上山砍竹。纸坊现在是从余姚购买竹料，买回后经过落塘、翻滩、堆蓬等工序加工后再使用。2018年时毛竹料的购买价格约为2.4元/kg，纸坊一年购买量约近5 000 kg。

2. 主料二：纸边

祝南书纸坊生产祭祀纸会在毛竹料的基础上添加一些纸边，总的纸浆比例约有一半以上掺加纸边。纸边通常是废弃的木浆纸边或是竹浆纸边，购自江苏，购买价格为3.2元/kg。纸坊废纸边一年需求量为1 500 kg左右，通常由商户根据需求将纸边送到村里。

3. 主料三：棉花

据祝南书介绍，制浆时加入棉花是为了让纸品更具韧性，纸张成品不易破损。纸坊所用棉花购买自绍兴，购买价格约为2元/kg，2018年纸坊棉花的需求量约为500 kg。购买方式为商户直接送往山基村各造纸作坊。

4. 辅料一：水

祝南书纸坊制作祭祀纸需要大量的水，纸坊就地取材，选择的是山基溪里流淌的溪水。实测山基溪水的pH为5.5～6.0。

5. 辅料二：姜黄粉

制浆环节中加入姜黄粉是为了让纸张成品颜

⊙1

⊙1 废纸边 Abandoned paper edge

⊙2

色更明艳，具体用量依据每次原料用量而不同。祝南书纸坊所用姜黄粉购买于山东，购买价格为330元/包，每包约25 kg，一般一年用量在50包左右。

⊙3

⊙5

⊙4

⊙2
山基村竹林
Bamboo forest in Shanji Village

⊙3/4
水源pH测试
Testing the water source pH

⊙5
姜黄粉
Turmeric powder

（二）祝南书纸坊祭祀纸的生产工艺流程

据祝南书口述以及调查组成员2016年9月28日、2019年3月5日两次实地调查，祝南书纸坊祭祀纸的生产工艺流程可归纳为：

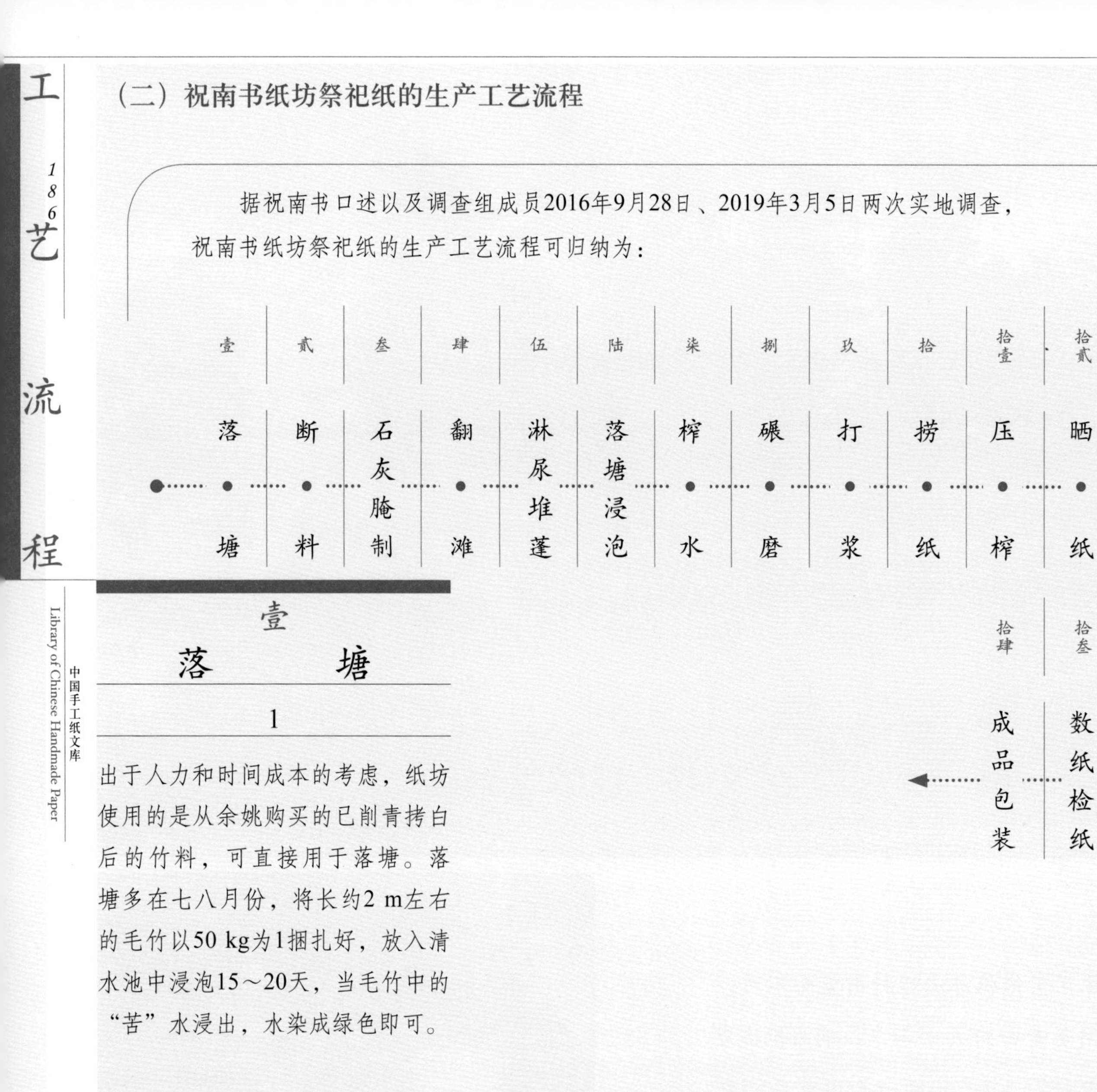

壹 落塘

1

出于人力和时间成本的考虑，纸坊使用的是从余姚购买的已削青拷白后的竹料，可直接用于落塘。落塘多在七八月份，将长约2 m左右的毛竹以50 kg为1捆扎好，放入清水池中浸泡15～20天，当毛竹中的“苦”水浸出，水染成绿色即可。

贰 断料

2

从清水池中捞出毛竹，使用断料刀将毛竹截成4～5段，每段40～50 cm，再将砍好的竹料以15 kg为1捆用塑料绳扎好。

叁 石灰腌制

3

将扎捆好的竹料放入石灰水中浸泡。按照当天所需浸坯的量放入石灰和水，约15 kg竹料配1 kg石灰。腌制通常由4～5人完成，整个腌制过程持续一个月左右。检验标准是看石灰水有没有均匀覆盖在每一捆料上，未完全覆盖的需要将料在石灰水中反复翻滚再腌制。

肆 翻滩

4

将浸泡在石灰水中的料捞出开始翻滩，用清水从上向下清洗，掰开每捆料，以保证可以清洗到内部的石灰，再将一捆料上下晃动，利用惯性将石灰水沥出。此过程反复进行3～4次，当沥出的水是清水时即表示翻滩完成。

伍
淋尿堆蓬
5

将清洗好的料放于尿桶内淋尿，促使竹料纤维软化并去除竹料表面黏附的石灰。再将淋过尿液的竹料堆放到一起，利用夏季的高温以及竹料与尿液的反应促进发酵。堆蓬的时间依据温度高低而有不同，一般需要15～20天。

陆
落塘浸泡
6

将堆蓬后的竹料放入未注水的池子，放料多少与池子大小相关，池子里装满料后注入清水浸满水池。整个浸泡过程持续约一个月，中途不需要换水。当料池中的水变成黑色时，即表示可进行下一步工序。

柒
榨水
7

将竹料从料塘中取出，使用压榨机榨干竹料水分，压榨时间依据榨料多少而有区别。

捌
碾磨
8 ⊙1

将压榨好的竹料用石磨细细碾磨，至竹料细碎用手可揉成碎末时为止。一般碾磨时间在半小时到1小时之间。

⊙1

玖
打浆
9 ⊙2

在打浆机中将废纸边用水浸湿放置约20分钟，再加入棉花和碾磨好的竹料，配比约为85%毛竹料＋10%废纸边＋5%棉花。打浆时间一般在1小时左右，通常一次会打好3～5天捞纸所用的纸浆。

⊙2

⊙1 碾磨后的竹料 Bamboo materials after grinding

⊙2 打浆机 Beating machine

拾 捞纸

10 ⊙3⊙4

祝南书纸坊由祝南书本人负责捞纸，每天早上5点起床，工作到下午三四点钟。一般在捞纸前一天的晚上加入纸浆和姜黄粉，以便为第二天的捞纸做好准备工作。捞纸过程中需要使用和单槽棍顺时针搅匀纸槽里的水和纸浆，每捞三五次纸就需要搅拌一次。将纸帘从上到下倾斜下到纸槽内，缓慢匀速抬起，纸帘移出水面时朝前倾斜形成湿纸膜。最后将抄好的湿纸放在压榨机上等待压榨。

⊙3

⊙4

拾壹 压榨

11 ⊙5

下午3点左右开始压榨，在待压榨的湿纸块上逐一放上垫板和千斤顶，压榨时力度从小到大缓慢增加，直至湿纸块无法挤压出水时，压榨完成。一般单次压榨5 000~6 000张纸，时间约1小时。

⊙5

⊙6

拾肆 成品包装

14 ⊙7⊙8

用塑料绳将成品纸张扎成捆，一捆纸为2 000张。再用磨纸工具将纸堆的四面磨光，使其整齐美观。

拾贰 晒纸

12

晒纸环节在家中完成。从湿纸块上每5张一叠将纸帖取出，阴天时将一叠叠纸放在用于晒纸的竹竿上自然阴干，晴天时放于楼顶阳台晒干。晒纸时间依据气温而定，夏季太阳强烈时一下午即可晒好，气温低时则需4~6天才能阴干。

拾叁 数纸检纸

13 ⊙6

将晒干的纸每5张为一叠收集好，这个过程中要将破损或褪色的纸品剔除。

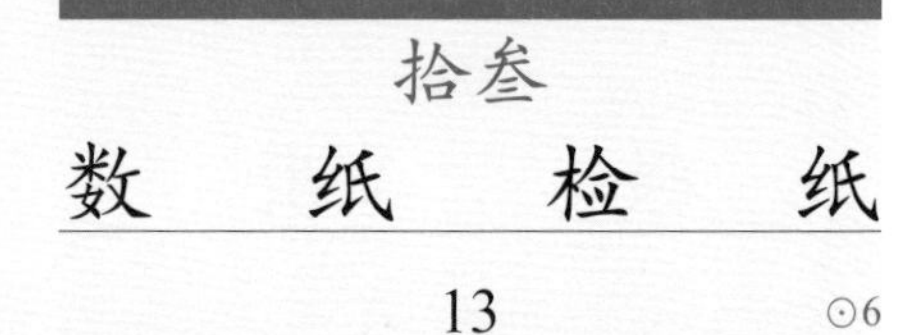

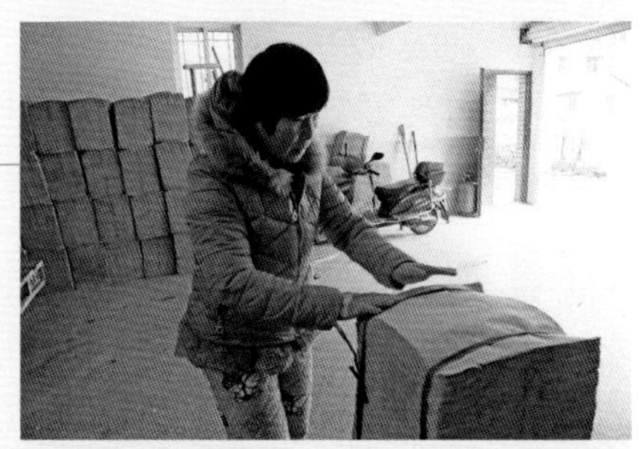

⊙7

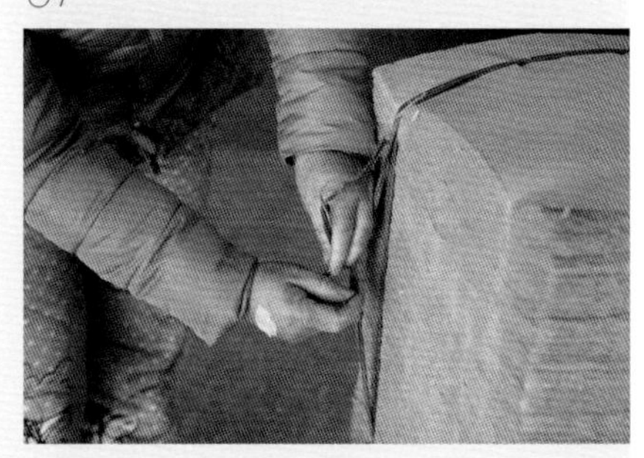

⊙8

⊙3/4 祝南书演示捞纸 Zhu Nanshu demonstrating papermaking

⊙5 压榨机 Pressing machine

⊙6 江春芹演示数纸 Jiang Chunqin demonstrating counting the paper

⊙7/8 包装捆扎 Packing and binding the paper

（三）祝南书纸坊祭祀纸的主要制作工具

壹 纸槽 1

盛放纸浆的设施，长方体容器，捞纸工站于其侧边进行工作。纸槽所在的屋子叫作捞纸坊。祝南书纸坊所用纸槽是清代祖传下来的，由几大块石板构成。古槽的石头和石头之间有扎口将其扎住，把石板拼起来，接口缝隙处用石灰、混凝土混合填充。

祝南书纸坊使用的古槽有200年以上的历史，类似这样的古槽在祝南书纸坊存有2个。祝南书纸坊的一个老纸槽用来捞纸，另一个用作储料池，放置打好的纸浆料。实测纸槽的尺寸为：外长213 cm，内长198 cm；外宽171 cm，内宽154 cm；高74 cm。储料池的尺寸为：外长124 cm，内长112 cm；外宽64 cm，内宽59 cm；高77 cm。

⊙9

⊙10

贰 纸帘 2

捞纸工具，用于形成湿纸张和过滤多余的水分，由细竹丝编织而成，表面刷有黑色土漆，光滑平整。实测祝南书纸坊所用纸帘尺寸为：长110 cm，宽43 cm。2019年回访时了解到，祝南书纸坊所用纸帘购自大源镇，购买价格为400元/张，一般一年更换一张。

⊙11

叁 帘架 3

木制的长方形框架，面积稍大于纸帘，纸帘可完全嵌于框架中，作用是捞纸时承载纸帘。祝南书纸坊所用帘架是请当地木匠所做，价格约为250元/张，一般使用2年换一次。实测纸坊所用帘架尺寸为：长110 cm，宽50 cm。

⊙12

⊙9 纸槽 Papermaking trough

⊙10 储料池 Pool for storing the materials

⊙11 纸帘 Papermaking screen

⊙12 架在纸槽上的帘架 Screen frame putting on a papermaking trough

肆
和单槽棍
4

用来搅动水和纸浆的工具，将纸槽内的纸浆搅均匀方便捞纸。实测纸坊所用和单槽棍的尺寸为：顶部宽22.5 cm，长30 cm，厚2 cm；杆长192 cm。

伍
温手锅
5

专门给捞纸人冬天工作准备的温手工具。做纸的人每天带一瓶热水倒在温手锅中，炉煌中生火以维持温度。以前用短的松末燃火加温，现在用木柴烧火。传统捞纸大都是露天操作，冬天气温低非常寒冷，需要温手锅时不时来帮助舒缓手部的冰寒，以便能长时间进行捞纸工作。实测纸坊所用温手锅尺寸为：长48 cm，宽36 cm，高77 cm。

⊙14

⊙15

⊙13

⊙13
和单槽棍
Stirring stick

⊙14/15
温手锅与炉灶
Tool for warming papermaker's hands and the stove

五 祝南书纸坊的市场经营状况

5 Marketing Status of Zhu Nanshu Paper Mill

据祝南书本人口述，他们夫妻二人每年除去因劳动强度大等原因导致身体不适而休息的天数外，基本上全年都在捞纸晒纸。祝南书每月工作15～20天，每天工作10小时左右，一日可捞近6 000张纸。夫妻二人一年的收入为8万元左右。祝南书说近些年还是有活可以做的，但是手工祭祀纸受机械祭祀纸的冲击越来越大，未来不知道什么时候纸坊可能就无法为继了。

2019年回访时，祝南书介绍自己不捞纸的时候会在浙江周边收购纸品到杭州进行销售。因此纸坊的收入分两部分：销售纸坊自产纸品和收购别家纸品再销售。2016年收购纸品利润约8万元，自家造纸收入也在8万元左右。2018年纸坊产出纸品500余捆，造纸毛收入8万元左右，从外面收购纸品销售的利润约为9万元。两年内纸坊的收入情况未有明显波动，收纸数量和收纸地区与2016年相比并未扩大。

谈及对外收购纸品的业务，祝南书表示最大的困难是收购的纸品质量参差不齐，有的纸户只讲价钱，不讲究质量。2018年时因纸品问题，在绍兴曾面临退货的困境。祝南书直言好的纸品最重要的是硬度，硬度不够的纸，大多在造纸过程中加入了过多的纸边，减少了竹料用量，但这种纸根本达不到买家的要求。其次是不能有破损，颜色要黄，张数要够。自己收购的纸品有时会因这些问题被退回，而退回的纸品很难再退回给纸农，只能自己买单。

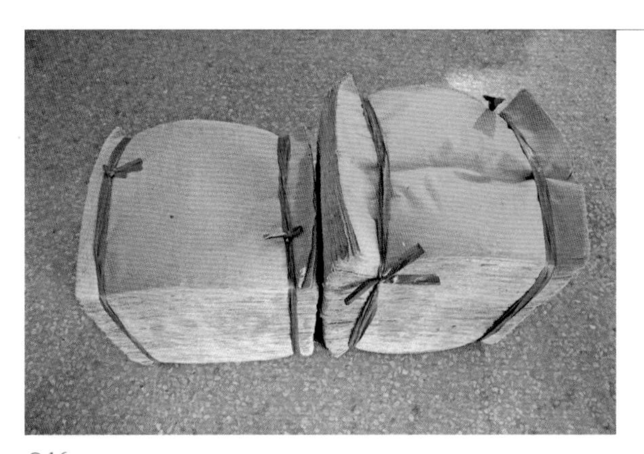

⊙16

⊙17

⊙16
收购的纸品（左边为质量稍次的褪色纸）
Purchased paper (slightly inferior faded paper on the left)

⊙17
祝南书家中堆放的纸品
Paper piled in Zhu Nanshu's house

⊙1

⊙1
背靠直坞水库的山基村造纸旧址
The former papermaking site of Shanji Village, backed by Zhiwu Reservoir

六 山基村手工造纸的文化和民俗故事

6 Culture and Stories of Handmade Papermaking in Shanji Village

（一）年初开工宴

据祝南书和山基村当地几位造纸的老人叙述，在山基村，每年年头开工造纸时有这样的民俗：东道主会请自家的工人吃顿丰盛的大餐，喝点好酒，在这顿饭中工人只需要把自己的嘴巴带来尽情享用美食美酒就好而不用付钱，寓意着开工利市，也算是给造纸工人提供的一个小福利，希望这年大家能够继续用心造好纸张。

这种民俗流行于雇人经营纸坊时期，在年三十或重要节气时会请工人吃饭，但随着造纸行业逐渐衰落，纸坊和纸工数量减少，便不再有此习俗。设立开工宴除了保证工人造纸的积极性，也有向山神祈求提高纸品产量、祝愿纸工平平安安的意思。村里有几十年捞纸经验的师傅表示，曾经在开工宴前会点上香和红烛，提醒山神有人向他发出祈愿，由东道主拜三拜后说出新年造纸的愿望，之后开工宴正式开始。

⊙2

⊙2 遗迹中仍在使用的料塘 Soaking pool still in use among the remains

（二）丰富的造纸现场遗存

山基村是富阳少数还留有较多传统造纸作坊遗迹的村落，可作为实物例证证明富阳是“土纸之乡”，而同其他传统造纸村落相比，山基村造纸遗存蕴含着其独有的特点。

首先是作坊遗迹保存较完整，大大小小约十多个料塘、石灰池、皮镬等传统造纸设施保存状况良好，而且大多集中于直坞水库前，基本保留了原有的格局。现今在每年6月左右，村内加工竹料时仍会用到部分造纸工具，一定程度上还原了当时的土纸生产场面，有利于遗产的展示。遗迹中部分造纸工具至今仍在使用，祝南书表示纸坊目前使用的是生料法，包产到户后几年村里的纸坊还以熟料法为主，后考虑到精简工序、节省成本等因素，熟料法渐渐被生料法所替代，用于蒸煮的皮镬就几乎不再使用了。

其次，山基村具有多种文化遗产，既有传统

造纸作坊遗迹，又有槽户老宅以及名人历史，这些遗迹将作坊、纸民、自然风光串联起来，成为留住文化记忆及山基村造纸转型的重要途径。山基村内留有如祝南书纸坊中纸槽这样的清代古法造纸遗迹，回访时村里的造纸户也表示，仅遗迹内的皮镬就有100多年历史，这一判断是从锅里面残存的木头构件推算而来的。村内多数遗迹已经荒废，地方政府计划修缮该地，将其作为文化传承体验观光景点，但2019年回访时，观光及修缮工作还未开展。

⊙2

⊙1

⊙3

七 祝南书纸坊的业态传承现状

7 Current Status of Zhu Nanshu Paper Mill

（一）面临已经走不下去的困境

2016年入村调查时，整个村子里尚有70户左右的造纸户，但从事造纸业的手艺人最低年龄也在40岁以上，更多的是60岁以上的老年人坚守在传统手工造纸的岗位上。而2019年回访时，村内仅剩下30户不到的纸坊仍在生产，目前现存造纸作坊最年轻的造纸人也近50岁了，各纸坊几乎都面临着后继无人的困境。祝南书表示，自家纸坊隔壁的纸户也是独立经营，经营者已经70多岁了，从事捞纸工作近40年，一个月有20天都在捞纸，平均每天工作11个小时左右，就在2015年年末，老人因年龄和体力问题，纸坊关停不再生产。

山基村传统造纸产业迅速没落、后继无人的现状，从根本上来说是市场原因所造成的。祝南书手工纸坊从原先做书画纸到如今改做祭祀用纸，

⊙1 皮镬遗迹 Remains of the papermaking utensil

⊙2 山基村近代土纸作坊遗存 Remains of a modern local laper mill in Shanji Village

⊙3 山基村自然风光 Natural scenery of Shanji Village

都是对现代纸业市场的无奈妥协和顺势改变，纸的种类和品质在变化，销路在变化。祝南书表示学习造纸技艺需要长年累月的积累、寒冬酷暑的历练并承受日常高强度的工作，这种高投入低回报的模式无法吸引到年轻人参与，面临的是祝南书等老一辈捞纸手艺传承人愿意免费教学生，但无人来继承学习的困境。

（二）可能的跨界拓展空间

山基村具有开展乡村旅游的优质先天条件，山清水秀，交通便利，距杭新景高速公路灵桥出口仅4 km，距杭州市中心40 km左右。[1]村子两面环山，背邻直坞水库。水库附近就是安顶山古道，回访时据当地村民介绍，时常有户外运动爱好者来这里爬山涉水，享受自然风光。但目前山基村的旅游还未完全开发，也未形成明确的旅游主题，游客大多来自周边地区。因此可以整合山基村原有造纸文化遗存与其周边自然风光，打造一条具有品牌特色的文化体验旅游线路，以旅游促进村民造纸的积极性和乡村经济发展。

谈及山基村传统手工造纸的发展，祝南书表示自己这几年收购纸品时发现，山基村造纸业整体呈慢慢消亡的状态。收购纸品的作坊是长期合作的，每年收购的纸品数量较为稳定，这两年收购和销售的价格都有所提高，但祝南书表示实际所得的利润却在减少。其很大一部分原因在于各纸坊提供的纸品质量未达标，这是纸农造纸积极性不强，重数量不重质量的结果，而这种结果更可能造成恶性循环。因此除了打造造纸文化体验游，更应在纸品质量和培养造纸匠人上下工夫。

关于纸品质量，祝南书表示从2018年开始，在收购纸品时他会明确地向纸农给出纸品合格的标准，给予优质纸品更高的收购价格，以此来提高周边纸农的造纸积极性。

在培养传承人方面，祝南书的一双儿女均未习得造纸技艺，也没有继承纸坊的打算。祝南书也表示自己想教，但因为造纸辛苦又没有较高的收入，没有人愿意来学。

⊙4

⊙5

⊙6

⊙4 68岁的祝波泉演示捞纸 68-year-old Zhu Boquan demonstrating papermaking

⊙5 山基村的休闲登山道 Recreational hiking trail of Shanji Village

⊙6 废弃的纸槽 Abandoned papermaking trough

[1]崔彪，刘小军．富阳山基村传统造纸遗存及其保护调研[J].杭州文博，2015(01).

祭祀纸

Joss Paper
of Zhu Nanshu's Paper Mill

染黄祭祀纸（毛竹+废纸边+棉花）透光摄影图
A photo of yellow joss paper (*Phyllostachys edulis*+waste paper edges+cotton) seen through the light

第十一章
富阳区皮纸

Chapter XI
Bast Paper in Fuyang District

第一节

五四村桃花纸作坊

浙江省
Zhejiang Province

杭州市
Hangzhou City

富阳区
Fuyang District

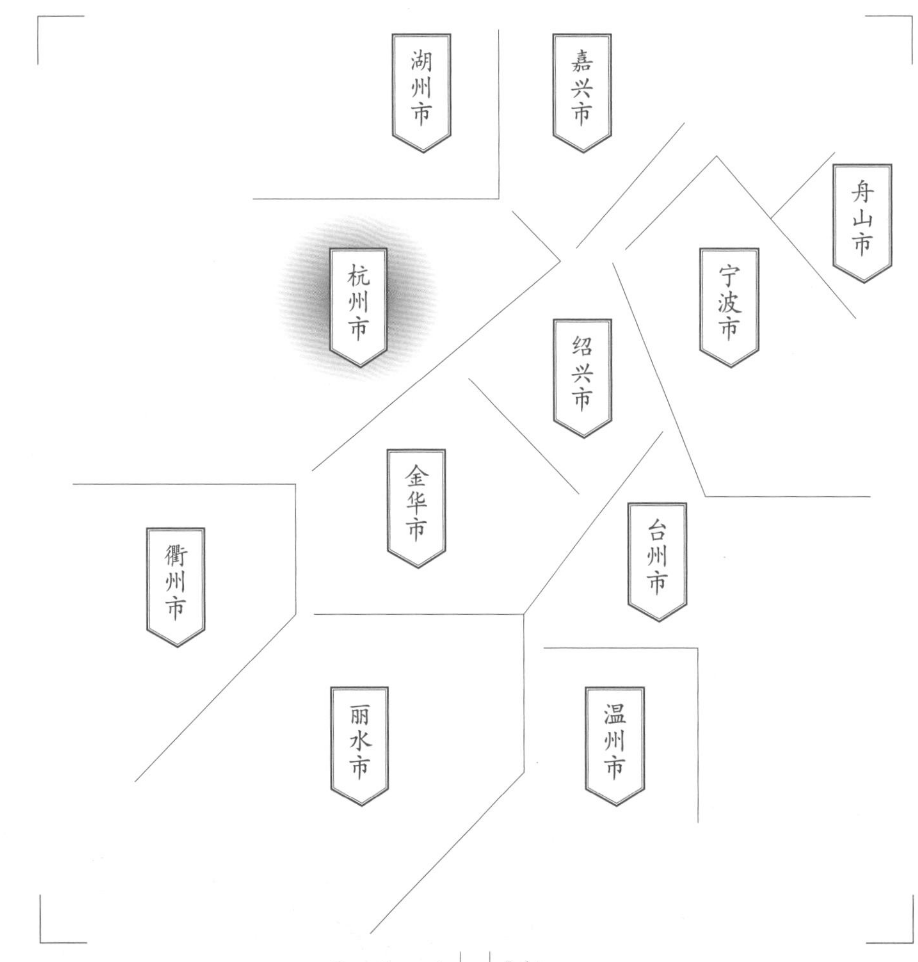

调查对象

富阳区鹿山街道五四村
桃花纸作坊
皮纸

Section 1
Taohua Paper Mill in Wusi Village

Subject

Bast Paper of Taohua Paper Mill in Wusi Village of Lushan Community in Fuyang District

一
五四村桃花纸作坊的基础信息与生产环境

1
Basic Information and Production Environment of Taohua Paper Mill in Wusi Village

调查时的桃花纸作坊，是鹿山街道五四行政村村民叶汉山在富阳区“非遗”保护中心支持下，于2018年下半年为恢复富阳历史名纸桃花纸生产工艺建成的小型手工纸坊，位于鹿山街道五四村，地理坐标为：东经119° 50′ 10.62″，北纬29° 59′ 1.79″。

2018年6月，富阳区“非遗”保护中心对鹿山街道原上里乡出产的桃花纸制作技艺进行抢救性记录，同时拨付5万元，由老纸工叶汉山负责恢复桃花纸生产工艺，至10月底生产出第一批1 500多张桃花纸。

2018年6月现场调查时，五四村属于鹿山街道上里片区。上里，民间习惯称之为“上里山”，位于鹿山街道西侧，320国道以西，南新线（鹿山街道蒋家杭春棉入口处至新浦埠头）贯穿整个上里山，沿途经过3个行政村，依次为三合村（原坞口村、三合村合并）、五四村（原五四村、上叶村合并）、新祥村（原桢祥村、新新村合并）。这一带山青水秀，毛竹自然资源丰富。五四行政村区域面积15.5 km^2，耕地面积0.46 km^2，竹山、柴山5.41 km^2，东面与鹿山街道三合村相邻，南邻新桐乡新浦村，西面与新登镇大山村、鹿山街道新祥村接壤，北面是富春街道柳溪村。五四村包括桃坑、华家、上叶、下叶、江家、山坑里、山坑坞里等自然村，以龙山为界分为南、北两坞，两坞绵延数公里。五四村的村名，据说是一位教师提出来的，源于“五四”青年节，充满活力，生机勃勃，带有很强的时代色彩。2018年五四村常住人口1 523人。

作为当代挂牌的“中国竹纸之乡”，实际上富阳皮纸生产的历史更悠久，起源更早。富阳区域造皮纸最早可以追溯至魏晋时期，当时，在相距不远的嵊州一带出现了历史名纸“剡藤纸”，近邻的余杭出现了历史名纸“由拳纸”，富阳在1 700年前就是造纸产区。余杭产纸的由拳村与

⊙1

⊙2

⊙1
桃花纸恢复现场
Recovery scene of Taohua paper

⊙2
鹿山街道上里山五四村山间盆地环境
Mountain basin environment of Wusi Village in Shangli Area of Lushan Community

路线图
富阳城区
↓
五四村桃花纸作坊
Road map from Fuyang District centre to Taohua Paper Mill in Wushi Village

五四村桃花纸作坊位置示意图

Location map of Taohua Paper Mill in Wushi Village

考察时间
2018年6月 / 2018年10月 / 2018年12月

Investigation Date
Jun. 2018 / Oct. 2018 / Dec. 2018

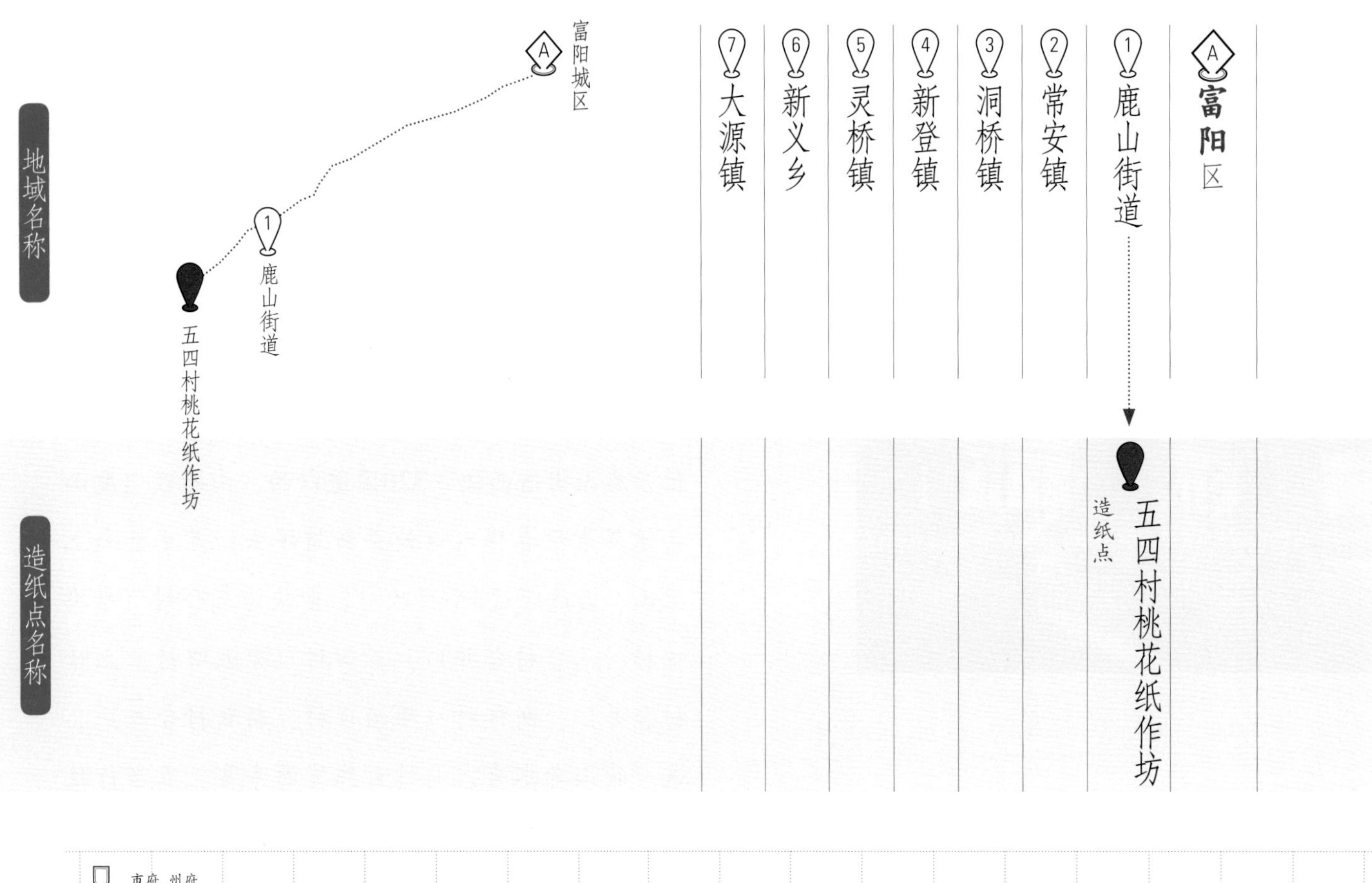

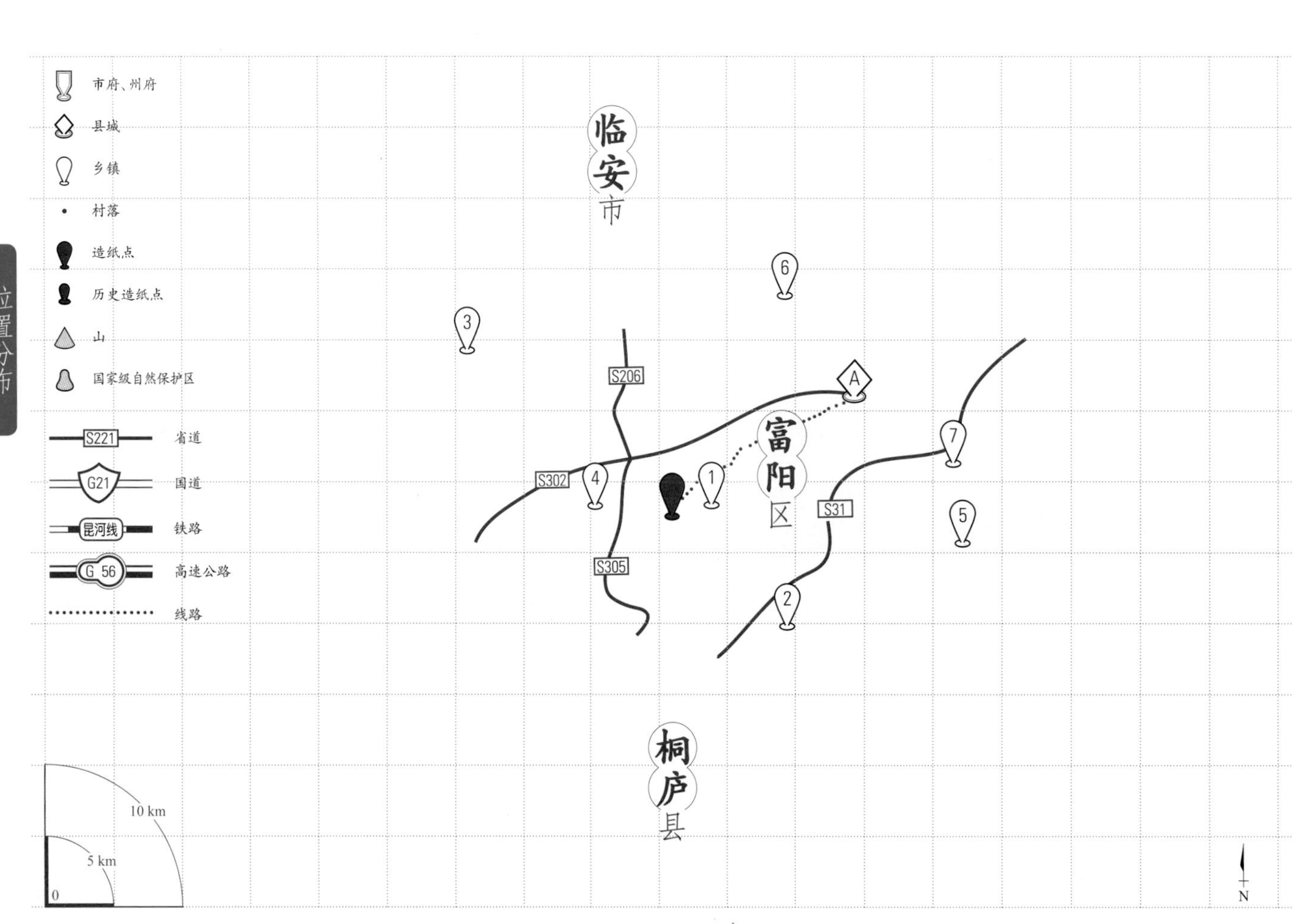

今日富阳银湖街道的铜岭村相距仅1 km，纸乡相连，技术融通，因此可以推断富阳地区造纸一开始是由皮纸制作技艺引领的。

明代富阳皮纸已经形成了成熟的制作工艺。明正统五年（1440年）吴堂纂修《富春志·贡赋》提到：富阳县岁办“桑穰六千一百三十”。桑穰，桑树的第二层皮，白色，为造皮纸的上等原料。初刊于明崇祯十年（1637年）宋应星著《天工开物·中篇·杀青》提到：“桑皮造者曰桑穰纸，极其敦厚，东浙所产，三吴收蚕种者必用之。”[1]《富春志》提到的“桑穰”不管是指代桑穰纸还是桑皮纸的原材料，因为是上贡的去处，说明富阳桑皮纸或桑皮纸原料在那个年代就是比较优质的。

历史上富阳上里山皮纸生产不仅历史悠久，而且质量上乘，曾经是富阳皮纸的重点产区，但并非仅止于桑皮纸，构皮等作原料的纸也相当有名。清光绪《富阳县志》载：“货之属——皮纸。以楮皮为之，西北乡造，为包裹银洋之用。桑皮纸以桑枝皮为之，西南乡造。”西北乡，指现在的银湖街道新义片、坑西片，西南乡，指现在的鹿山街道上里山片区。据《浙江之纸业》记载，1928年富阳生产的桑皮纸年产量达6 400件，每件168张，总量达107万张以上，位列全省同类产品第一位，并有名牌产品蜚声中外。

民国十九年（1930年），上里产的本色桑皮纸在西湖博览会上获特等奖。上里产的白绵纸，又名白皮纸，全桑皮制作，因纸质柔软洁白得其名，也获一等奖。以上里白绵纸为原料加工的富阳油纸，则在1929年的西湖博览会上获得特等奖。上里山面积不大，但竹纸、草皮纸、皮纸三

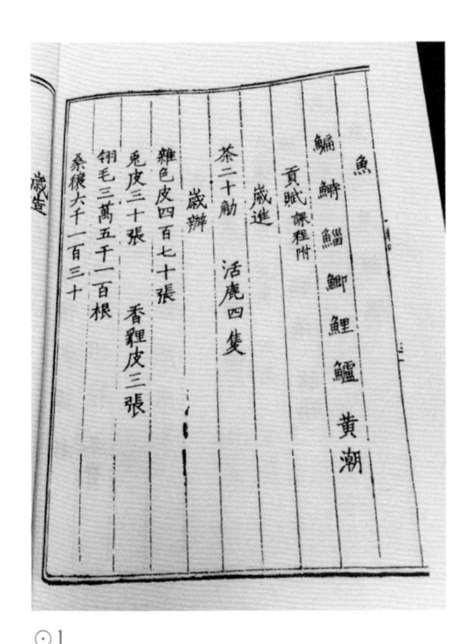
魚

鯿 鰣 鯔 鯽 鯉 鱸 黄 潮

貢賦（課程附）

歲進

茶二十觔 活鹿四隻

歲辦

雜色皮四百七十張

兎皮三十張 香貍皮三張

翎毛三萬五千一百根

桑穰六千一百三十

⊙1

⊙2

⊙1 明正统五年吴堂纂修《富春志》关于『桑穰』的记载
Record of "Mulberry" in *The Annals of Fuchun* edited by Wu Tang in the Ming Dynasty (1440)

⊙2 光绪《富阳县志》对地方产皮纸的记载
Record of local bast paper in *The Annals of Fuyang County* during Guangxu Reign of the Qing Dynasty

[1] 宋应星.天工开物[M].杭州：浙江人民美术出版社，2013.

类纸共存，技术融通，在富阳区也是一个较有意思的现象。20世纪50年代，上里一带所造的京放纸曾被浙江省评定为超级京放纸，名声远传国内京、津、沪等中心城市。上里山也是富阳地区最早从传统手工造纸走向机械造纸的重要纸产区。上里山手工皮纸生产业态中断大约是在80年代中期。

⊙1

⊙1 叶汉山生产的桃花纸
Taohua paper produced by Ye Hanshan

二
五四村皮纸的历史与传承

2
History and Inheritance of Bast Paper in Wusi Village

1963年富阳县统计局资料《全县土纸生产情况调查资料》显示：上里公社有竹纸槽45厂、草纸槽5具、桑皮纸框2 161只。在1974年富阳县土特产公司调查资料上，已不见上里公社，上里公社合并到三山公社。当时的三山公社有17个大队，其中16个大队造纸，有竹纸槽51厂、草纸槽405具、桑皮纸框1 955只、白皮纸槽60具。

2018年10月 8 日的调查中，据叶汉山介绍，20世纪六七十年代上里山的6个大队都建有皮纸厂，有3个村的皮纸厂是上规模的：五四造纸厂、三合皮纸厂、坞口造纸厂，这3个厂的规模与地位相当，其他3个小村祯祥村、新新村、上叶村的生产规模不大。而6个大队的每个生产小队都有竹纸和皮纸生产。抄制方式，不仅有用纸框的浇纸法，还有用纸槽的抄纸法。20世纪73年前后，五四皮纸厂、三合皮纸厂生产机制皮纸时，74年的统计资料还清晰记录着三山乡拥有数量不小的桑皮纸框和白皮纸槽，由此推知，传统手工皮纸生产业态到20世纪70年代一直在延续，并且还较兴旺。从访谈中得到，上里山手工皮纸生产完全消失大约是在80年代中期。五四村皮纸生产情况与上里山整体的皮纸发展是同步的。

据叶汉山介绍，五四村造纸历史悠久，皮纸、竹纸业态俱全。叶汉山祖上是用竹造纸，生产元书纸、“迷信纸”等。上里山生产皮纸历史最久的是三合村。五四村山坑坞里有几家是做皮纸的，夏申炳家是其中一户，他的侄子夏志仁原来就在他家做外场（注：外场，即在室外从事制浆方面工作），后来成为五四造纸厂皮纸生产的中坚力量。

2018年6月7日的调查中，叶汉山自述其造纸生涯是从皮纸生产开始的。叶汉山，1942年出生于化民乡第八保（今杭州市富阳区鹿山街道五四村下叶自然村）。父亲叶有礼，生于1921年10月，卒于1942年8月，享年仅23岁，死于伤寒病。母亲吴秋英，生于1918年8月，卒于1998年9月。有一位姐姐叶仙云，1940年出生，1967年离世。叶汉山是家里的独子，也是家里的顶梁柱。祖父有三个儿子，父亲排行老三，大伯很能干，二伯是残疾人不能说话，到叶汉山13岁时家里分了家。祖父当年有二厂竹纸纸槽，商号为"叶正泰"，因此后来被划为富农成分。二厂竹纸槽纸农大部分来自常绿镇，主要生产京放纸。叶汉山小学毕业后，第二年兼做生产队记工员，后来兼任生产队会计，兼职财会工作20多年。

1959年7月，当年的五四大队开始兴办集体所有制的"五四造纸厂"，实际上就是搭建了3间茅草房，面积170 m^2左右，添置木质纸槽5具、旱棚碓3个，而蒸煮用的皮镬、晒纸用的焙笼都是借用生产小队竹纸生产的设施设备。

当时五四大队由7个生产小队组成，生产小队仍然从事竹纸生产或浇纸法草皮纸（桑皮和稻草混料纸）生产，大队从各生产小队抽了一批年轻力壮、诚实、吃苦耐劳的小伙子作为制浆、抄纸主要力量。传授抄纸技艺的师傅是三合村的方关

⊙1

⊙1
五四村下叶自然村前的田畴道路
Field road in Xiaye Natural Village of Wusi Village

兰，制浆师傅是五四村的夏志仁。夏志仁以前在他小伯夏申炳的皮纸厂做过帮工，懂得制浆的整个工艺。制浆的主要原材料是桑皮，以野生洋桃梗汁（即野生猕猴桃枝）作为纸药，生产本色皮纸。制浆用石灰腌料，用皮镬蒸料，用旱棚碓碾压皮料，用摇臼桶分散纤维，抄纸架使用竹丝串架。1959年10月1日，“长东洋”雨伞纸正式出产。

据调查时叶汉山回忆，五四大队皮纸厂最初兴建时，外场制浆的有夏志仁（已故）、蒋荣奎（已故）、江明余（已故）、章文标（已故）、蔡阿元（已故）、叶丙根、夏生根；纸坊抄纸的有华季中、江仁富（已故）、叶宗云（已故）、叶根元、叶汉山；焙笼晒纸的有叶林胡、叶秋根；检纸是吴金娜；供销员是华锦胜（已故）。

⊙1

⊙2

⊙1 五四村境内造纸皮镬遗存 Remains of papermaking utensil in Wusi Village

⊙2 叶汉山（讲话者）讲述桃花纸制作历史 Ye Hanshan (speaker) telling the papermaking history of Taohua paper

1960年，随着市场发展需要，五四造纸厂开始生产绵纸。制浆辅助采用现代化工材料烧碱、硫酸、漂白粉，但仍是全手工制作。制作方式与2018年底新恢复桃花纸工艺相同。绵纸分正号绵纸和副号绵纸，正号绵纸是打字蜡纸的原纸，副号绵纸是工业产品的电池用纸。

1964年以后，五四造纸厂开发滤油纸，商标为“向阳牌”，主要原料是造纸棉。同时也生产绵纸。制浆部分采用比较现代化的工艺，蒸煮用蒸汽压力锅，浆料处理用打浆机，洗料用洗料机，等等。抄制部分还是纯手工，滤油纸和绵纸的抄制方式是一样的，都采用打浪法。产品销售旺盛，由五交化二级站总经销，产品源源不断销往全国，主要用于金矿、食品、化工、墨水厂等厂家各类过滤场合使用。那时候，全国大厂家，如沈阳造纸厂、北京滤油纸厂、温州打字蜡纸厂，生产的滤油纸采用国标为270 g/m^2，而五四造纸厂滤油纸标定为220 g/m^2，纸质松软，过滤速度快，很受用户欢迎。滤油纸和绵纸的利润率在35%以上。五四造纸厂纸槽最多时达16口，在富阳县的知名度相当高。

20世纪70年代，全国开展知识青年“上山下乡”运动。杭州华丰造纸厂、新华造纸厂部分“知青”来五四造纸厂落户。在造纸“知青”的帮助下，造纸机械化设备和生产技术有了很大变化，完全剥离了手工作业的工序。1973年引进全机械长网泼浪式造纸机1台，皮纸系列长网造纸代替了手工抄制。之后又增加了圆网造纸机（787 mm × 1092 mm）1台，原料用麦秆、竹壳，生产糖果纸及28～50 g书写纸。3条生产线，

⊙1

职工120人左右，年产值80万元左右。

20世纪80年代，五四造纸厂先后引进了1.5吨、4吨锅炉，每小时3吨切草机，蒸球3立方、8立方、14立方各一只。年产值增加至300万元，职工扩充至200人左右，解决了全村大部分的劳动力就业。至1987年，五四造纸厂主要产品有白绵纸、引线沙纸、滤油纸、书写纸等。

1995年5月，五四造纸厂改制为私营独资企业，更名为"康达绵纸厂"。圆网纸机生产卫生纸，长网纸机生产皮纸系列防风纸、电池绵纸等。2014年康达绵纸厂停产。

2018年12月23日调查时，鹿山街道五四村60岁以上村民尚有几位完全掌握手工桑皮纸抄纸法生产工艺的老纸工，叶汉山是其中代表者之一。1959年，叶汉山师从三合村方关兰，在新建的五四造纸厂学习桑皮纸抄制工艺。据叶汉山的说法，他抄了14年绵纸，每天约抄500张，成品率达90%～95%。期间他带出夏关水、叶中云两个徒弟。1973年以后，五四造纸厂开始用机械化方式生产，叶汉山作为技术人员一直坚守在生产岗位上。五四村是富阳最先发展起来的机械造纸村之一，叶汉山作为技术人员与其他村民一起外援其他七八个村兴建毛纸机械厂。1995年他受聘于康达绵纸厂，分管半机械滤油纸生产及销售，至2006年5月离职退休。前后共48年一直工作在村级造纸岗位上。

⊙2

⊙1 旧存滤油纸 Old oil-filtering paper

⊙2 叶汉山在抄造桃花纸 Ye Hanshan making Taohua paper

三

五四村皮纸的代表纸品及其用途与技术分析

3 Representative Paper and Its Uses and Technical Analysis of Bast Paper in Wusi Village

(一) 五四村皮纸代表纸品及其用途

据叶汉山介绍，五四村手工皮纸生产主要有两个系列：一是生产小队用浇纸法生产的草皮纸；二是五四大队用抄纸法生产的皮纸，包括绵纸（棉纸）、桃花纸、雨伞纸、防风纸等。

1. 正号绵纸（打字蜡纸原纸）

绵纸是五四造纸厂手工纸生产期间的主要产品。是用构皮、桑皮为原料制作而成的白色薄型纸，因其纸质如丝绸般软绵柔韧，洁白亮泽，故称白绵纸。桑皮、构皮因纤维表面有一层透明的胶质膜，若加工得法，纸张易呈现出丝质光泽。绵纸分正号绵纸和副号绵纸。正号绵纸，即打字蜡纸原纸，副号绵纸即电池用纸。正号绵纸比副号绵纸制浆技术要求更高。据叶汉山介绍，五四造纸厂生产打字蜡纸原纸的时间大约在1962年，规格为85 cm × 46 cm。

⊙3

⊙3 正号绵纸 Mian paper, a type of wax paper

2. 铁笔蜡纸原纸

铁笔蜡纸原纸是浙江省20世纪20年代开发的新产品，富阳从二三十年代起就已与上海挂钩，生产铁笔蜡纸原纸。铁笔蜡纸原纸生产原料主要为山棉皮及山桠皮，制作工艺相当精细。五四造纸厂铁笔蜡纸原纸生产时间并不长。据叶汉山介绍，浙江境内的铁笔蜡纸原纸所用以遂昌山棉皮为最好，但五四造纸厂当年用的是贵州买来的山棉皮。

3. 桃花纸

民间认为，桃花盛开时节的水最适宜生产皮纸，所以把质量精美的皮纸称为“桃花纸”。桃花纸又名桃花笺，洁白匀净，薄而强韧，主要用于裱糊纸屏、纸窗、纸阁，制作风筝、雨伞以及用于高档包装，亦可用于书画。富阳桃花纸的原料主要为桑皮与构树皮，主产于上里山的三合村、五四村、坞口村等6个村。历史上上里山桃花纸的产量曾占浙江全省桃花纸总产量的40%以上，是浙江省皮纸中的名品。

4. 油纸

油纸是一种以采用韧皮纤维为原料制成的纸张为原纸，经柿油或桐油等浸渍加工制成的防水纸。因白皮纸或桃花纸浸油后的油纸具有透明度较高的特点，常用于摹拓书画；也有利用油纸抗水性能较强的优点，用于制雨伞、糊灯笼。

5. 桑皮纸

富阳纸农习惯把以桑皮为原料抄造而成的本色纸称为桑皮纸。富阳生产的本色桑皮纸有两种：一种以全桑皮为原料制成，其具有较好的强韧性能，撕裂度、强度和耐破度都较高。杭州历史著名工艺品“杭扇”中的黑纸扇，就是采用富阳桑皮纸制作而成的，又称桑皮扇。富阳的本色桑皮纸在民国十八年（1929年）西湖博览会上获得过特等奖。另一种是用桑皮和稻草按不同配比制成的，也叫草皮纸。桑皮成分多的叫大桑皮，桑皮成分少的叫小桑皮。五四村生产小队曾经有较长时间生产草皮纸。

需要说明的是，正号绵纸（打字蜡纸原纸）、桃花纸、雨伞纸、防风纸都是由桑皮或构树皮为原料制成的，所不同的是制浆工艺的精细度不同，绵纸的制浆工艺要求最高、时间最长，桃花纸的品质比正号绵纸差一些，比雨伞纸要好些。桃花纸的浆料要在水棚碓上碾压四五次。雨伞纸有抄纸法制作的，也有浇纸法生产的，是两种性质完全不同的纸。防风纸是用80%的桑皮或构皮、20%的机制书写纸纸边与笋壳形成的混合料制作的，纸张规格为80 cm × 50 cm，拉力很好，主要用于糊窗户、包茶叶。

⊙1

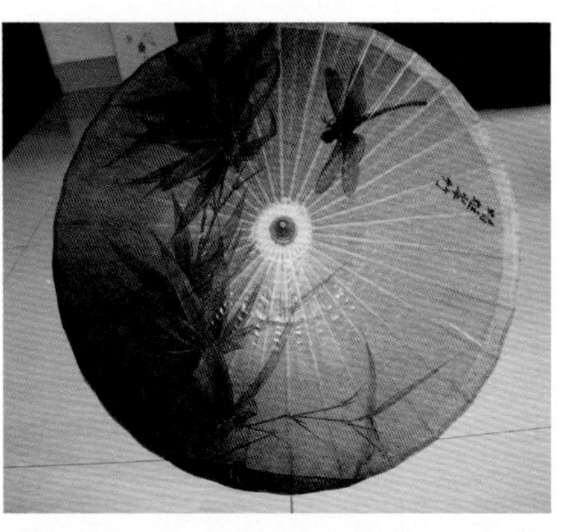

⊙2

⊙3

⊙1
灯影下的桃花纸局部
A part of Taohua paper under the light

⊙2
富阳境内至今仍在生产的油纸伞
Oil paper umbrella still in production in Fuyang District

⊙3
桃花纸的浆料要在水棚碓上碾压四五次
The pulp materials of Taohua paper pressed four or five times under the hydraulic pestle

（二）五四村新制桃花纸性能分析

测试小组对采样自五四村的新制桃花纸所做的性能分析，主要包括定量、厚度、紧度、抗张力、抗张强度、撕裂度、撕裂指数、湿强度、白度、耐老化度下降、尘埃度、吸水性、伸缩性、纤维长度和纤维宽度等。按相应要求，每一指标都重复测量若干次后求平均值，其中定量抽取5个样本进行测试，厚度抽取10个样本进行测试，抗张力抽取20个样本进行测试，撕裂度抽取10个样本进行测试，湿强度抽取20个样本进行测试，白度抽取10个样本进行测试，耐老化度下降抽取10个样本进行测试，尘埃度抽取4个样本进行测试，吸水性抽取10个样本进行测试，伸缩性抽取4个样本进行测试，纤维长度测试了200根纤维，纤维宽度测试了300根纤维。对五四村新制桃花纸进行测试分析所得到的相关性能参数如表11.1所示，表中列出了各参数的最大值、最小值及测量若干次所得到的平均值或者计算结果。

表11.1　五四村新制桃花纸相关性能参数
Table 11.1　Performance parameters of newly made Taohua paper in Wusi Village

指标		单位	最大值	最小值	平均值	结果
定量		g/m²				19.6
厚度		mm	0.079	0.065	0.073	0.073
紧度		g/cm³				0.268
抗张力	纵向	mN	16.5	12.9	15.0	15.0
	横向	mN	11.7	8.2	10.6	10.6
抗张强度		kN/m				0.853
撕裂度	纵向	mN	509.3	430.5	477.2	477.2
	横向	mN	624.1	545.0	580.7	580.7
撕裂指数		mN·m²/g				27.0
湿强度	纵向	mN	1241	1053	1146	1146
	横向	mN	923	780	865	865
白度		%	69.8	69.3	69.5	69.5
耐老化度下降		%				4.4
尘埃度	黑点	个/m²				92
	黄茎	个/m²				120
	双浆团	个/m²				0
吸水性	纵向	mm	30	23	26	19
	横向	mm	25	19	22	4
伸缩性	浸湿	%				0.25
	风干	%				0.50
纤维	长度	mm	5.9	0.7	2.3	2.3
	宽度	μm	32.4	6.2	14.5	14.5

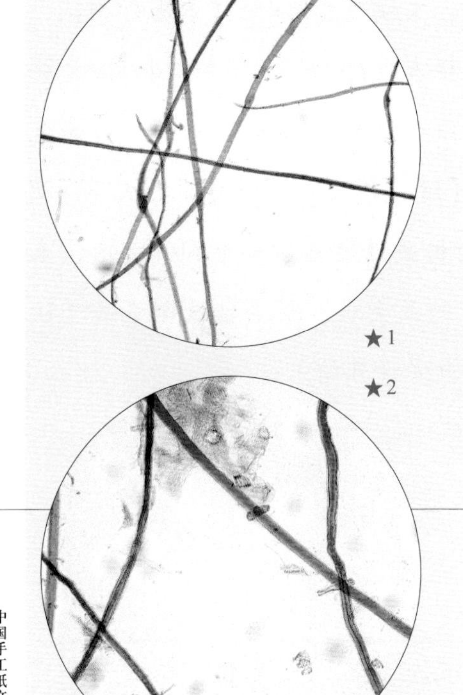

★1

★2

⊙1

性能分析

由表11.1可知，所测五四村新制桃花纸的平均定量为19.6 g/m^2。五四村新制桃花纸最厚约是最薄的1.215倍，经计算，其相对标准偏差为0.075，纸张厚薄较为一致。通过计算可知，五四村新制桃花纸紧度为0.268 g/cm^3。抗张强度为0.853 kN/m。所测五四村新制桃花纸撕裂指数为27.0 mN · m^2/g；湿强度纵横平均值为1 006 mN，湿强度较大。

所测五四村新制桃花纸平均白度为69.5%。白度最大值是最小值的1.007倍，相对标准偏差为0.003，白度差异相对较小。经过耐老化测试后，耐老化度下降4.4%。

所测五四村新制桃花纸尘埃度指标中黑点为92个/m^2，黄茎为120个/m^2，双浆团为0。吸水性纵横平均值为19 mm，纵横差为4 mm。伸缩性指标中浸湿后伸缩差为0.25 %，风干后伸缩差为0.50 %，说明五四村新制桃花纸伸缩差异不大。

五四村新制桃花纸在10倍和20倍物镜下观测的纤维形态分别如图★1、图★2所示。所测五四村新制桃花纸纤维长度：最长5.9 mm，最短0.7 mm，平均长度为2.3 mm；纤维宽度：最宽32.4 μm，最窄6.2 μm，平均宽度为14.5 μm。

★1
五四村新制桃花纸纤维形态图（10×）
Fibers of newly made Taohua paper in Wusi Village (10× objective)

★2
五四村新制桃花纸纤维形态图（20×）
Fibers of newly made Taohua paper in Wusi Village (20× objective)

⊙1
五四村新制桃花纸润墨性效果
Writing performance of newly made Taohua paper in Wusi Village

四
五四村新制桃花纸的生产原料、工艺与设备

4
Raw Materials, Papermaking Techniques and Tools for Newly Made Taohua Paper in Wusi Village

（一）五四村新制桃花纸的生产原料

1. 主料：桑皮、构皮

上里山桃花纸主要原材料为桑皮、构皮。桑皮主要来自本地，作为中国丝绸产业基地——浙江的一部分，富阳曾经也是种桑养蚕大县，曾有3万亩桑园，蚕茧质量之佳，向来与浙江嘉兴、湖州地区相媲美。随着国家“蚕桑西进”政策的推进，区内新登镇、胥口镇、万市镇、洞桥镇等地仍保持有一定规模的蚕桑养殖业。每年春蚕结茧后，四五月份便可将桑树枝剪下来剥皮。100 kg桑皮干皮通常最多能做16 kg成品纸。

2018年9月由叶汉山负责的恢复桃花纸生产错过了春蚕结茧后剪桑枝的合适时机，原材料采用的是构皮。虽然富阳的构树皮原料也很丰富，但叶汉山这次恢复制作桃花纸是从陕西西安地区采购到野生干构皮70 kg为原料的。

⊙2

2. 辅料：水、烧碱、次氯酸钠、硫酸、聚丙烯酰胺

在桃花纸的制作过程中，水的质量好坏对纸的质量影响很大。桃花纸恢复制作的第一轮试验选用的是自来水。据调查组成员在造纸引水现场的测试，制作桃花纸所用的水pH约为6.5，呈弱酸性，至于其他成分未能测试。

⊙3

调查组了解到的信息是本次在制浆过程中还用到了烧碱、次氯酸钠、硫酸等，采用了现代化学制浆的某些材料。

硫酸购买，公安局是管控的，为此叶汉山还委托方仁英代为起草了申请报告，原文如下：

⊙2 上里的桑树 Mulberry tree in Shangli Area

⊙3 富阳本地的构树 Native mulberry tree in Fuyang District

关于要求购买皮料生产需要化工原料的申请报告

富阳区公安局：

为协助富阳区非遗保护中心实施桃花纸抢救性记录，该项目工艺恢复由我主持。在桃花纸生产过程中，140斤皮料浸泡需用化工若干原料，其中硫酸6斤，液碱95斤，次氯酸钠50斤。为此特向杭州特种纸业有限公司购买。

希望贵局予以批准！

鹿山街道五四村村民叶汉山

二〇一八年九月二十五日

蒸煮前在皮料中加烧碱，目的是增加皮料的腐烂速度，更深度地分解皮料纤维；添加次氯酸钠主要是为了增加皮料的白度。在皮纸生产中使用稀硫酸，这在1930年《浙江之纸业》中就有明确记载：“用圆木耙放入袋内，在清水中搅成细柔纤维，然后提起滤干，置于缸中。用漂白粉液均匀浇入料内，每料六十斤，约用漂白粉一斤左右，逾一二时，再加适量稀硫酸，隔五六时之后料始洁白，捞至摇桶中，用人力往复摇搅，使其匀散，复纳布袋内在清水中洗去药汁，即可制纸。”（《浙江之纸业》，1930年版，第11章第254～255页）“亚硫酸法，其所以为今日最通行之法者，以其为木化质纤维素之最良分解法。”（《浙江之纸业》，1930年版，第20章第705页）。本次制浆叶汉山用的是浓硫酸，叶汉山后来告诉调查人，大约用了0.25 kg的量。

在抄纸过程中要用到纸药。据叶汉山介绍，纸药在上里山被称为“油水”，1975年以前，上半年清明节气以后用梧桐梗榨汁，下半年用野生的杨桃梗（猕猴桃梗）榨汁。杨桃梗是从湖源山里小章村收来的，上里山几个大队每年要收50 000～100 000 kg，割成2 m多长，按20～25 kg的重量一把把捆扎好，底部浸没在水里20 cm，上面用稻草盖好，可以保鲜，一直用到清明，接着用梧桐梗榨汁。叶汉山表示，以前湖源山里的山光秃秃的，杨桃梗比较容易找到，现在植被太好，反而不容易找到了。

梧桐梗要到桐庐县百江镇去收。把梧桐梗浸放在水里，上面抛洒点硫酸铜。硫酸铜为绿色固体，有发凉的效果，起着防止梧桐梗腐烂的作用。

以前用长网造纸机造纸时，光是油水，三合皮纸厂就要用10～20 m³水池储存，需要30多人专门做油水。1975～1976年，五四造纸厂成功使用聚丙烯酰胺，才解决了油水的问题。聚丙烯酰胺，原产日本，干粉状，呈白色，类似细盐。使用时在水中浸一夜，用机器打散、打匀，浓度为15%～20%，再过滤，不滤的话，会结成小块状。这次恢复桃花纸生产采用的纸药就是聚丙烯酰胺。

在浆液中融入纸药，纸药使浆液在帘面上的流动速度加快而滤水速度减缓，使优质纤维沉积成纸膜，纤维束及杂质却不易停留，抄纸师傅技巧性地一拨，将其连同多余的浆料一起拨回纸槽。这正是利用了纸药的滑性，使纸面平整、光滑和细腻。

⊙1

⊙2

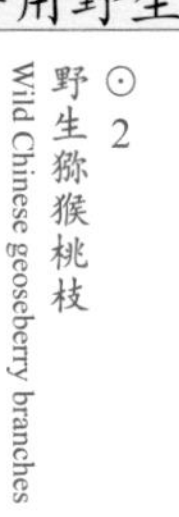

⊙1 做『油水』用的梧桐树枝 Chinese parasol tree branches used for papermaking mucilage

⊙2 野生猕猴桃枝 Wild Chinese gooseberry branches

（二）五四村新制桃花纸的工艺流程

据叶汉山的介绍及调查组现场观察，五四村新制桃花纸的制作主要包括以下流程：

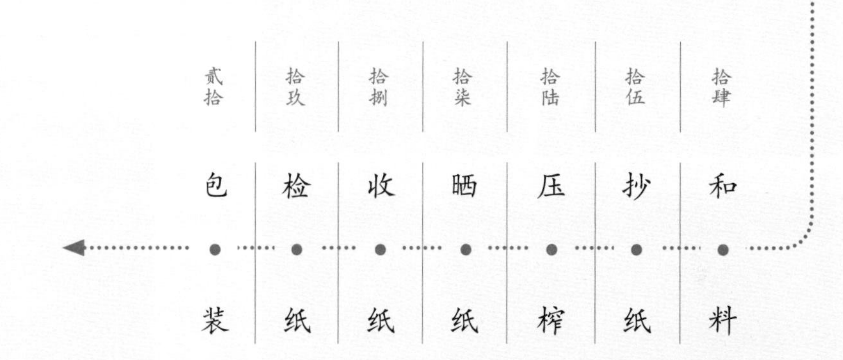

本次恢复制作桃花纸是从外地购买的干皮，因此省略了剥皮、晒干两道工序，直接从拣皮开始。

叁 拣皮

3

皮梢上的筋，即木质素密集部分不能造纸，要将其去掉，只保留韧皮。拣皮传统上是女人的活，一个人一天可拣50～100 kg。

肆 敲皮

4

一人将一捆重4～5 kg的皮料置于石板上，两人在其左右各拿一木榔头敲皮。边敲边拉动皮，约10分钟可敲好一捆皮。敲皮作用有二：一是把皮敲软，以便于吸收化工原料；二是通过敲皮去掉皮壳，使皮料纯净。现场调查时是在铁板上敲打皮料。

⊙3

⊙3 敲皮 Beating the bark

伍
切　皮
5　⊙4⊙5

将敲好的皮料置于切皮凳上，用刀将皮料切成长约15 mm的段。切得太长，纤维不均匀；太短，则淘洗时易洗掉，浪费原料。

⊙4

⊙5

陆
浸　皮
6

把切好的皮料放在清水里浸泡4小时以上，然后把水倒掉，沥干。

柒
蒸　煮
7　⊙6

先加水进锅，放入烧碱液后搅匀，再放皮料入锅，盖上锅盖，煮8小时以上。比例是50 kg干皮料，60 kg水，35 kg含量为30%的液碱。煮时要注意火候，先用中火烧滚（沸腾），再改用小火，使水一直滚（即沸腾）；如用大火，蒸汽太大，盖会被掀起来。

⊙6

捌
洗　黑　液
8　⊙7

将煮好的皮料放到洗皮笪（即竹篓筐）里，在溪水中清洗，如洗皮笪孔较大，需要在其内部铺一层纱布来防止漏料。将一定量的皮料置于洗皮笪内，用水洗掉70%的黑液。在交流中叶汉山特别强调，不能洗得太干净，要留有一部分，以在水棚碓碾压中保持润滑性，使纤维不受损伤。

现场调查时是在山边用水泵冲洗，一次可洗15 kg干皮，大约十几分钟即可。

叶汉山回忆，以前五四造纸厂一天用350～400 kg干皮，需要两个人在溪水里洗一整天。

⊙7

⊙4 叶汉山在切皮料
Ye Hanshan cutting the back materials

⊙5 切成15 mm长度的树皮小块
Bark cutting into small pieces 15 mm in length

⊙6 老技工吴金坤在野外煮料
Skilled mechanic Wu Jinkun boiling the materials outside the mill

⊙7 叶汉山冲洗黑液
Ye Hanshan cleaning the bark materials

玖 春料

9 ⊙8⊙9

一人将洗好的30～40克重量的小块皮料扔进水棚碓，另一人站在桥碓上踩碓。踩碓时皮料向四周炸开，溅到木圈壁上，积累到一定数量再回收。一般要舂5次，是否舂好的判断方法：拿一小团料放在水里，搅开，看看有没有筋（即纤维束），无筋即可。舂好的皮料叫黑料。

⊙8

拾 洗黑液

10 ⊙10

将舂好的细料放到料袋里，用料耙挑着在流动的河水里搅洗，将皮壳和第一次洗黑液时剩下的30%黑液都洗掉，直到完全洗清为止。

⊙9

⊙10

拾壹 漂料

11 ⊙11

将皮料放进水缸，50 kg水放3.5 kg干皮，加次氯酸钠溶液漂一个晚上，使皮料变白，50 kg干皮用14 kg 10%的次氯酸钠溶液。第二天再加浓硫酸，50 kg干皮用0.6 kg 98%浓度的浓硫酸，淘匀，漂一晚即可。叶汉山说，放硫酸这个环节叫“斗缸”，很危险，要十分谨慎。叶汉山认为，浓硫酸的作用，一是去污，二是去掉氯的残留物。漂得好的料，易保存，放几年再做纸都没关系。

叶汉山表示，集体造纸厂的时候是用30%含量的漂白粉和浓硫酸，现在根据漂白粉的有效含量转换成次氯酸钠含量。漂料容器以前有用水缸也有用池子的。

⊙11

⊙8 水棚碓舂料 Beating the materials with a hydraulic wooden pestle

⊙9 舂碎的细料 Fine beaten material

⊙10 老技工吴庆荣在河里洗黑液 Skilled mechanic Wu Qingrong cleaning the bark materials in the River

⊙11 水缸漂料 Bleaching the materials

拾贰 洗白料

12

将漂好的皮料再次纳入布袋中，内置圆木耙，在清水中漂洗，利用圆木耙捣搅，使皮料纤维细柔。洗好的皮料称白料。洗后取出少量料，加碘试剂，洗清了的呈绿色，洗得不清的呈红色。

拾肆 和料

14

将纤维疏解后的浆料倒入料缸内，注入水，加入适量聚丙烯酰胺，搅拌，使之混合均匀。按叶汉山的说法，上里山当地以前使用的纸药，春夏季用青桐梗汁，秋冬季用杨桃藤汁。

拾叁 摇白

13 ⊙12

把白料放到摇白桶中，加水稀释后，4个人站在两边摇十几分钟，纤维分散好即可。调查时用的摇白桶一次分散的料可抄纸约1 500张，叶汉山说也有小些的摇白桶。

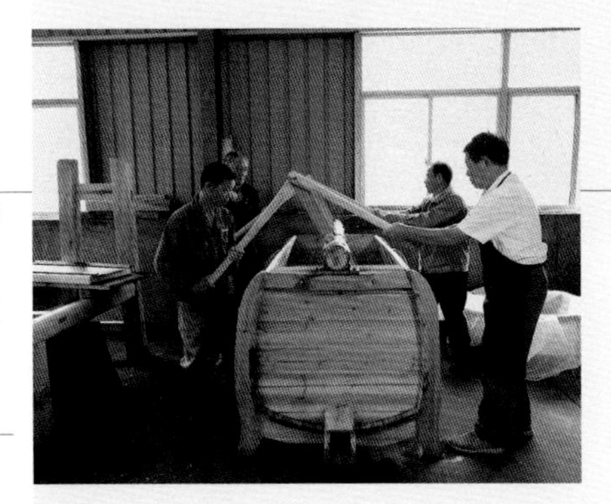

⊙12

拾伍 抄纸

15 ⊙13～⊙17

抄纸时先打前浪、抄后浪、再打前浪，共荡约20次，凭经验看纸面均匀与否而定；然后将纸帘翻盖于湿纸堆上（第一张扣放在铺于底面的旧纸帘上），用手将纸帘前面的额头（即木片）往上翻过来压一下，再将整个纸帘揭开。“好”的一槽料可抄15张885 mm×450 mm的纸，正劳力（青壮年劳力）一天可抄约500张。好的抄纸师傅成品率可达90%～95%，但叶汉山说能达到这个水准的不多。

叶汉山介绍说：抄纸好不好，主要看前浪。前浪大，把粗筋打下来，后浪使纸更均匀。抄15张后，就要放一些纸浆，再加一些纸药，然后用双手持方形木耙使劲在槽桶内搅拌，再用竹竿打槽，使纤维不结絮。打字蜡纸原纸抄制比桃花纸抄制要求技术更高，前三个浪头劲要大一点，抄一张就要搁置一下，看帘面是否有粗筋（纤维束），要把它拣掉。

上里地区桃花纸荡帘约20次的工艺在中国一般皮纸制作中是很少见到的，其技艺的独特性和难度都值得深入探究。

⊙13

⊙14

⊙12 摇白 Shaking the materials in a bucket

⊙13 打槽 Stirring the materials

⊙14 打前浪 Scooping the papermaking screen from the front side

拾陆
压 榨

16 ⊙18

抄完纸后，在湿纸堆上放一旧纸帘，其上放桶板，竖向放3根小木栅、几根大木栅、榨杆，用篾索将榨杆套好，篾索在滚筒钉处打结，将棒棍插在滚筒眼里，手压棒棍至压半干为止，一般需7～8分钟。判断是否压好的依据：一是没有水滴下；二是用手按纸堆按不进去，松手后，手按处呈白色，即可认为达到标准了。

据叶汉山介绍，制作铁笔蜡纸原纸时，压榨这一环节很特别，它不脱水，会像馒头一样胖起来，要慢慢压榨，榨完后把整个纸堆放在太阳下晒，晒干以后在水里浸涨、浸透，再压榨，再晒纸。

⊙15

⊙16

⊙17

⊙18

⊙15 抄后浪 Scooping the papermaking screen from the back side

⊙16 揭纸帘 Uncovering the papermaking screen

⊙17 把湿纸覆盖到纸桩上 Piling the wet paper on a board

⊙18 压榨 Pressing the paper

拾柒
晒　　纸

17 ⊙19⊙20

由晒纸工箝开纸坯右上角，揭开右上角的纸页后，慢慢地将纸页沿湿纸块额头从右到左慢慢揭开，然后从上到下，双手把一页纸竖直揭下来，用刷子将其刷到焙垅上。焙垅温度约60 ℃，纸张刷上去5～6分钟即干。纸张刷到焙垅上时必须要刷平整，不能有皱纹。高档白绵纸用羊毛排笔刷，一般的白皮纸用松毛刷。

叶汉山回忆：以前焙垅长10多m，高约1.8 m，上下横晒2张纸，左右10几张。1面墙2个人晒纸，1个焙垅4个人一起晒，1人晒完半面焙垅即可收纸。1人晒纸，2人抄纸。

⊙19

⊙20

拾捌
收　　纸

18 ⊙21

将烘纸墙上晒干的纸逐张收下来。

⊙21

⊙22

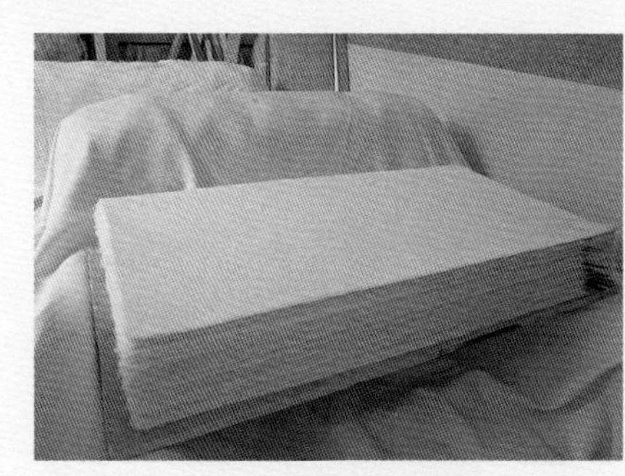
⊙23

拾玖
检　　纸

19 ⊙22

桃花纸验纸，主要是看整张纸是否有破损，有破损即要剔除掉。而打字蜡纸的验纸就复杂多了，需要将纸按质量分4种分别堆放：正号、双帘、单帘、副号。全好的为正号的三帘纸，如有部分损坏，则用木尺裁成双帘、单帘，裁不了的归为副号。三帘纸不是指一帘三张，而是指长度单位。三帘纸的长度是单帘的3倍，双帘纸的长度是单帘的2倍，而单帘的长度也就是打字蜡纸的长度。裁下的一帘、二帘可以拼起来成为一张全纸。副号绵纸用作电池绵纸。

贰拾
包　　装

20 ⊙23

100张为1小刀，1 000张即10小刀为1大刀，8 000或10 000张为1件。每1小刀用一纸条隔开，1大刀用纸包起来。正号绵纸按件用木箱装箱后销售；桃花纸、防风纸、雨伞纸等则用两块木夹板压起来外销。

⊙19 从湿纸块上揭下湿纸 Peeling wet paper down from wet paper stack

⊙20 将湿纸贴到烘纸墙上 Pasting wet paper on the drying wall

⊙21 收纸 Peeling the paper down

⊙22 叶汉山示范检验桃花纸 Ye Hanshan demonstrating how to check Taohua paper

⊙23 新制桃花纸成品 Newly made Taohua paper

（三）五四村新制桃花纸的生产工具

2018年在富阳区非遗中心支持下恢复桃花纸生产采用的生产工具，大部分是新添置的，为叶汉山请村里人按照旧日造纸工具制作，如蒸桑皮的铁桶、清洗皮料的竹制皮筤、舂料用的水棚碓、洗黑料的料袋与料耙、分散纤维的摇白桶、抄纸用的料槽、帘床、纸帘、榨纸简用的榨床，都是新请人制作的。晒纸借用了大源镇另一位造纸人庄道远在新桐乡纸坊的焙垅。漂料的料缸、放纸药的纸药缸，新买了几只。手工拷皮的榔头、切皮料用的切皮凳与铡刀是叶汉山家里自备的。各工具具体介绍如下：

壹 蒸锅 1

生产队时期用铁桶蒸桑皮或构皮，考虑到铁桶容易生锈，2018年这次用的是不锈钢板制成的。调查组实测尺寸为：直径0.9 m，高0.8 m，下面支架离地0.4 m。

贰 皮筤 2

清洗皮料用的圆形竹篓筐，用毛竹篾片编成。调查组实测尺寸为：直径1.1 m，高0.3 m。

⊙25

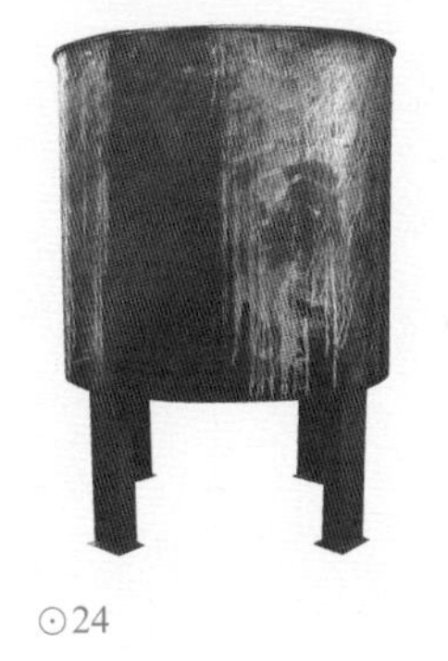

⊙24

叁 水棚碓 3

碾料用的碓，桥碓由桥碓棚、碓跳、木碓头和石碓座4个部分组成。桥碓棚为2根圆木柱中间架1竹制横杠。碓跳是指接碓头的长木条。碓头为长方体木棰。石碓座的底座是石块，调查时用的是直径为0.6 m的圆形大理石，大理石周围放一直径为0.58 m、高0.2～0.3 m的木圈。水棚碓木碓头撞击面和大理石平面，合拢时要求绝对平整。纸农站在桥碓棚上踩踏时，利用了碓的结构和动能原理,但不是舂捣，而是碓头下落时，两个平面在力的作用下产生强力碰撞，使投掷在大理石上的皮料炸开，溅向四周围桶，达到分散纤维的效果。

⊙26

⊙24 蒸煮原料的不锈钢锅 Stainless steel cooker for steaming and boiling the raw materials

⊙25 皮筤 Bamboo basket

⊙26 水棚碓（中间的木制工具） Hydraulic wooden pestle (the middle one)

肆 料袋 4

洗黑料时用的袋子，传统是用苎麻布做的，孔较大，一般为60～70目；2018年恢复制作桃花纸时用的是尼龙布缝制的。料袋的缝制是一个绝活，一块尼龙布要多次裁剪拼贴，底部成方形，中间为圆形，上部要升出长长的两片，以便于缚结。

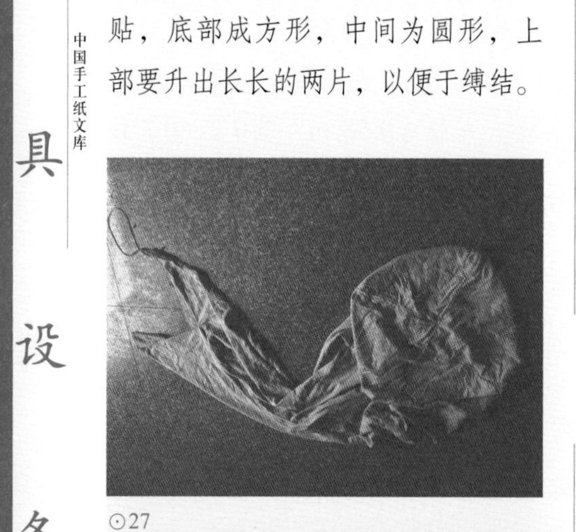

⊙27

⊙28

伍 料耙 5

洗料时用的耙子。料耙头为半圆球形，调查组现场实测尺寸为：直径0.18 m，高0.3 m；料耙柄是竹制的，长2.6 m。

⊙29

陆 料缸 6

一种上了釉的陶瓷水缸，富阳本地日杂店即可买到，产地为江苏宜兴和浙江长兴等地。最大的料缸称十石缸（意思是可盛放10担水的大水缸），最小的为一石缸，根据皮料的多少决定用缸大小。通常一处造纸作坊需要很多缸，分别用于配料、踏料、存放纸药等。

⊙30

⊙27 旧的草纸料袋（重新制作皮纸料袋仿制样本）
Old straw paper material bags (re-making imitation samples of bast paper bags)

⊙28 退休教师夏荣芳正在缝制料袋
Retired teacher Xia Rongfang sewing the material bags

⊙29 料耙（右起第一个工具）
Stirring rake (the first tool from the right)

⊙30 料缸
Material vat for storing and making the materials

柒
摇白桶
7

分散白料纤维用的木桶。生活用桶通常是圆形的，摇白桶却是长方形的。调查组现场实测尺寸为：长度上、下不等，上面长1.75 m，底面长1.65 m；宽度上、中、下也不等，上宽0.7 m，中宽0.9 m，下宽0.87 m；桶高0.70 m；底架高0.3 m。摇白桶有一出料口，匀散后的白料由此流出。摇白桶内等距离设置11条长0.9 m、宽0.05 m的竹片。摇白桶上方左、右两侧各设置两根上粗下细的长木条，用来牵拉。

如何更好地分解纤维，这是造纸人一直在思考的问题。调查中据富阳手工纸研究专家李少军分析：纸农们从用筷子搅动分散鸡蛋中得到启发，用棍子搅拌疏解纤维。后来纸农们又发现，在纸浆中撑开五指，手掌左右用力摇摆，同样能将块状原料很快疏解。根据这一原理，纸农发明了人工驱动的摇白桶，解决了生产中的问题。后来工程师们根据摇白桶的原理，发明了"摇摆式纤维疏解机"，解决了机制造纸中长纤维原料机械疏解的难题。

⊙31

⊙32

玖
纸　帘
9

以24号（指帘丝的直径）帘丝为纬、0.6 cm腈纶丝为经在木机上编织而成的帘子。帘丝以苦竹为材料，经多道工序制成竹丝。抄制桃花纸的纸帘比竹纸纸帘更加细密，1 cm宽有11根帘丝。实测纸帘尺寸为：长0.9 m，宽0.5 m。纸帘由富阳光明制帘厂汪美英制作。皮纸帘额头有3根帘丝，制作难度很高，是由上里山的鲍水珍穿制的。

捌
木纸槽
8

抄纸用的槽。抄制桃花纸的纸槽是杉木制成的，调查时实测尺寸为：长1.06 m，宽0.95 m，高0.33 m，底架高0.6 m。槽内设一隔板，分出两个区域，隔板两侧底端各有一小孔相通。抄纸区域宽0.57 m，另一侧区域宽0.3 m。纸槽上方设置一木架，木架的横档下缚一竹片，竹片两侧垂下两根绳索，与帘床相接，用来吊帘。

⊙33

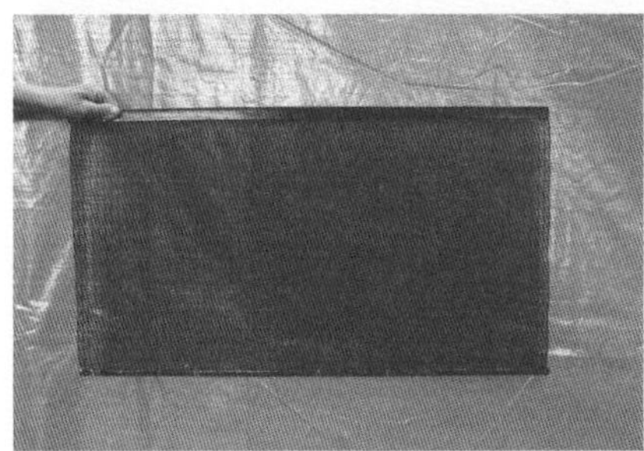
⊙34

⊙35

⊙31 吴金坤在制作摇白桶 Wu Jinkun making the shaking bucket

⊙32 摇白桶 Shaking bucket

⊙33 新制的纸槽与帘床 Newly made papermaking trough and the screen frame

⊙34 汪美英新制作的皮纸纸帘 Newly made bark papermaking screen made by Wang Meiying

⊙35 细密的纸帘局部 A part of fine papermaking screen

拾 帘床 10

托放纸帘用的架子，材质为杉木。长1.03 m，宽0.5 m，其中等距离设置15条细铅丝。桃花纸抄制用的是吊帘帘床，纸帘放在帘床上。

⊙36

拾壹 槽耙与槽棒 11

均为抄纸前用来打槽的工具。槽耙主要用于搅动底部浆料，使沉淀于底部的浆料向上浮动，耙头为长方形，现场实测尺寸为：长0.23 m，宽0.16 m，厚0.5 m；槽耙柄长1.55 m。槽棒主要用于划动槽内浆料，促使浆液均匀。槽棒由苦竹制成，长1 m，直径3 cm。

拾贰 压榨床 12

用于榨出湿纸帖水分的木榨，是手工造纸的重要设备。压榨时，用杠杆一端插在架子上"千斤横梁"的底部，这是固定在架子上的支点，杠杆离开支点4～50 cm的底面是湿纸块，将篾索的龙头套住杠杆的撬动端，篾索尾箍固定在木质转动轴上，篾索的中间部分在杠杆和转轴上套两个圈，由于绞动时杠杆和转轴上的篾索是在表面上滑动的，加上两圈篾索受力均匀，因此在压榨时会越绞越紧，不会自动松开，一旦尾箍松开转轴篾索会马上散开。上面提到的木质皮纸生产工具，如压榨床、帘床、木纸槽、摇白桶、水棚碓以及切皮凳、敲皮榔头、料耙、槽耙等，都是由吴金坤和叶军贤一起制作的。吴金坤，今年75岁，鹿山街道新祥村人，24岁开始做木工。叶军贤，今年70岁，鹿山街道五四村人。以前上里山的木质皮纸生产工具大部分都是他们俩制作的。

⊙37

⊙36 搁置纸帘的帘床 Frame for supporting the papermaking screen

⊙37 叶汉山手持槽棒打槽 Ye Hanshan stirring the materials with a stirring stick

⊙38

⊙39

⊙40

拾叁 焙　垅 13

晒桃花纸的焙垅（焙墙）用瓦窑里定制的专用青砖砌成，砖块尺寸为：长0.3 m，宽0.2 m，厚0.07 m。焙垅制作时底部的宽度大约是0.36 m，到1 m高时宽度缩小到0.3 m，和砖块的长度一样，然后按这个斜度向上打造砌砖。焙垅整体长5~8 m，高1.8 m。两面晒纸，墙表面用料筋或麻筋拌石灰（黏结状）粉抹，并用泥夹长时间研光，至表面硬化后才可试烧。每年需用桐油漆刷保养，用桐油漆刷的目的是减少纤维末梢在焙墙上的黏结，减少纸张焙疖的产生。

拾肆 松毛刷 14

晒纸时将湿纸刷上焙壁的工具，刷柄为木制，刷毛为松针。调查时桃花纸生产中使用的松毛刷借用自庄道远在新桐的晒纸坊，购自富阳区湖源乡，2018年约60元1把，实测尺寸为：长35 cm，宽12 cm。

⊙41

⊙38 压榨床（摇白桶与纸槽中间的设施）
Pressing device between the shaking bucket and papermaking trough

⊙39 叶汉山在压榨床上加放小木条
Ye Hanshan adding small wooden strips to the pressing device

⊙40 湖源乡颜小平正在制作松毛刷
Yan Xiaoping from Huyuan Town making brushes made of pine needles

⊙41 竹萁、敲皮木榔头、料耙、铁锹
Bamboo basket, wooden hammer for beating the bark, material rake and spade

五
五四村新制桃花纸的市场经营状况

5
Marketing Status of Newly Made Taohua Paper in Wusi Village

由于属于传统历史名纸的抢救性恢复，新制作出的第一批桃花纸主要都提供给了出资方——杭州市富阳区非遗保护中心，因此没有当前销售和市场信息。据叶汉山介绍，20世纪50年代至70年代末期的计划经济时期，五四村皮纸主要由供销社收购、销售，滤油纸主要由杭州五交化公司统一销售。1978年改革开放以后，防风纸主要运往延安、榆林地区糊窗户，后来销往西南地区包茶叶或用作裹尸纸。正号绵纸主要销往辽宁锦州蜡纸厂、北京腊纸厂、新华造纸厂等地。滤油纸主要销往全国各地金矿、电站、油化厂、食品厂、墨水厂、供电局等。蚕放纸主要销往杭嘉湖平原地区。桃花纸主要用于刺绣描图，也用于糊雨伞、糊窗户等，销往全国各地。

调查时富阳区境内手工制的桃花纸早已经被机器皮纸所取代，机制滤油纸也都已采用木浆生产，不再用植物韧皮纤维了。

六
上里山造纸文化与五四村的造纸故事

6
Papermaking Culture of Shangli District and Papermaking Stories of Wusi Village

（一）上里山造纸文化

1. 上里山皮纸的历史荣誉

上里山是富阳皮纸的重要产区。民国十九年（1930年），上里产的本色桑皮纸在西湖博览会上曾获特等奖。上里产的白绵纸，又名白皮纸，因纸质柔软洁白得名，也获一等奖。以上里白绵纸为原料加工的富阳油纸，则在1929年的西湖博览会上获得特等奖。

2. 上里山的抄纸名师

上里山的三合村造皮纸的历史悠久，从生产工具到制作工艺，形成了一套世代相传的精湛工艺。近代上里山有一大批皮纸抄纸师傅在杭州新华造纸厂、杭州蜡纸厂等地成为技术骨干。据叶汉山介绍，民国年间三合村有几个人在杭州松木场一带办皮纸厂。1956年国家提出“公私合营”，杭州的一些私营企业包括富阳人、奉化人开办的造纸厂都并到了新华造纸厂，当地公认的造纸能手姚阿元、吴明村就是这样并过去的。当时，新华造纸厂的工人，一部分是宁波奉化人，一部分是杭州本地人，其中40%～50%是富阳人，主要就是三合村人。

新华造纸厂当时在沈塘桥抄制正号绵纸。1949年后，上里山有一批做皮纸的人到龙游、遂昌、衢州做师傅。五四村山坑坞里的夏申炳到贵阳都匀蜡纸厂当师傅，后来整户人家搬到那里，就地取材，以当地构树皮为原料生产铁笔蜡纸原纸。杭州新华造纸厂、贵阳都匀蜡纸厂均为20世纪中后期中国著名的桑构皮纸生产大厂，可见上里山皮纸师傅群体对浙江皮纸产区和外省的技术输出，他们对浙江和外省韧皮纸业的发展起过重要的推动作用。

为欢度国庆，定于九月二十九日下午
二时正在大会议室召开先进人物座谈会，
敬请准时参加。
此致
吴明村 同志
新华造纸厂工会
一九八五年九月二十七日

⊙1

3.“爱国皮纸”的由来

民国十九年（1930年）出版的《浙江之纸业》一书记载：“此外尚有草皮纸，又名爱国皮纸，出于富阳、新登。其原料用桑皮六成，稻草四成，混合造纸。其制造和整理均在一起，即将原料处理完后，移往晒场，用长方式篾帘，帘用篾片编成，长约五口寸，阔约三尺九寸，将料浆和药汁少许调匀之后，用浇料杓将料浇于篾帘之上。藉日光晒干之后，逐张收回，堆叠成件，用篾捆缚，加盖牌号，即成市上之商品。”

这里明确地记载着“草皮纸，又名爱国皮纸，出于富阳、新登”。20世纪20年代以来，列强凭借政治特权和经济优势，对中国进行商品倾销，使中国民族工商业的发展面临严重危机。刚刚成立不久的南京国民政府，在各界舆论的强烈呼吁下，顺应民意，于1928年10月开展了提倡国货运动，全国各地国货界一致响应，这应该就是富阳、新登的草皮纸被称为“爱国皮纸”的由来了。

上里山三合村的吴明富收藏着一方木雕方印，方印上方刻着“三星商标”，左侧为“义泰厂督造爱国皮纸”，右侧为“开设浙江富阳大树镇”，下方刻着“请用国货”等字样。调查人咨询叶汉山，他说，这方印是吴明富的邻居吴耀桂（小名吴阿生，已故）家的，吴耀金生前办过皮纸厂，三合村以前就叫大树下。这方印究竟是何时启用的，目前还无法具体推断。据叶汉山介绍，1950年抗美援朝时期，上里地区大批生产草皮纸，供部队使用，那时也叫“爱国皮纸”。可见“爱国皮纸”的说法一直延续着。

⊙2

⊙1 吴明村保存的新华造纸厂请帖
Invitation card of Xinhua Papermaking Factory stored by Wu Mingcun

⊙2 吴明富收藏的『爱国皮纸』方印
Seal of "Patriot Bast Paper" collected by Wu Mingfu

（二）五四村造纸故事

1. 五四村引领了手工造纸到机器制纸的变迁

据叶汉山介绍，1973年新华造纸厂将长网纸机转让给五四造纸厂。长网纸机的生产速度是每分钟33 m，24小时不停机，可以生产很多纸。随后，完成机械化造纸改造的五四大队开始技术外援，支持兄弟大队。1975年前后，富阳县手工生产毛纸转为机械生产毛纸，来自五四造纸厂的叶汉山作为主要机修人员，他的职责是确保每个厂里的设备运转正常。五四大队陆续援建了环山乡西山村，鹿山街道长春村、祯祥村、坞口村、上叶村，新登南四村，新登新浦乡新杨村等多家乡村造纸厂。从一块平地开始，泥工、木工、机修工以及各类工程技术人员无私援助，直至产品出厂为止。1973年至1978年这一段时间，富阳县全县全部改为机制毛纸。

2. 皮纸生产中的伤人事件

叶汉山讲述道：蒸皮是一件很不容易的事，火过旺，水蒸气力度大，料就会向上喷，往外面涌，喷到锅外面，料就没有了。五四村后来改用一立方米立式压力锅蒸煮，上面盖上盖子，盖子上面装安全阀，没有出现过安全事故。上里山一带用锅炉蒸皮出过两次事故，坞口村弹死过一个人，祯祥村出过一次事故。有一次，1 m直径的压力锅铁盖高飞，冲破屋顶，飞向天空，远远望去像一个小笠帽。当时正是小麦收割季节，叶汉山和同村几个村民正准备去摆纸机、看地方、搞设计，家里将麦子碾成了粉，做好了麦塌饼，吃完饼才过去，才算幸运地躲过这一劫。

踩水棚碓劳动强度很大，也伤过不少人。下面抛料，上面踩料，一个不小心手会压伤。记得有一个人踩水棚碓被砸到，当时就晕过去了，准备抬到东梓关骨伤医院，快到医院时，抬伤者的一个人一个踢脚绊，脚伤掉了。那家纸厂就这样停掉了。

⊙1

⊙2

⊙1 曾任五四造纸厂厂长的何关元（左一）和夏明汉（右一）
He Guanyuan (first one from the left) and Xia Minghan (first one from the right) - former heads of Wusi Papermaking Factory

⊙2 叶汉山（站立者）讲述皮纸生产中的伤人事故
Ye Hanshan (standing one) relating the accidents happened in bast paper production procedure

七 五四村桃花纸的传承发展思考

7 Inheritance and Development of Taohua Paper in Wusi Village

2018年上里山桃花纸的复原，采用的是1963年后的工艺，并没有完全按照传统的桃花纸生产方法，制浆工艺中采用了现代化工材料与技术，但抄制方式是传统的抄制法。在中断数十年后，复原工艺“半传统”对于抢救性记录和进一步的传承来说多少还是有些缺憾。例如复原工艺中纸药的制作与使用环节缺了，这对桃花纸成纸的性能指标多少会有些影响。

谈到是否能恢复“古法”手工造纸时，叶汉山本人并不乐观，他认为，五四村手工皮纸（包括桃花纸）不可恢复，竹纸同样不可恢复。首先是完全没有劳动力。从上里山目前情况看，当年特种纸厂有300左右的技术工人，而现在上里各村的年轻人分布在全国各地工作，除了春节外，村里长年只剩一些老人，几乎看不到年轻人。其次是手工技艺无法谋生。长网纸机已经很流行，借助相关机器设备，原材料和辅料办理方便、简单，包括木浆、湿强剂、油水等的制作与使用。手工制作成本高，根本无法与机制纸抗衡。最后是环保的原因。桑皮、构皮手工制浆繁琐，成本高，污染也不小，在环保上无法妥善解决。所以要恢复难度很大。

仔细交流后，叶汉山提出了一个颇有创意的思路：“手工皮纸我来提供技术，叫大源朱中华生产。或者叫他儿子跟着我学，也许从非遗保护的角度还能培养出一个好苗子来。”2018年调查时，叶汉山已经78岁，他最年轻的徒弟夏关水也60多岁了，桃花纸生产作坊只能偶尔表演一下，仅依靠这几位掌握技艺的老人，生产性保护确实不具可行性。大源镇大同村的朱中华不到50岁，元书纸制作技艺娴熟，长年在手工造纸一线；儿子朱起扬20多岁，大学毕业后已经在家庭纸坊里学造纸近3年，是富阳区非常少见的在手工造纸生产一线“干活”的年轻人。如果两者开展协作，确实有可能恢复桃花纸手工造纸技艺，实现生产性传承。

⊙3

⊙4

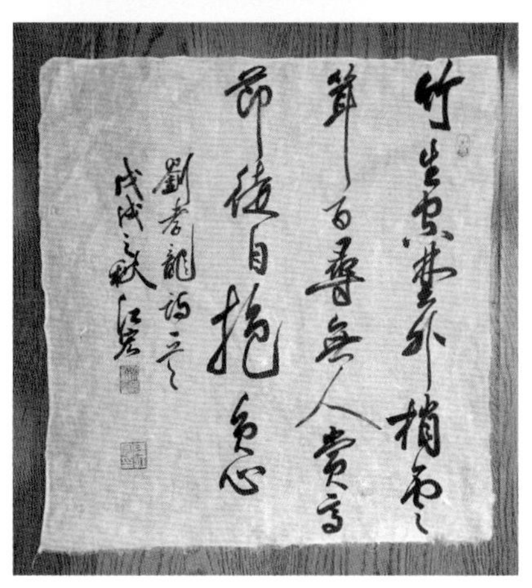

⊙5

⊙3 调研桃花纸制作工艺 Researching production procedure of Taohua paper

⊙4 展示叶汉山新制桃花纸用墨效果 Demonstrating the writing performance of newly made Taohua paper by Ye Hanshan

⊙5 新制桃花纸上的书写效果 Writing effect on newly made Taohua paper

231

五四村桃花纸作坊

桃花纸

Taohua Paper
of Taohua Paper Mill in Wusi Village

新制构皮桃花纸透光摄影图

A photo of newly made paper mulberry bark Taohua paper seen through the light

第二节

大山村桑皮纸恢复点

浙江省
Zhejiang Province

杭州市
Hangzhou City

富阳区
Fuyang District

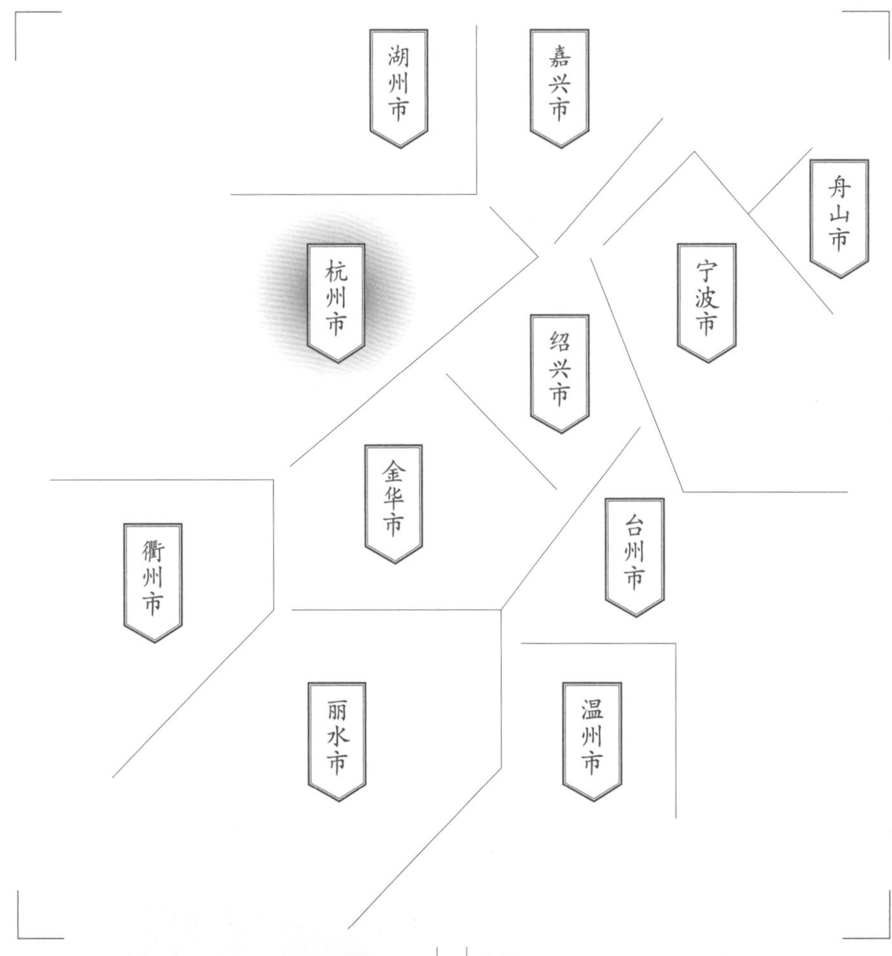

调查对象

富阳区新登镇大山村桑皮纸恢复点皮纸

Section 2
Mulberry Paper Recovery Site in Dashan Village

Subject

Bast Paper in Dashan Village of Xindeng Town in Fuyang District

一 大山村桑皮纸生产的基础信息与生产环境

1 Basic Information and Production Environment of Mulberry Paper in Dashan Village

大山行政村位于富阳区新登镇东边的大山坞深处，地理坐标为：东经119° 48′ 13″，北纬29° 56′ 28″。大山坞是一条自东北向西南，长达10余里的山坞。在它的东北方，有高兀的冬瓜岭俯视，周围峰峦连绵，山高谷深，源自东北群山间的宝山溪奔腾西去，活水常流，为手工造纸提供了充沛而优质的水资源。

宝山溪两侧的宝山村、大山村、胜总村3个村落首尾相连，这里的村民较长时期以生产桑皮纸为主要营生手段，尤其是大山村，曾经家家户户以造桑皮纸为业。

富阳皮纸生产的核心聚集区域以2016年6月、2018年3月、2018年12月调查时经过行政建制调整的鹿山街道上里片、新登镇大山村、渌渚镇新浦片*为主。从大山村往东有古道翻越村东的柏枝岭、冬瓜岭，越岭下行，即可到达现鹿山街道上里山。从鹿山街道的上里山翻过遥高坞、高坪岭就到达渌渚镇的新浦，高坪岭是上里与新浦两地的分界岭。在成纸方式上，新登的大山村、渌渚的新浦片以纸框浇注法为主，上里山则纸框浇注法和纸槽抄制法并重。

大山村位于新登镇东部，距新登镇行政中心约8 km，属半山区。调查时的2016年6月，村域范围东至鹿山街道，南至渌渚镇，西至双联村、上旺村，北至昌东镇，区域总面积9.78 km²。全村分布20个自然村，下属32个村民小组，总农户665户，总人口2 298人。

大山村历史上并非富阳县辖区，旧属新登县，随新登县于1950年并入富阳县成为新登镇后才纳入富阳造纸体系。新登县的桑皮纸生产业态历史悠久，并且形成了延伸产业链。如原新登县境内的官塘、湘溪等地借助桑皮纸加工的油纸，发展起制造油纸伞、折扇等产业。较为可惜的是，由于市场变化、消费萎缩的影响，大山村桑皮纸生产于20世纪80年代中期业态中断。

⊙1

⊙2

⊙3

⊙1 流经村边的大山溪（左下方） Dashan Stream running through the village (bottom left)

⊙2 大山村村口的路标 Road sign at the entrance of Dashan Village

⊙3 俯瞰大山村 Overlooking Dashan Village

* 上里、新浦曾是一个乡，后来上里并入鹿山街道，新浦并入渌渚镇。"片"是当地习惯上对几个行政村集合的一种称呼，比镇小，比行政村大。

路线图
富阳城区
↓
大山村桑皮纸恢复点
Road map from Fuyang District centre
to Mulberry Paper Recovery Site in Dashan Village

大山村桑皮纸恢复点位置示意图

Location map of Mulberry Paper Recovery Site in Dashan Village

考察时间
2016年6月 / 2018年3月 / 2018年12月

Investigation Date
Jun. 2016 / Mar. 2018 / Dec. 2018

地域名称

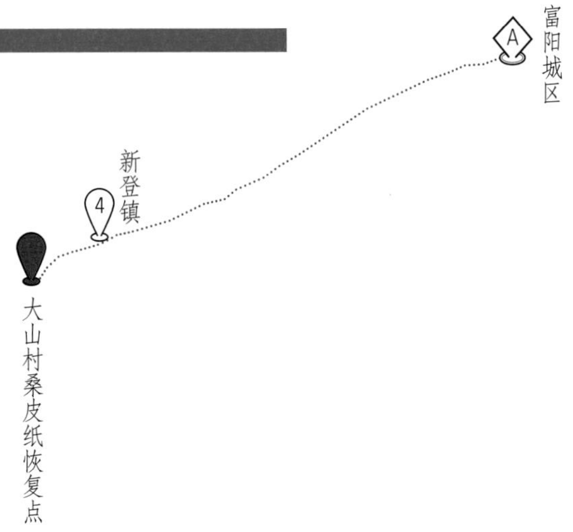

A 富阳城区
4 新登镇
大山村桑皮纸恢复点

A 富阳区
1 湖源乡
2 常安镇
3 洞桥镇
4 新登镇
5 灵桥镇
6 新义乡
7 大源镇

造纸点名称

大山村桑皮纸恢复点
造纸点

位置分布

市府、州府
县城
乡镇
村落
造纸点
历史造纸点
山
国家级自然保护区
S221 省道
G21 国道
昆河线 铁路
G56 高速公路
线路

临安市
富阳区
桐庐县
S206
S302
S305
S31
10 km
5 km
0
N

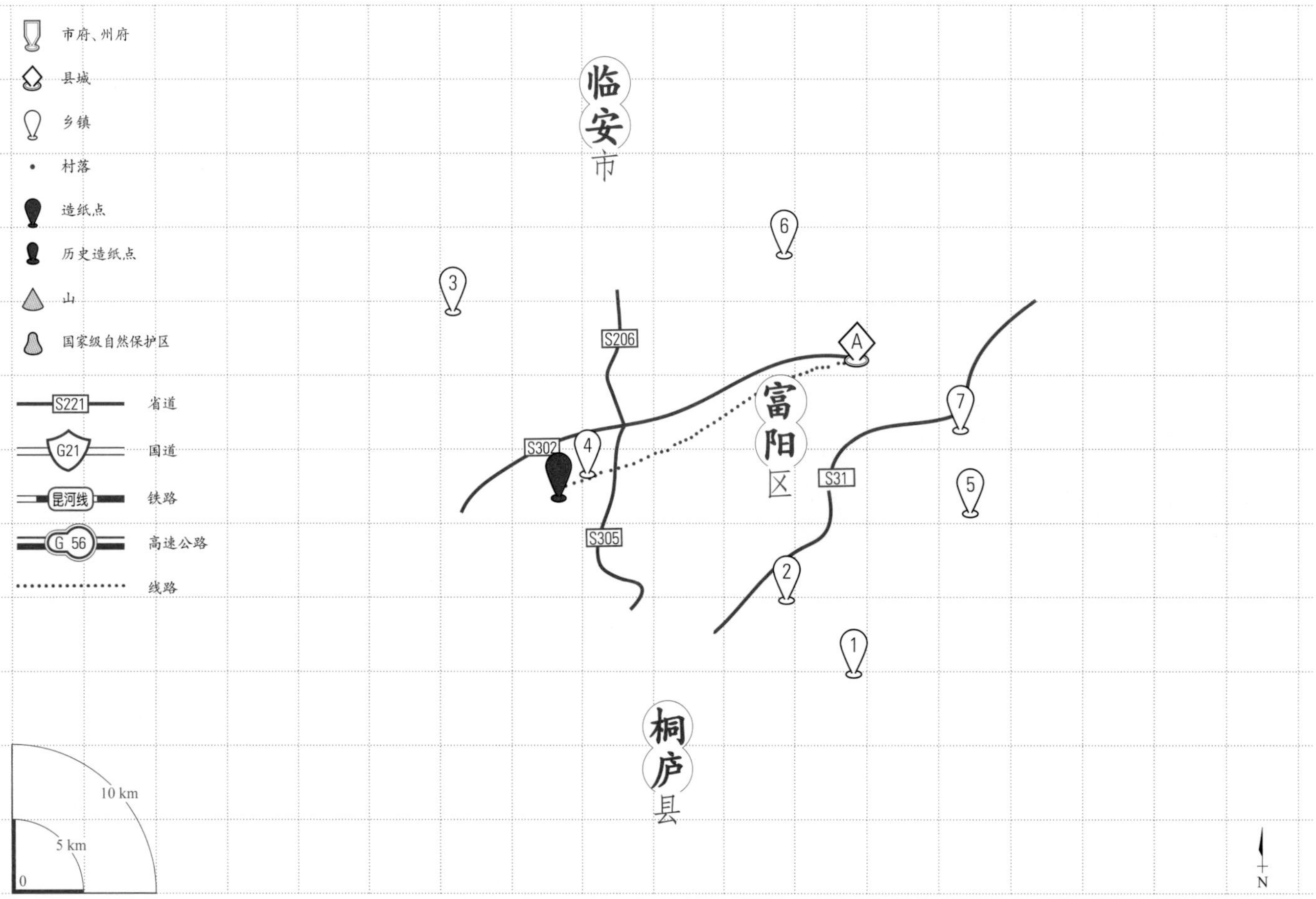

二 原新登县与富阳区新登镇桑皮纸生产的历史与传承

2 History and Inheritance of Mulberry Paper in Xindeng Town of Fuyang District and Former Xindeng County

新登地区的桑皮纸生产历史悠久。据地方文献记载，约650年前的明洪武二年（1369年），在新登县城区就已出现了专门卖油纸雨伞的店铺，由此推测，新登地区在那个年代或许自身已经在生产雨伞纸了。雨伞纸属于皮纸一类。宋应星的《天工开物》“造皮纸”一节记载：“又桑皮造者曰桑穰纸，极其敦厚，东浙所产，三吴收蚕种者必用之。”[2]桑穰纸，江浙收蚕种者必须用到，说明这是一种蚕放（生）纸，又称蚕种纸。蚕放纸一直是原新登县的特产，《浙江之纸业》明确记录其原材料是桑皮、稻草、笋壳，按一定比例配方[3]，用浇纸法制成。富阳、新登是浙西山区与浙东杭嘉湖平原的过渡地带，虽然无法明确断定明代宋应星所指的“东浙所产”，“东浙”是否一定包括新登县，但桑穰纸“极其敦厚”的特色，倒是与新登历史上所造桑皮纸质感很相似。

清道光年间吴墉纂修的《新登县志》（1822年）也明确记载：“纸，有绵白纸、竹帖纸、方高纸、元书纸、蚕生纸、桑皮纸、银皮纸、油纸、毛纸，又有名八百张者。”[4]绵白纸、蚕生纸、桑皮纸、银皮纸、油纸，都明显属于皮纸一类。

民国初期，新登皮纸进入一个新的鼎盛时期。皮纸作为新登造纸的3个大类之一，产量和质量都位于浙江全省前列。民国时期富阳、新登地区皮纸又分为3大类：一是绵纸类，取材于桑树皮、楮（构）树皮、山桠皮、山麻（绵）皮等，绵纸、雨伞纸、绵白纸属于这一类；二是白皮纸类，大多取桑皮为原料制成，以桑皮纸为大宗，桃花纸亦属于这一类；三是灰皮纸类，色尚灰，质粗劣，原材料不仅限于各种树皮，还掺杂了稻

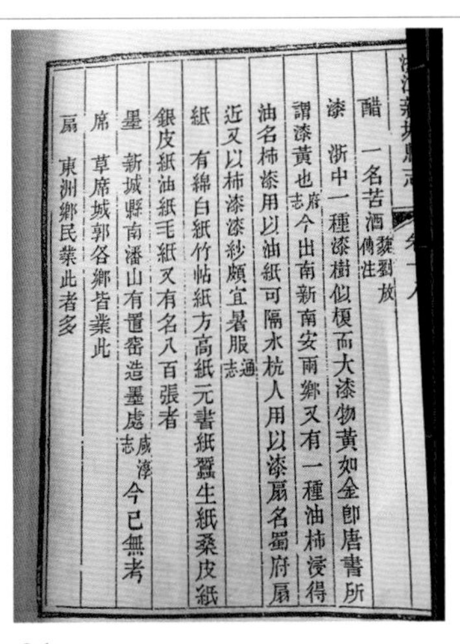
醋　一名苦酒劉攽傳注
漆　浙中一種漆樹似榎而大漆物黃如金卽唐書所
謂漆黃也府志今出南新南安兩鄉又有一種油柹浸得
油名柹漆用以油紙可隔水杭人用以漆扇名蜀府扇
近又以柹漆漆紗頗宜暑服通志
紙　有綿白紙竹帖紙方高紙元書紙蠶生紙桑皮紙
銀皮紙油紙毛紙又有名八百張者
墨　新城縣南潘山有置窑造墨處咸淳志今已無考
席　草席城郭各鄉皆業此
扇　東洲鄉民業此者多

⊙1

⊙1 清道光年间《新登县志》中的文字记载
Written records in *The Annals of Xindeng County* during Daoguang Reign of the Qing Dynasty

[2] 宋应星.天工开物[M].北京：中国画报出版社，2013:189.

[3] 任保全.近代纸业印刷史料：第3册·浙江之纸业 [M]. 南京：凤凰出版社，2014:322.

[4] 杭州市富阳区地方志编纂委员会.道光新城县志：卷十八·物产[M].北京：国家图书馆出版社，2016.

草、笋壳等，蚕生纸等属于这一类。民国四年（1915年），在巴拿马世界博览会上，新登皮纸曾获荣誉奖。

1937年抗日战争期间，杭城（杭州）一些纸商为躲避战火居于大山村，直接在此进行油纸加工，桑皮纸生意十分兴隆，小小山村有纸框（当地人称“戗架”）1 000张以上，桑皮纸行业发展迅猛。

1950年，新登县并入富阳县成为新登镇。20世纪五六十年代，富阳皮纸生产聚集地主要在上里、城阳、新浦、松溪、南安、新义、新关等乡村，年总产量维持在3 000～5 000担上下。据周关祥编著《富阳传统手工造纸》一书记载的1959年富阳供销社皮纸收购价，可以大略了解当时的皮纸生产情况：桑皮纸，长3.4尺，横4.5尺，重量70斤，504张一件，分一级25.8元、二级24.8元、三级23.6元、四级22.2元。洋皮纸，长1.1尺，横1.5尺，45斤，每件10 000张，97元。蜡纸坯，每件10 000张，正牌594元、副牌322元。加重桑皮纸，每件420张，收购价25.6元。真皮纸，每件51.2元，大皮纸每件20元。[5]从这里大致可以了解1959年前后富阳生产的皮纸种类有桑皮纸、洋皮纸、蜡纸坯、加重桑皮纸、真皮纸、大皮纸等。其中蜡纸坯、洋皮纸是上里山独有的，其他皮纸种类如桑皮纸、加重桑皮纸、真皮纸、大皮纸，大山村等村落也在生产。

表11.2　民国十九年（1930年）富阳县、新登县皮纸规格、产量数据
Table 11.2　Specifications and production data of bast paper in Fuyang County and Xindeng County in the 19th year of the Republican Era (1930)

县名	纸名	单位	重量	张数	规格（cm×cm）	产量（担）
富阳县	桑皮纸	件		168	117×155	6420
新登县	草皮纸	块	重30斤	168	111×145	2916
	雨伞纸	块	重63斤	16000	35×37	645
	蚕种纸	块	重60斤	500	44×66	370

注：资料来源自《近代纸业印刷史料：第3册　浙江之纸业》，2014：340，347。

⊙1

⊙1
浇纸纸帘（当地人称『戗』）
Papermaking screen (the locals calling it "Qiang")

[5] 周关祥.富阳传统手工造纸[Z].2010:320-323.

20世纪60年代富阳皮纸生产迎来新高潮。据富阳县统计局《全县土纸生产情况调查资料》的统计数据，1963年富阳县有竹纸槽厂1 116厂，草纸槽8 687具，桑皮纸框13 866只。其中桑皮纸框主要分布如下：松溪1 608具，城阳5 536具，六渚72具，新浦4 136具，上里2 161具。[6] 大山村当年属城阳乡，城阳乡的皮纸生产主要集中在大山村及其相连的宝山村、胜总村。

20世纪70年代手工皮纸依然保持一定规模，同时在上里乡出现机制皮纸这一新业态。富阳县土特产公司调查资料显示，1974年全县共585个大队，其中造纸大队有479个。竹纸槽厂1 515厂，草纸槽9 553具，桑皮纸框7 871只，白皮纸槽74具。皮纸主要分布在三山乡、松溪乡、城阳乡、六渚乡、南安乡。其中，三山乡（此时上里乡已经并入三山乡）桑皮纸框1 955只，白皮纸槽60具；松溪乡桑皮纸框104只；城阳乡桑皮纸框5 700只；六渚乡桑皮纸框112只；南安乡白皮纸槽14具。1972年以后上里开始成为机制皮纸的主要产地，以上里的三合皮纸厂为例，该厂机制皮纸的主要产品有白皮纸、“工农牌”滤油纸、打字蜡纸原纸、“长城牌”引线纱纸。1974年，属于城阳乡的大山村一带桑皮纸框有5 700只，手工纸生产规模有增无减。

在2018年3月的入村访谈中了解到，20世纪六七十年代，大山村全村拥有做纸戗场60余处，年产量约3 000担，产值约20万元。在当时物资供给匮乏的年代，大山村依靠桑皮纸产业富甲一

⊙2

⊙2
浇纸托架（当地人称『戗架』）
Frame for supporting the papermaking screen (the locals calling it "Qiangjia")

[6] 周关祥.富阳传统手工造纸[Z]. 2010:20-21.

Mulberry Paper Recovery Site in Dashan Village

⊙一
黄洪渭家附近的山地
Mountain field near Huang Hongwei's house

方，新登地区也因此一度流行着“嫁大山人穿毛线衣，娶大山媳戴金戒指”的俚语。

20世纪80年代中期，新登大山村传统手工皮纸业态中断，完全退出人们视野。

新登镇大山村世代以桑皮纸制作为业的家族众多，特别是在1949年中华人民共和国成立前后的一段时期，大山村几乎家家户户都开戗场制作桑皮纸。大山村桑皮纸制作技艺有比较复杂的工艺流程和操作要领，本节以本次抢救性恢复时负责技术指导的大山村黄家自然村黄洪渭家族可记忆的传承谱系为例介绍如下（以2018年3月和12月访谈时为时间节点）：

第一代，黄成贵（？—1922年），黄洪渭高祖父，开戗场制纸。

第二代，黄得胜（生卒年月不详），黄洪渭曾祖父，开戗场制纸。

第三代，黄昌荣（生卒年月不详），黄洪渭祖父，开戗场制纸。

第四代，黄玉铨（1925—），黄洪渭父亲，调查时93岁。黄玉铨父辈以造桑皮纸为业，建有造纸戗场1座，家庭比较富裕。黄玉铨少年时曾在学校读过书，20世纪40年代因父丧弃学回家，聘请同村方阿来负责家庭造纸戗场的生产，同时拜方阿来为师学习桑皮纸制作技艺，学成后自主经营造纸作坊，并将技艺传授给儿子黄洪渭。方阿来（生卒年月不详，约长徒弟黄玉铨30余岁）是家里请来的浇纸师傅。

第五代，黄洪渭（1966—），调查时52岁，1966年9月出生，杭州市富阳区新登镇大山村人，系黄玉铨儿子。黄洪渭16岁起师从父亲学习桑皮纸制作技艺，掌握全套桑皮纸制作程序和要领，是2018～2019年初调查时大山村能独立开场制作桑皮纸群体中相对较为年轻的造纸师傅。

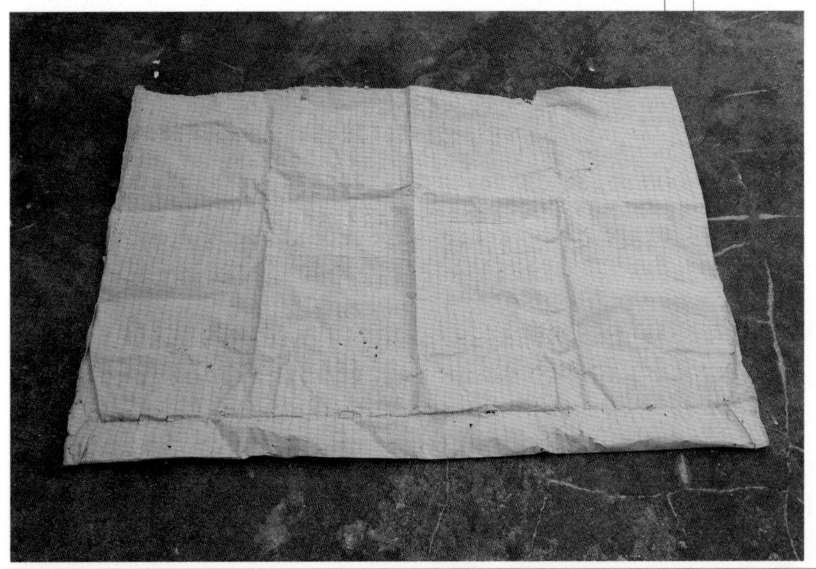

⊙1

⊙2

⊙1 大山村纸农保存的20世纪80年代初手工造的桑皮纸
Handmade mulberry paper produced in the early 1980s,preserving by papermaker in the Dashan Village

⊙2 访谈年老卧床的黄玉铨老人
Interviewing Huang Yuquan, the old man who stayed in bed

三

大山村桑皮纸的代表纸品及其用途与技术分析

3
Representative Mulberry Paper and Its Uses and Technical Analysis in Dashan Village

（一）大山村桑皮纸代表纸品及其用途

1. 草皮纸

大山村草皮纸的原料采用桑皮、稻草的混合料，配比根据客户的需求而定，一般为桑皮6份、稻草4份。传统草皮纸每件重15 kg，168张，规格是111 cm × 145 cm。大山村造的草皮纸历史上以厚实抗拉、柔韧性好著称，广泛用于包裹日常物品、设备，包装工业零件、器械、枪支弹药等，有很好的防潮防锈功效。

2. 矾红纸

据黄玉铨介绍，他们也曾生产矾红纸，原料构成为桑皮7份、稻草3份，是加了矾的红色纸，专项用于旧日典当铺里包裹高档衣服、银行里包裹白洋（银元）。

3. 油纸

经过进一步加工，草皮纸可以做成油纸。在塑料、尼龙等材料问世前，油纸的使用十分广泛，做油纸雨伞、油纸笠帽等是江南地区油纸的一大用途。油纸的另一个作用是在一些沙漠或平原地区，农民在地上撒上种子，为了防止鸟儿来偷吃，盖在上面当隔离层。2018年调查中了解到油纸还有一个特色性的用途：1937～1949年这一阶段，战争频繁，油纸的用量很大，军人行军打仗，每人配桑皮油纸一张，野外宿营时垫在地上阻挡潮气，即可随时可休息。据说抗战时期杭州城里一大批加工油纸的客商逃到大山村，大山村一下子出现了很多加工油纸的作坊。

⊙3

⊙3
大山村旧日造的草皮纸
Straw paper produced in the old days in Dashan Village

（二）大山村草皮纸性能分析

测试小组对采样自大山村2016年6月恢复新制

草皮纸所做的性能分析，主要包括定量、厚度、紧度、抗张力、抗张强度、白度、纤维长度和纤维宽度等。按相应要求，每一指标都重复测量若干次后求平均值，其中定量抽取5个样本进行测试，厚度抽取10个样本进行测试，抗张力抽取20个样本进行测试，白度抽取10个样本进行测试，纤维长度测试了200根纤维，纤维宽度测试了300根纤维。对大山村草皮纸进行测试分析所得到的相关性能参数如表11.3所示，表中列出了各参数的最大值、最小值及测量若干次所得到的平均值或者计算结果。

表11.3　大山村草皮纸相关性能参数
Table 11.3　Performance parameters of straw paper in Dashan Village

指标		单位	最大值	最小值	平均值	结果
定量		g/m²				88.7
厚度		mm	0.558	0.362	0.465	0.465
紧度		g/cm³				0.191
抗张力	纵向	mN	8.6	7.0	7.6	7.6
	横向	mN	6.6	5.3	6.0	6.0
抗张强度		kN/m				0.453
白度		%	12.2	10.3	11.2	11.2
纤维	长度	mm	2.4	0.3	0.8	0.8
	宽度	μm	26.9	4.5	11.4	11.4

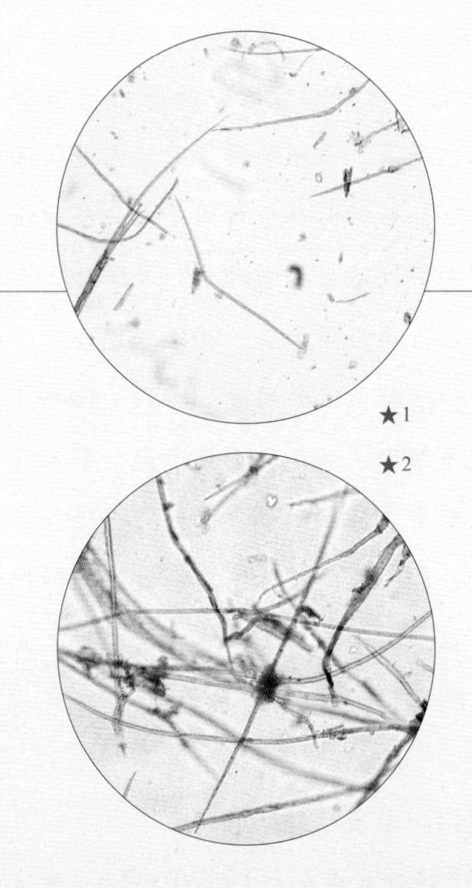

★1

★2

由表11.3可知，所测大山村草皮纸的平均定量为88.7 g/m²。大山村草皮纸最厚约是最薄的1.541倍，经计算，其相对标准偏差为0.148。通过计算可知，大山村草皮纸紧度为0.191 g/cm³。抗张强度为0.453 kN/m。

所测大山村草皮纸平均白度为11.2%。白度最大值是最小值的1.184倍，相对标准偏差为0.061。

大山村草皮纸在10倍和20倍物镜下观测的纤维形态分别如图★1、图★2所示。所测大山村草皮纸纤维长度：最长2.4 mm，最短0.3 mm，平均长度为0.8 mm；纤维宽度：最宽26.9 μm，最窄4.5 μm，平均宽度为11.4 μm。

★1
大山村草皮纸纤维形态图（10×）
Fibers of straw paper in Dashan Village (10× objective)

★2
大山村草皮纸纤维形态图（20×）
Fibers of straw paper in Dashan Village (20× objective)

四 大山村草皮纸的生产原料、工艺与设备

4 Raw Materials, Papermaking Techniques and Tools for Straw Paper in Dashan Village

（一）大山村草皮纸的生产原料

大山村草皮纸有纯桑皮制作的，也有草皮混合纸。通常涉及的原料和辅料包括干旱稻草、干桑皮、新鲜的滑涅（学名山苍）叶、腌料用的石灰和烧火用的柴草。2016年6 月开始的抢救性恢复生产，因为业态已经中断多年没有储存的干桑皮，采用的桑皮是新鲜的，从大山村田间地头砍来嫩桑树枝剥皮而得。

1. 主料：干桑皮、干旱稻草

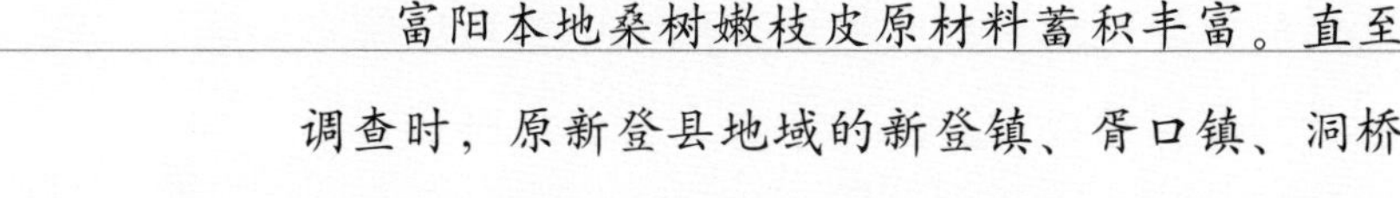

富阳本地桑树嫩枝皮原材料蓄积丰富。直至调查时，原新登县地域的新登镇、胥口镇、洞桥镇、万市镇还有很多农户种桑养蚕。在高峰阶段，随着桑皮纸生产的扩大，桑皮原料不足部分会向作为中国蚕桑丝绸基地的杭嘉湖平原购买。收购桑皮时，为了能够长久储存，原料有干燥度要求，水分含量不能超过8%。

⊙1

大山村造草皮纸对稻草的要求，以在含沙量高的山垅田里生长的早籼米稻草为优，有点类似安徽泾县宣纸的主原料之一沙田稻草的要求。

2. 辅料：滑涅叶、水

大山村造草皮纸使用的纸药当地土名叫滑涅叶，学名叫山苍子叶。有两种提取汁液方法：

⊙2

一种是用水浸渍，里面的黏稠汁水会自然析出，但是要将可用成分尽可能全部释放出来，就必须要人赤脚反复踩踏。人在缸里反复踩踏，汁液要形成像木莲豆腐一样的视觉与触觉质感才算达标，然后就可以用滑涅箩进行过滤。第一次过滤后，可以将叶子重新倒入水缸踩踏。再次过滤，过滤出来的汁液用浇纸桶作量具，以1.5桶的量，掺入十石缸（意思是能盛放10担水的大水

⊙1 刚刚砍下的新鲜桑树枝条 Fresh cut mulberry branches

⊙2 沙田早灿米稻草 Early rice straw in sandy field

缸）内，然后加入水和草、皮浆一同搅拌。

另一种制法需要打造一只堂锅灶（类似民间煮饭用的土灶），堂锅大小以能盛放2担水为标准。滑涅叶用石灰浆拌过，捞起后放入堂锅内，这样一层树叶一层末石灰（用水化开的粉石灰）码放。需关注的要点是火不能烧得太旺，以文火为主，水量少但要严防烧焦。煮至第二天取出来，堆成几堆，用棍子敲打，越打越黏，最后打到像“清明节用糯米和艾青做的青粿一样”。然后把敲打好的纸药材料团成篮球大小的药球浸在水中，这样可以储存2～3个月。需要用时从水中捞起，放入水缸里打开打散，用清水搅拌，过滤一下即可使用。

滑涅汁采集集中在每年夏至过后至寒露之前这段时间。据黄玉铨介绍，夏至过后夏风吹过3天，滑涅汁水就能浸出来。寒露之后滑涅的叶子就掉光了，掉落的叶子就浸不出来汁水了。

在草皮纸的制作过程中，水温对草皮纸原料中稻草纤维的提取影响很大。自然低温发酵，稻草纤维就好；天气炎热，稻草料在水中就会浮起来，草料就易过度腐烂，稻草纤维就不容易提取了。大山村选用纯净的山泉水制作草皮纸。调查人员在造纸引水现场测试，制作草皮纸所用的水pH约为6.5，偏弱酸性。

⊙1

⊙2

⊙1
大山村附近山间生长的滑涅树
Litsea cubeba tree growing on the mountains near Dashan village

⊙2
流经村里的山溪水
Mountain stream flowing through the village

（二）大山村草皮纸的制作工艺流程

2016年6月调查记录的桑皮原料不是传统上使用的干桑皮，而是从大山村田间地头的桑树上直接剪下来的新鲜枝条。根据对黄洪渭、缪柏堂、沈健根等造纸人的现场访谈和调查组的观察，归纳大山村草皮纸制作的主要工艺流程如下：

壹	贰	叁	肆	伍	陆	柒	捌	玖	拾	拾壹
剥皮	晒皮	腌料	煮料	漂洗	制浆	浇纸	揭纸	理纸	捆件	治印

壹 剥皮

1 ⊙3⊙4

用铁榔头敲打桑枝，将桑皮从桑枝上撕下。

⊙3

⊙4

⊙3 老纸工沈健根在敲打剥取桑皮 Shen Jiangen, an old papermaker beating and stripping the mulberry bark

⊙4 刚剥出的桑皮 Freshly peeled mulberry bark

贰
晒　皮
2

将鲜桑皮晒干备用。

叁
腌　料
3　⊙5⊙6

将50 kg石灰放入腌料塘中，加入半塘水，再把300 kg干旱稻草、25 kg干桑皮放入塘中并搅拌翻身数次，浸透泡匀后捞起堆放在皮镬灶旁。这些数量的混合料约能造3担（按照标准数字是3 024张）稻草桑皮混料纸。

⊙5

⊙6

肆
煮　料
4　⊙7

皮镬灶锅内倒进足量的水，先将腌制好的桑皮放在底层，再把腌制好的稻草放进去，一边烧煮一边添草料（自然下陷下去时则需添满），直至把腌好的草料全部放进锅灶内，然后上面压一层纸筋泥浆，像锅盖一样封在上面，一直烧到水蒸气冲出这一泥浆层后一定的时间才熄火。现场煮料的缪红潮师傅表示，一般要烧煮12个小时左右，为确保熟料质量，最好焖1天或1夜再出锅。

伍
漂　洗
5　⊙8

分别漂洗煮熟的稻草和桑皮。

出锅后的熟草料先搬运到放满清水的漾塘中，进行初步清洗漂汰，然后放干脏水，加满清水，浸泡熟草料，保持活水一头流进一头流出。接下去要每天拎翻一次浸泡中的熟草料，7天左右才能漂清。漂清后，在漾塘边上把草料堆放成长方形，外表糊光拍实，以免下大雨冲刷导致精料流失。

煮熟出锅的桑皮先用双脚踩踏至外皮脱落，再进行初步清洗漂汰，最后放到小漾塘用活水漂漾数日，直至漂清。

据李少军介绍，7天的翻洗绝对不能偷工减料。如果每次翻洗装装样子，出工不出力，7天过去了原料灰浆还是很浓，堆放后发酵的时间会延长，发热质量也会下降，含有大量有机质的料筋会成倍增加，不但原料质量下降，原料利用率也会下降，人家做纸有钱赚，你做却要亏本了。

大山村草皮纸质量、等级的好坏，很主要的工艺是这个漂的过程。有些漂在山坑水里5～6天就可以捞了，但有些漂在山坞的清水中，需要1～2个月甚至更长的时间。漂的时间越长，做出的纸拉力越好，韧劲越大；反之则会成为质量比较差的桑皮纸。

⊙7

⊙8

⊙5
沈健根在化石灰
Shen Jiangen melting the lime

⊙6
黄洪渭（左）和沈健根（右）在腌料
Huang Hongwei (left) and Shen Jiangen (right) fermenting the materials

⊙7
缪红潮在蒸煮桑皮和稻草
Miao Hongchao steaming and boiling the mulberry back and straw

⊙8
缪红潮、沈健根在活水漾塘里漾料漂洗
Miao Hongchao and Shen Jiangen cleaning the materials in the cleaning pool

陆 制浆

6 ⊙9～⊙15

用稻草、桑皮和滑涅汁配置制浆。取漂清后堆放数日的精料约30 kg放到十石缸中，人赤裸双脚在缸中反复踩踏，踩踏的技术要点是边踩边用脚翻动原料，把没有踩烂的原料用脚翻动到中间，踩烂到细糊状，然后加入水，再用一根小竹条或小木条多次搅拌，捡掉粗、长的料筋。

取漂清后的熟料桑皮约3 kg，放到石板上用硬木棒槌敲打制浆。棒槌的形状与棒球棍相似，握手部分小，呈圆形；槌打部分长，呈方形。反复敲打，把皮料打成“薄饼”，又叠起来重新敲打，如此反复，直至敲打成细糊状才算完成。用水疏解后，倒进草料缸内搅拌。

采摘来的滑涅叶需提前浸泡在小缸内数日，再放到专用的十石滑涅缸内，加上适量清水后用脚进行踩踏。踏得像木莲豆腐一样稠，就可以用滑涅箩进行过滤了。第一次过滤后，可以将叶子重新倒入水缸踩踏，再次过滤，过滤出来的汁液用浇纸桶作量具，以一桶半的量，掺入到纸浆缸中。

最后把十石纸浆缸加满清水，用一根粗竹竿多次搅拌均匀后，就配置好了可浇晒的纸浆。

⊙9

⊙10

⊙11

⊙12

⊙13

⊙14

⊙15

⊙9 沈健根在踩细稻草料 Shen Jiangen stamping fine straw materials

⊙10 用小竹条或小木条多次搅拌 Stirring the pulp multiple times with bamboo strips or wooden strips

⊙11 捡掉粗、长的纸筋 Picking out the thick, long paper residues

⊙12 罗忠火用棒槌敲打桑皮料 Luo Zhonghuo beating mulberry bark materials with a wooden mallet

⊙13 踩蹚滑涅叶 Stamping the leaves of *Litsea Cubeba*

⊙14 过滤滑涅汁 Filtering *Litsea cubeba* sap

⊙15 缪柏堂在调制皮水（纸浆） Miao Baitang making paper pulp

柒

浇纸

7

⊙16～⊙20

把加入草料、桑皮料、滑涅汁三种原料配制好的纸浆水，用2只浇纸桶，每只舀大半桶，约15 kg一桶，拎到晒场，倒入平放的戗内，1戗1桶，趁湿迅速用羽毛掸拂刷，将纸浆刷平刷均匀。水慢慢从戗笠片中滤出，待纸浆不会流动时，用竹梢戗柱撑起一档，然后干一点撑高一档，再干一点再撑高一档。不能操之过急，否则纸浆流下来就成了破纸废纸。若上半张干燥了，要把纸戗掉个头，把下半张掉到上面来。还要按太阳光移动调整方向，必须把纸戗置放倾斜，纸面朝向太阳，跟着太阳转向，这样才干得快（一般夏季1天可以浇3批纸）。据缪关宏介绍，浇纸中的戗每只约有40 kg重，每堂戗共有100多个，调戗头是个体力活，得有气力耐力，否则调不过来。

⊙16

⊙17

⊙18

⊙16/17 倒浆浇纸
Pouring the pulp to make paper

⊙18 用羽毛刷平纸浆
Flattening the paper pulp with feather brushes

⊙19

⊙20

⊙19
黄洪渭在支戗
Huang Hongwei raising the screen frame

⊙20
支戗晒纸
Raising the screen frame for drying the paper

捌 揭纸

8 ⊙21

待纸干燥后，将纸一张一张从戗上扯剥下来，折叠好背回家。折叠方式：以中间为基点，左边对折，右边再对折，然后在中间合拢。一张纸折成4层，4个角都裹在里面。

玖 理纸

9 ⊙22

大山村浇纸法生产的桑皮纸规格尺寸为155 cm × 117 cm，草皮纸规格尺寸为145 cm × 111 cm，由于尺寸大，不宜平张包装，因此将每张纸折叠后才能包装打件。在折叠前有一个整理检验产品的过程，质量要求：洁净，无泥沙，无破张。折叠桑皮纸一般是妇女干的活，男人在雨天也帮着干。在家里堂前，横放1张两人坐的长凳，上面搁一个竹篾编制的圆形晒箕，把1张晒干的纸摊放在上面，将破张剔除，放在一边，可用作一件纸顶部、底部包装用；将质量完好的桑皮纸、草皮纸横向折1次，然后纵向折1次，共2次折叠，每张成4层，折叠后的尺寸为72.5 cm × 55.5 cm。堆叠成件，每件504张大纸，共2 016层。

⊙22

⊙21

⊙21
缪柏堂揭下草皮纸（图中的纸戗是损坏的，在中间置一木条加固。揭下的是半张草皮纸）
Miao Baitang peeling down the straw paper (the screen frame in the picture was damaged, so a wooden strip was placed in the middle to reinforce. A half of the straw paper was peeled off)

⊙22
两次对折后的桑皮纸
Double folded mulberry paper

拾 捆件

10 ⊙23

每件纸折叠完成，要放进压纸架打件。将1件纸在厚木板里压实，再放入纸架压榨打件。压榨纸架和竹纸压纸架相同，为长方形木架，压纸板按折叠好纸的大小做成，很厚实的2块木板，木板中间凿有2道小木槽，用来穿嫩竹篾捆纸，1件纸是504张（其中包面4张是破纸，作包装纸）。

捆件过程如下：在压榨纸架上放1块压纸板，再放上504张折叠好的纸，上面再压1块压纸板。纸架一个横头做好相应的横档，另一横头有一个圆木滚筒，滚筒上有2个小圆洞，套上1根粗绳索，然后在上面压1根大粗木榨杆，一头伸入木架横档中，另一头套上绳索，用1根小木棍插入滚筒中的小洞内往下多次滚转，直至把纸压实。再穿入嫩竹篾捆缚好，一般先捆横向3道篾，放开榨后再捆纵向1道篾（叫关節）。这样就完成了1件纸的包装。

⊙23

⊙23 压纸架 Pressing device

拾壹 治印

11 ⊙3

压纸打件后，在每件纸上印上本坊专印，标注生产厂家，类似于现在厂家的商标，即可按担出售。

在与黄洪渭的访谈中了解到，大山村草皮纸制作有4个值得特别说的要点：第一要求水质清澈，没有污染，好水才能使料漂得干净；第二是要精确配料，一塘料用600 kg稻草加100 kg桑皮，生产3担桑皮纸（3 012张），桑皮料加得越少，草皮纸的质量就越差；第三是料要踩踏细腻，谷壳、桑皮壳等杂物要捞净，否则纸张会粗糙不堪；第四是浇纸要一气呵成，一遍成形，不能反复，否则这张纸就厚薄不匀而成次品了。

（三）大山村桑皮纸浇注工场选择和设施制作

1. 浇注工场

浇注桑皮纸要借助阳光晒纸。浇注工场要选择地面平坦、阳光充足、水源不断且水质清澈的溪边，戗架用地面积约需1亩大小。戗架的摆放角度，要迎合阳光照射方向，具体指向应该是正南偏西三四分（30°～40°），戗架下面需要挖一条小沟，并让小沟连通大沟，便于过滤后的废水排出（水沟用石块砌成）。戗架前后要留足走路通道，便于员工调整戗架与阳光的角度，东西的通道要宽于南北的走道。每个工场放戗架84～112具。

在浇纸工场一侧要挖1口水井，石块构筑，水井口要留1 m边沿便于做工时用水操作，水井边上摆放4只十石缸（2只配料缸，1只滑涅缸，1只踏料缸），水井横头放2只小缸浸泡滑涅用。还要放1块平整石板敲桑皮用，配1个1 m长檀树或青柴等硬木方榔头。

2018年恢复桑皮纸制作时，选择了宝山村戏台前较大的空地，大约可以放10具戗架。

2. 皮镬灶、腌草塘、漾滩等土建设备

原料加工时，还需要0.6亩地用于建造生产桑皮纸的必备设施：蒸煮原料的皮镬灶、腌草场地（当地人叫“腌草塘”）、漂洗场地（当地人叫“漾滩”）。

（1）皮镬灶。

用石块砌造一个大于皮镬灶外径的灶台，灶台中心要留一个镬堂（也叫火堂），底部直径大于1 m（上部小于铁镬口直径8 cm），并且留好外部到火堂的柴道和镬灶内部向外的排水口。在平整好灶台，安装好铁镬后，开始建造皮镬灶蒸

⊙1

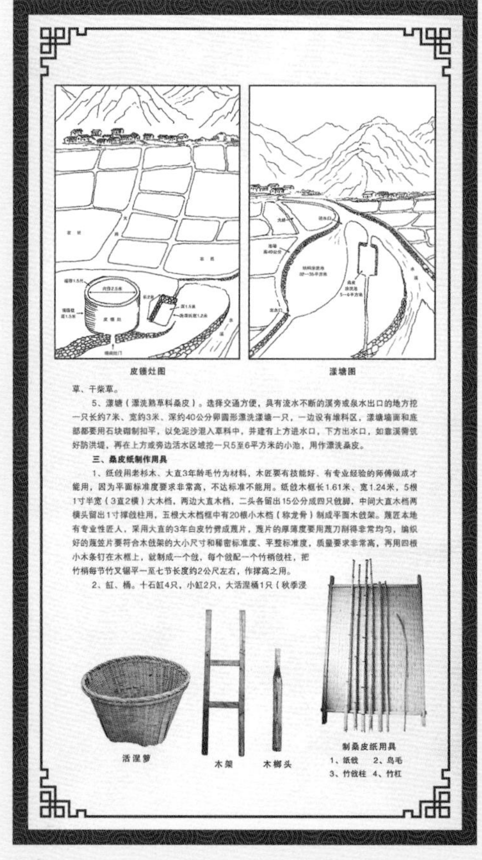

⊙2

⊙1 浇场示意图
Diagram of the papermaking

⊙2 皮镬、腌草塘、漾塘示意图
A diagram of papermaking utensil, and pools for soaking and cleaning the materials

料的本体（有储水功能的桶体结构）。首先以镬子为中心点，划两个圆圈，内圈半径为85 cm，外圈半径为115 cm，然后用竹爿编织内圈（篱笆），编织所用的竹爿都是用整根毛竹劈成的，由于每片竹爿根部宽度都超过1.5寸，因此编织到根部时十分坚硬，需要几个强壮劳力齐心合力才能编入。在完成竹圈编织后，将已经准备好的黄泥敲碎，并加上泥土量三分之一的末石灰用锄头拌匀。内圈与外圈之间空间用黄泥30%，末石灰30%，溪里的鹅卵石与细沙35%～40%，或加上5%的麻筋，用锄头拌匀成三合土，达到手捏不散即可，接下来一层层倒在两圈中间，用杵夯实。然后将黄泥加末石灰的泥坯加工成高黏土，将高黏土黏抹在竹圈表面和顶面，将竹爿全部包裹起来，每天派人用木板做成的木刀密密麻麻地在桶体上“斩”，目的是挤压泥巴中的水分，“斩”完以后又用木锤将粗糙的泥壁表面敲平，从木刀斩糙到木锤敲平反反复复，直到“斩无痕、敲无迹”，然后灌水，进入后期保养。这样的土建皮镬灶，日晒不裂痕，冰冻不脱壳，几乎坚不可摧。

非常可惜的是大山村的皮镬早已被清理拆掉了，一个也不剩。2016年6月恢复制作技艺现场，煮料用的是酿酒蒸料时的铁皮桶。

(2) 腌草塘。

在皮镬边需挖1只腌草塘，长2 m，宽1.5 m，深1.5 m，四壁用石块砌好，以黄泥石灰混合土加固防漏，一边留有空地堆料。由于大山村桑皮纸活态生产中断已久，2016年6月恢复制作技艺现场的腌草是放在1只大缸里进行的。

(3) 漂洗漾滩。

在腌草塘附近交通方便、流水不断的溪旁或泉水出口的地方，挖1只长约7 m、宽约3 m、深50～60 cm的椭圆形草料漂洗漾滩，一边设有堆料区。漾塘墙面和底部都要用石块砌制扣平，避免泥沙混入草料中，并设上方进水口、下方出水口。在草料漂洗漾滩上方或旁边活水区域挖一只5～8 m^{2}的专门漂洗桑皮的漾滩，要求底平面进水和出水有10 cm高低的落差，便于残渣清洗。

(4) 堆料场。

在腌草塘的一边，要留出一块约0.3亩平整的空地，用于堆放干旱稻草、干柴草。

（四）大山村桑皮纸的主要制作工具

壹 纸 戗 1

纸戗由戗架、篾帘与戗柱组成，是浇纸法成型的重要工具。2018年实测大山村制作草皮纸的戗架的木框长1.61 m，宽1.24 m。有5根（3直2横）约5 cm宽的大木档，5根大木档框中有20根小木档作龙骨，制成平面木戗架。

戗架用杉木制成，最基本的质量要求是“木架子平整”，因为浇纸时戗架平放在地上，如果四角不平，纸浆会产生厚薄。篾帘由篾匠采用大而直的3年生白皮竹劈成厚薄均匀的篾片编织而成，用4根小木条将篾帘固定在戗架上，以达到承接纤维、防止流失的效果。以大山村桑皮纸150 cm×114 cm的尺寸来说，篾笠的尺寸要略大于纸张的尺寸。李少军现场测得的篾帘经纬篾片数量为：经（纵）向篾片（从左向右）共90～91片，每片长152 cm，宽1.2～1.3 cm，篾与篾并列间隙0.1～0.2 cm；纬（横）向篾片（从上到下）共23～25片，每片间隔4～5 cm，长116 cm，宽1.2～1.3 cm，篾片厚度统一为0.10～0.15 cm。每个戗配1个竹梢戗柱，把竹梢按一定距离保留2 cm的竹叉，用于撑高。

2016年大山村恢复制作桑皮纸时，从全村搜集来了10张纸戗，大多破损，进行了修复。

贰 缸和桶 2

一处桑皮纸的浇纸工场需要4只十石缸，其中2只用于配料，1只用于踏料，另1只用作滑涅缸。其他还需要2只高60 cm、缸口直径55 cm的小缸，用于浸泡滑涅叶所用；大滑涅桶1只；浇纸桶2只。

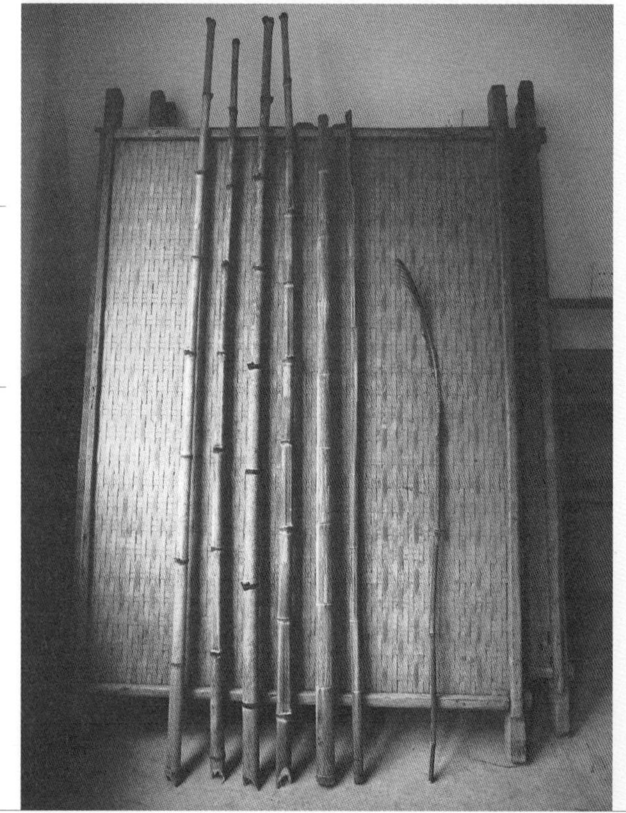

⊙1 纸戗、羽毛、竹戗柱、竹杠
Screen frame, feather, bamboo posts, bamboo bars for papermaking

⊙2 十石缸
Stone vat

⊙3 浇纸桶
Papermaking bucket

叁 小型用具 3

腌草铁捞钩2把，硬柴木榔头1个，烧火大铁火叉1个，大铁火铲1个，滑涅箩1只，滑涅架1个，海鸥鸟羽毛2支（羽毛大约50 cm长，4 cm宽）。

肆 压榨纸架 4

配件有压纸板2块，大榨杆1支，粗绳索1根，嫩竹篾若干。

⊙4

⊙4 滑涅箩、木架、木棒槌 Papermaking mucilage basket, wooden frame, wooden mallet

五 大山村桑皮纸的市场经营状况

5 Marketing Status of Mulberry Paper in Dashan Village

由于大山村桑皮纸生产业态中断已有30余年，2016年6月基于非物质文化遗产抢救性保护实现的恢复性生产并未面向市场产生销售，因此没有近期的市场数据。历史上大山村的桑皮纸主要用作油纸原纸，做雨伞、箬帽、扇子的用纸，也有包装枪支弹药、防雨防潮器物等用途。从销售方式说，调查中获悉的信息是，大山村桑皮纸过去是人力肩挑到渌渚镇下船，通过水路运至杭州、上海出售。20世纪30年代中期公路开通后，则挑到松溪乡昌东村众圃自然村通过汽车外运。抗战时期有杭州纸商到本地蹲点收购，直接做成油纸运回。当时小小山村纸商竟多达50～60人。1952年供销社收购站建立后，就直接挑到新登街上卖给收购站。

2018年调查时，据大山村的老人们回忆，约40多年前，杭州西湖沿岸有50多家出售桑皮纸的商店，他们都会上门到新登其他村或大山村来收购，主要卖到上海，以及浙江本省的绍兴、金华、兰溪等地。上海等地也有人到村里来收购的，他们买回去后，会用一把特制的榔头将纸面敲平整、紧实。大山村桑皮纸成品有明显的篾帘痕迹，篾条十字编织形成的篾帘凹凸感是显而易见的，所以纸张表面不太平整，经过敲打，使纸面平整后，再上桐油。

⊙1

⊙1 用大山村桑皮纸制作的官塘桑皮纸扇 Guantang mulberry paper fan made of mulberry paper in Dashan Village

六
大山村的造纸文化与造纸故事

6
Papermaking Culture and Stories of Dashan Village

（一）大山村造纸技艺与文化品牌

1. 大山村的造纸名人与桑皮纸品牌

据黄品耀撰写的《黄关银、方茂潮与桑皮纸》一文记载：

黄关银（1914—1995年），大山村人，他的纸槽建在大山村罗家桥的赤凉亭，取用绝无污染的山坞里的溪水。他的桑皮纸出名于20世纪40年代中后期，品名即为“黄关银”，共有80多张纸框。在大山村有个名叫方茂潮的师傅，生产“方茂林”品牌的桑皮纸，比黄关银还早一些，他家有102张框。

现代则有李金根三兄弟承继祖业，曾经拥有200多张纸框。李金根（1907—？）10来岁就做桑皮纸，1956年夏天，他创造出1个月生产“四担一头”（一担两头，一头一件504张，一担即1 008张，四担一头即4 536张）的记录，每天100框，要浇6缸纸料。[7]

2. 独特的成纸方式——旱地竹戗浇纸法

浇注法主要分布在原三山乡（上里山）、松溪乡、城阳乡（大山村）、六渚乡（新浦片）等地，以纯桑皮浆或按不同比例配比的稻草和桑皮浆为原料生产桑皮纸或草皮纸。方法是把处理好的原料和水在缸内先调成料浆，加进少许滑涅汁搅拌使之均匀，然后用浇料桶兜浆液浇注于事先准备好的一个个纸框上。帘面浇注上纸浆后，趁湿用羽毛掸拂平整，水则从篾垫下自行滤出。然后用支脚把纸框成斜角撑起，迎日光暴晒。一般夏季晴朗天气，每天可浇注2～3次，晒干后逐张收起，堆叠成件，用篾捆缚，加盖牌号，即可上市。大山村桑皮纸制作技艺保持了较“抄纸法”更为原始、简便的“浇纸法”，和新疆墨玉县维吾尔族桑皮纸的“水坑布帘浇纸法”相比较，新登大山村桑皮纸采用了“旱地竹戗浇纸法”，生产效率更高。这也是富阳区抄纸技法丰富性的一个体现。

⊙2

⊙2 旱地竹戗浇纸法
A method of papermaking by pouring the pulp onto the papermaking screen

[7] 浙江省富阳市政协文史委员会.中国富阳纸业[M].北京：人民出版社，2005：68.

（二）大山村的造纸故事

1. 桑皮纸浇纸靠天吃饭

访谈中据村里的文化员缪关宏介绍，大山村桑皮纸浇注在露天工场进行，因此浇纸必须要趁晴好天气，以利用阳光晒纸。产量根据一年四季气候而定，冬天太阳不旺，1天只能浇1批；夏天太阳光强，1天能浇3批。浇纸一般是两个人做的活，但灵快者一人也行。浇3批天蒙蒙亮就要开工，第一批浇好吃早饭，第二批、第三批紧跟着连续操作，遇着好天气，3天可晒干1担纸。如碰到雷阵雨天气，那就极有可能白忙一天了。

凡会浇纸的人，大多会看云识天气。早上起来看西边有没有红云，西边有红云预示今天要下雨。大山村是山区，山上树蛙很多，树蛙开始狂叫预示要下雨。浇注桑皮纸很辛苦，皮纸生产中的腌制草料、漂洗草料、踩踏料浆都要赤裸双脚，冬天也如此。一年四季风越大雨越大，越要出门。下雨要收合保护戗架，两张戗架搭成人字架，戗架反面朝外。刮风要放平戗架，防止吹破。戗架常年露天日晒、风吹雨打，腐烂损耗特别严重。

2. 大山村生产桑皮纸由来的传说

三合村村民吴明村曾被聘为杭州新华造纸厂抄纸师傅，据他介绍，大山村里的桑皮纸制作技艺是从上里山传过去的，有一个传说性的依据，大山村桑皮纸尺寸比上里山的皮纸要小，是因为大山村木匠做纸框时是按上里山桑皮纸大小做的，而不是按照框子大小做的。由于大山村的桑皮纸大量生产影响了上里山桑皮纸销售，两地为此还发生争斗，打起了官司，最后是上里山赢了官司。

坊间还传说，上里山赢了官司与清代乾隆年间富阳籍大学士董诰有关。董诰姑姑嫁在上里山三合村，董诰曾在三合村的私塾里上过学。上里山人去递状子时得到了董诰门人帮助，因而打赢了官司。如果这一传说确实有来源，那么大山村桑皮纸生产的时间就在董诰（1740—1818年）为官的同时或稍前，而上里山自然则要更早些。

⊙1

⊙1 手拿桑枝解说的缪关宏 Miao Guanhong introducing with a mulberry branch in his hand

七 大山村桑皮纸的传承发展思考

7 Inheritance Status and Development of Mulberry Paper in Dashan Village

从20世纪80年代中期大山村桑皮纸制作业态消失后，时感焦虑与可惜的地方政府和民间也曾下过大气力进行挖掘整理。2004年原富阳县政协文史委组织人员编撰《中国富阳纸业》一书，对富阳传统纸产区进行了颇为详实的实地调查，记录了一些珍贵资料，富阳皮纸方面内容涉及生产工艺、品名种类、代表人物等，但对桑皮纸浇纸法制作技艺的记录内容极简，只有约300字。此外，该书有竹纸、草纸生产工艺的系列图片，但没有收集到皮纸生产的任何图片。

⊙2

2016年6月，富阳区非遗保护中心组织申报镇级非遗名录，新登镇发动各村申报镇级非遗名录项目，大山村组织村民恢复桑皮纸生产并对桑皮纸浇注法技艺进行抢救性记录，此次工艺恢复留下了一批珍贵的照片和全程录像。大山村村民缪关宏整理了大山村桑皮纸的制作工艺，富阳区素质教育基地杨荣华撰写了申报书。2017年，大山村桑皮纸制作技艺被列入第七批富阳区级非物质文化遗产代表性项目名录。

2017年，富阳竹纸保护与传承发展研究会秘书长李少军及倪国萍、陈志荣、杨荣华等组成调查组采访了渌渚镇新浦村塘坞里自然村造纸人叶永棠（时年89岁）、新登镇大山村造纸人缪关宏，李少军执笔形成了调研报告《记富阳桑皮纸与浇纸法技艺的田野调查》，文中对大山村桑皮纸浇纸法工艺进行了详细记录并提出了5点思考。2018年3月，本调查组成员与上述人员再次走访大山村核实相关信息，探讨大山村桑皮纸生产的可能性。本节撰写是在田野调查基础上结合上述材料和相关文献资料完成的。

⊙3

⊙2
富阳市政协文史委员会编撰的《中国富阳纸业》
Fuyang Paper Industry of China compiled by Fuyang Municipal Political and Cultural Committee

⊙3
方仁英（左二）、李少军（左三）、杨荣华（右二）等采访黄玉銓父子
Fang Renying (second from the left), Li Shaojun (third from the left), Yang Ronghua (second from the right) etc. interviewing Huang Yuquan and his son

随着机制纸产品种类的日益丰富，传统乡土社会采用桑皮纸的手工技艺产品几乎都可以在市场上找到价格低廉的替代品，导致桑皮纸的市场需求大幅缩减，市场竞争力衰退，至20世纪80年代后期，富阳桑皮纸作坊陆续停产，皮镬、漾

塘、戗场等场地陆续改作农林用途而消失，料缸、料桶、纸戗、压纸架等专用工具陆续损毁。2017～2018年调查获得的信息是：整个富阳区仅有数十位60岁以上的老人掌握整套或部分浇纸法制作桑皮纸技艺。同时，由于制作工艺的复杂，在停产以后年轻人失去实践体验，不大可能掌握这门技艺，大山村桑皮纸制作技艺面临着完全失传的风险。

2018年12月座谈时，大山村村主任方金勇表示：他一直想把种桑养蚕业与桑皮纸浇注的产业链发展起来。他说，现在小学生科学课上有养蚕这一内容，淘宝上每斤桑叶要卖到250元。我们村建一个桑树园，与杭州的学校签约，定期采摘桑叶，卖桑叶、卖蚕苗，还可摘桑果。桑枝用来造桑皮纸，最好用浇纸法浇注的桑皮纸能供书画家创作，再建一个桑皮纸制作研学基地，把学生引进来体验，这样还可以带动农家乐的发展。通过发展种桑养蚕业与桑皮纸浇注这样一个文化产业带动村级经济发展。

⊙1

⊙1
大山村文化礼堂陈列的桑皮纸制作工艺图板
Illustrations of the mulberry paper papermaking techniques displayed in the Cultural Hall of Dashan Village

草皮纸

Straw Paper in Dashan Village

草皮纸（桑皮+稻草）透光摄影图
A photo of straw paper (mulberry bark + straw) seen through the light

第十二章
工具

Chapter XII
Tools

第一节

永庆制帘工坊

浙江省
Zhejiang Province

杭州市
Hangzhou City

富阳区
Fuyang District

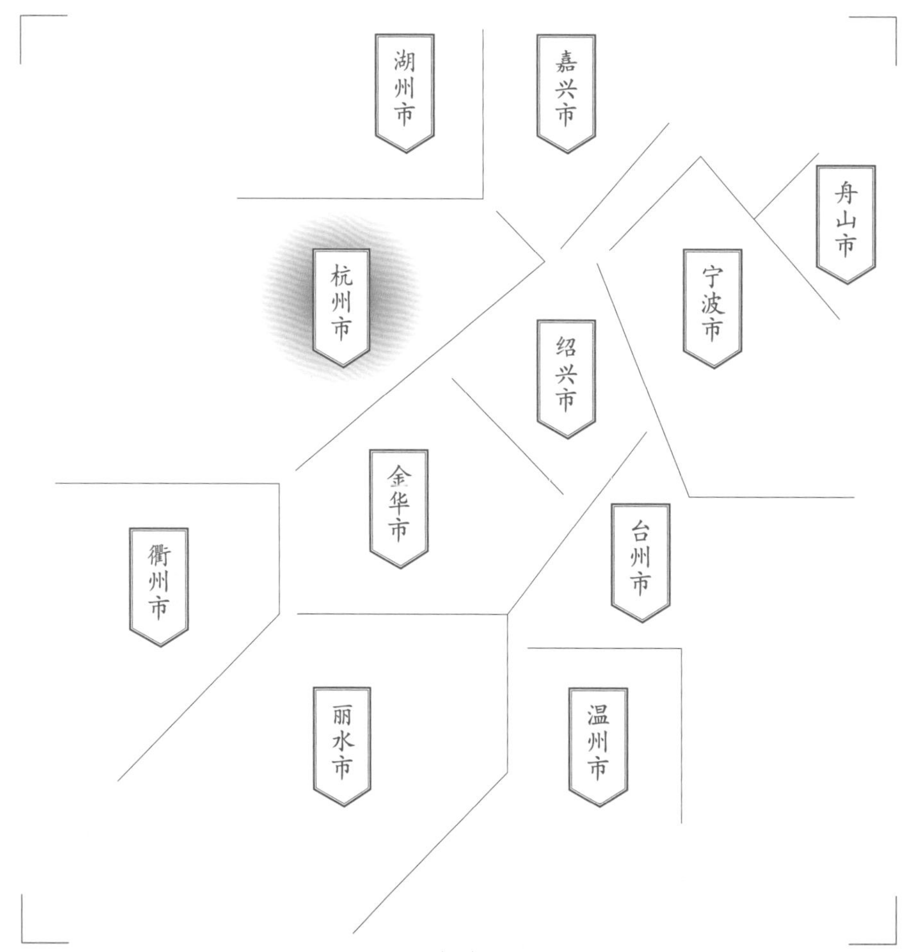

调查对象
富阳区大源镇
永庆制帘工坊
纸帘

Section 1
Yongqing Screen-making Mill

Subject
Papermaking Screen of Yongqing Screen-making Mill in Dayuan Town of Fuyang District

一 永庆制帘工坊的基础信息和生产环境

1 Basic Information and Production Environment of Yongqing Screen-making Mill

大源镇位于富春江南岸，距富阳城区中心7 km，距杭州市区42 km，东邻萧山市，南与上官乡、常绿镇相连，北接灵桥镇，西面靠富阳城关镇，镇区域面积10.5 km^2。地理坐标为：东经120° 0′，北纬30° 0′。大源是一个交通、通信便捷，工、农、商、贸综合发育的大杭州市区的卫星镇。

2018年，杭黄高铁驶进了大源镇，这条全长265 km的铁路，185 km在杭州域内，设杭州东、杭州南、富阳、桐庐、建德东、千岛湖6座车站，其中富阳站就设在大源镇亭山村，为两台四线侧下式。同时，富阳站预留了地铁出入口。富阳站站房共有三层，地下一层，地上两层，地面一层主要是购票厅、出站口、候车厅等。富阳站的开通不仅为大源镇居民、游客的出行和旅游提供了极大的便利，还给大源的经济发展、乡村建设带来了新的可能。

⊙2

⊙1

同样在2018年，“杭州绕城高速西复线大源互通”项目也正式启动，线路全长98 km，双向6车道，设计时速100 km/h，其中富阳境内48.54 km，占全线总长的近一半，待2020年西复线通车后，从东前互通到杭州只要半个小时，将有力改善大源的交通运输能力，推动包括大源镇在内的富阳和杭州主城区的快速联动，极大地带动大源镇的经济发展。

⊙1 杭黄高铁富阳站 Fuyang Station of Hangzhou-Huangshan High-speed Railway

⊙2 大源高速公路 Expressway of Dayuan Town

路线图
富阳城区
↓
永庆制帘工坊
Road map from Fuyang District centre to Yongqing Screen-making Mill

永庆制帘工坊位置示意图

Location map of Yongqing Screen-making Mill

考察时间
2016年9月 / 2019年3月

Investigation Date
Sep. 2016 / Mar. 2019

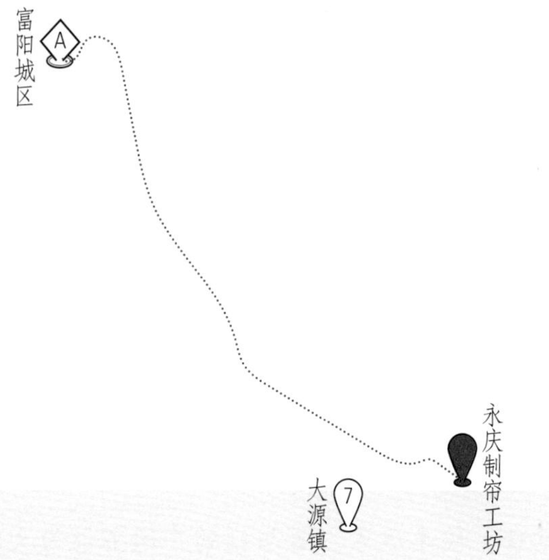

地域名称

富阳城区
永庆制帘工坊
大源镇

A 富阳区
① 湖源乡
② 常安镇
③ 洞桥镇
④ 新登镇
⑤ 灵桥镇
⑥ 新义乡
⑦ 大源镇

造纸点名称

生产点 永庆制帘工坊

位置分布

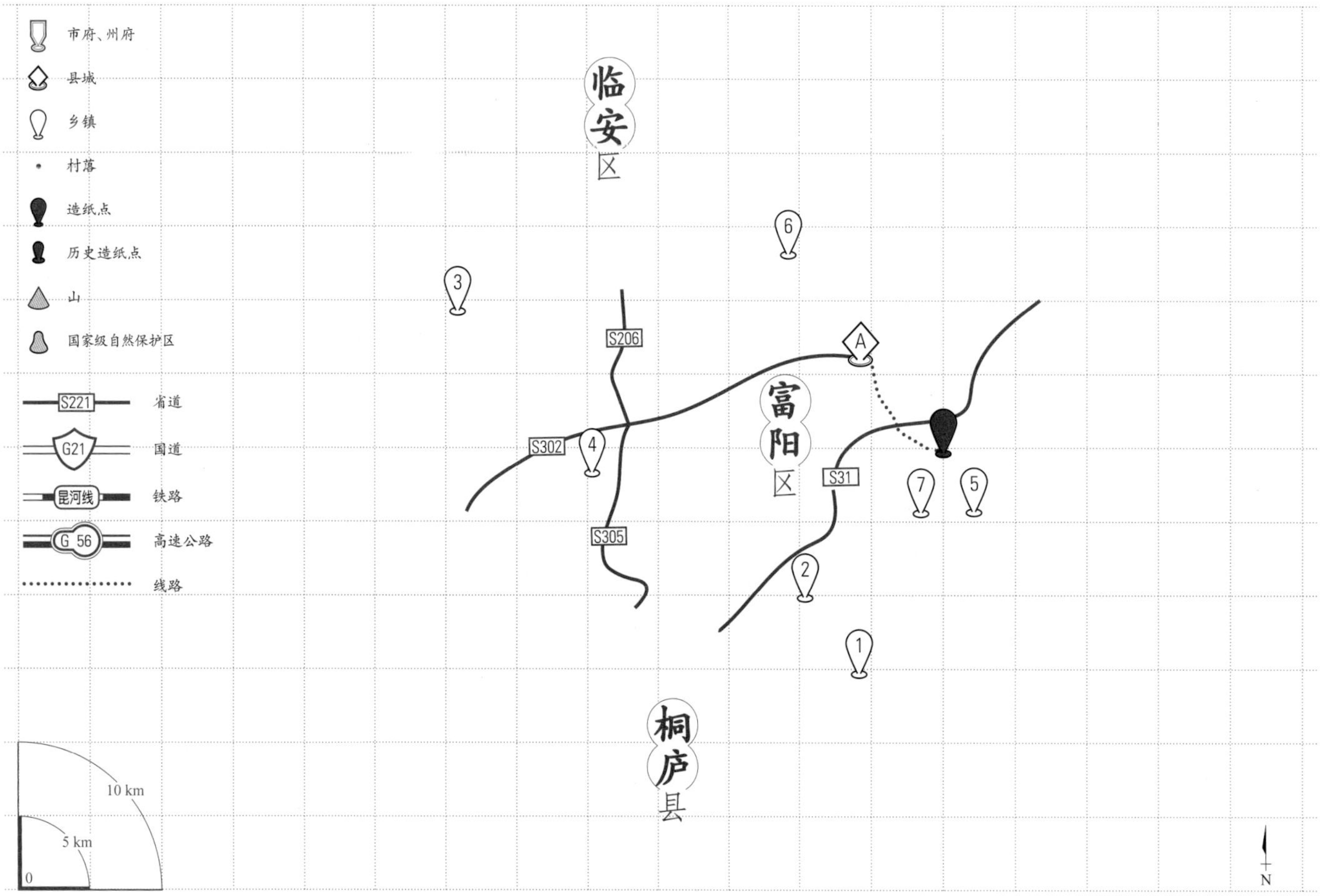

市府、州府
县城
乡镇
村落
造纸点
历史造纸点
山
国家级自然保护区
S221 省道
G21 国道
昆河线 铁路
G56 高速公路
线路

临安区
富阳区
桐庐县
S206
S302
S305
S31

10 km
5 km
0
N

⊙1

2019年3月调查组回访时得知，近两年来，大源镇启动了史上规模最大的集镇搬迁，杭黄站场区域基本拆净，为下一步产业转型、城市建设留足了空间。围绕高质量打造宜居宜业新大源的发展定位，区里将进行杭黄一平方公里城市建设，这块区域就在大源镇，今后将打造成高铁商务区。

大源镇是富阳历史上著名的竹纸之乡，手工造纸业态一直很发达，永庆制帘工坊就坐落于大源镇的永庆村。永庆村本为行政村，2007年行政村规划调整时，将其与原来的浦东、潘塘、望仙、下郎五村合并调整为大源村，此后永庆村属于大源镇大源村下的一个自然村。现大源村区域面积4.95 km^2，有26个村民小组，农户1 159户，常住人口3 780多人，经济收入主要依靠金属门窗制造、加工、造纸等。

永庆制帘工坊是一个家庭式的手工制作纸帘的作坊。2016年9月29日，调查组对永庆纸帘工坊进行了实地考察。永庆制帘工坊地理坐标为：东经120° 0′ 20″，北纬30° 0′ 7″。

⊙2

⊙1 大源风光图 Landscape of Dayuan Town

⊙2 永庆制帘工坊外观 Exterior view of Yongqing Screen-making Mill

二
永庆制帘工坊的历史与传承

2
History and Inheritance of Yongqing Screen-making Mill

富阳作为可考造纸历史近千年的造纸之乡，其传统一直使用原始的帘果掼织织帘法编织抄纸的帘子。帘果掼织织帘法，是纯手工的编织纸帘方法，先担好架子，用长短不一的腈纶线挂上一串串的“帘果”，编织纸帘的时候利用“帘果”一上一下绕，把竹丝编上去。据王增福介绍，“帘果”是用泥烧制而成的编帘小工具，两头凸出，中间凹陷的部分可以缠绕腈纶线。“帘果”的作用一是为了防止腈纶线的相互缠绕，二是更加便于受力，可以将竹丝编得更紧一点。

⊙1

⊙2

经过长时间的演变，到现代出现了和织布相似的木机制帘方法，20世纪70年代通过针对性的技术革新，木机制帘技艺得到完善，并成为主流被沿用下来。

富阳的纸帘制作历史悠久，但作为一个手工造纸的附属行业，相关文献记载稀少，其起源时间也没有断代的支撑资料。但富阳制帘的工艺水平曾经是非常高的，1926年在北京国货展览会上，富阳竹帘得到特等奖牌；20世纪60年代，富阳制帘厂开始生产竹帘工艺品，1983年，以竹帘编制技艺为主要工艺手法的竹丝画帘屏风，获中华人民共和国对外经济贸易部颁发的荣誉证书。

⊙1
王增福展示帘果架
Wang Zengfu showing a tool made of clay for screen-making

⊙2
帘果
A tool made of clay for screen-making

长期以来，富阳抄纸竹帘的制作技艺依靠家族传承和师徒传承，其技法各有小变而不离其宗。制帘工序相当复杂，有50多道，在制作时对手工技艺和手感把握要求很高，难以形诸文字及语言，言传身教是主要的传承方法，要凭个人的悟性以及长期实践才能掌握。

2016年9月第一次入坊调查时，永庆制帘工坊是富阳地区极少的一家还在生产的竹帘工坊。作坊主王增福和妻子章爱苏负责制作竹帘，女儿王华负责销售。王增福于2013年申请成为杭州市纸帘制作技艺非物质文化遗产传承人，工坊则在2015年12月被评为杭州市富阳区非物质文化遗产生产性保护基地。

⊙3

⊙4

王增福，1954年出生于永康市，年幼的时候随着父辈来到了大源永庆，2016年第一次调查时为62岁。1978年进入富阳制帘厂（后改名富阳工艺美术总厂）工作，在富阳制帘厂里拜老师傅陆余良为师，学习漆帘等制帘技艺，师祖是陆余良的父亲陆柏松。学了两三年之后，便可以独立做帘了，于是在厂里担任漆帘工（手工竹帘上漆）并参与制作画屏风形式的竹帘。1996年企业破产，王增福回到大源镇永庆村创办王增福竹纸帘手工作坊，以制帘为业。据王增福的说法，他家祖上六代都是篾匠，由此才有了做抄纸竹帘的基础功底。

⊙5

⊙3 杭州市富阳区『非遗』生产性保护基地授牌
Award of the Intangible Cultural Heritage Production Protection Base in Fuyang District of Hangzhou City

⊙4 『非遗』传承人荣誉证书
Honorary certificate of inheritor of Intangible Cultural Heritage

⊙5 王增福工作图
Wang Zengfu at work

王增福的妻子章爱苏，1956年出生于常绿镇，1982年，26岁的章爱苏嫁给了王增福，也进入了富阳制帘厂工作，在厂里学会了织帘的技术。1996年，与丈夫王增福一起办起了永庆制帘工坊。

⊙1

王增福有两个女儿，大女儿王华，1983年出生，12岁的时候就开始跟着母亲章爱苏学织帘技术，大学毕业后一直在外从事会计工作，不再织帘。小女儿王洁，2001年出生，2019年3月调查组回访时，王洁正在读高二，没有学习过织帘技术。大女婿施银能，1983年出生，2013年时跟随王增福学习过漆帘的入门技艺。但是据王增福说，女婿从事的是软件行业，工作比较好，“学这个只是为了玩玩，并不会从事这个行业”。

⊙2

⊙1 织帘的章爱苏
Zhang Aisu making the screen

⊙2 与调查员交流制帘技艺的王增福（右）、章爱苏（左）
Wang Zengfu (right) and Zhang Aisu (left), exchanging techniques of making screen with the researchers

三
永庆制帘工坊的竹帘品种

3
Varieties of Bamboo Screen in Yongqing Screen-making Mill

富阳地区竹帘品种并非仅有造纸所用这一类，还包括其他多类用途。依据2019年3月回访调查的信息，永庆纸帘工坊制作的竹帘用途分为以下3种：

1. 造纸用帘

帘面平整如绸，滤水快，造出的纸张均匀、光洁。尺寸可以根据来单定制，永庆制帘工坊纸帘订单最多的是四尺纸帘。

⊙3

王增福透露，纸帘的价格一般根据尺寸来定，约700元/m^2。一个纸帘的使用时长和捞纸师傅的手艺相关，一般情况下半年一换，好的师傅是一年一换。

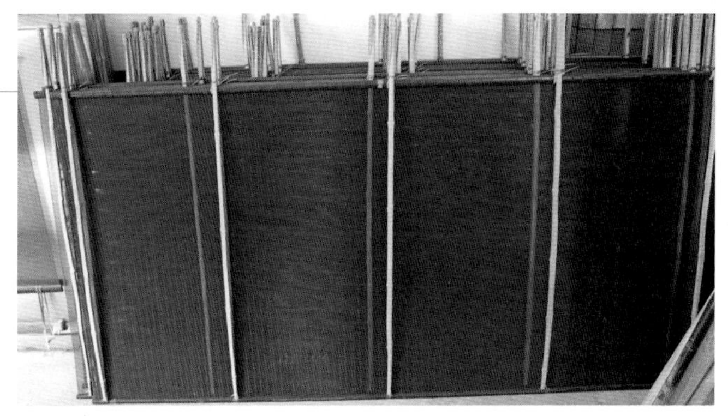

⊙4

2. 门窗帘

竹丝编制，通常漆成紫色，悬挂于门、窗，可筛挡日影光波，轻巧，透风，遮阳，隔视线，挡雨水。门窗帘的尺寸不一，根据订单定做。价格也是不一，如果自己去上海推销，价格会高一点，约750元/m^2，若是客户自己上门，价格会比较便宜，约400元/m^2。王增幅讲述，这种门窗帘不易坏，可以用好几年。

⊙3 四尺纸帘 4-chi papermaking screen

⊙4 八尺纸帘 8-chi papermaking screen

⊙1

3. 茶帘

2016年调查中王永福介绍，工坊于2014年开始制作竹丝编茶席用途的茶帘。帘子分大小，价格约为400元/m²，最小的尺寸为3 cm×3 cm，最大的为30 cm×300 cm。

茶席竹帘是北京的客户专门定做的，但是从2016年开始订货量意外大幅下降。2019年3月回访时，王增福介绍，2017年这种茶帘销量很低，到了2018年，销量有所回升，去年约有80 000元的销售额。

⊙2

据王增福介绍，早年间在富阳制帘厂的时候，还做过一种画帘，竹丝细如发丝，可编织成不同的画面，悬挂于厅堂、书房，精细雅观，古香古色，这种画帘多用于出口。但是要制作这种画帘十分费劲，首先需要专用的竹丝，这种竹丝400元/kg，再来需要请专业的师傅去作画，十分

⊙1 门窗帘 Screen for door and window

⊙2 半成品茶帘 Semi-finished screen for tea

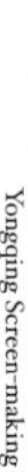

⊙3

麻烦，用章爱苏的话来说，“一般来说小作坊是吃不消的”。

此外，王增福还介绍了另一种帘子——屏风帘，其中最知名的便是竹贴屏风。这种屏风是先做好普通的屏风帘，然后将帘子上好白漆，用胶水贴上修剪成各种形状的紫竹片，利用紫竹片贴出屏风图案。王增福说他的师傅陆余良会设计竹贴屏风，早些年的时候都是从王增福这里买成品的屏风帘回去加工成竹贴屏风。如今陆余良年岁比较大，早已经退休，不再做了。

⊙4

⊙3 画帘 Painted screen

⊙4 紫竹 Black bamboo

四 永庆制帘工坊的制帘材料、工艺与设备

4 Screen-making Materials, Techniques and Tools of Yongqing Screen-making Mill

（一）抄纸竹帘的生产原料

1. 苦竹丝

制作抄纸竹帘的原料以苦竹为首选，因为苦竹茎节长，易取出无竹节的较长竹丝，对抄纸时湿纸形成平整有利。苦竹秆基的节间较长，竹根少，两侧有芽眼2～6枚，既可以发育成竹鞭，在土中横向生长；也可抽笋长成新竹，成丛生长。1年生的成竹为幼龄竹，2～3年生竹为壮龄竹，4～5年生竹则为老龄竹。

据王增福、章爱苏夫妇介绍，他们家工坊的苦竹主要是从温州瑞安购买的半成品竹丝，大约20年前用的是富阳当地的苦竹，后来由于当地苦竹林被破坏导致品质不好而弃用，从1996年自己创建工坊开始就直接购买竹丝了。苦竹丝选用的竹料以5年左右的老竹为最佳。成品的竹丝2016年市场价约为140元/kg，更细一些的每千克要200元以上，价格趋势是越细的越贵。2019年3月回访纸坊，得知苦竹丝的价格是年年增长，2017年约为150元/kg，2018年约为160元/kg。

2. 土漆

与中国另一大竹纸聚集地四川夹江地区的制帘工艺不同，王增福制帘刷漆只用生漆即可，不需要进行二次加工。生漆是从陕西购买回来的，单价约为240元/kg。一次未使用完的生漆要用塑料薄膜尽量密封，放在阴凉处，不能暴晒。2019年回访得知，土漆目前的价格约为280元/kg。

⊙1

⊙2

⊙3

⊙1 购买的半成品竹丝 Purchased semi-finished bamboo filament

⊙2 购买来的土漆 Purchased raw paint (not requiring secondary processing)

⊙3 盛漆的木质桶 Wooden bucket for holding paint

3. 腈纶线

腈纶线是织帘时的重要辅料，用来将竹丝编织固定。王增福工坊内的腈纶线还是1996年工坊建立之初在上海购买的腈纶丝，自己去杭州的萧山区加工成腈纶线。由于当年买的量大，加上每年的纸帘订单也不多，用了20年还没有用完。

⊙4

⊙5

（二）抄纸竹帘的制作工艺流程

因造纸生产所需，富阳竹帘制作已有千年的传承，具有一整套成熟的制作工艺。当代富阳抄纸竹帘的制作技艺，借助木织机，以线为经，竹丝为纬，编织成松紧相宜、结边挺直的竹帘，涂以优质土漆，具有帘丝细、匀、圆、滑、韧的特点。

根据调查中王增福的介绍，主要制帘工艺可归纳为：采竹（砟竹、锯竹）、劈篾（对开、劈条、去黄、晒燥、运回、贮藏）、抽丝（水浸、笃篾、刮青、去节、削尖、缚把、脚踩去毛、抽丝、晒燥、蘸漆、搓匀、摊地阴干、熟帘丝分长短、再缚把）、织帘（蚕丝打线、调线、剪线、帘果绕线、线挂帘架、放帘丝编织、缚帘、上脚篾、摊开、钳开口、修边、撑帘）、漆帘（帘棒竹挑直、帘棒竹划线、帘部竹烧直、帘棒竹钻洞、帘棒竹刮节、吊帘、撑帘敞、漆帘、摊地、盖笠、蘸帘漆、漆二遍、量角测正、号字“作坊名、油漆日期”、直靠放、再量角测正）。通常漆后阴凉处晾一个月后可使用。

漆帘工艺要求高，线路和帘丝必须成90°直角，帘部用火烤时要直；刷漆均匀，不能有杂质和漆块，不能两丝结块。

调查中了解到，王增福对传统工艺进行了改进和简化。

⊙4 腈纶线 Acrylic Line

⊙5 王增福、章爱苏展示腈纶线团制作工艺 Wang Zengfu and Zhang Aisu demonstrating the papermaking techniques of acrylic thread

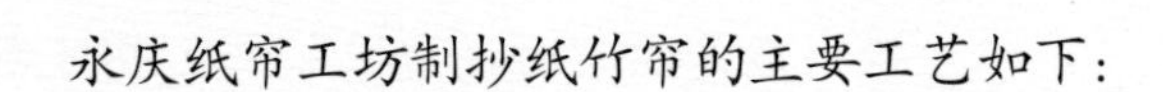

永庆纸帘工坊制抄纸竹帘的主要工艺如下：

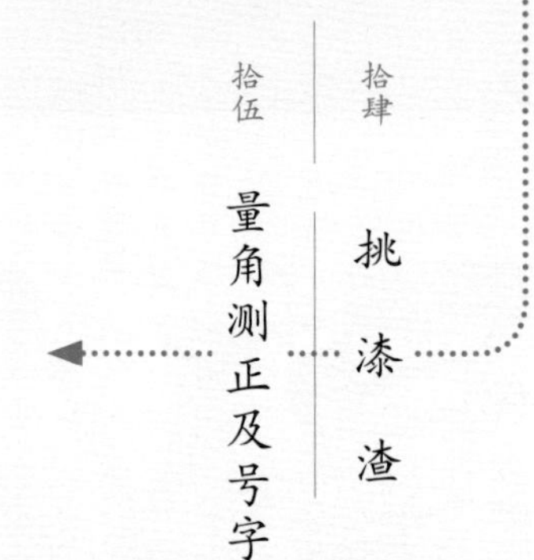

壹 锯竹

1 ⊙1

将苦竹料锯成长竹节。

⊙1

⊙3

⊙4

贰 去节刮青

2 ⊙2⊙3

用绞结刀将竹节部分刮平，并将竹节剖开，刮除青色部分，再将竹节剖成宽约1 cm的竹条，用冷水浸泡1天。

⊙2

⊙5

叁 抽丝

3 ⊙4～⊙6

将竹条根据客户定制竹帘帘纹密度的需求抽成不同细度的竹丝，调查时王增福的工坊可以完成最细直径0.58 mm的竹丝。

⊙6

⊙1 王增福在锯竹
Wang Zengfu sawing the bamboo

⊙2/3 王增福在去节刮青
Wang Zengfu scraping the bamboo joint and the cyan part

⊙4/6 王增福在抽丝
Wang Zengfu threading

肆

帘丝上漆

4 ⊙7～⊙9

将抽好的竹丝，用漆刷均匀地刷上土漆，自然阴干后，脚踩帘丝去毛，这样可以防止毛躁的竹丝在织帘时伤到手。

⊙7

⊙8

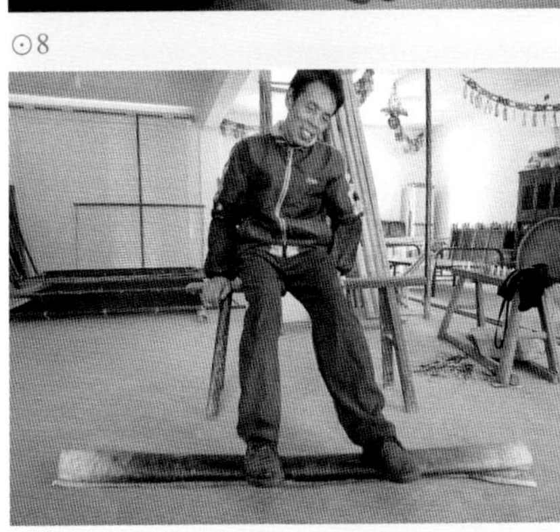

⊙9

⊙10

伍

煞帘丝

5 ⊙10

即将上完漆的帘丝按长短分类，便于织帘时准确取用。

⊙11

⊙12

陆

织帘

6 ⊙11～⊙13

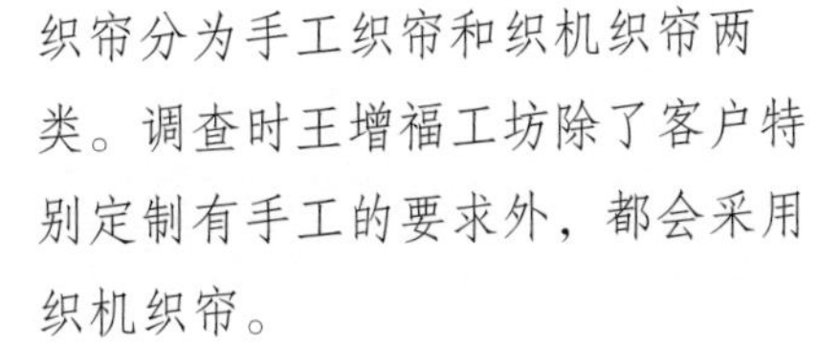

织帘分为手工织帘和织机织帘两类。调查时王增福工坊除了客户特别定制有手工的要求外，都会采用织机织帘。

⊙13

⊙7 王增福在给帘丝上漆 Wang Zengfu painting the screen filament

⊙8 帘丝上漆后阴干 Drying the screen filament silk in the shade after painting

⊙9 王增福脚踩帘丝去毛 Wang Zengfu stamping the screen filament to remove the rough edges

⊙10 章爱苏在煞帘丝 Zhang Aisu classifying the screen filament

⊙11 章爱苏在用织机织帘 Zhang Aisu making the screen with a loom

⊙12 工人在用织机织帘 Workers making the screen with a loom

⊙13 章爱苏示范纯手工织帘 Zhang Aisu demonstrating the screen making by hand

⊙14

柒

打结头

7 ⊙14

将编制好的竹帘的每两根帘丝的接触部分用腈纶线固定，防止散开。

捌

帘棒竹烧直

8 ⊙15

将用来固定悬挂竹帘的帘棒弯曲的地方烧直。

⊙15

玖

吊帘

9 ⊙16⊙17

在帘棒上打孔，并将竹帘固定在帘棒上。调查时王增福已使用打孔机打孔。

⊙16

⊙17

拾

钳开口

10 ⊙18

用普通的眉毛钳手动对每排没有对齐的帘丝进行矫正对齐，使得每一排每一列的竹帘帘丝的接口处相互对齐。这项工作比较耗费时间和精力，对焦时间久了眼睛也会酸涩疼痛。

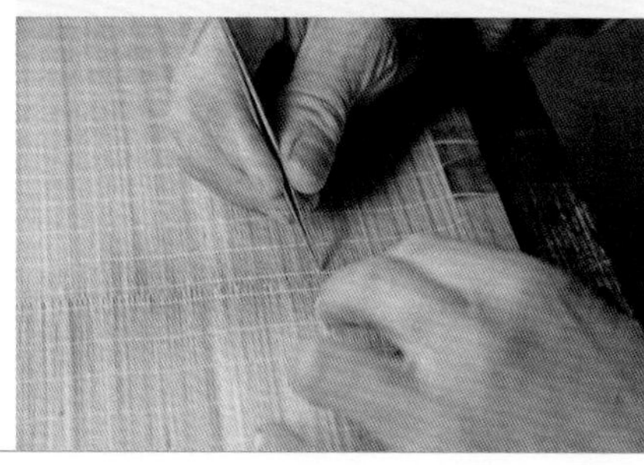

⊙18

拾壹

排稀密

11 ⊙19

对竹帘竹丝之间疏密不一的间隔做调整，保持间隔距离等距并固定。

⊙19

⊙14 工人在给竹帘打结头 A worker knotting a bamboo screen

⊙15 王增福在烧直帘棒 Wang Zengfu making the screen rod straight through burning

⊙16 王增福在给帘棒钻孔 Wang Zengfu drilling a screen rod

⊙17 王增福将帘棒与竹帘固定在一起 Wang Zengfu fixing the screen rod and bamboo screen together

⊙18 王增福在钳开口 Wang Zengfu aligning the junction of the screen filament

⊙19 王增福在排稀密 Wang Zengfu arranging the screen filament

拾贰
竹帘上漆

12 ⊙20

将编织好的竹帘均匀地刷上土漆。王增福口述的技艺要领是：抄纸竹帘在阴雨天上漆最好，茶席竹帘晴天上漆最好。王增福工坊只需给竹帘上一次漆即可。据王增幅介绍，上漆的次数多了，虽然帘子看上去会比较漂亮，但是对于纸帘来说，其捞出的纸的质量会不好，容易破损。

拾叁
掸清

13 ⊙21

刷漆过程中，同时需要将产生的杂质用毛刷清扫干净。

⊙21

⊙22

拾肆
挑漆渣

14 ⊙22

竹帘干燥后，用尖门针把刷漆过程中夹在帘丝之间的漆渣挑除干净。

⊙20

拾伍
量角测正及号字

15 ⊙23⊙24

检测刷漆线路和帘丝是不是成90°直角，算是一种质检的方式，检查无问题后在竹帘上写上作坊名称和刷漆日期。

⊙23

⊙24

⊙20 王增福在给竹帘上漆 Wang Zengfu painting the bamboo screen

⊙21 王增福在给竹帘掸清 Wang Zengfu dusting the bamboo screen

⊙22 王增福在给竹帘挑漆渣 Wang Zengfu picking out the residues

⊙23 王增福在量直角 Wang Zengfu measuring right angles of bamboo screen

⊙24 号字 Sighing after checking the quality

（三）抄纸竹帘的主要制作设备和工具

壹 绞结刀和夹子 1

用来刮除竹节的突出部分。实测永庆制帘工坊绞结刀尺寸为：长20 cm，宽5.3 cm，柄长10.5 cm。竹夹尺寸为：长26 cm，直径0.5 cm。

自制的竹夹子是用细竹段剖开而成的，用来固定刀具，便于受力。

⊙1

⊙2

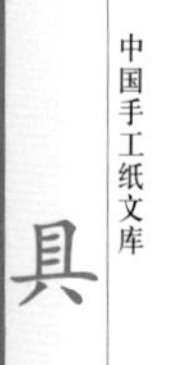

贰 抽丝工具 2

用来抽丝的工具，主要由支架和刀片构成。实测永庆制帘工坊的抽丝工具尺寸为：刀片长4.3 cm，宽2 cm，上面布满大小不一的洞；支架高18 cm，宽4 cm，厚1.7 cm。

⊙4

叁 扁刷 3

掸清时使用。实测永庆制帘工坊所用扁刷尺寸为：刷头长12.8 cm，刷毛长3 cm。

⊙3

⊙5

⊙1 绞结刀和夹子 Twisting knife and clip

⊙2 王增福示范刮竹节 Wang Zengfu demonstrating scraping the bamboo junction

⊙3 抽丝刀片 Blade for threading

⊙4 抽丝支架 Frame for threading

⊙5 扁刷 Flat brush for dusting

肆 漆刷 4

竹帘上漆、帘丝上漆时使用的刷子。实测永庆制帘工坊漆刷尺寸为：长32 cm，圆筒直径6 cm。

⊙6

伍 自制量线工具 5

用来测量腈纶线的长短与多少的自制工具。

⊙7

⊙8

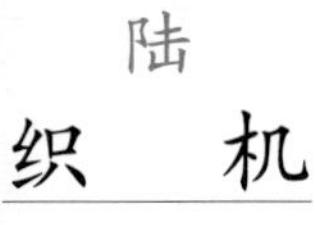

陆 织机 6

编织竹帘类似织布机的织机。

⊙9

柒 尖门针 7

王增福用来挑漆渣的工具。

⊙10

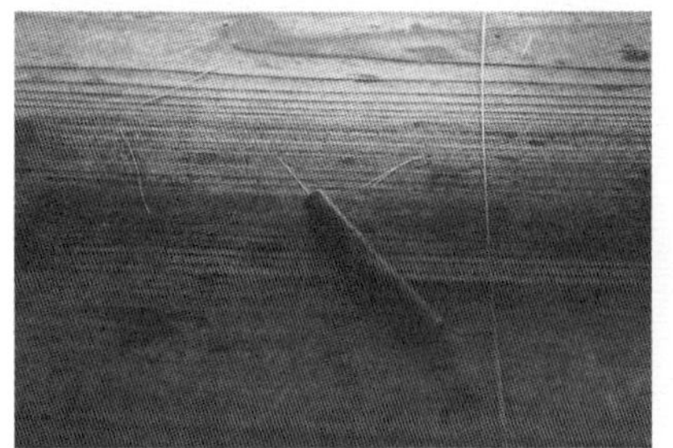

⊙6 漆刷 Brush for painting

⊙7 王增福展示量线工具的用法 Wang Zengfu demonstrating the usage of the measuring tool

⊙8 量线工具底部特写 Bottom of the measuring tool

⊙9 编帘织机 Loom for screen-making

⊙10 尖门针 Sharp needle for picking out the residues

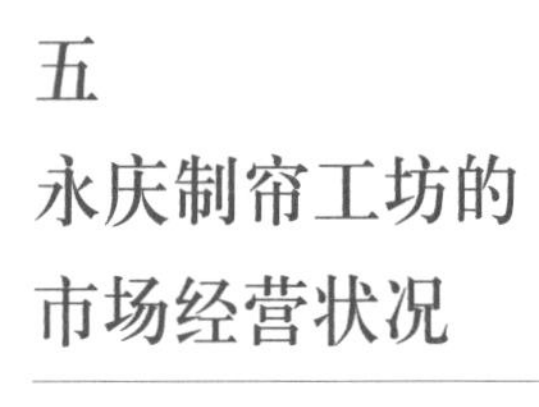

五 永庆制帘工坊的市场经营状况

5 Marketing Status of Yongqing Screen-making Mill

2016年入工坊调查时，永庆制帘工坊共雇佣了织帘工人2人，王增福和妻子章爱苏负责生产，以往订单多的时候也会临时加人生产，但近年来订单不多已经不再雇佣工人了。2019年3月回访工坊时，永庆制帘工坊只有一位王姓织帘工人，王增福告诉调查组，有一个织帘工人因为工坊的活不多不做了，但偶尔订单多的时候会通知她过来帮工。织帘工人的工资是多劳多得，一般而言，一张四尺的纸帘付70元的工资，按照四尺的标准来说，织帘工人一天可以织一张多，平均下来一天收入100多元。

2014～2016年，工坊年产竹帘约4 000 m^2，年销售额约20万元。根据质量不同，售价400～600元/张不等。生产的主要规格有四尺（153 cm × 85 cm）和小五尺（173 cm × 77 cm）两种，其余规格产品根据客户要求定制。2015年时，新产品茶席竹帘一度占据工坊销售额的一半，但在2016年迅速下降。2019年回访得到的信息是：纸帘是工坊的主要产品，2018年卖了300～400张不同规格的纸帘，销售额约为28万元；门窗帘的订单较少，只做了几张帘；茶帘的销售渠道比较窄。王增福透露，一张帘基本上只能赚一两百元，利润比较低。

工坊竹帘的销售区域主要集中在富阳周边地区，广西、安徽、北京、嘉兴、江西等地也有少量订购；2016年6月有法国客户上门参观订购，小批量销往法国。

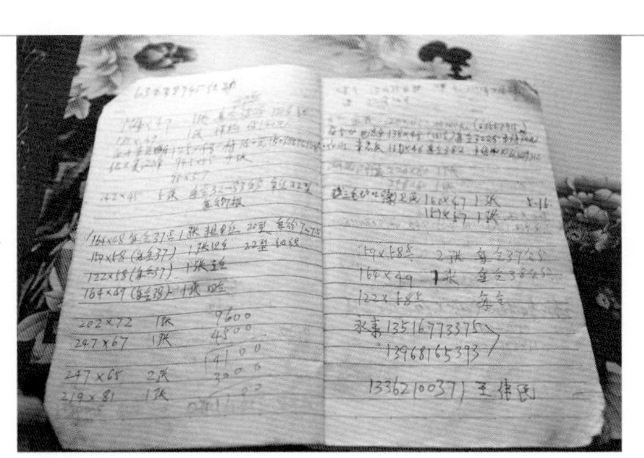

⊙1

⊙1 永庆制帘工坊的来往账本 Account book of Yongqing Screen-making Mill

⊙2

⊙3

对于线上的销售，其实早在2009年11月，永庆制帘工坊就由女儿王华注册了淘宝网店——娃娃手工艺小铺。王华认为工坊竹帘的销量主要还是线下，线上销售很少。她表示店铺主要由她经营，但是由于自己的工作比较忙，所以没有付出很多的精力和时间在网店上。2014年她稍微为网店做了点宣传，当年的线上销量就比较好，销售额达到四五万元。因此，王华认为，一旦能好好打理网店，做好店铺宣传，线上的销路还是可以打开的。

⊙2 女儿王华与法国客户交流
Wang Zengfu's daughter Wang Hua communicating with French clients

⊙3 法国客户赠送的皮纸
Bast paper given from a French clients

六
富阳区大源镇制帘工艺相关民俗及文化事项

6
Folk Customs and Culture of Screen-making Techniques in Dayuan Town of Fuyang District

(一)分工明:男漆女织

富阳竹帘在制作过程中,流传着一些习俗。富阳竹帘编织这道工序一般是由妇女做的,而漆帘等工序是由男人做的,是为“男漆女织”。这种分工约定俗成,似乎漆帘的只能是男性,织帘的只能为女性,分工明确,不容混淆。

但是凡事也有特例。20世纪中期,还在用帘果掼织法织帘的时候,织纸帘的工人中有男有女,其中很多是男性,并且大多是腿有残疾的男性。那时候有腿疾的男性没办法做别的活,而织帘却是可以只用手就能完成的工种,因此在当年很多织帘厂特意设立了这种“福利岗”,让有腿疾的人也有活可做。

⊙1

⊙2

⊙3

⊙1 女掼织 Female making the screen

⊙2 男漆帘 Male painting the screen

⊙3 永庆制帘厂厂牌 Brand plaque of Yongqing Screen-making Mill

(二)声名起:一次偶然的评比

据章爱苏讲述,以前自家的纸帘名气不够大,生意也不够好,生意的转机源于五六年前一次偶然间的宣纸纸帘评比。当时双溪有个造纸老板金祖民(音译),因为一时间需要多张宣纸纸帘,怕一家制帘厂来不及做,便定了三家做纸帘,分别是永庆制帘厂、大源的一家汪姓制帘厂和光明制帘厂。三家纸帘厂陆续交工之后,试用过三家纸帘的捞纸师傅便感觉永庆制帘厂的宣纸纸帘用起来更舒服,捞出的宣纸很光滑,纸的质量好。自此之后,永庆制帘厂的名声便传了出去,渐渐地客户量就多了。

七

富阳区制帘业及永庆制帘工坊的业态传承现状与发展思考

7

Current Status and Development of Yongqing Screen-making Mill

（一）面临的问题和困境

1. 环保风暴与机制纸流行双重夹击下的生存危机

近年来，一方面随着国家环保政策的出台与执行趋严，家庭式的造纸作坊因为排放污染无能力上环保设施被关闭或者面临关闭威胁的数量较多；另一方面，机制祭祀竹纸与书画纸对富阳手工造纸形成了很强冲击，从事手工造纸的厂家难以为继的也不少，一批造纸槽厂（坊）停办，基本客户萎缩严重，明显影响了作为配套产业的制帘业的生存。

2. 习艺从业艰苦与收入难保双重夹击下的后继无人

手工制帘上漆工艺复杂，技术要求高，土漆气味难闻且有致敏性，对漆工的要求高，习艺周期长，流程繁多，学习有一定的难度，劳动强度大，产量少，利润薄，因此年轻人不愿意学。如今师傅们大都年事已高，技术乏人传承，手工制帘技艺正面临着失传的危险。

⊙4

（二）传承与发展之路探索

（1）在保持较高水平制帘工艺的基础上，扩大客户与市场推广渠道，设法通过协作资源平台，建立网上与微信社交群的信息传播与行业人群交流通道，争取承接更大范围的潜在需求。

（2）积极争取政府与工艺组织推动的"非遗"研学体验计划的对象人群，如在一些非遗研学基地让学生、游客学习体验手工艺产品和技术，让制帘技艺与产品通过新的消费方式获得新的消费者。

（3）由富阳当地"非遗"中心等机构启动保护项目，全面收集、记录、整理富阳手工制帘上漆的制作技艺；收集、保护手工制帘上漆的相关文献、实物；成立大源手工制帘上漆保护机构；开展相关学术研究；保护健在的传承人，并探索新的传承方式，培养新的传承人。

⊙5

⊙4 独自织帘的章爱苏 Zhang Aisu making the screen alone

⊙5 学生体验用的纸帘 Papermaking screen for student to experience

第二节

光明制帘厂

浙江省
Zhejiang Province

杭州市
Hangzhou City

富阳区
Fuyang District

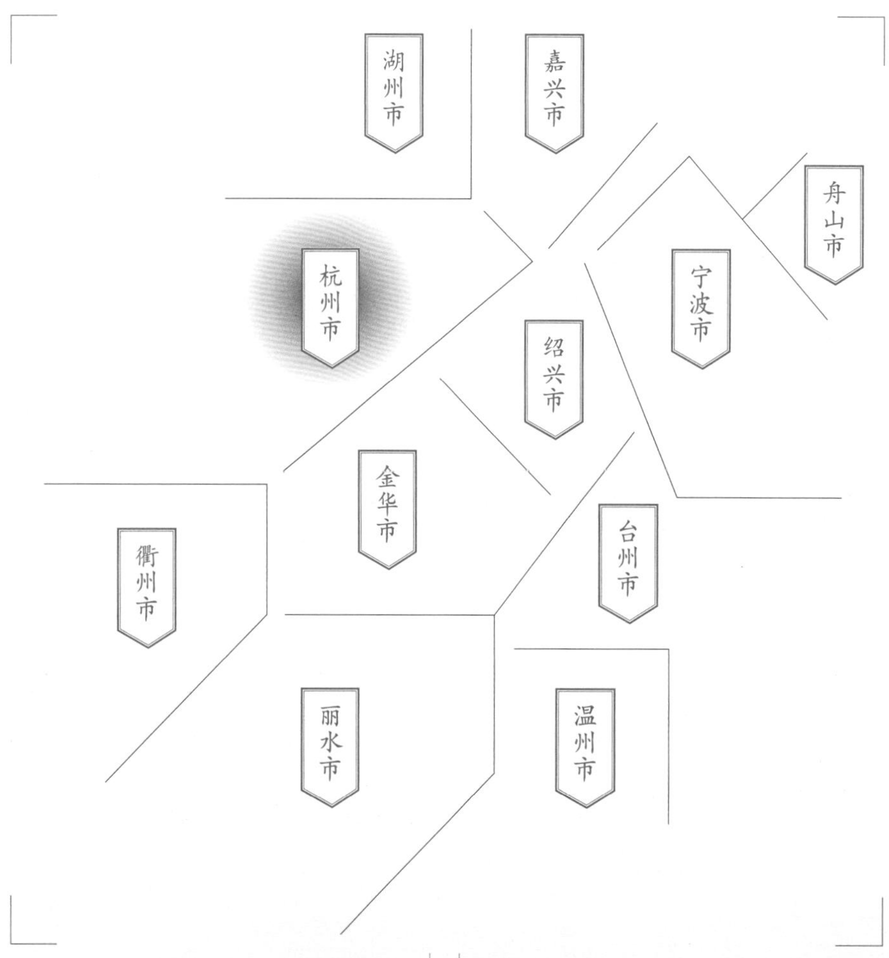

调查对象
富阳区灵桥镇光明村
光明制帘厂
纸帘

Section 2
Guangming Screen-making Factory

Subject

Papermaking Screen of Guangming Screen-making Factory in Guangming Village of Lingqiao Town in Fuyang District

一

光明制帘厂的基础信息与生产环境

1

Basic Information and Production Environment of Guangming Screen-making Factory

浙江富阳手工造纸已有千年历史，其竹纸制造技艺作为中国第一批国家级非物质文化遗产之一，享有“京都状元富阳纸，十件元书考进士”的美誉。富阳竹纸曾经长期成为富春江南岸山区及青云、龙羊、新登等地的重要传统手工支柱产业。2016年富阳区荣获中国文房四宝协会评审的“中国竹纸之乡”称号。富阳竹纸和安徽宣纸一样采用床架式抄纸法制作，为典型的蔡伦造纸法成纸方式，抄纸用的竹丝帘子作为生产制作竹纸的重要工具，也伴随着手工竹纸的制作在富阳当地得以催生发展。

《富阳日报》2014年曾载文介绍富阳的纸帘制作行业：

富阳的竹丝帘子，一出现便用于造纸。富阳山区生产竹纸，平原做草纸，富春江两岸都有土纸槽产，做纸少不了竹帘，所以，那时竹帘制作遍及富阳城乡。清光绪《富阳县志》载：“竹纸帘，丝细如发，漆如紫色，供造纸用。悬于门

⊙1

⊙1 富阳区境内的富春江两岸 Both banks of Fuchun River in Fuyang District

路线图
富阳城区
↓
光明制帘厂
Road map from Fuyang District centre to Guangming Screen-making Factory

光明制帘厂位置示意图

Location map of Guangming Screen-making Factory

考察时间
2016年8月 / 2019年3月

Investigation Date
Aug. 2016 / Mar. 2019

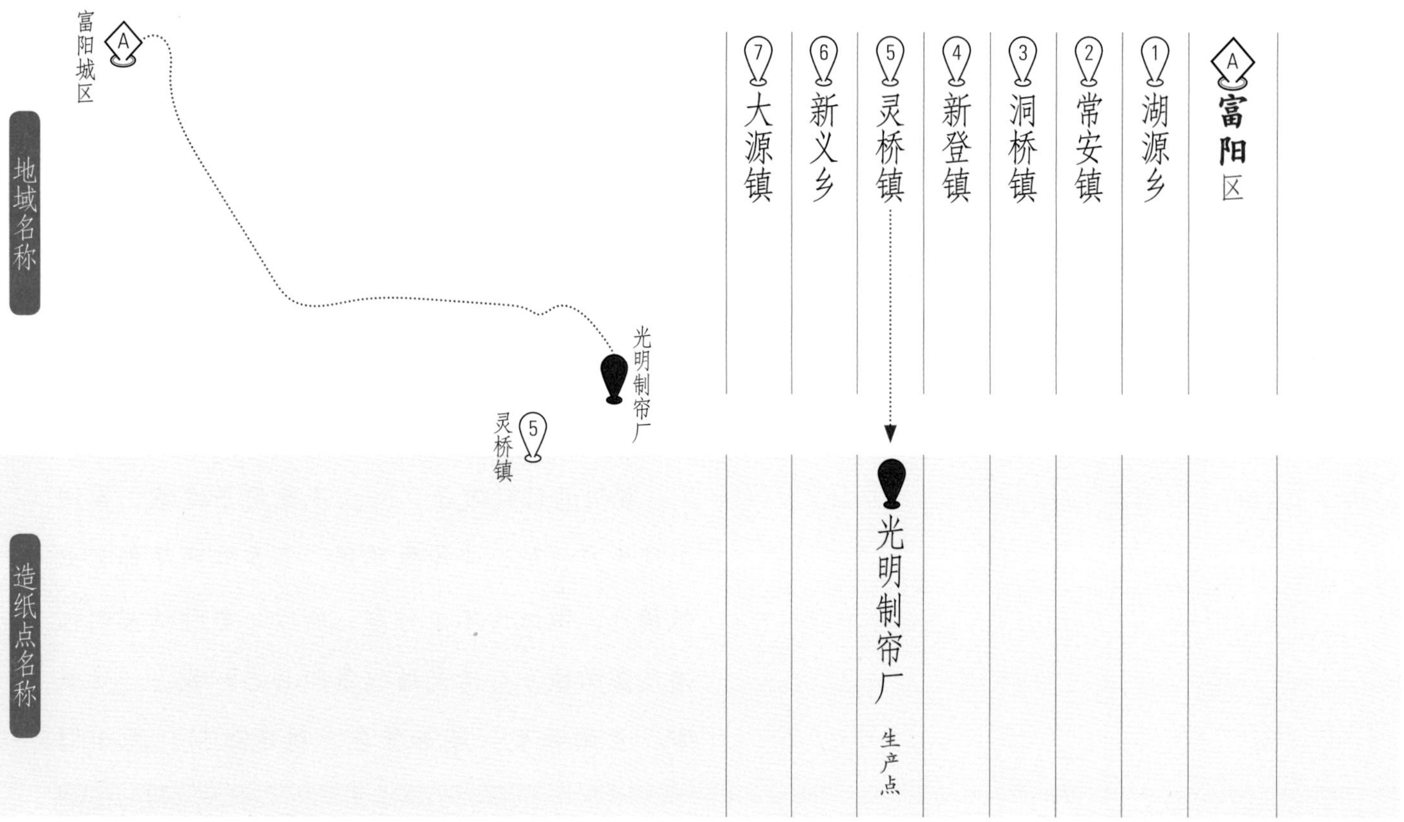

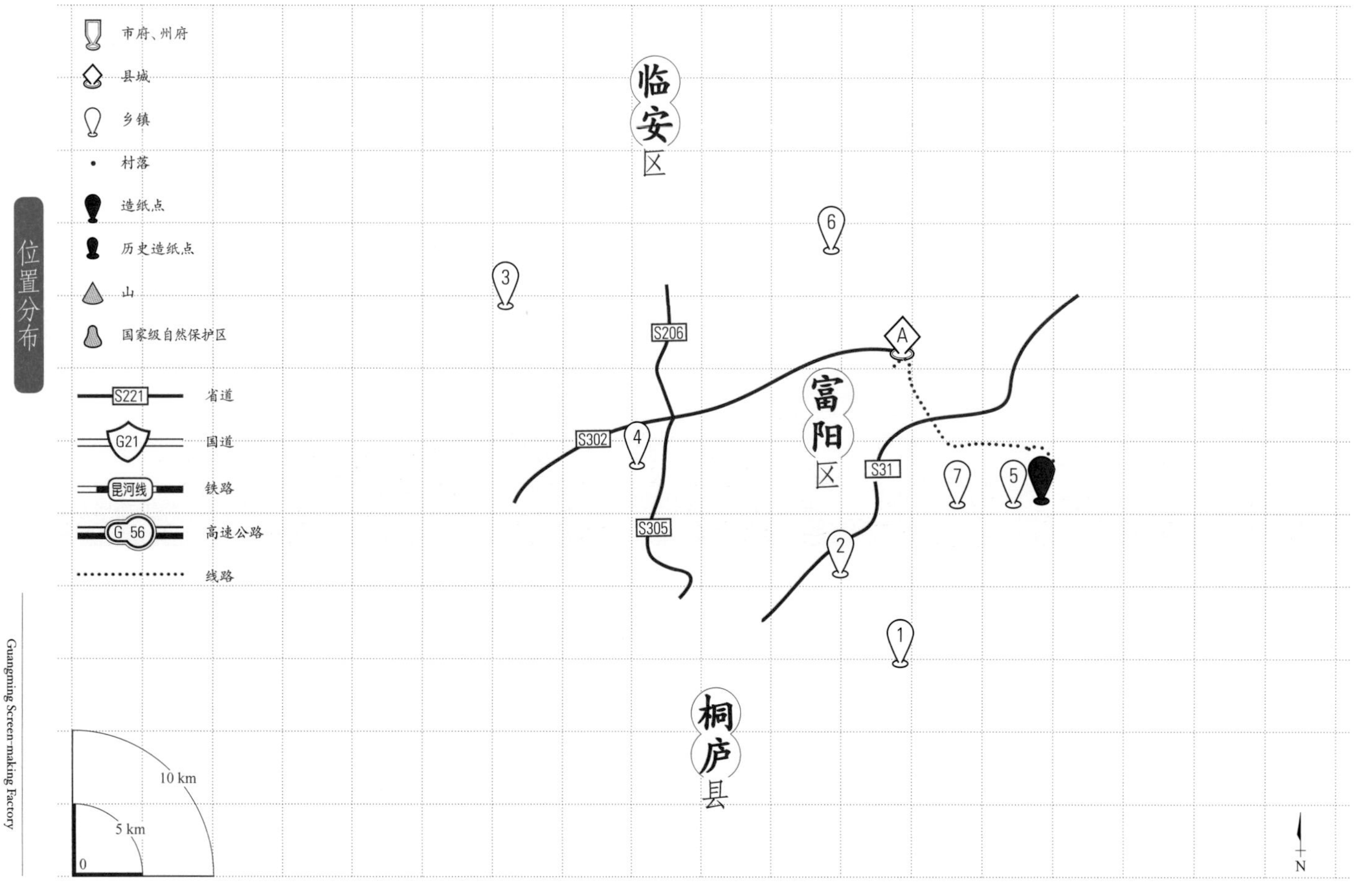

⊙1

窗，日影筛波，极为雅观，它处不能及也。”据《浙江经济略》记载，民国十五年（1926年），在北京国货展览会上，富阳之竹帘得特等奖牌。1983年，富阳工艺美术总厂生产的竹丝画帘屏风，获中华人民共和国对外经济贸易部颁发的荣誉证书[1]。

2016年8月23日，调查组一行对富阳区灵桥镇光明制帘厂进行了实地观察，通过访谈得知的基础信息为：光明制帘厂调查时的负责人汪美英，是第三批杭州市级非物质文化遗产手工制帘项目代表性传承人；光明制帘厂20世纪80年代末建厂，制帘厂从事竹帘制作加工已有30余年历史，加工场所占地面积约300 m^2，年产捞纸竹帘数量汪美英没有详细统计过，大概每月制作各类帘子20～30张。

光明制帘厂现厂区位于富阳区灵桥镇光明行政村，地理坐标为：东经120° 2′ 17″，北纬30° 0′ 25″。光明村属于富春江南岸的山区村，在2007年浙江省行政村规划调整中由原光明村和梓树村合并而成，2015年的统计数据是下辖5个自然村，12个村民组，有在籍村民583户2 028人，地域面积8.7 km^2。光明制帘厂厂区紧邻杭千高速公路富阳出口处，方莒线县道从其门前经过，交通便捷。

⊙1 灵桥镇光明村村碑石
Name stone of Guangming Village in Lingqiao Town

[1] 陈志荣.汪美英的织帘生涯[N].富阳日报,2014-08-16:12版.

二 光明制帘厂的历史与传承

2 History and Inheritance of Guangming Screen-making Factory

制帘虽然是富阳的传统流行手工业，但纸帘的制作以及传承历史相关的文献资料十分匮乏，几乎不见记载。据2016年8月调查中来自汪美英本人的叙述，汪美英1957年出生于灵桥镇，今年59岁，从16岁开始学习织帘，从事纸帘制作行业已有40余年，从未间断，称得上是制帘的“老师傅”。2012年5月，汪美英被认定为第三批杭州市非物质文化遗产项目代表性传承人。她与丈夫方如堂一起创办的光明制帘厂实际上就是家庭作坊，平日都是夫妻俩负责生产，只有当织帘订单较多忙不过来的时候，才会加雇人手帮助织帘。

据汪美英回忆，1972年5月底初中毕业后，刚刚走出校门的她就跨进了富阳制帘厂，应该说运气不错，一毕业就有“吃商品粮”的稳定工作，那时是很让人羡慕的。当时她住在今天的外汪村，当时叫灵桥公社光明大队，离外汪村1 km的地方有座灵岩寺,富阳制帘厂就坐落在那里，后来改名叫富阳工艺美术总厂。

⊙1

⊙2

访谈中，汪美英向年轻的调查员讲述了她是怎样“吃上商品粮”的：20世纪70年代初，制帘厂从原来单一的造纸用帘扩大到画帘、画帘屏风制作，业务范围逐渐扩大，手工织帘一时人手紧

⊙1 光明制帘厂厂区
Guangming Screen-making Factory

⊙2 汪美英正在织帘
Wang Meiying making the screen

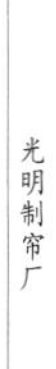

⊙3

缺，需要招募新工人。光明村的大队干部听到制帘厂要招工的消息，就前去联系了，希望能够获取一些名额。光明大队和制帘厂平时就有很好的业务关系，名额自然要来了。汪美英时年16岁，是学习织帘最好的年龄，然而制帘厂总共只招20人左右，可仅仅是大队的1个生产小队，符合年龄要求的姑娘就有20多人，在大家都想去的情况下，采取了抓阄的方式，每个生产队3人，8个生产队共24人。汪美英通过抓阄成为幸运者，成功进入纸帘厂当上了织帘女工。

一开始，和厂里正式工人一样的待遇，早上7点上班，中午吃饭1个小时，星期日休息，工资每月27元，还发放一些劳保用品。汪美英最初从事的工种是抽帘丝，师傅是负责生产的姜荣生，抽了3年的帘丝后，调换了工种，开始学织帘，跟随织帘师傅大源镇蒋家村人蒋阿金学习织帘。

到1980年，富阳工艺美术总厂在富阳城里的西堤路造了新厂房，各个车间陆续搬到了富阳城关镇上，但制帘车间还是留在灵岩寺。富阳工艺美术总厂把目标定位向绣衣、电子玩具等稍高档的产品发展了，在总厂下面设分厂。留在灵岩寺的制帘分厂由于产值低，利润薄，一定程度上牵制了总厂的发展步伐，于是在20世纪80年代末，纸帘厂被划分给了属地的光明村（系原光明大队改名），生产场所也迁移到了外汪村，灵岩寺转给灵桥中学做了校舍。

⊙4

⊙3 光明村周边环境 Surrounding environment of Guangming Village

⊙4 灵桥中学内遗存的灵岩寺旧水池 The old pool of Lingyan Temple in Lingqiao Middle School

光明村接过帘厂的业务后，聘请技术娴熟的汪美英担任培训师傅，带领七八个人学习制帘技术。3个月后她与丈夫方如堂合计不如自办制帘厂，于是懂一点制帘技术的方如堂辞去原地方化肥厂的工作回家办起了制帘厂。汪美英负责织帘等技术活，做帘棒竹、漆帘的事归方如堂。后来，外汪村承接的帘厂办不下去了，汪美英与方如堂创办的帘厂就成了灵桥镇光明村制帘厂。

三 光明制帘厂纸帘的制作原料、工艺与设备

3 Screen-making materials, Techniques and Tools of Guangming Screen-making Factory

（一）纸帘的制作原料

1. 主料：苦竹

据汪美英口述，制作竹帘的竹丝多采用苦竹为原料。苦竹呈圆筒形状，通常高达4 m以上，下部数节间距长达25～40 cm，直径约15 mm，比一般竹节明显要长，这是制作抄纸竹帘很重要的指标。不过1～2年生的苦竹不易拉丝，一般都采用3年左右的苦竹，其韧性和长度均达到制帘所需的最佳状态，拉出的竹丝弹性好，性质刚强，使用中耐磨损。根据纸张的品种与厚薄要求选定竹丝的粗细，富阳传统抄造元书纸一般选用的竹丝直径在0.6～0.8 mm，一张帘子选用的竹丝粗细要求尽可能一致，否则会影响抄纸时纸的质量。

汪美英介绍，光明制帘厂以前生产用的帘丝都是用苦竹自己加工成的竹丝，由于加工过程耗时耗力，现在已经不自家抽丝了。当然，自己加工抽的帘丝比买的成品要细。现在生产所需的帘丝全部为外购加工好的成品，主要来自本省内的温州瑞安地区，2015～2016年采购价格为每千克220元左右，据汪美英估计2019年的价格在每千克240元左右。

2. 辅料：涤纶线

涤纶线主要起到固定竹篾丝的作用，使得竹

⊙1

⊙2

⊙1 抽好的苦竹丝 *Pleioblastus amarus* filament after threading

⊙2 涤纶线 Polyester thread

篾丝紧密地结合在一起，不易散开，从而形成一张完整的帘幕。汪美英介绍，其使用的涤纶线购买价格为80多元/kg。调查时据说富阳当地对这种线的需求很好，很难在当地购买到，所以光明制帘厂生产所需的涤纶线都是从上海百货商店购买的。2019年3月4日回访时了解到，现在光明制帘厂所用的涤纶线都是早前购买的，目前已经买不到了。

3. 生漆

生漆即天然漆树漆，是从漆树上采割的乳白色胶状液体，接触空气后氧化变为褐色，数小时后表面干涸硬化而生成漆皮。生漆具有耐腐、耐磨、耐酸、耐溶剂、耐热、隔水和绝缘性好、富有光泽等特性，常用于漆制工艺品、木家俱、古建筑，不仅漆面光洁，能防腐抗热，而且特别经久耐用。光明制帘厂所用的生漆采购自富阳当地经销商，2016年采购价格约为300元/kg。2017年采购价格也在300元/kg左右，当时汪美英购买了500斤，差不多能用到2019年。

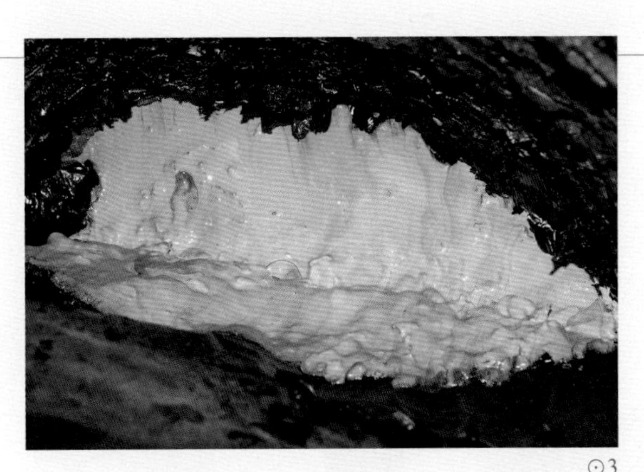
⊙3

⊙3 生漆 Raw paint for painting the screen

（二）纸帘的制作工艺流程

抄纸竹帘的制作工艺分为采竹、劈篾、抽丝、织帘、漆帘5个过程，从上山选材到剖竹、撕篾、油漆、下架，整个竹帘的生产可以细分为

50多道工序，全部采用手工方式，几乎无法用机械操作。当然，到21世纪初，光明制帘厂的这些工序在保留原有传统手工工艺的基础上也有部分被机器所代替。

根据调查组成员对光明制帘厂进行的实地调查和访谈，综合汪美英的叙述，归纳其纸帘制作的主要工艺流程如下：

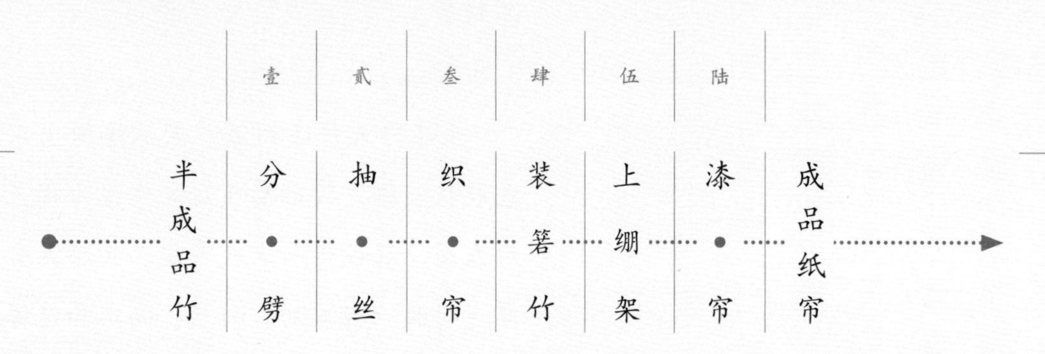

壹 分劈

1 ⊙1

将半成品竹子在水中浸泡一定时间，根据干燥竹片的浸水渗透情况，用固定在板凳上的两片刀片进行分劈，两刀片之间间隔1～1.5 cm。用刀片将竹片劈裂后用手掰开，再用特制的刀具把半成品竹片切黄、去芯，剖成比竹丝略粗的若干根竹篾。

贰 抽丝

2 ⊙2

将分劈好的细竹篾穿过一个特制的有一定大小的圆孔，圆孔的直径根据竹帘帘丝的粗细要求而决定。穿过圆孔后，用钳子夹住露出头的部分，左手水平直线抽出竹丝，一气呵成将整根竹篾从圆孔里穿过，使得每根竹丝都一样粗细。

⊙1

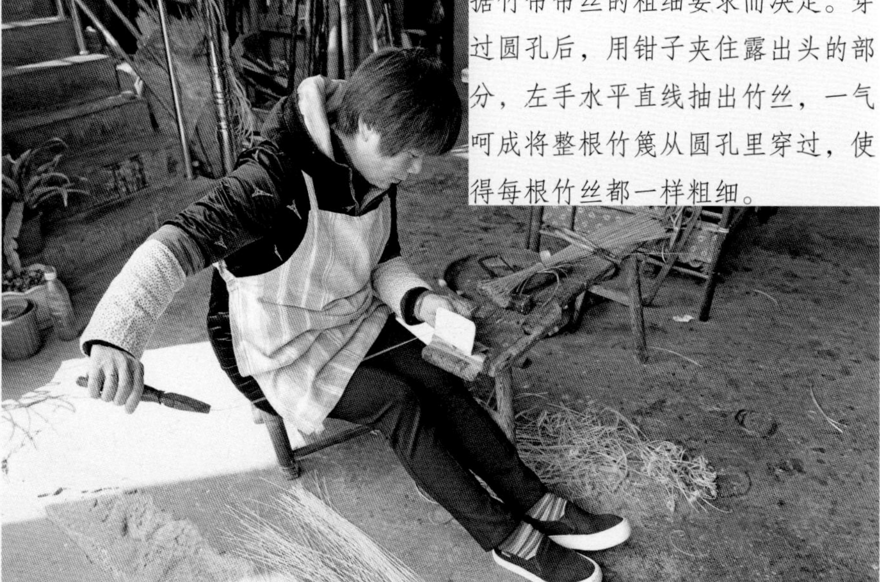

⊙2

⊙1 竹片切黄、去芯 Cutting the bamboo and removing the core

⊙2 汪美英演示抽丝 Wang Meiying demonstrating the threading procedure

叁
织帘

3 ⊙3⊙4

调查时光明制帘厂沿用的是20世纪70年代富阳制帘厂改造后的编织方法——木机编织法。木机编织沿袭了织布的方法，以线为经，竹丝为纬，在木机上编织竹帘。木机的大小不一，一般上、下由12块纵板组成，下面有3块踏板，用脚踩踏后带动挡板上下活动，形成相互交织状态。帘丝放入交叉的两线之间，再来回踩踏交织，使得帘丝之间缝隙变小编织成网，线与线之间的距离根据捞纸质量、厚薄决定。编织时，眼睛、手、脚都要配合：脚踏后，丝线形成相互绞织，放入帘丝。脚踏时须用力匀称，使帘丝密度均匀，这样织出来的竹帘才会松紧适宜。

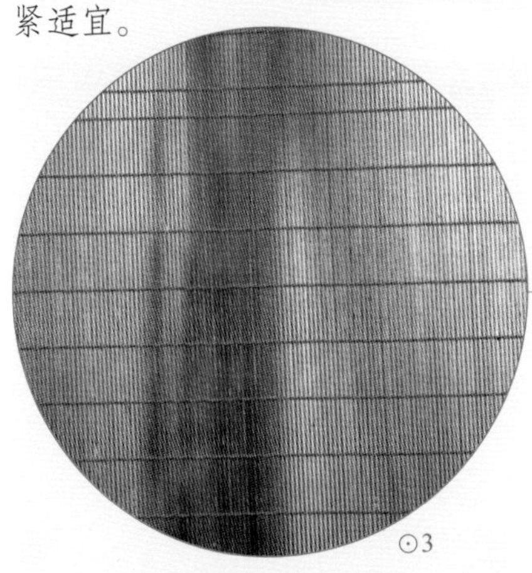

⊙3

调查时了解到，富阳造元书纸所用纸帘以前1 cm宽度有10根竹丝，现在基本保持在9根；造民间祭祀用的火烧纸所用纸帘1 cm宽度有7根竹丝，后期纸张厚度增加后1 cm为6根竹丝。

汪美英介绍，木机织帘比原先全手工织帘提高了至少10倍的效率，编织1张面积较小的帘子只需要4～5个小时，将近2天时间可以编织好1张四尺帘子。

⊙4

⊙6

伍
上绷架

5 ⊙6

将装好箬竹的帘子平整并四角腾空地固定在一个由4根木棍围成的长方形木床中间，上下两边用力均匀使得竹帘完整打开。

陆
漆帘

6 ⊙7

把漆分段刷在帘丝上，漆的用量需要适当多一点，然后双手轻轻搓揉，使得漆汁在搓揉的过程中均匀分摊在帘丝上。等第一遍刷完后半干燥时，再用不加漆的刷帚重新刷一遍，这时要求不能有上下帘丝粘黏的情况发生。刷完的竹帘一般需要在室内阴干，切不可在太阳下暴晒。刷漆时需要保持一定的温度和湿度，一般经过一个月的保养后方可下水抄纸。

⊙5

肆
装箬竹

4 ⊙5

编好的帘子需要在横向的两边各装一根箬竹作为固架，保持竹帘不变形。固架杆选用富阳当地深山里面的箬竹，箬竹砍伐下来一般是曲折弯曲状的，需要用火烤拗直再用水冷却。在上下两根拗直的箬竹上每隔2 cm左右钻有0.8 cm的小孔一排，用针、线将织好的帘网固定住，并将两边多余部分的帘丝剪齐。

⊙7

⊙3 帘丝密度特写 Close-up of density of the screen filament

⊙4 编帘 Making the screen

⊙5 汪美英在装箬竹 Wang Meiying packing the bamboo for fixing

⊙6 绷架 Stretcher

⊙7 方如堂在漆帘 Fang Rutang painting the screen

（三）纸帘的主要制作工具

壹 剖竹刀 1

把竹片劈成竹篾，同时给竹片切篾、去芯的铁质工具。

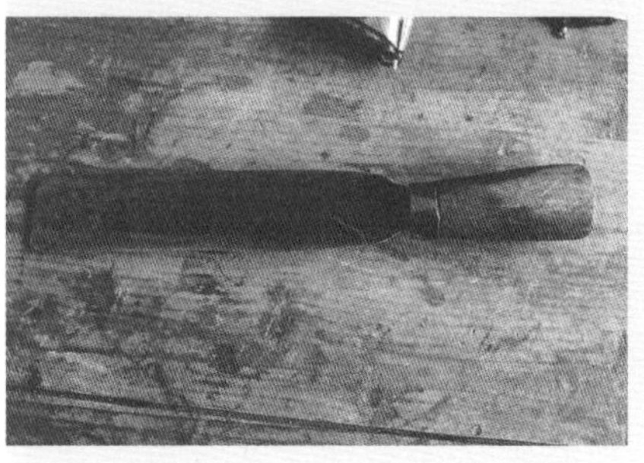

⊙1

贰 织帘木机 2

编织竹帘的工作台，主要由一个木架组成，工作原理与原始织布机略有相同。

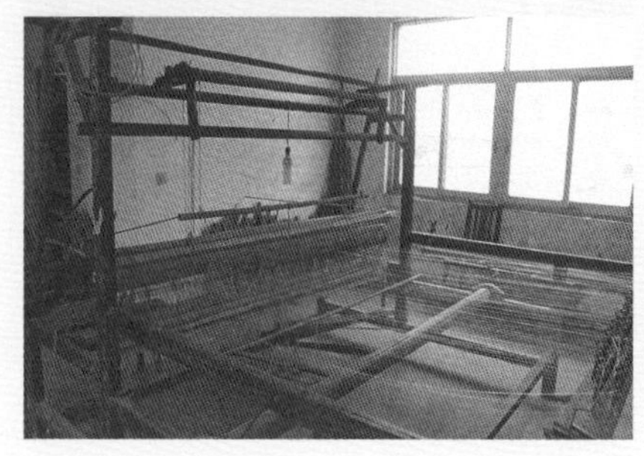

⊙2

叁 木钻 3

给箸竹开孔时所用。

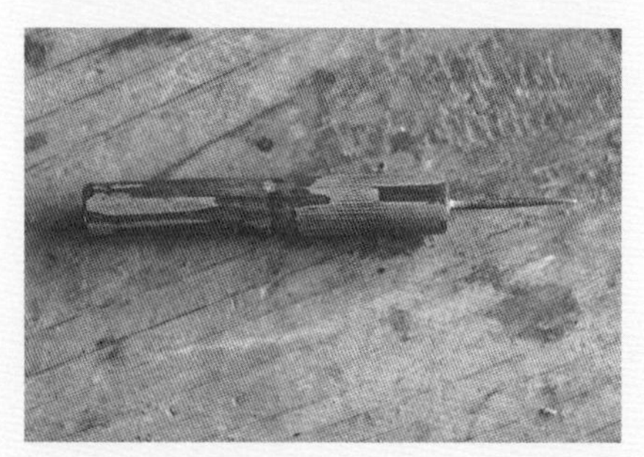

⊙3

肆 绷架 4

用于漆帘的木质框架，主要起到固定竹帘的作用，使得竹帘充分展开，便于漆帘与快速晾干。

⊙4

⊙1 劈竹刀 Iron tool for cutting the bamboo
⊙2 织帘木机 Wooden loom for making the Screen
⊙3 木钻 Wooden drill
⊙4 绷架 Stretcher

四
光明制帘厂的市场经营状况

4
Marketing Status of Guangming Screen-making Factory

手工造纸业态的兴衰直接决定着抄纸竹帘的兴衰，光明制帘厂制作的竹帘主要用于富阳当地的元书纸以及祭祀用途“迷信纸”抄纸所用。调查时的情况是，21世纪初特别是2010年以后，富阳及周边的传统手工造纸厂坊数量急速减少，纸帘销售受到的影响很大。但同时，纸帘厂家在全国范围内也消失了很多，因此坚持下来的手工制帘厂家依然有活路，用汪美英的话说：“现在，虽然手工造纸不多，用帘也少，但我们厂小，这点业务还是有的，富阳本地零散做纸的槽户需要，连云南、贵州、四川、山东、安徽、北京都有来订帘的。”光明制帘厂也在积极开发拓展新产品，转变竹帘的用途，向家居装饰方向发展，作为杭州市的非遗项目，杭州、北京等地的博物馆会向他们订做竹帘，用于馆内的布展装饰。

目前光明制帘厂所生产的帘子主要有纸帘、门窗帘和茶帘等多个品种，年产量300张左右。价格上“宣纸”帘按照张数和尺寸来计算，四尺的700～800元/m^2，六尺的约1 000元/m^2。其他种类按竹丝粗细和平方计算，细的500～600元/m^2，粗的250～300元/m^2。所有产品都是订单式生产，先付50%的定金，交货后付完全部款项。总体年利润约10余万元。

⊙5

⊙6

⊙5 作坊内刷漆间场景 Scene of the paint room in the workshop

⊙6 光明制帘厂生产的茶帘 Screen for tea produced by Guangming Screen-making Factory

五 光明制帘厂的业态传承现状与发展思考

5 Current Status and Development of Guangming Screen-making Factory

（一）纸帘制作技艺传承情况

1. 新的传承人很难获得

光明制帘厂是富阳当地规模不大的制帘家庭作坊，平时主要是夫妻俩生产，业务多的时候，才雇人帮助织帘。这种纯手工的活，流程复杂、工序多、工作量大、产量小、学习难度大，愿意从事织帘上漆这类工艺的工人也越来越难找到。2016年调查时汪美英也表达出担忧：“织帘的技艺，不知是否能够传承下去。”

⊙1 ⊙2

2. 积极参加竹帘制作技艺的公众普及

从2010年起，光明制帘厂传承着的竹丝帘制作技艺成为富阳区级非物质文化遗产名录项目，后来又申报成为杭州市级“非遗”名录项目。现在，每逢与竹、纸搭界的文化艺术展，汪美英都会在富阳区非物质文化遗产保护中心的号召下积极参加。如2011年6月在原富阳市影剧院开展的富阳竹文化艺术展、2014年6月在原富阳市体育馆开展的富阳纸文化艺术展等，汪美英都曾带着笨重的木机到现场进行表演，向公众展示传统竹帘制作技艺。

感谢状

汪美英 同志：

为庆祝第六个“文化遗产日”，6月11日至13日在市影剧院举办“富阳市竹文化艺术展”，贵厂生产的竹纸帘参加展出，您在现场展示了织帘技艺。

特此感谢！

富阳市非物质文化遗产保护工作委员会

⊙3

⊙1/2 工作中的方如堂和汪美英 Fang Rutang and Wang Meiying at work

⊙3 汪美英参加富阳市竹文化艺术展的感谢状 Wang Meiying's certificate of appreciation for participating in Fuyang Bamboo Culture Art Exhibition

（二）保护和发展竹帘的若干思考

1. 打破单一用途，开拓新的产品线

在富阳等传统手工造纸地区，竹帘编制主要用于抄纸，产品用途较为单一，受手工造纸行业景气度影响较大。同时，目前的竹帘制作多是家庭式作坊，规模小，市场销售面窄。秉承保持传统、不失其本，在改良制作过程中保持与提高品质的宗旨，可以尝试拓展多种产品类型来扩大销路。比如同竹制品厂家合作，开发新产品，走竹帘精加工，制作家居竹帘、书画帘、竹帘艺术品等。销路打开，传统行业才会成为有源之泉。

2. 发挥政府作用，变革发展思路

“非遗”项目的创新由于个人力量有限以及某些“非遗”产品受众面小等原因，其效果并不理想。因此，对纸帘制作这样很小众定向消费的“非遗”产品的传承和保护，政府相关部门应起到牵头作用，组织开展项目关联人群的研修研习、工艺旅游与展览体验活动，促进跨界思考与产品、渠道培育。与此同时，让深陷小地域埋头生产的纸帘技艺持有人吸纳其他“非遗”项目跨界运营的经验，变革一成不变祖传发展思路，结合新技术手段，在产品创新、市场运营等方面做文章，这样小众定向消费的“非遗”项目才会走得更长远。

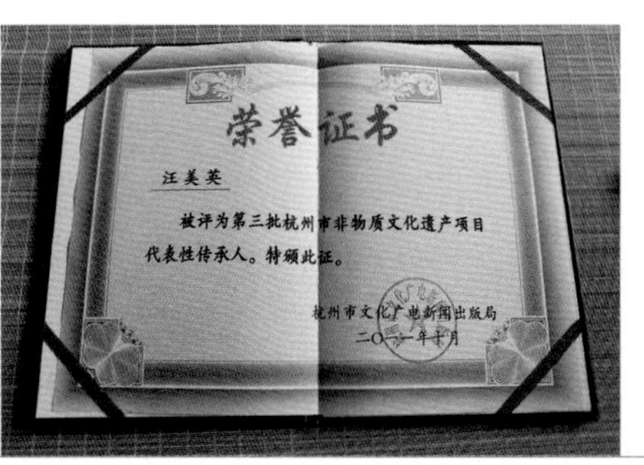

⊙4

⊙4 汪美英被评为第三批杭州市非物质文化遗产项目代表性传承人证书
The certificate of Wang Meiying as the third representative inheritor of Hangzhou Intangible Cultural Heritage Project

第三节

郎仕训刮青刀制作坊

浙江省
Zhejiang Province

杭州市
Hangzhou City

富阳区
Fuyang District

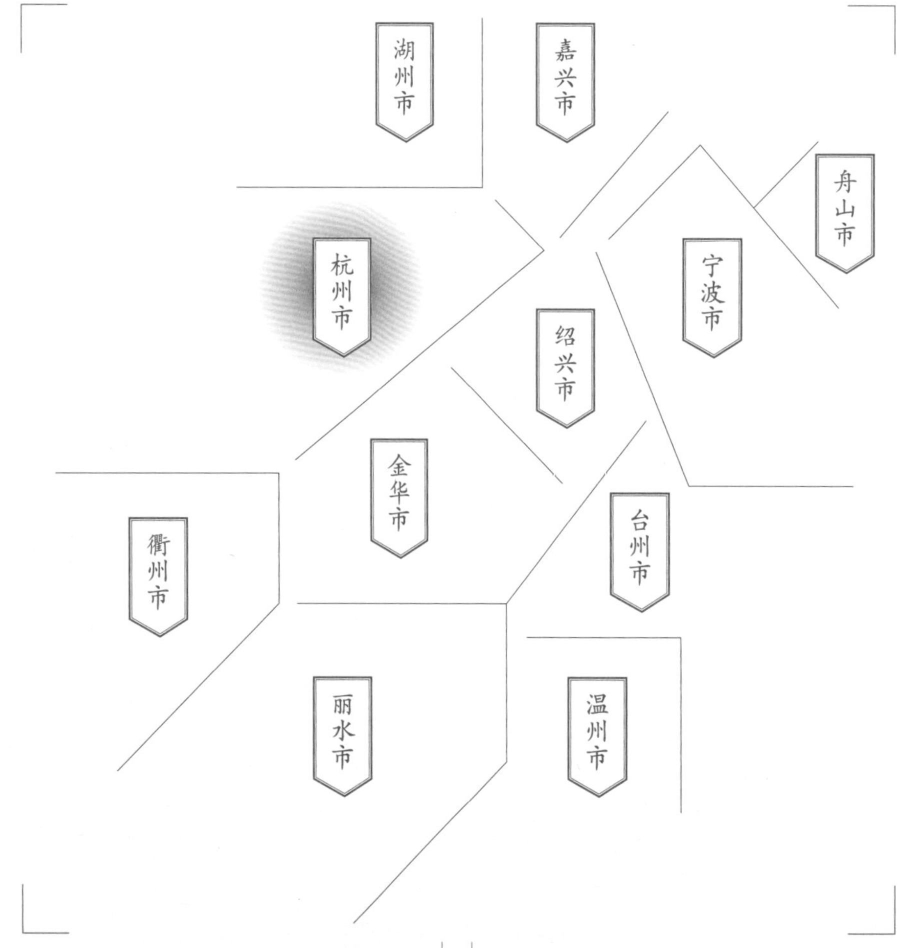

调查对象
富阳区大源镇
郎仕训刮青刀制作坊
刮青刀

Section 3
Lang Shixun Scraping Knife-making Mill

Subject

Scraping knife of Lang Shixun Scraping Knife-making Mill in Dayuan Town of Fuyang District

一
郎仕训刮青刀制作坊的基础信息与生产环境

1
Basic Information and Production Environment of Lang Shixun Scraping Knife-making Mill

郎仕训刮青刀制作坊位于富阳区大源镇朝阳南路二弄，地理坐标为：东经120° 0′ 26″，北纬30° 0′ 9″。刮青刀为富阳元书纸原料制作的重要工具之一，用于将竹子上的青皮刮去，方便后续原料处理。

2016年9月29日和2019年3月8日，调查组两次前往郎仕训刮青刀制作坊进行田野调查，获得的基础信息是：郎仕训刮青刀制作坊长年制作人员仅有2人，也就是郎仕训本人及其妻子庄美华；作坊占地面积40 m^2左右，基本上是个住家外搭起的铁皮棚子。据郎仕训的说法，他1个人制作1把刮青刀花费3小时即可完成，当然，这与郎仕训已经从事40年打铁造器行当技艺娴熟是分不开的。访谈中向郎仕训了解到，2016年富阳区只有郎仕训作坊和高桥镇2家打铁铺还在打制铁器工具。

⊙1

⊙1 郎仕训刮青刀制作坊 Lang Shixun Scraping Knife-making Mill

路线图
富阳城区
↓
郎仕训刮青刀制作坊
Road map from Fuyang District centre
to Lang Shixun Scraping Knife-making Mill

郎仕训刮青刀制作坊位置示意图

Location map of Lang Shixun Scraping Knife-making Mill

考察时间
2016年9月 / 2019年3月

Investigation Date
Sep. 2016 / Mar. 2019

地域名称

造纸点名称

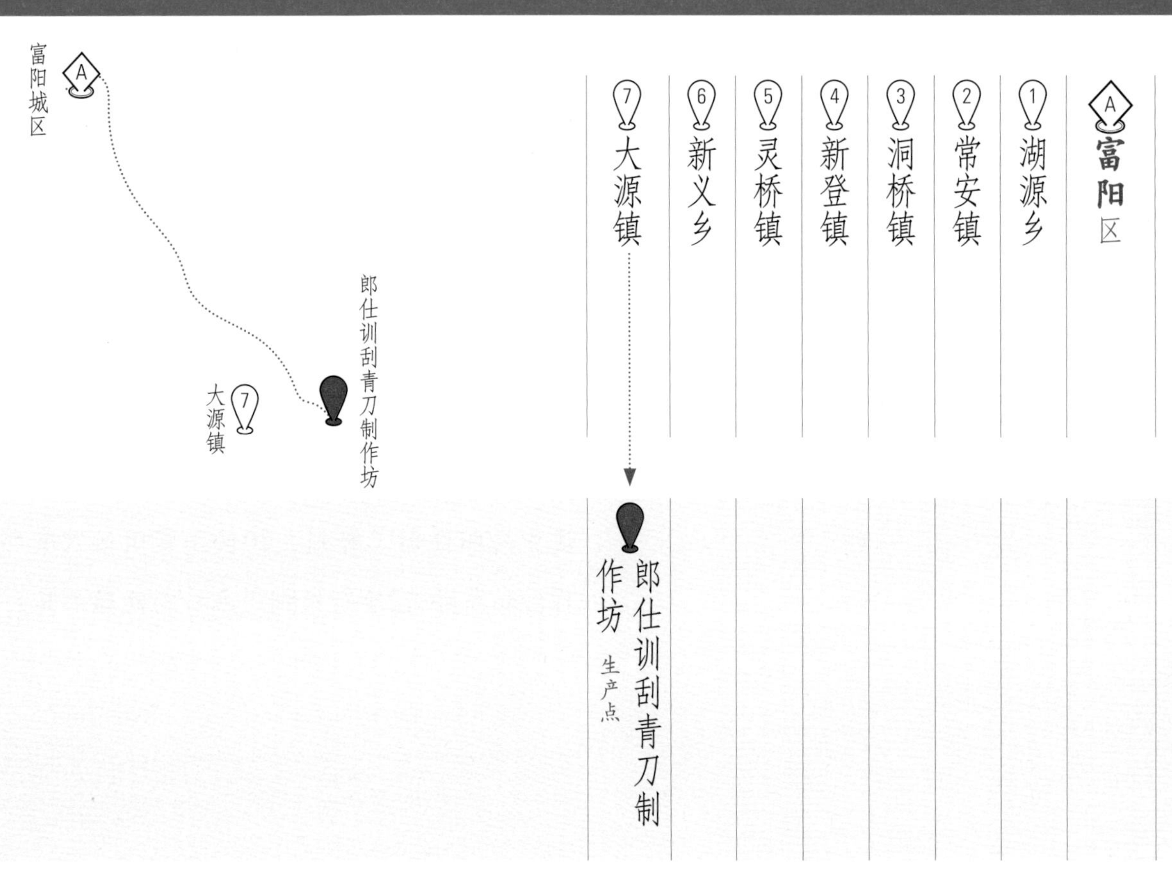

位置分布

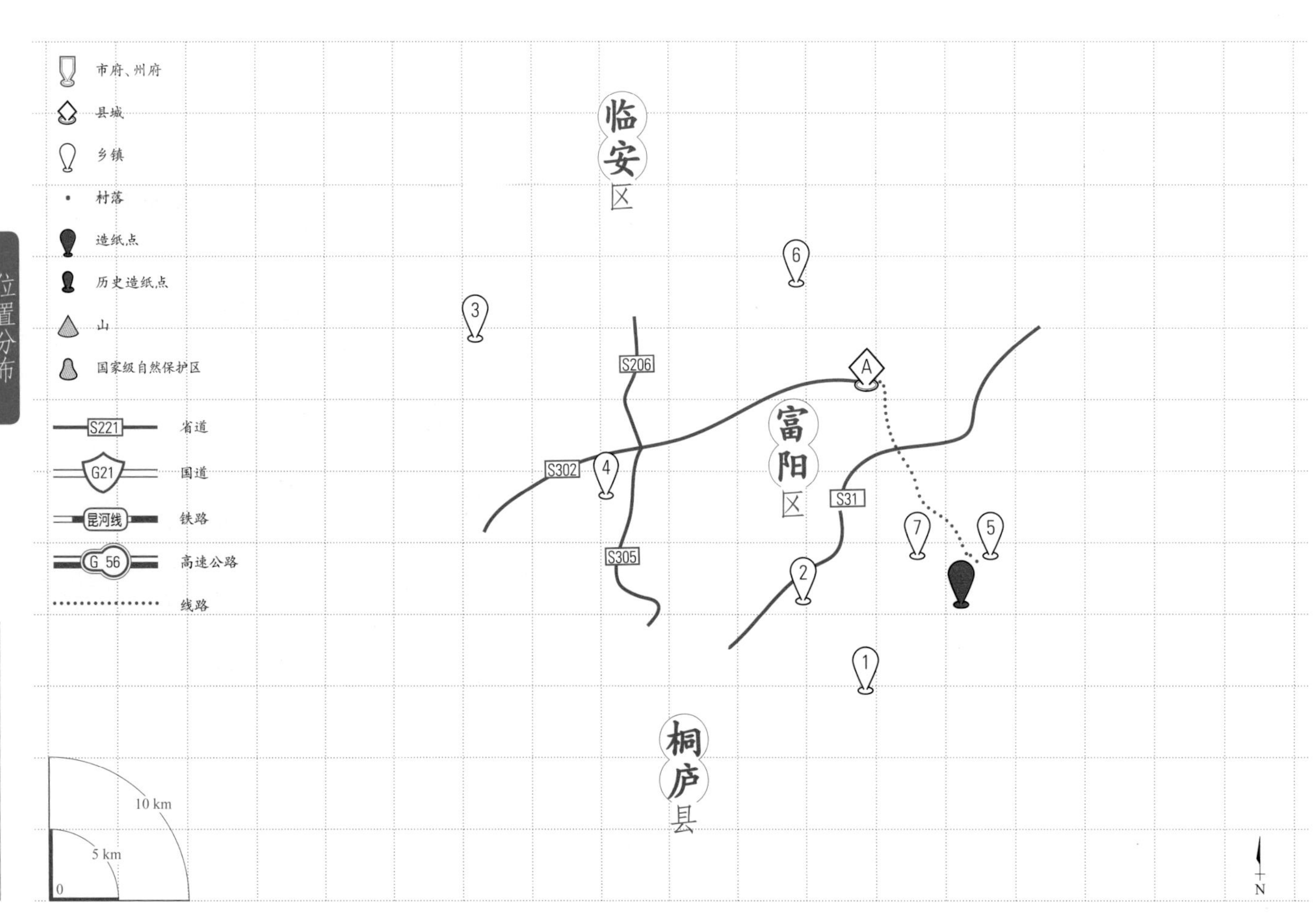

二
郎仕训刮青刀制作坊的历史与传承

2
History and Inheritance of Lang Shixun Scraping Knife-making Mill

⊙1

郎仕训出生于1956年，出生地为永庆村。19岁开始在大源镇农机厂当学徒，师从出生在富阳的一位永康籍师傅程茂根（出生及生长在富阳区小源镇）学习打铁。郎仕训记得以前大源镇农机厂都是永康的师傅在打铁。永康打铁师傅的作业习惯是2人合作锻打，但现在很难找到合作对打的那个人，因此现在几乎没有永康师傅在富阳区打铁，也不知道这些师傅去哪里了。大源镇的农机厂于1985年倒闭，被迫下岗的郎仕训在厂子倒闭后去了福建做卷帘门生意，但老实的郎仕训并不擅长做生意，卷帘门的贷款收不回来，苦苦支撑了3年后只能回家，重拾打铁旧业。郎仕训回来后，办起了家庭打铁铺，因为没有能力雇人，妻子庄美华（1958年出生）开始给他打下手，一起打铁，刮青刀就是打铁作坊的主要产品之一。

⊙2

⊙3

⊙1/2 在打铁铺与郎仕训交流
Interviewing Lang Shixun in the iron shop

⊙3 在打铁铺与庄美华交流
Interviewing Zhuang Meihua in the iron shop

三
刮青刀的用途

3
Usage of Scraping Knife

刮青刀是造竹纸时使用的工具，主要用于富阳造中高端元书纸原料——嫩毛竹的青皮处理。毛竹肉是造高档元书纸必不可缺的原料，嫩毛竹在砍伐后，需要在当天将表面青色的皮去掉。此时，削竹者面朝扶桩，左腿前跨成左弓步，右手在前，左手在后，两手握刮青刀一前一后平行向前削竹。削竹皮是原料准备中重要的工序之一，特制的刮青刀在削竹工序中至关重要。

⊙2

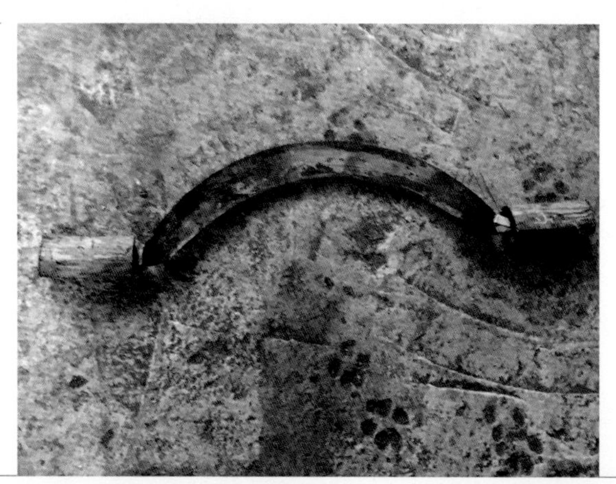

⊙1

⊙1
刮青刀
Scraping knife

⊙2
演示刮青刀削竹动作
Demonstrating scraping the bamboo with a scraping knife

四
郎仕训刮青刀制作坊的制作原料、工艺与设备

4
Knife-making Materials, Techniques and Tools of Lang Shixun Scraping Knife-making Mill

（一）刮青刀的制作原料

1. 主料一：A3钢

刮青刀必不可缺的重要材料是A3钢，A3钢是一种含碳量偏低的碳素钢。据郎仕训介绍，制作刮青刀的A3钢是在富阳当地的钢铁市场购买的，没有具体的尺寸要求，最贵时5 000元/吨，便宜的时候2 000元/吨。2016年时属低谷，购买价为2 000元/吨；2019年时价格约为4 000元/吨。刮青刀毛铁需要0.75 kg，打好后大概缩到0.6 kg左右。

2. 主料二：45号钢

制作刮青刀刀锋部分的钢叫45号钢。45号钢为常用中碳调质结构钢，该钢冷塑性一般，退火、正火比调质时要稍好，具有较高的强度和较好的切削加工性，经适当的热处理后可获得一定的韧性、塑性和耐磨性，材料来源方便。适合于氢焊和氩弧焊，不太适合于气焊。焊前需预热，焊后应进行去应力退火。据郎仕训介绍，45号钢4 000多元一吨，但钢废料价格便宜些，其制作刮青刀的45号钢是旧汽车零件回收的，价格为3 000多元一吨。

⊙3

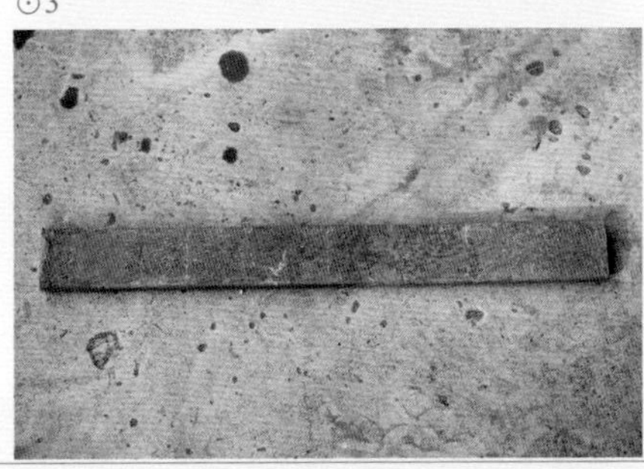
⊙4

⊙5

⊙6

⊙3/5 A3钢 A3 steel

⊙6 45号钢（夹在中间的为45号钢）No. 45 steel (middle)

3. 辅料：煤

制作刮青刀重要的辅助原料是煤。郎仕训介绍，其制作刮青刀使用的是在富阳当地购买的煤，2016年购买价为1 000元/吨，2019年无烟煤购买价为1 500元/吨。

（二）刮青刀的制作工艺流程

据调查中郎仕训的介绍，其打制刮青刀的工艺流程为：

壹	贰	叁	肆	伍	陆	柒	捌	玖	拾	拾壹	拾贰
裁料	反复烧热锤打	冷却调整	砂轮打磨	铲口造刃	砂轮打光	锉刀打磨	砂轮打磨	淬火定性	冷轧定型	上销	冷却

壹

裁料

1

⊙1⊙2

用切割机锯下0.75 kg的A3钢和45号钢备用。

⊙1

⊙2

⊙1/2 裁料 Cutting the materials

贰
反复烧热锤打

2 ⊙3～⊙19

第一步，用炉灶旁边的鼓风机鼓风，用废纸引燃煤炭，燃烧约4分钟可达1 000多度。

⊙3

⊙4

第二步，把切割下的钢放在一个支承物上，不然容易烫手，入炉烧4分钟。

⊙6

⊙5

第三步，用空气锤锤打铁坯，然后放回炉子继续高温烧4分钟左右。

第四步，用锤子继续手工锤打，在中间敲出一条缝，以便于后面往缝里放钢。

⊙10

⊙11

⊙7

⊙8

⊙9

⊙3 炉灶生火 Making fire in a stove

⊙4 鼓风机鼓风 Blowing with a blast furnace

⊙5 打钢时夹住钢片的钳子 Pliers holding the steel sheets while hitting the steel

⊙6 烧红铁块 Burning iron block

⊙7/9 空气锤第一轮锤打 Air hammer first round hammering

⊙10/11 铁锤第一轮锤打 Iron hammer first round hammering

第五步，放入炉中烧5分钟左右再进行第二轮锤打，打出刀的弯度。

⊙12

⊙13

第六步，先用空气锤敲打45号钢，再转为敲打A3钢。此时经过空气锤的反复锤打后，已经初具刀形。锤打时先竖着打，再正反两面横着打，再竖着打，3次一循环。

⊙14

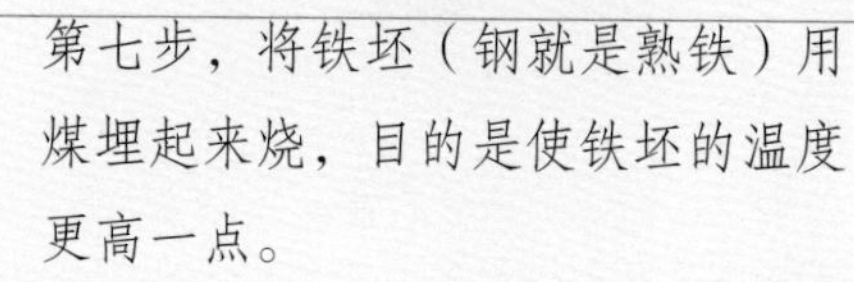
第七步，将铁坯（钢就是熟铁）用煤埋起来烧，目的是使铁坯的温度更高一点。

第八步，用空气锤锤打，进一步锤成自己想要的形状。

第九步，拿着一个定型的模具先竖着打，再横着打。敲打完后接着用小锤手锤固形，固形后再用空气锤敲打。

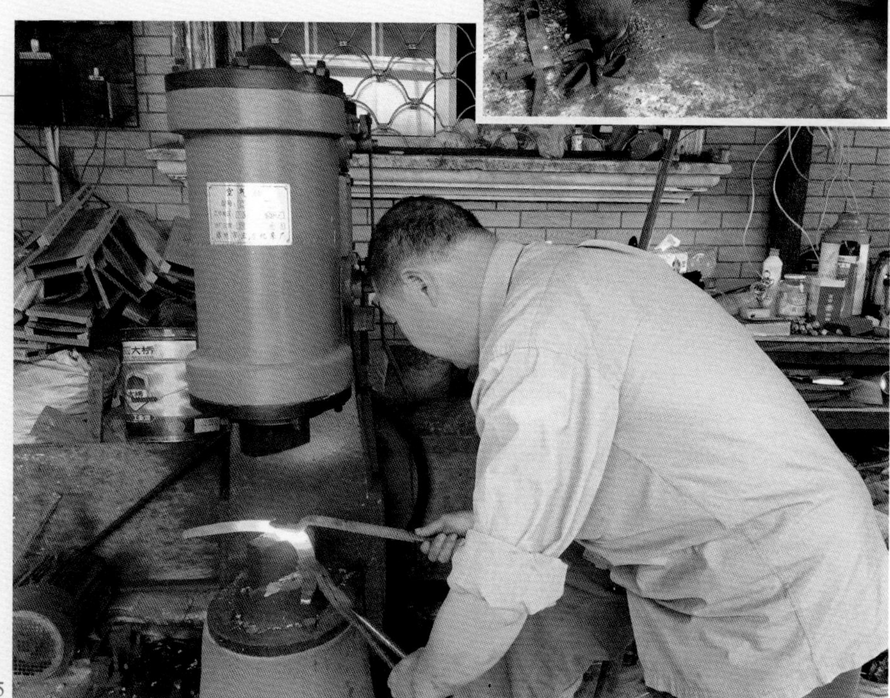
⊙15

⊙16 ⊙17

第十步，入炉煅烧后继续用小锤整形；再次入炉煅烧后，先用小锤锤打出刀柄，再用空气锤锤击刀柄，锤出相应的形状。

第十一步，锤击完刀柄后入炉煅烧2分钟，然后用小锤敲打，敲击另外一边的刀柄。

⊙18

⊙12/13 铁锤第二轮锤打 Iron hammer second round hammering

⊙14 空气锤第二轮锤打 Air hammer second round hammering

⊙15 空气锤第三轮锤打 Air hammer third round hammering

⊙16 铁锤第三轮锤打 Iron hammer third round hammering

⊙17 空气锤第四轮锤打 Air hammer fourth round hammering

⊙18 铁锤第四轮锤打 Iron hammer fourth round hammering

第十二步，敲打完成后煅烧30秒，然后专门针对手柄的位置用空气锤锤打。手持刀身，不停转换刀柄方向，使得刀柄变尖。每次锤完放在冷水里面冷却，该水每年换1次。现场观察，郎仕训经过四轮次空气锤和小锤锤改，打出刀具毛坯一共用了44分钟。

⊙19

叁 冷却调整 3

将锤打好的刀具毛坯放置在地上散火，让其冷却。冷却到手可以拿起就可以用砂轮磨打精加工了。冷却的时候铁匠可以做其他活，让其自然冷却。冷却打磨后再用小锤锤打来调整刀具的形状，敲打的目的主要是为了使刀具的刀锋两边平整。

肆 砂轮打磨 4 ⊙20～⊙22

敲打完成后，用电砂轮进行刀锋的打磨，以使刀锋锋利。打磨约10分钟后形成半成品。

伍 铲口造刃 5 ⊙23⊙24

用锉刀进行打磨，约需锉10分钟。用锉刀打磨能清晰地看到钢分布在哪里，是否均匀。有光影发亮的就有钢，不发亮说明只有铁没有钢，这个时候就要用锉刀打磨，使钢显现出来，如果不用锉刀直接用砂轮打磨，打制出来的刀在使用的时候有些地方“快”有些地方“钝”。

⊙20

⊙21

⊙22

⊙23

⊙24

⊙19 初具刀型 Original knife form
⊙20 砂轮 Grinding wheel
⊙21/22 砂轮打磨 Grinding the blade with a grinding wheel
⊙23/24 铲口造刃 Sharpening the blade with a shovel

陆

砂轮打光

6 ⊙25

继续用砂轮打磨约5分钟。

柒

锉刀打磨

7 ⊙26

继续用锉刀锉约13分钟。

捌

砂轮打磨

8 ⊙27

继续用砂轮打磨约5分钟。

⊙25

⊙26

⊙27

玖

淬火定性

9

上炉淬火之前在刀具两面各刷一层泥，然后入炉，这样刀烧红后不会变弯。淬火即将刀烧到像杨梅那样的红，此时可以根据自己的经验和需要对刀进行微调。

拾

上销

10 ⊙28

将刀柄烧热，趁热插入松木刀把。

拾壹

冷却

11 ⊙29

用冷水将其冷却，制作一把刮青刀的全部工序就完成了。

⊙28

⊙29

⊙25 砂轮打光 Polishing the blade with a grinding wheel
⊙26 锉刀打磨 Grinding the blade with a file
⊙27 砂轮打磨 Grinding the blade with a grinding wheel
⊙28 上销 Fitting the knife into a pine handle
⊙29 冷却 Cooling

（三） 刮青刀的主要制作工具

壹 空气锤 1

锻打工具。郎仕训刮青刀制作坊使用的空气锤是2016年1月从山东省滕州市三力机床厂购买的，购买价11 000元。此前使用的为1985年从大源镇农机厂以1 500元购买的空气锤，该空气锤1977年生产，时间久不太好用了，因此调查当年购买了新的空气锤。

⊙30

⊙33

贰 铁 锤 2

用来锤打调整刀具形状的工具。实测郎仕训刮青刀制作坊使用的小锤尺寸为：柄长28.5 cm；锤头长17 cm，宽3 cm，最厚4 cm；锤子两侧一侧厚2.5 cm，一侧厚1.5 cm。大锤尺寸为：柄长65 cm；锤头长17.5 cm，宽一头4 cm、一头6 cm，厚一侧3 cm、一侧4.5 cm。

⊙31

⊙32

叁 钳 子 3

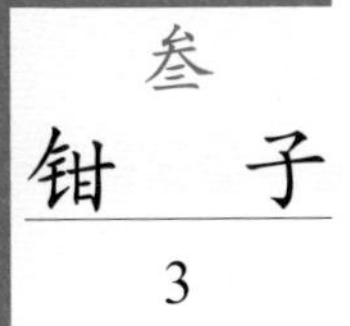

用来夹住铁坯及在炉灶里面支撑铁坯。郎仕训刮青刀制作坊使用了多种钳子，长度从42 cm到57 cm不等。

⊙34

肆 锉 刀 4

用来打磨刀坯的工具。实测郎仕训刮青刀制作坊使用的锉刀尺寸为：刀身长20.5 cm，宽2 cm；柄长10 cm。

⊙35

⊙30 空气锤 Air hammer

⊙31/32 铁锤（上边为大锤，下边为小锤） Iron hammer (sledgehammer on the top and small hammer on the bottom)

⊙33/34 各种用途的钳子 Pliers for various purposes

⊙35 锉刀 File

伍

铲子

5

用来打磨刀坯（用在铲口造刃工序）的辅助工具。实测郎仕训刮青刀制作坊使用的铲子尺寸为：长47 cm；铲刀长14.5 cm，宽2 cm。

⊙36

陆

锉刀容器

6

铁制，用来盛放锉刀的容器（锉刀是用来打磨刀坯的工具，手握的地方为木头制作，圆柱形，锉也为圆柱形，铁制）。实测郎仕训刮青刀制作坊使用的锉刀容器尺寸为：高20 cm；直径4.8 cm；底座宽10.5 cm，长16.5 cm。

⊙37

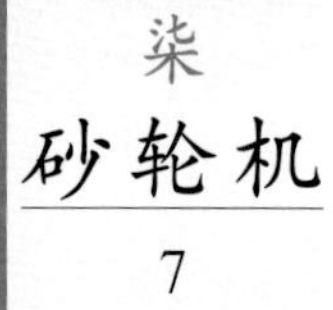

柒

砂轮机

7

用来打磨刀具的工具。实测郎仕训刮青刀制作坊使用的砂轮机尺寸为：砂轮直径29 cm，厚4 cm；底座高61 cm，长51 cm，宽25 cm。

捌

墩头

8

用铁锤锤打刀具时下方的支承工具。实测郎仕训刮青刀制作坊使用的墩头尺寸为：长52 cm（尖头部位长18 cm，中间长20 cm，尾长14 cm），宽20 cm，加上底座一起高约72 cm。

⊙38 ⊙39

⊙40

⊙36 铲子 Shovel

⊙37 锉刀容器 File container

⊙38 砂轮机 Grinding machine

⊙39/40 墩头 A support tool when hammering the knife with a hammer

五
郎仕训刮青刀制作坊的市场经营状况

5
Marketing Status of Lang Shixun Scraping Knife-making Mill

郎仕训刮青刀制作坊一直为家庭式作坊。刮青刀是富阳地区以传统工艺制作中高端竹纸（以元书纸为主）的必备工具，但到20世纪90年代以后，市场与消费发生了剧变：第一，富阳地区手工造纸厂坊萎缩明显，数量少了很多；第二，现在还在生产的厂家有很多不刮青皮了，刮青刀用户急剧萎缩；第三，刮青刀是耐用品，一把刀可以用很多年，以上三方面的原因导致刮青刀的需求目前已近乎于无，郎仕训表示已经10多年没打过这个刀了，因为没有客户。即便偶尔有零星需求要打刮青刀，一来成本太高客户不接受，二来他也挣不着钱，所以大都以谈不拢而告终。

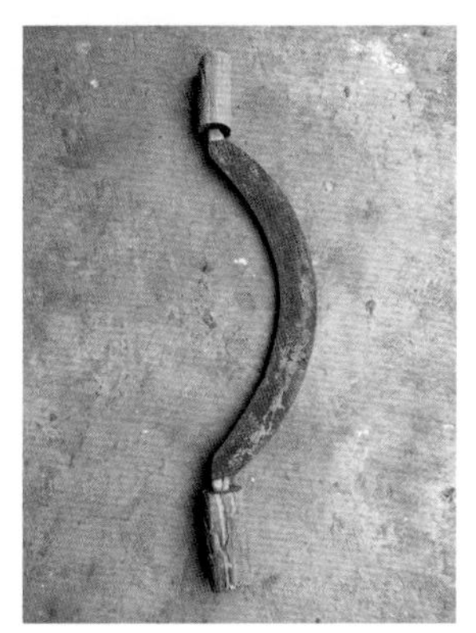
⊙41

郎仕训介绍，刮青刀制作坊近10年主要以做锄头、镰刀、菜刀、砍柴刀这类农具为主。他本人信奉基督教，现在除了每周日去教堂做礼拜，其他时候都生产，但是经营状况一直处于下滑态势，因为农具的需求量也在不断下降。

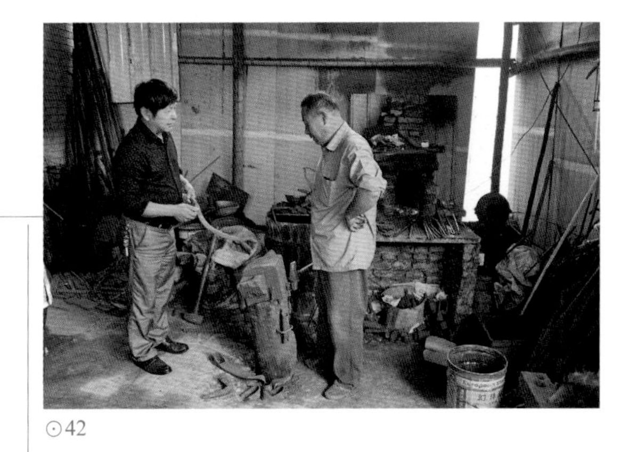
⊙42

⊙41 刮青刀 Scraping knife

⊙42 作坊里谈到市场显得无奈的郎仕训 Helpless Lang Shixun when talking about the marketing status

六
富阳区大源镇打铁习俗

6
Customs of Ironing in Dayuan Town of Fuyang District

调研中郎仕训提到，在打铁做刀的时候，他们行业内一直流传着这样一句话：“长木匠短铁匠”。意思是取料的时候，木匠做木工原料要取长一点，这样即便原料长了还可以切短一些继续制作需要的木具，如果取短了，哪怕是短一点，都无法制作出需要的木具，这样整个原料就浪费了；而他们打铁取原料时要取短一点，因为如果长度不够可以再将原料打长一点，但是原料取长了再将原料切断就会造成不必要的浪费。

⊙1

七
郎仕训刮青刀制作坊的业态传承现状与思考

7
Current Status and Development of Lang Shixun Scraping Knife-making Mill

刮青刀一直是富阳地区制作中高端竹纸（以元书纸为主）传统工艺中原料制作的重要工具，其制造工艺非常复杂，工作十分艰苦，需要在扬尘、噪音、炎热的艰苦环境中连续工作3个小时以上才能打出1把刀。虽然有一整套的工艺流程，但是每一步的具体操作完全凭打刀师傅的经验，因此，这是一个很辛苦而且技艺要求还不低的手艺活。

由于手工造纸锐减、造纸工艺简化，富阳目前使用刮青刀的人很少，即便有些纸坊保留了刮

⊙1
郎家湾公园
Langjiawan Park

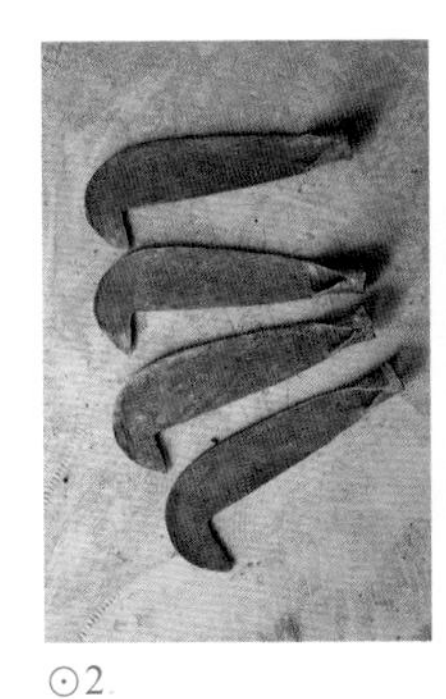
⊙2

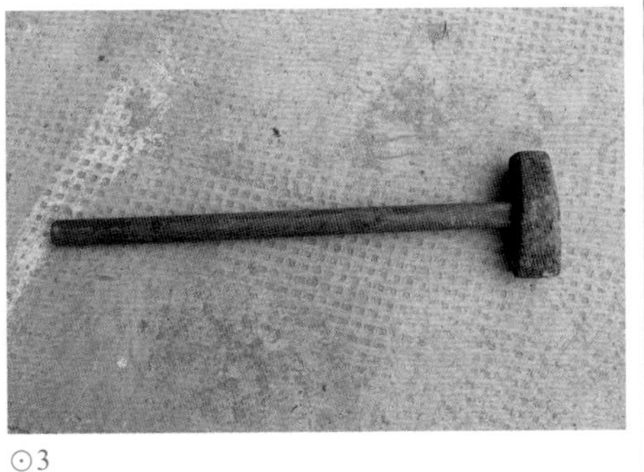
⊙3

青皮这一工序，往往也会使用土法制作的电动刮竹皮设备。郎仕训刮青刀制作坊目前以生产锄头上的勾刀、菜刀、锤子为主，已经有十多年没有生产过刮青刀了。没有需求，没有人愿意学，没有继承者，刮青刀制作的技艺传承目前看来后继无望。

刮青刀是为元书纸量身打造的，因此其营销渠道一直很单一——只供应富阳当地企业和当地供应商。而且由于需求较少，也一直无法机械化。在手工纸市场不景气的大环境下，刮青刀已经10多年没有生产，下一步的传承看来只能从影像保护+实物采集的途径加以实施， 即尽快完成全部技艺与材料、用途的视频拍摄及数字化保存，尽快采集刮青刀系列实物进入专门的“非遗”展馆或集中保护场所。面对这一无法生产性保护的技艺，迫切需要富阳地方文化遗产保护机构实施行动，否则技艺的完全消失无可避免。

⊙4

⊙5

⊙2 锄头上的勾刀 Hook on the hoe

⊙3 锤子 Hammer

⊙4 郎仕训为调研组制作的刮青刀 A scraping knife made by Lang Shixun for the researchers

⊙5 富阳区大源镇郎仕训刮青刀制作坊周围环境 Surrounding environment of Lang Shixun Scraping Knife-making Mill in Dayuan Town of Fuyang District

Introduction to Handmade Paper in Zhejiang Province

1 History of Handmade Paper in Zhejiang Province

1.1 Natural and Cultural Characteristics of Zhejiang Region

As a provincial administrative region, Zhejiang, referred to as “Zhe”, is located in the southern side of the Changjiang River Delta in the southeast China. The name comes from the zigzag of the longest river in this territory, Qiantang River (the upstream of which connects Fuchun River and Xin’an River which is the main source of Qiantang River, originating from the Liugujian mountain at the junction of Xiuning County of Anhui Province and Wuyuan County of Jiangxi Province), which was accordingly called Zhijiang, Zhejiang or Zhejiang River (three different Chinese characters meaning zigzag are used).

Qiantang River (photo provided by Xu Jianhua)

Technically speaking, the natural and cultural connotation of Zhejiang has witnessed three major changes since ancient times before it eventually formed:

(1) As the name of an administrative region, eastern Zhejiang region and western Zhejiang region were set up in the Tang Dynasty. The eastern Zhejiang region dominated seven prefectures. They are Yuezhou, Quzhou, Wuzhou, Wenzhou, Taizhou, Mingzhou, Chuzhou areas. Western Zhejiang region covered the southern part of Changjiang River in Jiangsu Province, the eastern part of Maoshan Mountain and the northern part of Xin’an River in Zhejiang Province today.

(2) The area was set up as a province in the Yuan and Ming Dynasties: during the Yuan Dynasty, Zhongshusheng Institute was set up, with the provincial center transferred from Yangzhou to Hangzhou. In 1366, Zhu Yuanzhang’s army occupied Hangzhou where he set a local government there. With the development through the Ming and Qing Dynasties, its jurisdiction gradually laid the foundation for the administrative area of today’s Zhejiang Province.

(3) Modern Zhejiang Province is among the 34 provinces, municipalities, autonomous regions and special administrative regions, set up on the basis of previous jurisdiction after the founding of the People’s Republic of China. It is between 118°00' to 123°00' east longitude and 27°12' to 31°31' north latitude. The westernmost point of Zhejiang Province is the west line of Chetianban Village in Suzhuang Town of Kaihua County, while the easternmost point is the reef (now Tongdao Island) to the east of Shengsi Islands, the east-west longitude difference being 5°5'; the southernmost point is Xingzai Island, which is to the south of Nanguan Island in Cangnan County, and the northernmost point is on the north side of Changwei Village in Meishan Town of Changxing County, with a latitude difference of 4°19' between north and south.

As a province, the history of Zhejiang Province is not too long, more than 600 years. However, as a place harboring civilization, it enjoys a long history. As early as the Middle Pleistocene 450 000 years ago, ancient human activities could be traced in the present Anji County in the north of Zhejiang Province, which is the prelude of the currently known archaeological history of Zhejiang. In the middle and late Paleolithic period about 100 000 years ago, a group of primitive men named “Jiande people” lived in the mountains of western Zhejiang. If the remarkable Hemudu Culture in the Neolithic Age is also counted, Zhejiang has enjoyed at least 7 000 years of civilization, forming a unique commensalism between human and nature, and showing the evolution of regional civilization.

Zhejiang Province covers an area of 101.8 thousand square kilometers, yet the natural environment enjoys diverse features. In this geographical space, the local customs vary from place to place. Several sub-regional subsystems coexist and flourish together in the Hangjiahu Plain, Ningshao Plain, Wentai Coast, Jinqu Basin etc., through the Ming, Qing Dynasties and the Republican Era to present. Wang Shixing, a human geographer from Zhejiang Province in the Ming Dynasty, once described the characteristics of local customs in Zhejiang Province in vivid language: “Hangzhou, Jiaxing, and Huzhou harbor plains, rivers and lakes; Jinhua, Quzhou, Yanzhou, Chuzhou are mountainous and dangerous, residents there can live in harmony with the difficult environment; Ningbo, Shaoxing, Taizhou, Wenzhou sit between the mountains and the sea so inhabitants there are living by the sea. People living in the three different areas have their own peculiar customs. People living by the lakes and rivers make a living by transporting goods. So the local residents are rich, and tend to pursue luxury and extravagance. The officers there have much higher status than the average people. People who live in the mountains tend to be strong and forthright due to their rigid living environment. They look down on the penal laws and are in pursuit of the frugality, but they often assemble to fight against the officers on account of their unyielding spirit; people who live by the sea earn their living by fishing, their life is quite arduous. They won’t be impoverished for their fishing practice but they can’t be wealthy on account of the fact that they have no contact with the vendors and business men. Officers and common people there get along well with each other. Half of the people there are pursuing extravagant life style while the rest are prone to be frugal due to the fact that the officers can be from the rich family or the common household.”

According to the geographers, Zhejiang Province can be divided into six major areas. The northern plains: which mainly consists of Hangjiahu Plain and Ningshao Plain and serve as the chief Zhejiang populous center, with interconnected waterways and paddy fields. The northwestern middle mountains and hills: there are numerous mountains and hills in the northwest Zhejiang; the famous Qiandao Lake scenic spot sits in the mountain area; Tianmu Mountain and Baiji Mountains lie along the border of Zheiang Province and

Anhui Province. The central basin: in this basin which is also called Jinqu Basin, two famous cities, Jinhua and Quzhou, embrace a long history and prosperous culture. The eastern hills: agricultural civilization presents a unique rhythmical image of beauty along the ordered rise and fall of the gently rolling hills in Cao'ejiang River Basin. The southern middle mountains: Lishui and west Wenzhou are crisscrossed by mountains with its 1 921-meter-high dominant peak Huangmaojian (also the highest peak in Zhejiang Province) in Donggong Mountains of Wuyi Mountain System. The coastal area: plains intermittently connected from north to south in the southeast coast, harboring coastal cities like Taizhou and Wenzhou etc., forming the coast line with echos from thousands of sea islands of different sizes, all of which constitute an image of maritime civilization in Zhejiang Province.

View of islands and mountains

It is generally held by Chinese historians that before the Tang and Song Dynasties, the Central Plains (Zhongyuan) was the center of the country in terms of politics, culture and economy. Zhejiang and other regions in the south bank of Changjiang River were relatively marginalized. From the Tang and Song Dynasties to the Ming Dynasty, the center of economy and culture in China gradually moved to the south, meanwhile the economy and culture in Zhejiang experienced its position from margin to center, from a relatively blocked environment to a leading role in opening up.

During the Sui and Tang Dynasties, communication between the Central Plains and Zhejiang intensified, and followed a comprehensive development of economy and society in Zhejiang promoted by the excavation and opening of the Grand Canal. From the Jin Dynasty, after about one-thousand-year upheavals and changes of dynasties in the Central Plains, the transformation of economic center from the Central Plains to regions in the south of the Changjiang River was quietly achieved. The peak of this transformation was finally accomplished by the Southern Song Dynasty establishing its capital in Hangzhou, which symbolized the transcendence of Jiangnan area (southern part of Changjiang River) over the Central Plains in culture, economy and living standards. In the Song Dynasty, Zhejiang was second to none in prosperity in Jiangnan area. The rise of Hangzhou City was the symbol of Zhejiang civilization in the Song Dynasty, and even today this city is still glowing like a "paradise". Zhejiang in the Song Dynasty bloomed with a flourish not only in overseas trade but also in its economy, politics and culture. Even in the relatively remote city Yongjia (Wenzhou) in the southern Zhejiang, there arose the famous Yongjia School in the South Song Dynasty, which was one of the major schools of academic thoughts and knowledge center together with Cheng-Zhu school and Lu Jiuyuan's mind theory.

The economic interaction and prosperity of Jiangnan area in the Ming Dynasty was as thriving as that in the Song Dynasty. Stable social status and advanced economy contributed to the culture and education prosperity of the region, and helped Zhejiang earn the name of "state of cultural relics". During the Ming and Qing Dynasties, Zhejiang witnessed emergence of cultural celebrities who enjoyed a national fame. Although Zhejiang was neglected in the planning and execution of Westernization Movement in the late Qing Dynasty and was comparatively late in its modern industrialization, it was the birthplace of "republican revolution" in overthrowing the monarchy and establishing the republic system.

In the Republican Era, a group of industrial and commercial capitalists from Zhejiang rose in China, who were called "Ningbo Faction" and "Jiangzhe Consortium" (or called "Zhejiang Consortium"), and exerted the most tremendous influence on Chinese industry and commerce. In contemporary era, Zhejiang is still attracting wide attention by the power of its private capital from local merchants.

Zhejiang Province enjoys a profound culture with a long history, which harbored over 100 relics of the Neolithic Age. Among them, the following are the representative ones: Hemudu Culture of southern China in the early Neolithic Age, formed between 5 000 and 7 000 years ago, centered in Hemudu Town of Yuyao City; Majiabang Culture in the Neolithic Age, enjoyed a history about 6 000 years, which was named after Majiabang Relics in Nanhu Town of Jiaxing City; Liangzhu Culture, centered in Pingyao Town of Yuyao District in Hangzhou City, formed between 4 500 and 5 300 years ago, whose rice farming, jade-making, silk and fiber spinning were well developed.

Liangzhu Ancient City Relics Park
(photo provided by Cai Yongmin)

Perhaps due to the geographical features, ancient residents in Zhejiang Province lay special attention on production and marketing of handcraft products. Therefore, this place abounds with diverse intangible cultural heritage today. In 2005, it was the first province in China to release provincial lists of intangible cultural heritages. Then in 2006, it ranked top in numbers of national level intangible cultural heritages. In 2007, it published the second batch of provincial lists totaling 225 kinds and 10 categories, which, to some extent, pioneered the trend of China holistically preserving intangible culture heritage in the inception of 21th century.

1.2 The Origin and Development of Ancient Papermaking Industry in Zhejiang Region

1.2.1 The Basic Clues of the Papermaking History in Zhejiang Region

Handmade paper was immersed in the unique local history, terrain and culture of Zhejiang region. Compared with other provinces of southern China, handmade paper in Zhejiang emerged much earlier. In this place, handmade paper was recorded as early as the Jin Dynasty. Since then, the migration from north to south brought in scholarly culture and advanced technologies, which accelerated Zhejiang's rapid development, making it home of literati and a place prominently pooled by cultures and civilization.

In Zhejiang, the early handmade paper practice was mainly originated from Shanxi area of Yuezhou Prefecture and Kuaiji Prefecture. Plenty of rattan gave birth to Teng paper (Teng, Chinese name for rattan), which then contributed to the local scholarly activities. Teng paper in Shanxi area and bamboo paper in Kuaiji Prefecture were favored by the literate giants and officers, including Wang Xizhi, Xie An and Huan Wen.

Tang Dynasty had witnessed the boom of handmade paper. During that time, Teng paper was mostly made in Hangzhou (now Yuhang District in Hangzhou City), Yuezhou (now Shaoxing City), Wuzhou (now Jinhua City) and Quzhou in Zhejiang. Besides the Shanteng paper (Shanteng means Teng in Shanxi area)from Shanxi area, Teng paper from Youquan Village of Yuhang also enjoyed a good reputation. Then in the Song Dynasty, with the economic center moving southward and the capital of Southern Song Dynasty set in Lin'an (now a distict of Hangzhou City) particularly, the political and cultural center in Jiangnan area had been thriving rapidly. The groundbreaking headway of the handmade paper in Zhejiang, to a large extent, was benefited from the tribute paper system formed by the Official Paper Office in Hangzhou and Shaoxing areas at behest of the government. There were 11 places designated to produce the tribute paper, including Hangzhou, Wuzhou and Quzhou.

The handmade paper encountered its important transitional point after the Northern Song Dynasty. The local rattan was over-exploited and gradually exhausted in Shanxi area, because of the large demands of the famous Teng paper. Teng papermaking practice hence withered and the paper gradually faded away from the public after the Southern Song Dynasty. Jiangnan area teemed with bamboo, so in the Tang and Song Dynasties, bamboo paper emerged and flourished in Fujian and Zhejiang regions. The new kind of paper significantly increased the quantity of paper output in Zhejiang, with Muzhou (now Chun'an County), Yuezhou and Hangzhou (now Fuyang District) as the bamboo papermaking center in the Song Dynasty. Of them, bamboo paper in Shaoxing City (also named Yue paper) was the best in quality, making Shaoxing the most prominent origin of high quality bamboo paper.

During the Song and Yuan Dynasties, Juan paper in Wenzhou City and straw paper made in different places were broadly put into use. Hence, more details of handmade paper in Zhejiang Province were recorded in literature. In the Song Dynasty, as China's papermaking center moved southward, the output, diversity and quality of handmade paper in Jiangnan area began to surpass that in the North and consolidate the leading role of Zhejiang in papermaking practice.

In the Yuan Dynasty, numerous books and textbooks of Directorate of Imperial Academy were printed in Hangzhou City because the place possessed great carving and type setting skills, and high quality paper with reasonable price. In the Ming and Qing Dynasties, papermaking flourished and blossomed almost everywhere in the whole province, the productivity and papermaking techniques of which reached the peak in the end of the Qing Dynasty, forming many papermaking centers such as Fuyang, Quzhou Changshan areas etc. *The Annals of Chinese Industry* recorded that papermaking industry in Zhejiang Province was extremely prosperous in the Ming and Qing Dynasties, among the paper types the Zouben paper (used for memorials to the emperor) produced in Changshan and Xianju areas, the Zhushao paper (used to hold a memorial ceremony for the dead) produced in Yuhang and Longyou areas and the Rili paper (used to make calendar) produced in Tonglu and Changshan areas were all well-known at that time.

1.2.2 Wei, Jin, Southern and Northern Dynasties: the Origin of the Handmade Paper in Zhejiang Province

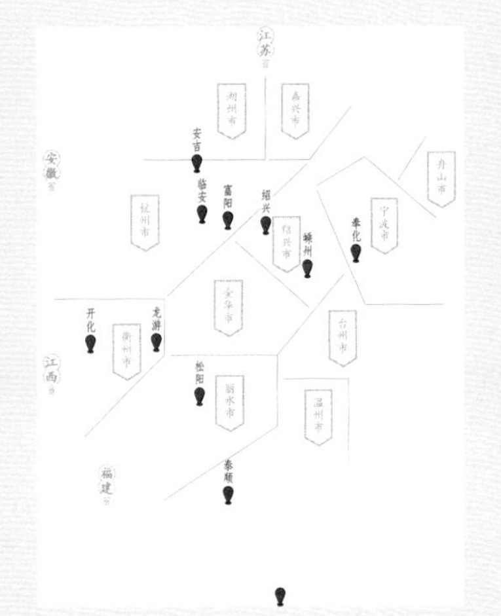

Distribution map of the papermaking counties in Zhejiang Province under the investigation

According to the current archaeological findings and the records in *The Book of Later Han Dynasty*: *The Life of Cai Lun*, some sporadic historical paper relics were discovered which can be traced back to the early Western Han Dynasty about 2 200 years ago. And in the Eastern Han Dynasty, Cai Lun presented the Emperor Hanhe the

Portrait of Wang Xizhi

"Caihou Paper" (105) he invented and then it was spread by the imperial court all over the country, which symbolized the formal completion of the first national recognition of papermaking invention. Historical literature failed to ascertain the period of the Han Dynasty as the beginning of handmade papermaking practice in today's Zhejiang Province. In some local tales, however, many traditional handmade papermaking areas such as Fuyang all dated their papermaking origins back to the Han Dynasty, which can also be regarded as an identification of regional tradition by the local culture.

In the Eastern Jin Dynasty, definite records emerged about handmade paper consumption in Zhejiang Region represented by the great calligraphy master Wang Xizhi (303—361) in his calligraphy and poems. Accordingly, the medium of calligraphy and poetry at least underwent the phases of bamboo, silk and paper, or co-existing stage in about 250 years since Cai Lun's papermaking techniques were recognized and promulgated (105) in the Eastern Han Dynasty. In the Three Kingdoms period and the late Han Dynasty, with the advancement of papermaking techniques, there were relatively great improvement in the production capacity and cost performance of papermaking compared with that of Cai Lun period. For example, the calligrapher and papermaker Zuo Bo (165—226) who was born in Donglai County of Shangdong Region and neighbor of Wang Xizhi from Langya County of Shangdong Region, created the outstanding "Zuobo Paper", subsequently "Zhangzhi Writing Brush" "Zuobo Paper" "Weidan Ink Stick" were celebrated as "Three Best Things in the World". We assumed the Eastern and Western Jin Dynasties as the emerging period of handmade paper in Zhejiang region owing to the ambiguous records of papermaking practice in this region, even though at that time paper had become the important writing medium.

Legend has it that the earliest record about the origin of handmade paper is "abundant rattan is available in Shanxi area to make paper" said by Zhang Hua, a famous person in the Western Jin Dynasty. From the quote, we learned that Shanxian County in Kuaiji Prefecture (now Shaoxing City and Ningbo City) had utilized rattan in papermaking. In the Eastern Jin Dynasty, papermaking in Kuaiji Prefecture has taken shape with a large scale, and the paper made in Yuezhou Prefecture was also well-received. Written record mostly concentrated on the papermaking center of Kuaiji Prefecture affiliated to Yuezhou City area. According to the literature study, the paper-involved descriptions were closely connected with local cultural activities. Famous literati and men of letters, pooled in Kuaiji Prefecture at that time, so the area was in large demand of paper, which inevitably stimulated the progress of the industry.

"In the Eastern Jin Dynasty, when Wang Xizhi served as the Kuaiji administrator, Mr. Xie (Xie An, also called Huan Wen, a well-known politician) asked for Zhili paper (also called Celi paper, which was said to be made from a kind of fern grown on dank rocks).Wang, then, gave all the stored paper to him, totaling 90 thousands pieces (or 500 thousands pieces based on a different version). Hence, it is known that paper was made in Kuaiji Prefecture." from *Punctuation and Collation of Kuaiji Erzhi*, authored by Shi Su, et al. in the Southern Song Dynasty. As for the amount of paper mentioned, no matter 90 thousand or 500 thousand pieces, it can be seen that Yuezhou enjoying a large-scaled paper production. Yu Bing, a politician at that time, also mentioned in his *A Letter to Wang Xizhi*, "Paper may amount to one zhang or qian (Chinese units of measurement), so my admiration only increased". The two specific units of paper at the time unveiled the large quantity of the paper made locally.

Papermaking practice shows typical regional features. Its development is bound to the abundance of local raw materials. Jute, rattan, paper mulberry, bamboo and mulberry from Hangzhou and Shaoxing areas are high quality materials to make paper.

Documents show that Tengpi paper made by wild rattan was the first to be recorded. During Xian'an Reign of the Eastern Jin Dynasty (371—372), Fan Ning, the Yuhang County magistrate, required "Tengjiao paper as local official document paper, while local paper was banned". It offered a glimpse of the great number of Tengjiao paper. Shan Qianzhi in the Liu Song Dynasty also mentioned in his Wuxing Records (a book mainly about the events happened in Zhejiang) that "Youquan Village is the home for Teng paper".

In early days, besides Tengpi paper, bamboo paper also emerged early in Zhejiang region. In *Collection of Earthly Paradise*, a book written by Zhao Xique in the Southern Song Dynasty, he recalled the local handmade paper when Wang Xizhi was still alive that "the paper in the south region was made by vertical-stripe papermaking screen. Hence the authentic works by two giant calligraphers, Wang Xizhi and his son Wang Xianzhi, must be written on bamboo paper in Kuaiji Prefecture, for the two spent most time over there since the southern migration (in order to avoid war and seek safety), and the paper from the north region was hard to come by". Combining the above two details, it can be seen that Teng paper stood the mainstream while bamboo paper was rare.

Yuluan (fish egg) paper from today's Dongyang area in Zhejiang Province was famous in the Southern and Northern Dynasties. Legend has it that Mrs. Wei (272—349), Wang Xizhi's calligraphy teacher said in her book of calligraphy theory, the Diagram of Writing, that "Yuluan paper in Dongyang is soft and smooth". As the book was written in 347 A.D. (the third year in Yonghe Reign of the Eastern Jin Dynasty), back to then, the so called Dongyang, just a county at that time, was equal to today's Jinhua City. Yuluan paper was claimed as white as newly born cocoon when it was put against the sun and its dots like fish egg (or fish line), which explained the source of the name. From previous description, the paper was highly similar to the advanced cocoon Jian (a kind of expensive paper for writing), which was very likely made from plant bark.

In early records, several kinds of celebrated paper from Zhejiang can be found, but no detail of papermaking had ever been referred to. Through "paper and ink both were made by themselves" in *Zhang Yong's Biography in the Book of Song*, a historical book recording the history of Liu Song Dynasty in the Southern Dynasties written by Shen Yue (441—513), we knew that, at that time, some landlords were engaged in papermaking and ink-making too. The lack of the paper processing records was due to the fact that more concern had been given to paper related stories of the celebrities and calligraphy culture than the despised craft skills of papermaking.

Zhejiang paper, at that time, was served as commercial products. In addition to supplying local needs, it was also exported to other places for storage and sales. In Shen Yue's *The Book of Song*, from "politician Kong Ji's younger brother, Kong Daocun and Kong Hui saw dozens of vessels loaded with silk and paper on their way from Kuaiji to the capital city, Jiankang (today's Nanjing City)", we can see that in Liu Song Dynasty (420—479), Kuaiji paper was already a vital commodity product in circulation.

（南宋）会稽二志点校

A photo of the book *Punctuation and Collation of Kuaiji Erzhi (Southern Song Dynasty)*

It should be pointed out that in the Wei and Jin Dynasties, with the initial development of papermaking, public and private paper mills were built from north to south, even in some ethnic minority areas. With materials obtained locally, many papermaking centers had gradually formed. The north centered in today's Luoyang City in Henan Province, Xi'an City in Shaanxi Province, Shanxi, Hebei, Shangdong and other places, where majorly output hemp paper, and mulberry paper. In the south, the industry took off after the southward migration in the Jin Dynasty, and centered in Kuaiji of Zhejiang, Jianye (today's Nanjing), Yangzhou, Guangzhou and other areas. On the whole, the center of Chinese papermaking mainly lied in the north.

1.2.3 Tang Dynasty: the First Developmental Peak of Handmade Papermaking in Zhejiang Region

Tang Dynasty had witnessed the first peak of papermaking in Chinese history: the producing area, on the one hand, was spread nationwide, and on the other hand, production center transferred to the south. Cai Xiang, a noted calligrapher in the late Northern Song Dynasty said "most celebrated paper came from the south, like Wutian, Gutian, Youquan, Wenzhou and Huizhou areas, paper made in Jixi (a county of Anhui Province) cannot compare with paper made in these places".

Zhejiang region was a main paper producer in the Tang Dynasty, which in turn witnessed the first boom of Zhejiang papermaking practice. During the period, papermaking covered more areas in Zhejiang and enjoyed an increasing diversity. From *Record of Yuanhe County*, *The Geographical Record of the Tang Dynasty* and *General Code*, 11 places were designated to produce tribute paper, and Zhejiang enjoyed one third of them, namely, Hangzhou, Yuezhou, Wuzhou, Quzhou areas. In Kaiyuan Reign of Emperor Xuanzong in the Tang Dynasty, Quzhou produced "tribute Mian paper" to the amount of six thousand pieces annually in the period of Yuanhe Reign of Emperor Xianzong, Wuzhou made tribute Teng paper in Kaiyuan period and white Teng paper in Yuanhe period, Dongyang County yielded six thousand pieces of tribute paper, and Kuaiji, Yuezhou areas contributed Teng paper too in the same period.

In the Tang Dynasty, Teng paper made in Shengzhou area along the Shanxi area in Kuaiji County (hosting Yuezhou General Fu, chief military command organization in the early Tang Dynasty) was more than famous, with a reputation of "the most wonderful paper". In addition, Xihuangzhuang paper (paper used for filing a lawsuit), An paper and Ci paper (a kind of fine, yellow paper) made in Yuhang County, Wuzhou, Quzhou, Yuezhou and other areas also represented famous tribute paper, which was also true for another paper variety, Taijian letter paper.

New History of the Tang Dynasty recorded that "Teng paper served as tribute in Wuzhou", a distinguished Teng paper from the Eastern Jiangnan area, today's Jinhua City in Zhejiang Province. *Old History of the Tang Dynasty* recorded that when Du Xian (unknown—740), a famous prime minister in the Tang Dynasty, left his position in Wuzhou, he "planned to return home and was given ten thousand pieces of paper by the local administrator but ultimately only received one hundred pieces of them". At that time, Du Xian was called as "military officer taking only one hundred pieces of paper", in honor of his integrity. It also shows that Wuzhou was a paper production site with large quantity.

Shanxi River flowing through the Shengzhou Region

With accumulated papermaking experience, Teng paper in Shanxi area had a great improvement in quality in the Tang Dynasty, and was well received then. Gu Kuang, a prestigious poet in the Tang Dynasty composed a poem Eulogy of Shanxi Paper to extol the superior paper.

In the Tang Dynasty, transcription of classic scriptures was in a vogue. As a serious work, the paper chosen was also very demanding, so the population of Teng paper in Shanxi area revealed the excellent quality. Besides Gu Kuang, some other scholars also preferred the paper. Tao Gu in the Five Dynasties said in his Qingyilu, an ancient Chinese classic novel recording anecdotes of ancient culture and society, to the effect that Shanxi paper was so white and smooth that a dot of yellow seal can be spotted on the reverse side. It can be said that the paper used by Bai Juyi (a giant poet in the Tang Dynasty), collected by Tao Gu's ancestors, was given to the emperor by minister Yan Gu before given to Bai Juyi.

Thanks to its unique geographical resources and superior quality of paper, Shanxi, Shan paper and Shanteng, were important cultural symbols in the poems of the Tang and Song Dynasties. Table 1.1 presents a collection of Shanxi paper related poems (poems are omitted here).

Youquan Village in Yuhang County (belongs to today's Hangzhou City) is another famous producing place of Teng paper in the Tang Dynasty. Li Zhiyi, a Songci (poems) writer described the processing techniques of Teng paper in Youquan area, which is a valuable literature on local papermaking:

"The papermaking method used by Youquan workers followed Chengxin papermaking practice. The material was various, such as in northern Zhejiang and Jiangsu areas, bamboo was involved, so the paper was inferior in quality and purity. In the end of this summer, because of the pouring rain and violent tide, the water in fields failed to let out. Therefore, I could only walk around in my clogs within my hamlet. When a farmer came to tell me the harvest, the rain started again and I barely could move a single step. Thus, I sat idly at my table all day long. Finally, the sky cleared up. When I came out to relieve myself, I bumped into a neighbor who immediately invited me to write something on the paper. And I was delighted to do that and unwittingly used dozens of paper."

From Li Zhiyi's record, it can be seen that the papermaking method in Youquan Village was close to that of Chengxintang paper (a famous paper in the Southern Tang Dynasty), but unlike the latter one made from the refined bark of paper mulberry, the former one was inferior comparatively in quality.

With excellent quality, fair outlook and thick texture, Shanteng paper was selected to wrap up precious tea then and enjoyed a great popularity as packaging paper, which also enhanced its scale of production. The tea saint Lu Yu said in his grand work *The Classic on Tea* "Shanxi paper is fair and thick, and using it to package tea can maintain the aroma".

With regard to Teng paper in Shanxi area, the most systematic and complete literature is *Lament for the Rattan in Shanxi Area*, composed by Shu Yuanyu in the Tang Dynasty.

The poem, at least, revealed two things: firstly, the production scale of Teng paper in Shanxi area in the middle and late Tang Dynasty was indeed huge. Secondly, cutting rattan excessively to make paper had damaged the environment and ecology. To meet the need of papermaking, rattan was improvidently chopped from its root. Thus, the growth of rattan was impaired for its growth cycle is longer than that of hemp, bamboo and paper mulberry tree. The resources were limited, so Shu Yuanyu, as a Jinshi (a successful candidate in the highest imperial examination) born in Dongyang County of Wuzhou Prefecture in the Tang Dynasty, advocated cherishing the paper, otherwise "Shanxi can't produce sufficient paper" and "probably no rattan will be found in Shanxi". As visionary as Mr. Shu was, what he said had just been ignored. The truth is, the extinction of paper later in the Song Dynasty could be attributed to the previous ignorance.

In addition to Teng paper, Chu paper (or mulberry paper, a kind of paper made from the bark of paper mulberry tree) in Zhejiang region emerged in the Tang Dynasty too. In a famous work, *Mao Ying Record* (a prose written by Han Yu, a giant literati of the Tang Dynasty), showed that Chu paper came from Kuaiji and said "Mao Ying (refers to Chinese writing brush) has a good relationship with Chen Xuan of Jiangzhou area (refers to ink), Tao Hong of Hongnong area (refers to inkstone) and Mr. Chu of Kuaiji area (Chu paper)". By using personification, it called Chinese writing brush "Mao Ying", Chu paper in Kuaiji area "Mr. Chu" and assumed them friends. Judging from Han Yu's position as a literate master, we could assume the public admiration for Chu paper in Kuaiji area.

In the same period, the emergence of bamboo paper was recorded in Zhejiang region. *Bei Hu Record* (a geographical book of Lingnan area), authored by Duan Gonglu of the Tang Dynasty, mentioned "Zhumo paper", annotated by Cui Guitu (a scholar in the late Tang Dynasty) "produced in Muzhou Prefecture". Muzhou Prefecture is today's Jiande City in Zhejiang Province. And in the inception of the Song Dynasty, Su Yijian also said in his *Records of Four Treasures of the Study* (Systematic exposition of bibliographic works on the production and anecdotes of writing brush, ink, paper and inkstone) that "most paper was made from tender bamboo in Zhejiang and Jiangsu" (basically before the Song Dynasty).

Portrait of Han Yu

Apart from the core production areas, papermaking in remote areas like Wenzhou had also flourished. For Wenzhou area, "there was Juanfu paper (tribute handmade paper made from mulberry bark)" in the Tang Dynasty, and for those who made the paper were free from service (a preferential policy made by the emperor), that's why it was named Juan (meaning exemption). Now the paper made by Yongjia area (a prefecture in Zhejiang region) was favored by scholar bureaucrats, who were vying to buy it, for its ability to carry out the noble spirit of Chinese literati. As a good one, the paper can be populous both in previous and present times, almost as good as Chengxintang paper (one of the three treasures of the study in the Southern Tang Dynasty, and it enjoyed the fame of the best paper in Chinese history). Su Yijian also extolled the Juan paper in Wenzhou area, proving the high prestige of it.

Tang Dynasty is an important phase in the history of handmade paper in Zhejiang region, which mainly served for official usages, literati, scholar bureaucrats and classic transcribing. And from the literature study, packaging paper was just limited to high-end items like tea and basically not for everyday use.

1.2.4 Song Dynasty: the Unprecedented Boom of Handmade Paper in Zhejiang Region (the Yuan Dynasty also included)

Along Chinese history, the Song Dynasty is renowned for its civilization-oriented governance, prosperous culture, and exquisite life style. During the Song Dynasty, papermaking industry unprecedentedly prevailed in Zhejiang region. It surpassed former dynasties not only in papermaking expertise, varieties, and output, but also in the widespread use in daily life. Especially after the Song royal family moved its capital to Hangzhou area, with the development of politics, economy, culture and education driven by capital center area effect, the need for paper increased and papermaking industry gradually played an increasingly important role in history, forming the coexistence of three papermaking centers, i.e. Zhejiang, Bashu (Sichuan and Chongqing) and Fujian regions.

The scale and papermaking techniques of handmade paper in Zhejiang region in the Song Dynasty surged enormously. It was reflected in the following aspects.

Initially, China completed its transfer of papermaking centre. The south centered around Zhejiang and Fujian regions exceeded the north in papermaking in all respects with the appearance of a great deal of famous paper. Official system played an increasingly significant role both in supervising papermaking and in imposition of paper-related taxes, which grew into an important factor promoting the expertise and industry development of handmade paper.

Moreover, bamboo paper from the famous central papermaking area in Zhejiang gradually substituted Teng paper, which was popular in the past centuries. Due to the inexhaustible resources and low price of bamboo, bamboo paper became the newly-crowned dominant paper.

Finally, paper was highly rich in its raw materials and varieties since the Northern Song Dynasty. Su Yijian in the Song Dynasty described traits of Chinese paper from representative papermaking areas at that time in this way: "Paper is mainly made from hemp in Central Sichuan region, which was also called Yuxie or Xiegu. While in Jiangsu and Zhejiang regions, tender bamboo was generally used to make paper. Mulberry bark was usually used in the northern areas, and rattan in Shanxi, sea sedge for fishermen. People in Zhejiang region believed that paper made from wheat stalk and rice straw is easy to break and thin, and that paper made from wheat straw and rattan is the best."

From the late Tang Dynasty to the Five Dynasties, famous quality paper produced in Zhejiang region represented by Teng paper from Shengxian County and Youquan Village of Yuhang area, was still renowned in the Northern Song Dynasty, 50 000 pieces of which in compliance with the standard paper specifications were offered to the royal court every year. In the Song Dynasty, Sun Yin in his *Yue Wen*: *Yue Paper* praised Teng paper from Yuezhou this way: "Its luster and color can be seen through the golden sheet; it's anti-worm and anti-erosion." The outstanding literary master, Su Shi mentioned Yuban paper in Shengxian County in his work, *Four Poems to Sun Shenlao*, "Yuban paper is produced in Shanxi area". Shanteng paper was famous throughout the country, fueling a tremendous need and high price, which spurred the local papermakers to mass produce paper to pursue profits. Consequently, rattan as the raw material was almost exhausted, thus the concern of Shu Yuanyu in the Tang Dynasty was justified. By Jiatai Reign of the Southern Song Dynasty, output of Shanteng paper was scarce. By Hongzhi Reign of the Ming Dynasty, according to the record in *The*

A photo of *The Annals of Lin'an Prefecture During Xianchun Reign of the Song Dynasty*

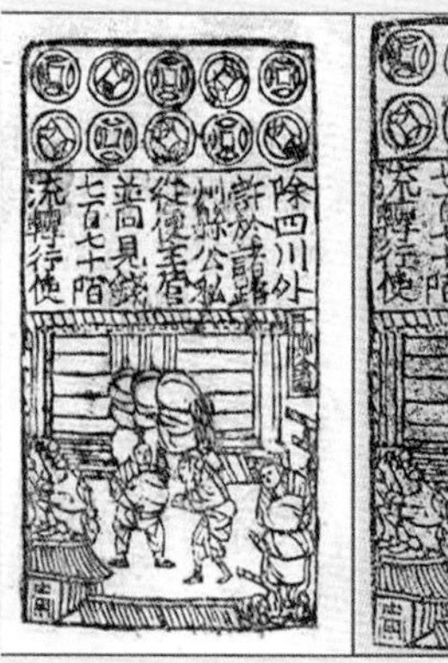
Jiaozi, paper money made in Chengdu area during the Song Dynasty

Annals of Shengxian County, "there is no papermaker of Shanteng paper." Since then Shanteng paper disappeared and remained only in the memory of history.

According to the documents in the Southern Song Dynasty, there were plenty of other types of famous paper in Zhejiang region during the Song Dynasty besides Teng paper. *The Annals of Lin'an Prefecture During Xianchun Reign of the Song Dynasty* carried a description of Hangzhou area, "Hangzhou offered Teng paper as the yearly tribute to the royal court, made in Youquan Village of Yuhang area. According to books, in order to save the paper for tribute, people use Xiaojing paper made in Fuyang area, Chiting paper made in Chiting Mountain as its substitutes". Outside Hangzhou area, there were other types of tribute paper, e.g. Huangtan paper in Linhai area, Juanjiang paper in Wenzhou area, and Yigong paper in Wuzhou area.

One of the major traits of the Song Dynasty was that both civilians and the government were engaged in papermaking practice. In the beginning, the authorities levied taxes from producing and selling paper. Later to print Jiaozi (paper money), the governments were in need of paper with special quality, so established their own papermaking factories. For instance, from the fifth year of Xining Reign (1072) in the Northern Song Dynasty, Chengdu Fu (Fu, ancient name for city) established papermaking factories to produce paper of Jiaozi in Sichuan; the officials in the Southern Song Dynasty made mulberry bark paper successively in Huizhou and Chengdu areas, and then transported it to Lin'an for printing money. Because Chengdu was too remote and the freight was too high, during the Southern Song Dynasty, Lin'an Papermaking Office moved papermaking factory from Jiuquchi in Lin'an Fu to Chishan area. This papermaking factory had papermaking workers and apprentices up to 1 200; then the office built Anxi Factory in Hubin; in the second year of Xianchun Reign (1266), these two factories were merged. In the late years of the Song Dynasty, Ma Duanlin wrote in his *Wen Xian Tong Kao (Literature Study)*, in the Southern Song Dynasty "paper was from Huizhou and Chizhou areas, then made in Chengdu Fu, later in Lin'an Fu". Moreover, in the Song Dynasty, Yuan Fu also mentioned paper in his work that in the Southern Song Dynasty, "Each year each papermaking site made ten million pieces of paper; till next summer, there came one hundred million pieces of paper from seven papermaking sites". Huizhou Fu (Jixi area as the center), Yanzhou (now Jiande City in Zhejiang Province) and other 5 local governments made ten million pieces of paper each year. The output was rather considerable under the operation of the official papermaking offices.

In Yuquanyuan of Beishan in Hangzhou area, a papermaking workshop was built, as well as four papermaking bureaus in Tangpu, Xinlin, Fengqiao and Sanjie of Shaoxing area.

In terms of papermaking center in Zhejiang region, in the early Northern Song Dynasty, papermaking center and main producing areas

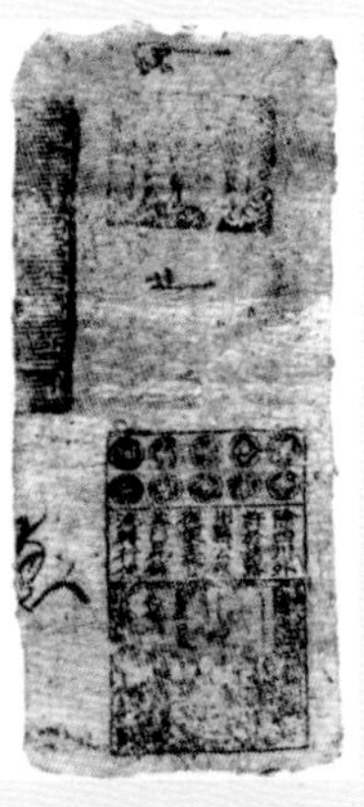

Huizi, paper money made in Lin'an area during the Southern Song Dynasty

of famous paper were clustered in Yuezhou/ Kuaiji (in the Northern Song Dynasty, Yuezhou and Kuaiji overlapped, and the two names were alternatively used; in the fifth year of Jianyan Reign of the Southern Song Dynasty (1131), Shaoxing Fu became the official name and was used ever since).

Shaoxing area as the traditional papermaking area, expanded its influence broader in the Northern Song dynasty. In *Fuxuan Zalu*, the papermaking practice nationwide was recorded as "China has mulberry paper, Teng paper in Shuzhong area, bamboo paper in Yuezhong area, mulberry bark paper in Jiangnan area". In the Ming Dynasty, Wen Zhenheng wrote in his *Zhang Wu Zhi*, "In the Song and Yuan Dynasties, colorful pink paper, wax paper, yellow paper, flower-decorated paper and ribbed paper were all from Shaoxing area".

Apart from Teng paper in Shanxi area still in production in Shaoxing area in the Song Dynasty, meantime there was another famous paper, Qiaobing paper, made with icy water in winter, which was welcomed due to its pure white color and smooth texture. According to *A Sequel to the Annals of Kuaiji Prefecture in Baoqing Reign*, "Qiaobing paper was from Shanxi area." Zhang Bochang described in his *Penglai Pavilion Poem*, "The expertise of making this Qiaobing paper shows how skillful the papermakers were." In *The Annals of Xin'an Area*, "Paper, especially Qiaobing paper was the best; in the westernmost areas of Shanxi, water was pure, with rattan ample in the mountain, therefore, winter was the best time for papermaking for its icy water". In the Northern Song Dynasty, poet Lv Ben said: "When holding the scrolls made by Qiaobing paper, I can feel the fragrance around the desk". All of the above mentioned quotes illustrate the extraordinary value of Qiaobing paper made with icy water.

Bamboo as the raw material in bulk production of bamboo paper emerged no later than the Tang Dynasty and was mass produced in the Song Dynasty. Bamboo paper had high cost performance as bamboo was easy to plant, fast in growing, and in China it was widely distributed with large reserves. Bamboo paper revealed a great advance in papermaking history of ancient China. It was even regarded as starting a new era in Chinese papermaking history.

Shaoxing area teemed with many varieties of bamboo, harboring a profound tradition of Teng paper, which made it one of the most active areas of bamboo papermaking in the Song Dynasty. In the Southern Song Dynasty, Ning Zongchao compiled *The Annals of Kuaiji Prefecture in Jiatai Reign*, in which he said: "Bamboo in Kuaiji Prefecture was such a beauty; there was one special variety of bamboo to make paper. Local families chopped bamboo and made a living by it. Dan bamboo and tender bamboo could both be used to make paper. Ku bamboo *(Pleioblastus amarus)* could be used to make joss paper." After the Song Dynasty moved its capital to the south, bamboo paper of Shaoxing area was more advanced in production with greater reputation, even surpassed the previous Teng paper, as the location of Shaoxing was inside the capital cultural circle. *The Annals of Kuaiji Prefecture in Jiatai Reign* recorded that, "Nowadays only bamboo paper was renowned across the country, other paper simulating it but only to find it impossible, thus Teng paper was gradually overshadowed". Jiatai Reign was in the mid of the Southern Song Dynasty, from which it could be concluded that before Ning Zongchao's time bamboo paper in Shaoxing had thrived and replaced the dominance of Teng paper.

However, in the Song Dynasty bamboo paper underwent a developing shift from immature to mature in expertise, from underrated to populous among people. Initially, it was repulsed by scholars and officials owing to its poor quality and lack of durability. In the Song Dynasty, prime minister Su Yijian said: "Bamboo paper was mostly made from young bamboo in Jiangsu and Zhejiang regions; but nobody dared to use the paper in writing secrets; because it was easily torn apart and hard to restore." Cai Xiang, a well-known calligrapher, put his words, "I banned the use of bamboo paper in my department; how could we fulfill our job, when the lawsuit was pending, but the paper was broken into pieces?"

As of the middle and later period of the Northern Song Dynasty, bamboo paper was generally acceptable by scholars and officials, writers and calligraphers, and became popular socially. Especially bamboo paper produced in Yuezhou area embraced affection and admiration from celebrities like Wang Anshi, Su Shi and Mi Fu. According to *The Annals of Kuaiji Prefecture in Jiatai Reign*, "Wang Anshi was fond of small pieces of bamboo paper which was shorter than what Shao used now; scholars also pursued the habit; before the Jianyan and Shaoxing Reigns, messages were mostly written on this kind of bamboo paper. When Su Shi returned from overseas, he wrote to Cheng Defan, "You help me buy one hundred pens of Cheng Yi writing brushes in Hangzhou area, Yuezhou paper two thousand pieces, half Changshi paper and half Zhanshou paper. Zhanshou paper is wider than Changshi paper." It means that it has the same length with ordinary paper, but double the width. Mi Fu, a calligrapher, had a unique admiration for bamboo paper in Yuezhou area. He started to use bamboo paper when he was 50 years old, holding that bamboo paper in Yuezhou area was the best, surpassing the famous Youquan paper of Hangzhou. In his poem, To Xue Shaopeng and Liu Jing, he praised, "Bamboo paper in Yuezhong area was shining like golden sheets, even better than Hangyou paper and Chijian paper". Hangyou paper in this poem referred to Youquan paper from Yuhang District of Hangzhou area.

In the Song Dynasty, calligrapher, Xue Daozu, praised bamboo paper in Yuezhou area, "It was as smooth as moss as a result of repeated beating of the raw materials; Yuezhou complimented itself as the hometown of writers and calligraphers; one piece of work was worth the output from one thousand papermaking households." According to Xue, the key tip for such high quality, was the repeated hammering of the materials in production, making the paper even more compact. Bamboo paper in Yuezhou area featured an immense popularity among scholars in the Song Dynasty; Sushi once purchased two thousand pieces of it for someone else. In the Southern Song Dynasty, Chen You said evidently in his *Discussion on Paper*, "In ancient times Shanteng paper was well-known, while currently Yuexi bamboo paper overwhelmed other paper types". From the Southern Song Dynasty, Teng paper in Yuezhou area was replaced by bamboo paper as a representative of Yue paper.

Portrait of Mi Fu

According to *The Annals of Kuaiji Prefecture in Jiatai Reign*, there were multitudinous bamboo paper types produced in Shaoxing Fu during the Southern Song Dynasty, of which Yaohuang, Xueshi and Shaogong were top paper types. Because they featured five major merits: smooth touch, excellent capability in showing ink color on the paper, desirable writing feeling and ink on it can last a long period, anti-worm and anti-erosion, they were especially favored by the calligraphers.

Besides Shaoxing area, as the capital city of the Southern Song Dynasty, Lin'an area was also thriving in papermaking industry. In the Northern Song Dynasty, papermaking industry marched hand in hand with the printing industry. Kaifeng area possessed the same amount of book-printing types as Hangzhou area, however its paper quality was inferior to that used in Hangzhou, thus its brand was not that influential as Hangzhou. Papermaking industry saw an early development in Hangzhou, for there was a record of local tribute paper in *The Annals of Jiuyu in Yuanfeng Reign*.

According to *A Sequel to History as a Mirror*, in June of the seventh year of Xining Reign, "the emperor ordered that a new specification of Xuan paper is supposed to be made in Hangzhou area, fifty thousand pieces each year. From now on, it is designated to be the official paper of local authorities, with the specifications fixed." This fact illustrated that Hangzhou in the middle of the Northern Song Dynasty became the papermaking base of the government.

Buddhism prevailed in the Song Dynasty, so temples and disciples transcribing Buddhist scriptures called for high quality of paper. Scripture paper of Jinsushan Mountain in Haiyan County of Jiaxing area in Zhejiang region was famous. There was a Jinsu Temple at the foot of the mountain in the southwest of Haiyan County. This temple stored scripture paper with good quality of the Northern Song Dynasty, with red seal of "Jinsushan Mountain Scripture Paper" on it. In the Ming Dynasty, Dong Yi wrote in his *A Sequel to the Annals of Ganshui* (1557), "In the Great Mercy Pavilion, there were two sets of the tripitaka, over ten thousand volumes. All volumes had the exactly same handwriting, which were likely to be written by one same man. On every piece of paper, there was the small red seal of "Jinsu Mountain Scripture Paper". Some of them were printed with the reign title of Yuanfeng which showed the paper history of 500 years. Both sides of paper were waxed with no decoration."

Hu Zhenheng (1567—1634) of the Ming Dynasty said in *Pictures of Haiyan County: Miscellaneous Knowledge*, "There was thousands of scrolls of scripture in Jinsu Temple, written on hard yellow cocoon paper, waxed both sides, smooth and shining, with red silky lines. The calligraphy was the regular script in a full way, as if all volumes were written by the same person. The ink was glossy and dark. On the back of each piece of paper was small red seal of "Jinsu Mountain Scripture Paper". The scripture paper of Jinsu Mountain was thick, and could be divided into layers as Song paper. Wen Zhenheng said in his *Zhang Wu Zhi*, "There was yellow and white scripture paper which could be separated into pieces." Zhou Jiawei (Qing Dynasty) said in *Zhuang Yan Zhi*, "When I framed something, I used Jinsu paper with paste made from bletilla, which never peeled, and looked extremely elegant." By the time of Qianlong Regin of the Qing Dynasty, it has been handed down over generations. During this period in Haiyan County, the famous collector and bibliophile, Zhang Yanchang, created On Jinsu Mountain Paper particularly for this paper.

In the Song Dynasty, Yuluan (fish egg) paper produced in Dongyang area was quite prestigious. A poem named *Xiegaoji Yihui*

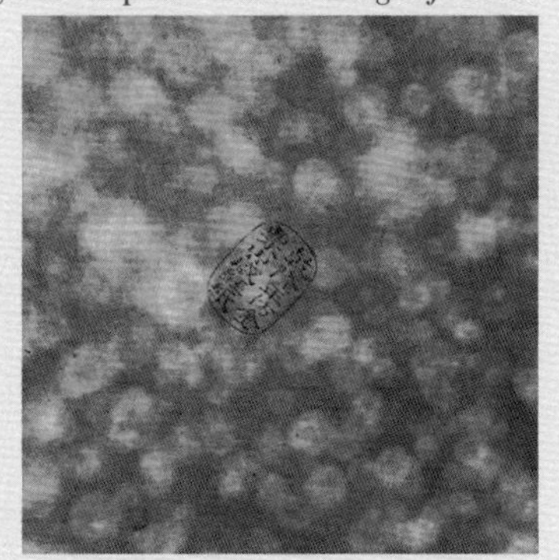
Jinsu Temple in Haiyan County and Jinsu paper (imitated during the Qianlong Reign)

Paper by Chen Zao in the Song Dynasty wrote: "Everyone adores the Yuluan paper shining with silver, only the paper used by Xue Tao (a female poet in the Tang Dynasty) to compose poems could be comparable." The origin of Yuluan paper was the Dongzishan District in Dongyang area. According to *The Annals of Dongyang County*: *Local Products* in the year of Daoguang Reign, "Among all kinds of paper, Yuluan paper is soft, smooth and clean; to the east of Dongyang County, that made in Baixi area was the best."

In addition to bamboo paper, Teng paper and mulberry paper, Zhejiang region also created one type of paper made from wheat stalk and straw in the Song Dynasty. Su Yijian in the Northern Song Dynasty, wrote in his book *Four Books of the Study* : "People in Zhejiang region think that the wheat stalk and rattan are quite superior among all the crisp and thin materials for making paper." This is the earliest record of making paper by grass fiber in the world. The papermaking method is as following: first, smash the straw and stalk; then soak with plant ash, bury in a pit or pond to ferment the material; later, put it into a permeable bag; at last, clean the material with flowing water. The emergence of straw paper has further expanded the source of papermaking materials due to the abundant sources of rice straw and wheat stalk. In the Ming Dynasty, a eunuch named Liu Ruoyu noted in his book *The Annals of Zhuozhong (Court Anecdotes)*: "The toilet paper used by the emperor was made by the paper production workshop which was supervised by the Grand imperial eunuch. It was light yellow, soft and thick, and the square cut of it could be over three cun (Chinese unit of length). It was collected by the eunuch who exclusively served the emperor and specialized in the management of the clean room. When it came to the former emperor (Wanli Emperor of the Ming Dynasty), on account of the fact that he thought the handmade paper made in the Imperial Court was not so qualified, he specially sent attendants to buy the superior handmade paper from Hangzhou area." From this book, it can be seen that the most well-known handmade paper in Zhejiang region was made in Hangzhou area, which served as tribute paper.

In the Song Dynasty, the papermaking industry in Siming area of eastern Zhejiang region, rose in a rapid speed. According to *Siming Annals in the Baoqing Reign*, paper for printing salt coupon was exclusively made in Fenghua County, Yinxian County and Xiangshan County, and then issued by the Imperial Court after printing. Every year the imperial court would purchase 79 300 pieces of Salt coupon paper, expending 1 019 *guan* (ancient Chinese unit of currency) and 568 *wen* (ancient Chinese unit of currency) copper coins every year. It was depicted in *Song Dynasty Chronicles: commodities and foods*. The size of the salt coupon paper is one chi seven cun long and one chi wide, but so far no relics have been found yet. Since the paper is often circulated, it should be durable and tough. Therefore, it is inferred that salt coupon paper was either exclusively or mainly made from tree bark with long fiber.

What's more, the "Huanggong paper" produced in Ninghai County also owned a certain popularity. It is said that the paper was invented by Huanggong, one of the "Shangshan Four Recluses" (four recluses in the late Qin Dynasty), made from Huang bamboo from Tongbai Mountain in Xixi area of Ninghai County. Su Dongpo (Su Shi, a great poet in the Northern Song Dynasty) once praised Huanggong paper and compared it to jade, from which we could infer the good quality of it. The development of the paper industry in the eastern part of Zhejiang region such as Mingzhou area laid the foundation for the prosperity of the book printing industry.

Juan paper in Wenzhou area as tribute paper started from the Tang Dynasty and thrived in the Song dynasty, embracing a great reputation for high-end purpose. The Volume 99 of *Taiping Geographical Annals*: *Jiangnan Dongdao 11-Wenzhou*, records that during Zhihe Reign of the Northern Song Dynasty (1054—1056), tribute Juan paper in Wenzhou area as its local product was one of the nine tribute paper types all over the country, together with Shanteng paper from Shengxian County, and Youquan paper from Yuhang listed as the three major tribute paper varieties in Zhejiang region, which manifested its superior quality at that time. Zhao Yushi in the Southern Song Dynasty quoted *Yuanfeng Nine Regions Annals* in his volume 10 of *Bing Tui Lu (Historical Anecdotes)* that during Yuanfeng Reign of the Northern Song Dynasty (1078—1186), "four thousand pieces of paper from Yuezhou, Shexian, and Chizhou areas with one thousand from each, and from Zhenzhou and Wenzhou areas with five hundred pieces from each, were served as tribute paper every year."

Juan paper was fine and white. Why it was called "Juan paper" was explained in *Qinbo Biezhi* written by Zhou Hui in the Song Dynasty, "in the Tang Dynasty, any household who made Juanfu paper was exempt from servitude. Thus, it was named "Juan". The paper was made in Yongjia area. Scholar officials loved it very much for its writing performance, so they competed to acquire it with a favorable price. It was even as precious as Chengxintang paper." This document revealed that firstly, Juanfu paper was named for the fact that the government exempted papermakers from hard labor who made this quality paper; secondly, in the Song Dynasty, Juan paper made in Wenzhou (Yongjia) area in the south part of Zhejiang region was already well-renowned. In *Guang Yu Ji (Atlas of Ancient China)*, Juanjiang paper, pure white and smooth, was better than Gaoli paper (a strong and spotless white Korean paper made from silk). In the state of Wuyue, who made this paper could be exempted from hard labor, hence the paper got the name. Chengbo in the Song Dynasty said in his *Sanliuxuan Zashi*, "Juan paper was made in Wenzhou, pure white and smooth, largely similar to Korean paper. In the southeast of the region it was produced most. Even Youquan paper was not as good as it and was produced less. Since the period of Zhihe Reign, it became a tribute paper. Bigwigs of the time demanded more and more, while the papermakers could not manage it." From the above description and analogy with Korean paper, Juan paper should be similar to mulberry paper. In 1961, *The Sutra on Contemplation of Amitayus* printed in the third year of Daguan Reign of the Tang Dynasty was discovered in White Elephant Tower in Ouhai District of Wenzhou area, which was detected as Juan paper relics from Wenzhou area by the researchers.

In the Song Dynasty, Fuyang, as one of the papermaking centers later in China, witnessed a rapid development of bamboo paper. According to *The Annals of Lin'an Prefecture During Xianchun Reign of the Song Dynasty*, "there was Xiaojing paper in Fuyang area, Chiting paper in Chiting Mountain." The famous eunuch, Xie Jingchu (1020—1084), from Fuyang in the Song Dynasty, processed "Ten-color letter paper", which was called "Xiegong paper" at that time, based on the

Portrait of Xie Jingchu

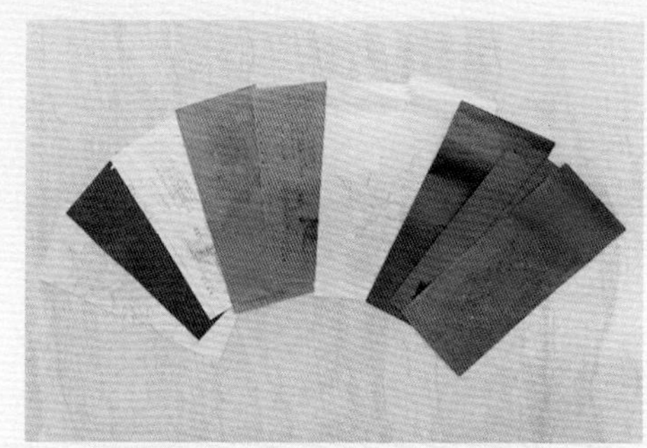
"Ten-color Xiegong Paper"made in Fuyang District nowadays

exquisite bamboo paper in Fuyang area and known around the country together with the distinguished "Xuetao paper". During Guangxu Reign, according to *The Annals of* Hangzhou, "paper was named after people, e.g. Xiegong and Xue Tao. Xiegong from Fuchun area was raised as "Xiansi" (in charge of judicial department), creating different types of letter to make it convenient for writing, hence called Xiegong paper. There were ten colors of it: crimson, pinkish white, apricot red, bright yellow, deep cyan, light cyan, deep green, light green, aerugo, light blue."

In addition to documentary records, Fuyang area in the Song Dynasty served as a vital papermaking base in China has been proved by archaeological researches. In 2008, archaeological workers discovered a papermaking site ruins of the Song Dynasty at Sizhou Village in Gaoqiao Town of Fuyang District, which was run by the government and mainly made bamboo paper. By far, it is the largest ancient papermaking workshop ruins enjoying the longest history and containing the most complete technological process. The ruins could be divided into work and living area, totally covering an area of 16 000 square meters. One seventh of the total area, about 2 400 square meters was excavated by the archaeologists in 2008. The working area was separated by a west-east canal into two parts. North side of the canal was mainly used as drying yard and the south was papermaking area, including pools for fermenting materials, cooking woks, pools for making pulp, grinding area, papermaking rooms and paper drying room. And there were stone paths, wells and ash pits. All those functional areas mentioned above are in accordance with those recorded in *Heavenly Creations* (a comprehensive scientific and technological masterpiece) and those of bamboo paper workshops in today's Fuyang District.

Almost 30 000 pieces of cultural relics have been unearthed at the site, and 300 of them can be restored. Besides stone-made tools, like grinders, stone mortar, duitou (the head of pestle used for rice hulling), inkstone, there are also ceramic building components, porcelain and copper money. The first exhumed relics were mainly in the Southern Song Dynasty, yet the discovery of bricks in "second year of Zhidao Reign of the Song Dynasty" and "second year of Dazhongxiangfu Reign of the Song Dynasty" can push the site era to the early Northern Song Dynasty. We know surely that Fuyang papermaking industry in the Song Dynasty had reached a large scale and was dominated by the government. In 2013, Sizhou Papermaking Site was listed in the seventh batch of national key cultural relics protection units.

Different from the Song Dynasty, little

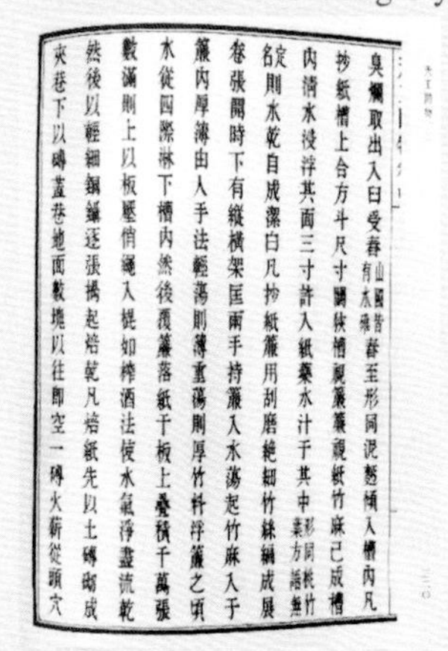
臭爛取出入臼受舂山國皆有水碓舂至形同泥麵傾入槽內凡
抄紙槽上合方斗尺寸闊狹槽視簾簾視紙竹麻已成槽
內清水浸浮其面三寸許入紙藥水汁于其中形同桃竹葉方語無
定名則水乾自成潔白凡抄紙簾用刮磨絕細竹絲編成展
卷張開時下有縱橫架匡兩手持簾入水蕩起竹麻入于
簾內厚薄由人手法輕蕩則薄重蕩則厚竹料浮簾之頃
水從四際淋下槽內然後覆簾落紙于板上疊積千萬張
數滿則上以板壓俏繩入棍如榨酒法使水氣淨盡流乾
然後以輕細銅鑷逐張揭起焙乾凡焙紙先以土磚砌成
夾巷下以磚蓋巷地面數塊以往即空一磚火薪從頭穴

Records in *Heavenly Creations*

Sizhou Relics Protection Exhibition Hall

Archaeological Excavation Report of Sizhou Relics in Fuyang District

information about papermaking in Zhejiang region can be found in existing records of the Yuan Dynasty. According to scholar Gao Lian in the Ming Dynasty, "In the Yuan Dynasty, there was some special paper, like Bailu paper, Qianshan paper, Changshan paper, Yingshan paper, Qingjiang paper and Shangyu paper and so on", among which, at least Chanshan paper and Shangyu paper was definitely produced in Zhejiang region. In paper part of *Collection of Ancient and Modern Books* (a reference book covering many fields and arranged according to subjects) in the Qing Dynasty, "There was Huangma paper, Qianshan paper, Changshan paper, etc. in the Yuan Dynasty". Here, Chanshan paper was referred to again. However, the ancient Changshan area is not equivalent to today's Changshan area. For instance, in the Tang Dynasty Changshan covered Kaihua County and main part of Yushan County in Jiangxi region while in the early Song Dynasty it just included Kaihua County. Nonetheless, it's doubtless that papermaking in Changshan area was well developed and in a large scale in the Yuan Dynasty.

Yuanshu paper made in Fuyang area in the Ming Dynasty

1.2.5 Ming Dynasty: Evolution of Types and Papermaking Areas of Handmade Paper in Zhejiang Province

In the Ming Dynasty, Zhejiang region became the important center of knowledge dissemination and printing in ancient China with the development of economy and culture. Mass paper demand required by the printing industry stimulated the great development in output, cost performance and specific uses of handmade paper in Zhejiang region. In the early Ming Dynasty, large amounts of writing and printing paper was levied by the government on Zhejiang region , which stimulated and boosted local papermaking practice. After the Ming Dynasty, increasing records show more daily uses besides to supply the government.

In the beginning, rattan and paper mulberry bark were mostly used to make official document paper and calligraphy and painting paper in Zhejiang region. Its cost and price were both high though quality is good. In the middle and late Song Dynasty, bamboo papermaking in Zhejiang region began to develop in a relatively large-scale and replaced the high-cost Teng paper and mulberry paper. In the Ming Dynasty, Yuhang in Hangzhou area was famous for its Zhushao paper and Pichao paper. The local name of

Zhushao paper was Shao paper or Qianzhang paper. It was square and could be cut into strips without break. Shanggao and Xiekeng areas, in the south part of Yuhang area, "Local people put bamboo into limewater and pound it to make paper. The paper replaces silk to be burned in sacrificial rituals. People in the south of Changjiang River all use this kind of paper and think it is great." "The local name for Pichao paper is Mian paper, also called Chu paper, which is produced in Nanjian area of Yuhang County. Cook mulberry bark and grain shell with limewater and pound the material to make pulp for making Pichao paper."

From the Eastern Jin Dynasty, Yuezhou-Kuaiji papermaking center was gradually surpassed by many emerging places in the area. Changshan and Kaihua of Quzhou area and Fuyang of Hangzhou area were all famous for papermaking practice. Songyang and Xuanping of Chuzhou area (ancient name of Lishui) located in remote mountain areas and suffered from poor transportation, also gained a reputation for papermaking. In Fuyang County, "people make a living by papermaking and even the elderly and children are busy with papermaking day and night." In the late Ming and early Qing Dynasties, Yuanshu paper, Xiaojing paper and Chiting paper in Fuyang County were designated as paper for writing official documents, memorial to the emperor, and for the imperial examination, which showed that these types of paper were all of high quality at that time.

Another great transition of Zhejiang papermaking in the Ming Dynasty was the decline of Juan paper in Wenzhou area from prosperity. *Qihai Trivial Talk* written by Jiang Zhun in the Ming Dynasty recorded that: "Juan paper made in Wenzhou area is pure white, fine and smooth almost like Gaoli paper (a strong and spotless white Korean paper made from silk). When the Qians ruled the Wuyue state, this kind of paper was used to exempt tax so it was called JuanpPaper." In Hongzhi Reign of the Ming Dynasty, *The Annals of Wenzhou Fu* (an encyclopedia of Wenzhou) recorded the method of Juan papermaking: "Put strong acids, flour and mirabilite into water and boil it until it becomes thick and cool the liquid. Put paper into the mixture of glue and alum and dry it. Then, brush the liquid over two sides of the paper and dry the paper. Wax the paper, then smooth the paper with a rough cloth bag." The papermaking techniques of this kind of paper was so complicated that it was not very productive. However, literati adored it very much for it was appropriate for calligraphy and painting.

The decline of Juan paper industry was relevant to the incessant demand of the feudal officials. At the beginning of the Ming Dynasty, the government opened a papermaking office in the old street of Quxi Town in Ouhai District in Wenzhou area in order to supervise the process of papermaking, which imposed a heavy burden on both the local authority and the inhabitants. In the fifth year of Xuande Reign of the Ming Dynasty (1430), He Wenyuan, the administrator of Wenzhou, perceived the complaints of local residents. As a result, for the sake of showing solicitude for the local people and giving them a break, He Wenyuan submitted reports to the throne that it was no longer suitable to produce Juan paper due to the increasing turbidity of local water. Jiang Zhun said in his book Qihai Trivial Talk, "He Wenyuan, whose style name was Dongyuan 'furtively schemed to change the quality of local water so that the Juan paper made turned black, in terms of which he claimed it was on account of the migration of the fortune of this area'. He declared that the quality of the paper was influenced by the deterioration of the local water, which was caused by the papermaking industry. In light of this situation, the central government of the Ming Dynasty decided to suspend the operation of Papermaking Office, which accordingly ceased the production of Juan paper.

Wenzhou was still required to deliver other type of paper as tribute to the imperial court despite of the fact that the production of Juan paper had been halted by the central government. According to the third volume of *The Record of Wenzhou Government*, "The number of Nanjing calendar yellow paper delivered by Wenzhou area is 6 040 every year while the number of white paper delivered by Wenzhou is 98 357 every year (including 3 950 pieces yellow paper and 48 091 pieces white paper manufactured in Yueqing as well as 2 085 pieces of yellow paper and 50 246 pieces white paper made in Rui'an area)", from which we could infer the general scale of Wenzhou handmade paper industry in the Ming Dynasty.

The second major change of Zhejiang handmade paper industry in the Ming Dynasty was that Changshan and Kaihua counties in Quzhou area became the main handmade papermaking centers nationwide, which laid the foundation for the place to become a production center of tribute paper in the early Qing Dynasty. *The Annals of Changshan in Guangxu Reign* once quoted the record from *Wanli Chronicles*: "paper mulberry tree, the raw material of one type of paper, did not grow in Changshan County. As a result, only residents in Qiuchuan area were good at producing such kind of paper... official paper was made in large scale every year, making tens of millions of silver money." In accordance with *The Record of Quzhou Annals* written in Tianqi Reign, Quzhou area produced "Teng paper, Huangbai paper, paper bed-curtain and so forth." Lu Rong (1436—1494), an officer of Zhejiang region in the Ming Dynasty, recorded the papermaking techniques of Quzhou area in detail in his book *Miscellaneous Notes of Shuyuan Garden*: "Residents in Changshan and Kaihua counties in Quzhou earned their living by making paper. The steps of making paper are as following: To begin with, steam the bark of paper mulberry tree, peel the rough solid bark, and then ferment in lime water for three days. After that, stir and boil the materials, remove the impurities, immerse the materials in water for seven days and boil once again. Subsequently, wash the materials and dry in the sun for a long time. Beat to make pulp, rinse by water, add in papermaking mucilage such as walnut cane, and then fill the pulp in the papermaking screen. By the time it condenses, peel the wet paper and dry with fire on a table made of bricks, and fire was set under the table." Lu Rong once was an official in Changshan County, which made his records quite reliable. It is worth noting in the description of Lu Rong that in Changshan and Kaihua Counties in Ming Dynasty, paper was drying on a flat table, instead of using regular fire walls in later practice.

Changshan papermakers took advantage of bamboo to produce paper as well, and the way they processed paper was quite typical, which was depicted in detail in *Heavenly Creations* written by Song Yingxing (1587—1661), the technical expert in the late Ming Dynasty: "The diameter of the pot to boil bamboo is about four *chi* (Chinese unit of length). The clay is used to adjust the edge of the pot with lime, so that its height and width are similar to those for boiling salt in the coastal areas of the middle areas of Guangdong region, carrying water more than ten *dan* (Chinese unit of weight). It is covered with a bucket that has a circumference of 15 *zhang* (Chinese unit of length) and about four chi in diameter. After bamboo is added into the pot, boil for eight days. Then stop for one day, take out bamboo and wash in the clear water pond. The bottom and the periphery of the pond should be blocked with wooden boards to prevent contamination (not necessary when making rough paper). After the bamboo and linen are washed, soak with firewood water, then put it into the pot and flatten it, and spread one cun thick of straw ash. After boiling, transfer the material another bucket and continue to rinse with the ash water. After the grass ash water is cooled, it should be boiled and rinsed again. After more than ten days, bamboo naturally rot and stink. Take it out and put it into a mortar and pulverize it into mud (there are hydraulic pestle in the mountainous areas) and pour it into the papermaking trough."

It can be seen from literature that the papermaking industry in Changshan County made great progress in the Ming Dynasty (1551—1602) in terms of the quality and the amount of the paper, compared to the Song Dynasty and the Yuan Dynasty. Hu Yinglin, a litterateur in Ming Dynasty, wrote in his book *A Series of Stories of Shaoshi Mountain*: the Mian paper produced in Yongzhou area (one of the counties in Jiangxi Province) was the best for printing books, then Jian paper produced in Changshan area (one of the counties in Zhejiang Province), and then book paper produced in Shunchang area (one of the counties in Fujian Province) while bamboo paper produced in Fujian region was the most inferior one among all."

Most of the official paper in the Ming Dynasty came from western Zhejiang region and eastern Jiangxi region. Lu Rong wrote in his book *Miscellaneous Notes of Shuyuan Garden*: "Residents in Quzhou area of Zhejiang region made their living by making paper. Countless official paper was wasted no matter for government use or personal use. However, at the beginning, ministers of Imperial House cared nothing for this phenomenon. It was heard that in Tianshun Reign, after an old minister in Imperial House came back from Jiangxi region, confronting the official paper used to paste the wall, he broke into tears. As a result, ministers of Imperial House all understood the hardship for making paper and lamented the tremendous waste of official paper. Here is another story told by a venerable old man: During Hongwu Reign, the textbooks used by the students of the Imperial College and the paper that had been used for practicing calligraphy were sent to the Ministry of Rites monthly. The copybooks for calligraphy were sent to the Guanglu Temple to pack the dumplings, and the textbooks were sent to the judicial department, with the backside of which used as draft paper. During the period from Yongle Reign to Xuande Reign, the annual fireworks and firecrackers of Aoshan Mountain were also made of used paper and this custom was suspended a long time later. In the year of Chenghua Reign, the consumption of paper was tremendous due to the fact that all the stuff like fireworks and firecrackers were made of top-hole paper. No old minister of Imperial House would stand up to share the hardship of papermaking."

From the above description of the waste of official paper, we can tell that the consumption of official paper in the Ming Dynasty was very large, and the demand of the government had in turn triggered the development of the private papermaking industry.

Record of papermaking in Quzhou City in *Miscellaneous Notes of Shuyuan Garden*

A photo of *The Annals of Changshan County* (published in 1990)

In the late Ming and early Qing Dynasties, papermaking Industry in Changshan area entered its heyday, and in addition to being part of the tribute, it was also sold to Beijing, Zhili (ancient name for Hebei), Henan, Shandong and other places. According to *The Annals of Changshan County in Kangxi Reign*: "The types of paper are numerous. Paper for compensating tax in Zhili region and Henan region and the official paper in Hunan, Guangdong and Fujian regions were all from our county and Yushan County."

Paper produced in Quzhou was mainly used for compiling Yellow book and land register book by feudal officials. Most of the official paper used in Jiangnan area, Henan, Hunan, Guangdong and Fujian in the Ming Dynasty were purchased from Quzhou.

In the Ming Dynasty, the gold, silver and tinfoil industries in Hangzhou, Shaoxing, Suzhou and Jinling (ancient name for Nanjing) areas reached considerable scale, supporting the development of special-purpose processed paper industry such as Luming paper and Wujin paper.

With the boom of folk tinfoil industry in the Ming Dynasty, Luming paper (a kind of paper used as the raw material for producing tinfoil) used for lining the tinfoil was in tight supply. Bamboo was the raw material of Luming paper. Due to the abundant bamboo in Kuaiji area, the great consumption of Luming paper and the high profit, many paper mills in Kuaiji area gave up the production of traditional bamboo paper at that time and turned to more profitable joss paper for sacrificial rituals. The absolute slump of the famous bamboo paper in Yuezhou area may directly resulted from this sort of market transition. *The Annals of Shaoxing Fu in Wanli Reign* (an encyclopedia of Shaoxing area in Wanli Reign of the Ming Dynasty) recorded that: "In the past, paper production in Yuezhou area was in large quantity. *Mao Ying Zhuan* (a biography of Mao Ying who was the personification of brushes) written by Han Changli (another name of Han Yu, a famous literati in the Tang Dynasty) called the paper Mr. Zhu." By the 1960s, there was still production of Luming paper in some areas of Kuaiji area such as Pingshui Town of Shaoxing City and the paper was used to make tinfoil. However, at that time Luming paper production was scaled down a lot and then was forbiden because of the superstition opposition tide in the period of the Great Cultural Revolution.

Wujin paper (a kind of black gold and glossy paper), a local product of Hangzhou, was specially used to make gold and silver foil. The raw material of this kind of paper was Ku bamboo (*Pleioblastus amarus*). The producing process of Wujin Paper was very

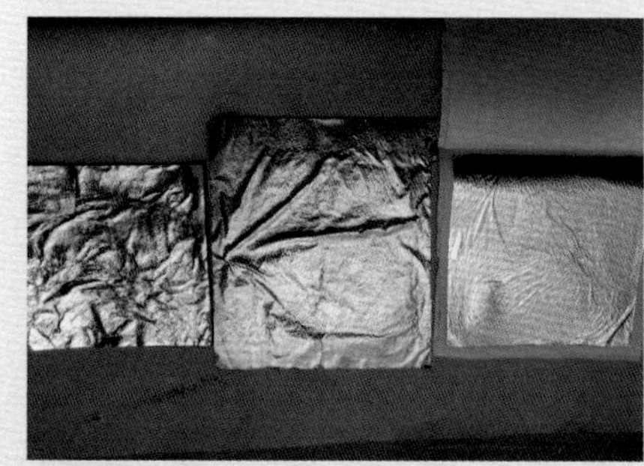
Modern tinfoil products in Zhejiang Province

Wujin paper made during the late Ming and early Qing Dynasties

complicated with many steps as follows: beating the paper repeatedly in "White House", gathering smoke from boiling soya-bean oil in "Smoke House", and painting, boiling, beating, smudging and dying the paper in "Black House".

Physics (an encyclopedia) was written by a scholar Fang Yizhi in the late Ming and early Qing Dynasties, the volume seven of which recorded that: "Put gold between two pieces of Wujin paper and beat repeatedly with a huge hammer to make gold foil. The Wujin paper remained intact. The color of the Wujin paper changes from brown at first to black gold over a long time." "Wujin paper can be only found in Hangzhou area and it can only be made with the water below Chunyou Bridge in the east of the area. The producing steps of Wujin paper are as follows: make Wujin water (a kind of ink made from coal) and spread the Wujin water on the paper, smoke the paper after it has become extremely black, and polish the paper with a special stone mill. After the process the tenacious and strong Wujin paper is made. Wrap small-size sheets of gold in Wujin paper and beat repeatedly to make the gold foil. When the small-size sheets of gold have become thinner sheet called gold foil, the Wujin paper is still intact. Wujin paper is widely sold at a pretty high price for perhaps only people in Zhejiang region are able to produce this kind of paper." said Wei Liangzai.

In the volume of *Gold* in *Heavenly Creations: Metal*, written by Song Yingxing (a scientist) in the late Ming Dynasty also recorded that: "The method of making gold foil is to wrap small-size sheets of gold in Wujin paper and beat repeatedly with full strength. All the Wujin Paper is made in Suzhou and Hangzhou areas. It is made from membrane of the Ju bamboo by the East China Sea. Burn soya-bean oil in airproof space with only pinholes to ventilate and smoke the paper to produce Wujin paper. Each Wujin paper after being used to make gold foil for fifty times will be sent to herb shops as wrapper of cinnabar. The Wujin paper is still intact perhaps because great techniques make wonder."

In the Ming Dynasty, there were already records about papermakers and paper mills in Jinhua area. In Yongle Reign of the Ming Dynasty, Jinhua paper was listed as a tribute to imperial court and the court levied yellow and white sacrificial paper from brands of "Tian" and "Shu" on Jinhua area. Emergence of paper brands showed that the mechanism of designating paper mills to produce different types of paper had been formed at that time, which was a feature worthy of attention in handmade paper history in Zhejiang region.

In the early Ming Dynasty, Wenzhou, a city located in southern Zhejiang region, preserved the recordings on the use and quantity of tribute paper in Leqing County. It is related that: "calender paper was included in the yearly purchase list of the local government as tribute to the royal family; the Ministry of Rites in feudal China offered 3 500 pieces of yellow paper, 500 pieces of white paper, while the administrative commissioner's office offered 53 000 pieces of white paper." The recordings show the diverse types and uses of tribute paper of the time.

In conclusion, based on literature review of the tribute paper in Zhejiang region in the early Ming Dynasty, a detailed industrial chain had already been formed then.

1.2.6 The Qing Dynasty: Great Prosperity Period of Handmade Paper in Zhejiang Region in Ancient Times

In the Qing Dynasty, handmade paper in Zhejiang region entered the stage of great prosperity, highlighted by:

Firstly, literature reveals that handmade papermaking sites are scattered throughout the province. Papermaking materials, production process, as well as categories, all reached the peak in the traditional era. In general, papermaking is associated with the origin of the raw materials, which was located in the mountainous area. Plant fiber was used as raw material, such as *Phyllostachys edulis*, *Pleioblastus amarus*, sympodial bamboo, *Phyllostachys heteroclada* Oliver, mulberry bark, paper mulberry bark, *Wikstroemia monnula* Hance, *Edgeworthia chrysantha* Lindl. and straw, covering a wide range.

Papermaking is more developed in the following areas: Hangcheng, Fuyang, Yuhang, Xiaoshan, Lin'an of Hangzhou, Ouhai, Rui'an of Wenzhou, as well as Anji, Xiaofeng of Huzhou. Needless to say, Fuyang has become a papermaking center. Huangbai paper, mulberry paper, straw paper, yellow paper produced in the mountainous areas under the jurisdiction of Huzhou area, such as Xiaofeng, Anji, Wukang, Gui'an, together with Meili letter paper produced in Jiaxing, have been well-known in the Qing Dynasty. With an intense focus on papermaking industry, Changshan, Kaihua, Jiangshan and Longyou areas of Quzhou Prefecture have become an important base for Chinese papermaking industry in the Qing Dynasty.

According to *Zhang's Testament* written by Zhang Xuecheng in the late Qing Dynasty, paper at Hubei market was partly from Hangzhou and other places. Even merchants in Huizhou area (the papermaking center from China's late Tang and Five Dynasties to the Song and Yuan Dynasties) came to Zhejiang region for paper. Due to the fact that Hui paper, which is branded as "Chengxintang paper", has a great reputation in history, the amount of tribute paper has increased every year, making the paper industry in Huizhou area unbearable with declining quality. In the middle and late Qing Dynasty, Hui paper was not good any more. Even the tribute paper had to be purchased from outside. According to *The Annals of Huizhou Fu* during Daoguang Reign, the tribute paper of Huizhou area was often purchased in neighboring places such as Changshan and Kaihua Counties in Zhejiang region. This also reflects that the production of Changshan and Kaihua paper is considerable, and even the tribute paper has a surplux for sale.

The use of hudraulic pestle as a tool for pounding the bamboo in papermaking appeared in the Song Dynasty and prevailed in the Qing Dynasty. In the records of *The Annals of Huizhou Fu* during Tongzhi Reign, "In Dongshen Qianjiabian area, streams divert to several directions, and the water force can be used to pound bamboo as papermaking raw material. Stone stove is placed to boil materials with dust. Pound them with a pestle to break fiber and boil them. As a result, the material will float in water and be exposed to sun for days. Then comes the yellow paper."

In Zeya Town of Ouhai District in Wenzhou City, there still exists the Monument in Memory of the Hydraulic Pestle jointly built during Qianlong Reign of the Qing Dynasty. And its contents include: "Ziyu and other seven farmers jointly invested in the construction of the hydraulic pestle and agreed that the property would not be allowed to be transferred casually. No quarrel is allowed or they will be fined, which rule was set at Pan's house in February 1790, during Qianlong Reign of the Qing Dynasty.

Propaganda material of Kaihua tribute paper

Old hydraulic pestle still being used in Zeya Town of Wenzhou City

Monument in memory of the hydraulic pestle

Secondly, it has continued until the contemporary papermaking center and the highly concentrated areas of the business came into being, with a rich variety of paper products.

In the Qing Dynasty, Fuyang County (including Xindeng County, which was later incorporated into Fuyang) began to become one of the national central production areas, with a significance of business concentration. For a long period of time, handmade paper is the largest commodity of the county. According to *The Annals of Fuyang Fu*, "paper is produced in all counties and the one in Fuyang ranks first". By the end of the Qing Dynasty, the trading amount reached no less than one million taels annually. Among them, bamboo paper enjoyed the largest output. "Bamboo paper is produced in southern village, using the raw material of *Phyllostachys edulis* and *Dianthus chinensis* L..There were various kinds of paper such as Yuanshu paper, paper of Gaobai, Shiyuan, Zhongyuan, Haifang, Duanfang, Changbian, Luming, Cugao, Huajian and Biaoxin, etc. It was the biggest papermaking site.

At that time, due to the fact that Fuyang area enjoyed well-developed paper business, even the large-size local paper and Kengbian paper were of top quality, sold throughout eighteen provinces across China. Yuanshu paper, a kind of bamboo paper produced in Dayuan Town, was claimed "best of all". Also named white paper, "it is specially made in the mountain in the south of Fuchun River, using hydraulic pestle in the papermaking process" "The quality of the paper lied half in people's work and half in the water, which would not be copied anywhere else."

Straw paper is the second largest paper product. Taking straw as its raw material, it was quite common in the place, with a large output. All the towns in the county could produce it, with the one made in Beixiang area ranking the top.

Bast paper is the third largest type of paper. One category is produced by residents of northwestern township, with paper mulberry bark as raw material; while the other category is produced by residents of southwestern township, with mulberry bark as raw material. *The Annals of Hangzhou Fu* records that "Small mulberry and wild mulberry trees are the best in Fuyang County. Moreover, the mulberry trees planted in other county are also called Fuyang mulberry."

It was recorded in the Qing Dynasty, due to the large scale of the industry, high circulation and concentration, some people got rich in papermaking. For example, during Kangxi Reign and Qianlong Reign, there was a household head named Shi Yaochen in Shijia Village of Dayuan Town in Fuyang County. He had more than 100 papermaking workshops and employed up to 900 people. The emergence of large papermaking households also manifested the scale of Fuyang's handmade papermaking and the prosperity of the paper industry.

Changshan County of Quzhou area (perhaps including several parts of the old Kaihua County) is another major papermaking center in Zhejiang region in the Qing Dynasty. *The Annals of Changshan County* in the Qing Dynasty records that Changshan in the Qing Dynasty made different varieties of paper. While the Guangxu version made similar descritpion, and claimed that 72 steps were involved in the papermaking process. Different from Fuyang area, its prosperity failed to continue. In the book *On Papermaking*, it is recorded that Qiuchuan Town in Changshan area has the reputation of "Capital of Paper". When making paper, "water should be taken from Qiuxi of Qidu area (an old name of Changshan County). Majinxi River runs across which belongs to Fuchun River system. On the west of Qiuchuan, Qiuxi brook flows through Qidu Village and finally flows into Majinxi River. Because of clear water, the paper mill was established here. While the floating Qiuxi water was used for soaking, boiling and washing the materials.

Yuanshu paper made in the Qing Dynasty preserved in Dayuan Town

According to *The Annals of Changshan County* during Guangxu Reign, at that time, there were more than 500 paper mills in Qiuchuan area alone. More than 3 000 people were engaged in papermaking, making Mian paper, tribute paper, calendar paper, exam paper, book paper, letter paper, three-color paper, nineteen-color paper, etc. In particular, there was a high-end paper named "Bang paper" (paper used by the government to publish information) with delicate papermaking techniques, it enjoyed the reputation of Qiuchuan Official Paper, as white as snow. When drying the paper, inside and outside the village, top and down the mountains, especially around the ancestral temple of the Xus, white paper could be seen everywhere, as it's snowing and became one of the ten scenic spots in Changshan area.

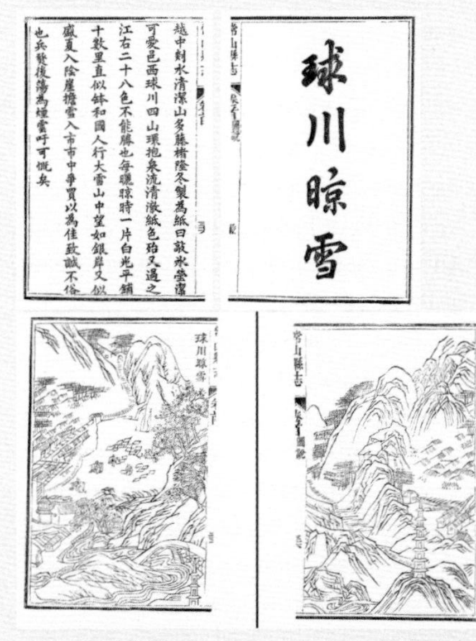
越中剡水清潔山多藤楮陰冬製為紙曰敲冰瑩潔
可愛邑西球川四山環抱泉流清澈紙色殆又過之
江右二十八色不能勝也每曬晾時一片白光平鋪
十數里直似鉢和國人行大雪山中望如銀岸又似
臘夏入陰崖擔雪入市市中爭買以為佳致誠不俗
也兵燹後殤為煙雲吁可慨矣

球川晾雪

Discription and pictures of *Qiuchuan Paper-snow Scene* in local annals of Changshan area

Thirdly, different areas in Zhejiang region have records of their specific papermaking practice and rich products.

(1) In North and northwest part of Zhejiang region, including Hangzhou, Huzhou, Jiaxing areas:
In Lin'an, all the mountainous towns were involved in papermaking. In Zhinan area, they made Huangshao paper by "putting bamboo in limestone water and smashing them to

make paper"; In Nanxiang area, they took use of straw to produce Chabai paper. Both Huangshao paper and Chabai paper were used in sacrificial rituals and sold to Songjiang and Shanghai areas. Yuanshu paper was produced in Taoyuan, Juren, Baigua, Qingcaowu of Nanxiang area; Cansheng paper was produced in Qianzhai Bridge of Zhinan area; Fanggao paper was produced in Dahuangwu of Zhinan area. In addition, there were Dapi paper, mulberry paper and so on.

Most places in Xincheng County produces Yinpi paper with a local plant bark.

Huangbai paper, Straw paper and Mulberry paper were produced in Xiaofeng, Anji and other mountainous areas with a large output. *The Annals of Xiaofeng County* during Qianlong Reign records: "There are Huangbai paper, straw paper, mulberry paper, etc., mostly from southeast township." Taking bamboo as the raw material, Huangbai paper was produced by the hydraulic pestle. "The hydraulic pestles are used in Xiaofeng county and Kuntong area. The local people take advantage of water of the river outlet, and put millet crops into the river to let the water hammer smash them mud-like. Then they get the crops out of water and steam the material. Finally, the mud-like crops turn into paper." In Anji area, "papermakers, from Mugan of Tianquan, to Huanghui of Gaowu, all gathered in Sanqiao County sell paper." Moganshan villagers were all involved in "papermaking... the straws rot in deep holes, then they will be put into water. After washing, paper will be made. People sell them at markets." Mulberry paper was mainly produced in Dipu of Anji area, while yellow paper was mainly made in Wukang and Gui'an areas.

The Annals of Huzhou Fu during Tongzhi Reign records the situation of bamboo papermaking in Xiaofeng area: "East of Shenqian's house, by the river, people use hydraulic pestle to smash bamboo...People then boil the material, and dry under the sun. Finally, the yellow paper is made." On the hillside, pieces of paper are scattered for drying all over the mountain. In the event of showers, women will rush out to put away paper up down the hillside, running as fast as monkeys, or paper would get wet and turn into mud very soon. There is a local proverb that vividly describes the case. "Near the Qians house, hydraulic pestle spin quickly, and every family make paper. Women work hard in wild wind."

Meili letter paper in Jiaxing area is famous at the time. *The Annals of Meili* during Qianlong Reign quoted "Spring Breeze" and said: "(Meili letter paper) is popular nationwide, among the celebrated papermakers are The Zhous from Jianzhai area, The Gus from Youtang area, The Wangs from Chubo area, The Chens from Beifang area."

(2) In Wenzhou Fu, south of Zhejiang region: After the revocation of Quxi Paper Office in Wenzhou area, the production of precious paper was shocked, but that of low-end paper was not affected. Instead, it was greatly developed, such as mulberry paper, umbrella paper, and red flower letter paper, etc. *Dong'ou Zazu* recorded the tribute paper in place of Jun paper: "Dagong paper was made in Xia'ao area, one of the fifty cities in Nanxi today. With root bark as raw material, it was white, two chi three or four cun long against two chi wide. It was sold near the countryside, which was thin and small. At present, people only know Dagong paper, but fame of Juan paper die out."

There is Baiqian paper in Kuixi area. In the Qing Qynasty, The Past recorded that "In July of the fourteenth year of Daoguang Reign, a visitor of Kuixi area said that due to the scarcity of white joss paper that year, bamboo shoots were used as food in the spring. Coupled with the severe famine, almost half of local people died, and the survivors also went out of business because they had no money to support papermaking. (Journal written on July 25)." A poem written during Tongzhi Reign of the Qing Dynasty described the scene of a paper dyehouse. *Local Story of Rui'an County* recorded that Nanping paper was also produced in Rui'an County in the late Qing Dynasty. "Nanping paper produced income of tens of thousands of gold per year and was sold to other provinces."

Hydraulic pestle on Zhishan Mountain in Zeya Town of Wenzhou City

The papermaking techniques have also been improved. In addition to the traditional ways, such as boiling in lime water, beating and smashing, and sun-drying, water was also commonly used. Hydraulic pestles were built in the turbulent river, and the bamboo was placed in it. The water smashed bamboo into mud. Then, the bamboo was picked up and stacked for steaming to make paper. The bamboo paper was produced in a city. In the 21st year of Kangxi Reign, Luo Yuansen, a local person, traveled to Lizhuang Village and saw many bamboo forests. Thus, the paper was made. Later, papermaking became popular among the neighborhood, while its profit was very little. "Gu paper, commonly known as Mian paper, was mostly produced in Yayang area." recorded in the volume II Local Products of Taishun Annals during Tongzhi Reign in the Qing Dynasty.

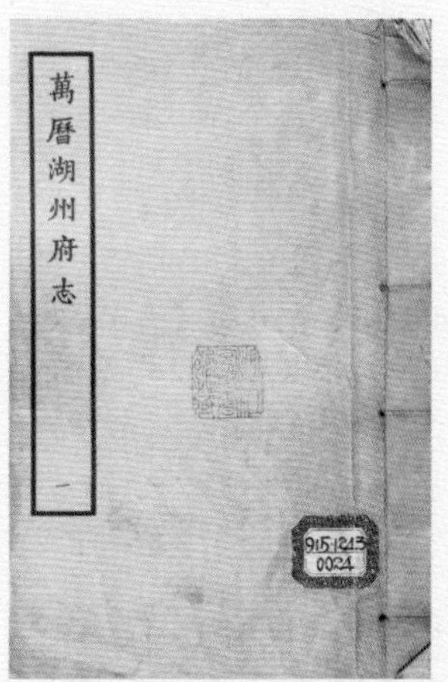

The Annals of Huzhou Fu compiled in Wanli Reign

(3) In Longyou County of Quzhou area in the west of Zhejiang region, the development of the paper industry relying on Longyou Merchants Group has its own characteristics: Although papermaking was mostly commodity production, it was mainly based on the family handicraft by the farmers. The professional papermaking workshops that were independent appeared in Longyou County in small numbers, and this was a rare phenomenon. The Longyou Merchants Group was named after Longyou. In fact, it included merchants from Xi'an, Changshan, Kaihua, Jiangshan and Longyou areas. The number of merchants from Longyou County accounted for the majority of group, and they were the most shrewd merchants.

Meanwhile, Longyou paper merchants were also very famous. As the raw material of papermaking, bamboo was plenty in the mountainous areas of Longyou County. The county enjoyed a long history of papermaking. Many merchants were engaged in paper trade, purchasing paper for export in local market. Such paper shops were scattered throughout city center and towns. The largest paper trade market was located in Xikou Village. Crowded with merchants, the village was even more prosperous than the town.

There are three types of Longyou paper in the Qing Dynasty: yellow letter paper, white letter paper, and Nanping paper. Nanping paper had two drying ways, that is, in the oven and in the sun. Xikou Village in this county was a paper trade center. Nanxiang folks relied on papermaking for a living. The villagers wholeheartedly cultivated bamboo in the mountains to make paper. It can be concluded that the paper industry was of great importance in the local economy.

In Longyou County, there were nearly 20 paper shops run by paper merchants in the 20th year of Guangxu Reign. In general, the paper merchants had a shop for paper trading. Some paper merchants moved to the country to do business because Longyou County was a paper trade center. For example, three brothers of Lin Qiongmao all moved from Shanghang County of Fujian region to settle down and engage in paper trade; other paper merchants sold Longyou paper outside. For example, in the early Qing Dynasty, Minting Business Groups sold paper to Jiangsu region, and then, Ningshao Business Groups took over the paper trading after the mid Qing Dynasty.

There are many historical records about the production and sales of Longyou paper in the Qing Dynasty, including local paper merchants in Longyou County, those from Fujian and Anhui regions, as well as those from Jiangxi and Dongyang and Jiangshan areas of Zhejiang region.

龙游商帮的产生和发展

龙游地处浙江西部，历史上一直是浙、闽、赣、皖四省和省内金、衢、严、丽四府之交通孔道，物产丰饶，粮食、竹木等大量外输。南宋建都杭州，对木材和纸张需求巨大，带动龙游木材与造纸行业开发。凭借地理优势和江南经济发展契机，依靠自身文化素养和开拓进取精神，龙游商人走出山沟，逐鹿中原，远征边关，漂洋出海，终成中国十大商帮之中唯一以县域命名的龙游商帮。

龙游商帮崭露头角于南宋，迨至明代已蔚为壮观，明万历时已有“遍地龙游”之谚。旧志中也有“龙丘之民，往往糊口于四方，诵读之外，农贾相半”，“龙游之民，多向天涯海角，远行商贾，几空县之半，而居家耕种者，仅当县之半”的记载。至清朝康熙、乾隆年间“其居家土著者，不过十之三四耳。”可见直至清前期，龙游商帮的势头依然有增无减。

Introduction to Longyou Merchants Group

Longyou paper boasts a long history. Qianshan in the neighboring Jiangxi, Jiangshan and Changshan areas in Zhejiang region were all important papermaking places, with superb techniques. Papermaking played an extremely important role in economy of Longyou County. In the Qing Dynasty, local official repeatedly banned access of the outsiders to Longyou County to cut bamboo shoots, for "most Nanxiang residents make a living by papermaking". Twelve bamboo trees could make 1 dan paper materials, each worth 2 yuan (Chinese unit of currency). When the cost was reduced, the papermakers would gain profit of 1 yuan. Selling paper material instead of making paper could get 3—4 *jiao* (Chinese unit of currency) to 7—8 *jiao*, while 12 bamboo shoots would only worth 30 to 40 *wen* (Chinese unit of currency). The gap of profits between was very large.

In the middle and late Qing Dynasty, Lin Julun, a celebrated paper merchant in Longyou County, was very famous for making a fortune by doing paper business and charity. Lin Pinmao, his grandfather, who was originally from Tingzhou area of Fujian region, came to Louyou during Qianlong Reign, together with three brothers. Later, the son of Pinmao (father of Junlun) lived in Youtou of Nanxiang area, taking papermaking as his career. "When his father passed away, Lin Junlun inherited his business and made a fortune by working hard. He was also known for being kind-hearted and once donated more than 11 000 *liang* silver ingots (ancient Chinese unit of currency) to build Tongsi Bridge. As a resident of Longyou County living in Xikou of Nanxiang area, Lao Jinrong, "liked to go to paper mill and took supervision as pleasure because Lao's father was in charge."

The papermakers must be employed by the households to produce paper. Most of them were from Qianshan of Jiangxi and Dongyang of Zhejiang. In Nanxiang of Longyou County, there were many artisans from other places. In the 24th year of Guangxi Reign, more than 400 workers in the paper mill went on a strike. The number of free laborers (raftsmen) was abundant in the production of Longyou paper, so it can be inferred that there was already a sprout of new production method in the paper mills.

Longyou paper merchants turned to the front-end papermaking industry, which was a commodity operation model that controled the complete industrial chain instead of pursuing self-sufficiency. Thus, commercial capital and industrial capital were combined, which was very meaningful in the economic structure, worthy of the attention of researchers in the history of handicraft management.

(4) In Ningbo Fu of Zhejiang region:

Yuyao was a key area for paper production in Zhejiang region in the Qing Dynasty. The paper produced would meet local needs and be sold to other places across the country. According to Summary of Ningbo Port Trade in 1902, an article published by Taxation Department of Zhejiang Customs on March 9, 1903: In 1869, the value of paper, wine, and medicine exported by Ninbo amounted to 90 000 *liang* silver ingots. In the 28th year of Guangxu Reign, there was a record of paper export in Ningbo Fu, "paper, porcelains and bamboo were transported to the northern area; while tinfoil, paper, grass mats and mellow wine were transported to the cities along the Changjiang River."

(5) In Jinhua Fu located in the middle of Zhejiang region, Dongyang County outshined others in papermaking:

The Annals of Dongyang County during Daoguang Reign records: "Rattan with texture can be used to make paper in Baixi Village of Yongning County." Dongbai Mountain area and Yushan Mountain area (Pan'an area) were the source of Shanxi area. With dense forests and ancient rattan, it used to be one of the main areas producing durable Teng paper. Baixi Village of Yongning County was the only place where Teng paper could be produced just because it was close to Shanxi area. On the tenth year of Kangxi Reign, there were only 3 households and 19 people involved in papermaking in Baixi Village, producing bast paper and Mian paper. One of the special products is Yuluan (fish egg) paper made of mulberry tree bark and bamboo shoots, among which the quality of Pu paper and Yuluan paper were the best. "Jinsan" paper was a famous-brand paper in the middle of the Qing Dynasty. In the bookstores of Hangzhou, Ningbo, Shanghai and other areas, it was exempted from quality check. Many bookstores also came to Baixi Village to purchase paper. Due to the thriving market transactions, more than 200 people in the village of Baixi Village had been working on papermaking till the end of the Qing Dynasty and the beginning of the Republican Era.

Fourthly, there appeared famous processed paper and paper-related products.

In addition to the paper products mentioned above, processed paper had also formed a brand. The letter paper of Hangzhou area in the Qing Dynasty was very famous. According to *Dream Record* by Xu Yong, which is a book that records treasures in the study, paintings and calligraphy of gold and stone, "My old friend Chen Bojun once got several pieces of Jianlian Luowen paper during Kangxi Reign. All pieces of paper with dark decoration were about 6 *chi* long. He entrusted Wang Chengzhi, a skillful papermaker in Hangzhou to make and dye the paper, making it look the same as the ancient paper." The most famous flower letter paper is Xiegong letter paper. According to *The Annals of Hangzhou Fu* during Guangxu Reign, "Some paper names are from famous person, just like Mr. Xie and Xue Tao. Mr. Xie refers to Xie Sifeng (also Xie Jingchu), whose literary name is Xie Gonghou. He created patterns of letter paper and that's how the paper got its name. There are ten different colors."

In the Qing Dynasty, Wujin paper made in Hangzhou area also developed. According to *Supplement to Compendium of Materia Medica*: "There is a type of paper in Jiangsu and Zhejiang regions which is black on both sides while smooth and brittle, not used for writing, but used to wrap treasures and Chinese medicine in the shops. It was also called Xunjinpaper, which was dyed black and well calendered." In the Qing Dynasty, Shanyin, Kuaiji, Shangyu and Fuyang areas all produced Wujin paper, which was regarded as the top one in Zhejiang region. In Shanyin area, only Hua family possessed the skill of making Wujin paper. According to *The Annals of Shangyu County* in the reign of Emperor Guangxu, "Wujin Paper was made in Cailin in the northwest, and nowadays only Zhao family in Jiudu County can make this type of paper. It is said that the craft can be passed to the daughter-in-law rather than daughter."

In the Qing Dynasty, the development of Luming paper in Shaoxing area was more prosperous due to the progress of the tin foil industry. The tin foil was processed against Luming paper. Through a number of steps, the small tin sheet was beaten into a thin piece, which was smashed as well as pressed. It is a kind of high-end processed paper used for worshipping Buddha and honoring ancestors.

In the late Qing Dynasty, Shaoxing Fu, Shanyin and Kuaiji Counties became the

Original Wujin paper made in Shangyu area during the late Qing Dynasty

Wujin paper used to make gold foil during the late Qing Dynasty and the Republican Era

Original Luming paper

large tin foil production base in the country. The number of people engaged in tin foil processing industry in Shaoxing was quite impressive, and it was known as "Half-tin City". According to the statistics of the year 1911, there were more than 31 000 employees, including 8 000 foil workers, and 2 000 backmen (including ingot moulding, paper flattening, paper cutting, blocks making, etc.). 400 workers are involved in powdering and most of them are child laborers. More than 20 000 male and female workers worked on other steps like calandering, peeling paper down, plus more than 1 000 relatives of the owners of foil shops. Except for domestic sales, Shaoxing's tin foil was mostly exported from Shanghai Port. According to the *Report on Employment in Kuaiji County*, during the period of 1909 to 1911, the annual output of tin foil was 1.6 million blocks, of which 1.4 million for export, worth 1.3 million silver *yuan*. In the late Qing Dynasty, tin foil produced in Shaoxing began to be exported to Indonesia.

Another aspect reflecting the development of handmade paper is the development of the paper-related products. For example, Shaoxing paper fan was made of fine mulberry paper and repeatedly painted with bituminous coal powder and extraction of jelly from unripe persimmons. Paper fan production was one of the important rural side businesses of farmers in Shanyin and Kuaiji Counties in the Qing Dynasty. According to *Report on Employment in Shanyin County* in the second year of Xuantong Reign (1910): "The silk, wine, tinfoil and folding fans are important Shanyi products." "The sale of folding fans is 5 500 000, worth188 000 *yuan*." *Report on Employment in Kuaiji County* (1911) records, "For the fan industry, about 60 000 fans are transported to other places, worth 3 000 *yuan*." During the late Qing Dynasty, the paper fan industry in Shaoxing was concentrated in Zhoujiaqiao, Yingjiaqiao and Zhanglan areas. All the local farmers and residents were involved in paper fan making.

The paper industry in Zhejiang region in the Qing Dynasty was still in the stage of household handicraft industry, closely integrated with local raw material production. Although it was a commercial production, there existed different situations and forms in terms of production format and commercialization degree.

The first form was the papermaking practice in the mountainous areas, which was based on bamboo, and served as the main source of income for local people. Here, the entire production process of villagers, from raw material growing to paper production, was completed in the family. According to *The Annals of Huzhou Fu* during Tongzhi Reign, in Dongshen Qianjiabian area of Xiaofeng County, every house produced yellow paper. "Drying in the sunlight, paper could be seen everywhere. In the case of storm, women rushed up and down the hill to move the paper into the house like flexible monkeys. After a while, the paper was piled like mounds." It is a form of independent household commodity production associating commercial agriculture with commercial handicrafts.

The second form was represented by letter paper and other related paper products. It was paper processing industry requiring higher technical standards. As an independent small commodity production, separated from agriculture, its commercialization level was obviously higher than that of the former local farmer commodity production.

The third form was a commercial production as a complementary sideline of the family. For example, the villagers of Mogan Mountainous area in Deqing County of Huzhou Fu, "rely on farming and wood cutting for a living and make paper in their spare time." If there was leisure time to make paper, it was only a necessary supplement for self-sufficient food production and collecting local natural products in the mountain.

The fourth form was a paper town like Fuyang. "Everyone is involved in

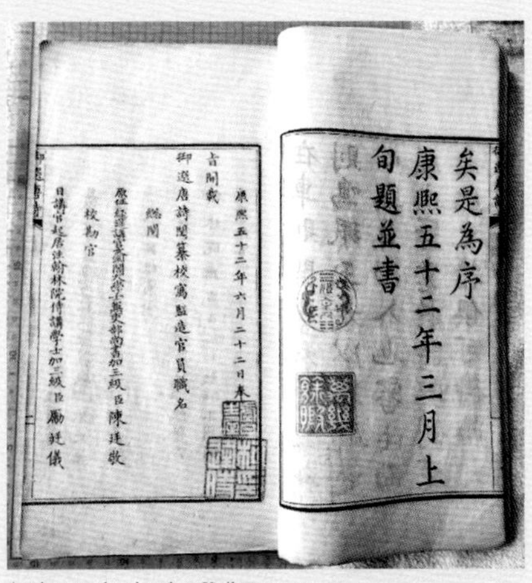

Ancient work printed on Kaihua paper

papermaking, young and old, working day and night" "mostly are women from the northern part. They have nothing to do but help their husband dry the straw paper, earning so little." Here, the paper industry was the economic pillar of many families, so the whole family devoted almost all energy to papermaking throughout the year.

The fifth form was paper sales and transporting like Longyou Merchants Group did, which was basically a pure business. In the later stage, considering taking control of front-end production, several merchants became production households at the same time.

There were many cases before the Ming Dynasty when we examined the influence over handmade paper industry in Zhejiang region from policy and mechanism supply. However, as a concise account of the ancient industrial policy, the summary here is limited to the period of the Ming and Qing Dynasties.

The first aspect is the industrial promotion caused by the tribute paper system.

In the Ming and Qing Dynasties, many varieties of handmade paper in Zhejiang region, such as Ping paper in Quzhou area, Juan paper in Wenzhou area, Yuanshu paper in Fuyang area, Bang paper and Kaihua paper in Changshan area, had such characteristics as strict material selection, fine workmanship, good water quality and excellent paper quality. For a long time, as a tribute to the royal court and official document paper as well as examination paper, it was under the supervision of the government, ranging from quality control to production scale, so that they must pursue excellence continually.

"In the early Ming Dynasty, the court assigned each paper production area to pay tribute. Task of 250 000 pieces of tribute paper was assigned to Zhejiang region, second only to Hebei region, ranking second in the country. As for the royal books and official seals from the late Ming Dynasty to Qianlong Reign of the Qing Dynasty, Kaihua tribute paper took priority for a long period of time. Due to the large demand of the royal family, "Kaihua Paper Version" became famous in the field of ancient book collection. Even several volumes of 7 sets of *Complete Library in the Four Branches of Literature* were printed on Kaihua paper. The amount of paper used was huge.

Although Juan paper in Wenzhou area, which was supplied as tribute paper from the Tang Dynasty, was suspended at this stage, and the tribute paper as a system came to a halt temporarily. However, Wenzhou paper was still used as a tribute to the royal court every year. According to the tribute section of the volume 3 of *The Annals of Wenzhou County*: "Wenzhou area sends six thousand and forty copies of Nanjing calendar yellow paper and ninety-eight thousand three hundred and fifty seven sheets of white paper every year." The quantity was still quite large.

The second aspect is the thriving of the paper market for book printing. In the Ming and Qing Dynasties, bamboo paper became the mainstream paper for book printing because of its excellent performance. Zhejiang region was not only one of the largest bases for bamboo paper production in China, but also a prosperous place for cultural events such as reading, imperial examinations and a gathering place for intellectuals. The local photoxylography and book-collection industry centered on Hangzhou, Ningbo and Jiaxing areas were extremely advanced. During the Ming and Qing Dynasties, the number of books printed in workships or privately in Zhejiang region was enormous, and the demand for examination and practice paper was also considerable. Paper was not only supplied to the local market, but also sold to the whole country, thus promoting the development of the paper industry in Zhejiang region.

The third aspect is the increase in foreign sales. Ping paper and bast paper produced in Wenzhou area were the major exports of Wenzhou Port. According to *The Annals of Fuyang County* during Guangxu Reign, which has the earliest county-wide data, that bamboo paper in Fuyang area could be exported to other countries in 1906, earning about six or seven hundred thousand gold per year, and approximately three or four hundred thousand gold by selling straw paper. The main export destination was Southeast Asian countries.

1.3 Development of Modern Papermaking Industry in Zhejiang Province

1.3.1 From the Early Period of the Republican Era to the First Half of 1937: Coexistence of the Rapid Development of Production Capacity and the Disappearance of Famous Paper Types

Since the late Qing Dynasty, due to the input of "foreign paper" and introduction of machine-made papermaking techniques, the relationship between Zhejiang handmade paper industry and local political, economic changes had become closer. On the one hand, with the shift of China's economic center from the core areas of Jiangnan area, such as Hangzhou, Ningbo, Nanjing and Suzhou Cities to Shanghai City, the largest metropolis in modern China, which had only begun to rise in modern times, the overall economic structure of the Changjiang River Delta has undergone tremendous changes. On the other hand, with the continuous fluctuations in the industrial economic development in Zhejiang Province, local handmade paper industry experienced several ups and downs after 1860. Although it still showed its peak in terms of output, there was a looming crisis. The main influencing factors include: after the Republican Era, traditional famous paper types quickly faded. Most of the handmade paper production was for daily necessities and sacrificial ritual supplies; the tribute paper system that had lasted for more than a thousand years disappeared, and there was a lack in both motivation and pressure in papermaking techniques innovation to meet higher requirements. Entering the modern times, both domestic and international situation were turbulent. Competition from foreign paper had caused the market of traditional handmade paper to shrink; the exploitation and control of domestic bureaucratic capital had led to the compression of industrial profit. Accordingly, most of the handmade papermaking companies or workshops in Zhejiang Province were in difficulty.

At the end of the Qing Dynasty, Zhejiang Province was one of the main battlefields between the Qing Army and Taiping Army. As a result, the population of Zhejiang plunged into a sharp decline from the year 1860 to 1866. Under the double impact of Qing Army and Taiping Army, there were many households burned and looted in cities and rural areas. Coupled with the natural disasters around 1867, Zhejiang handicraft industry and local economy suffered from the catastrophe. Not until the end of the Guangxu and Xuantong Reigns of the late Qing Dynasty, did it come back to life. The subsequent political situation remained stable for nearly 20 years. Under the influence of the "Five Ports of Commerce", the handmade paper industry came into being and developed. However, the period for such development was very limited, because then The Revolution of 1911 broke out, followed by warlordism. The political situation became chaotic and turbulent. The civil war continued throughout the years, traffic interrupted, business and transportation isolated, which seriously hindered the development of industry and commerce.

From the successful cooperation between Kuomintang of China and the Communist Party in the Northern Expedition (1928), to the "September 18th" Incident (1931), Zhejiang handmade paper industry at this stage was in another prosperous period, and the output value of most paper-producing counties in the province had reached a historically high level recorded in the literature. According to the statistics of the Ministry of Agriculture, Industry and Commerce in 1916: "The national output value is only 42 million *yuan*, with Jiangxi 8 million *yuan*, Fujian 7.5 million *yuan*, Jiangsu 6.5 million *yuan*, Zhejiang only 4 million yuan." In 1924, paper industry in Zhejiang Province was not weak at all, although it could not keep pace with Fujian Province. There were paper products worth 3 million *liang* annually. The top papermaking sites consisted of Yanzhou, Quzhou and Jinhua Cities. In 1930, the output value of handmade paper in Zhejiang Province amounted to 20 million yuan, accounting for about two-fifths of the country.

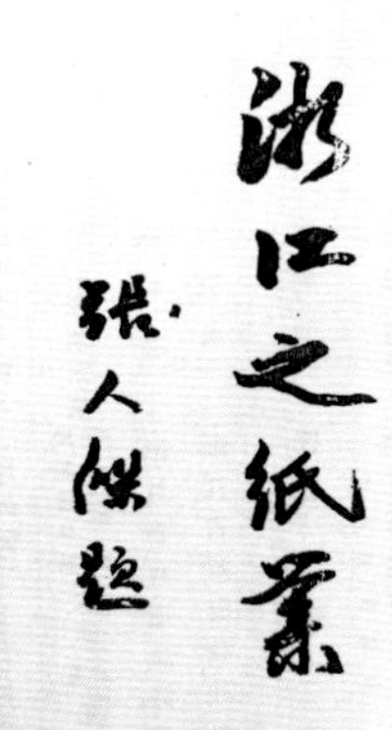

First edition of *Zhejiang Paper Industry*

It can be seen from Table 1.2 that in the middle of the Republican Era, the output value of handmade paper in different counties of Zhejiang Province varied greatly, and the regional distribution was extremely uneven. Fuyang had outshined all other papermaking sites. In the first year of the Republican Era (1911), paper produced in Fuyang accounted for 44% of the total; in the 19th year of the Republican Era (1930), Fuyang became the county with the highest output value all over the country, with a production value of 8 667 912 *yuan* (silver dollar), accounting for 41.56% of the province's total output value of handmade paper, ranking first. In the peak year, Fuyang handmade papermaking once accounted for 25% of the total national paper output.

Table 1.2 Handmade paper output value and proportion of cities and counties in Zhejiang Province in 1930

City/county	Output value: *yuan*	Proportion (against the total province)	City/county	Output value: yuan	Proportion (against the total province)
Fuyang	8 667 912	41.56%	Wuyi	243 456	1.17%
Xiaoshan	1 360 620	6.53%	Sui'an	239 778	1.15%
Quzhou	1 034 783	4.96%	Jingning	202 350	0.97%
Jiangshan	75 336	3.61%	Shaoxing	179 880	0.86%
Zhuji	714 450	3.47%	Fenghua	169 050	0.81%
Taishun	685 200	3.19%	Shouchang	158 400	0.76%
Yuhang	671 100	3.22%	Tangxi	155 420	0.75%
Ning'an	548 028	2.63%	Shangyu	127 224	0.61%
Yongjia	529 644	2.54%	Suichang	110 783	0.53%
Huangyan	518 722	2.49%	Songyang	108 913	0.52%
Rui'an	506 070	2.43%	Wenling	100 980	0.49%
Longyou	454 910	2.18%	Shengxian	87 966	0.42%
Tonglu	405 347	1.94%	Yongkang	85 194	0.41%
Jinhua	308 640	1.49%	Qingyuan	71 156	0.34%
Xiaofeng	283 520	1.36%	Pujiang	61 848	0.30%
Xindeng	276 307	1.33%	Yuqian	54 921	0.26%
Jinyun	274 807	1.32%	Changhua	52 248	0.25%
Changshan	274 800	1.32%	Pingyang	48 536	0.23%
Linhai	248 040	1.20%	Xinchang	16 836	0.08%

Fuyang handmade paper industry had a high output value, and the number of papermakers and output also topped in the province. Papermakers accounted for one-fifth of the county's total population and one-third of the male labor force. *Zhejiang Paper Industry* records: In the same period, Fuyang County had 10 069 families engaged in papermaking with 10 864 papermaking troughs, and the production output of handmade paper reached 6 411 100 pieces (59 002 tons). According to the *Construction Monthly* of Zhejiang: "Price of Fuyang local paper has been relatively stable since the 15th year of the Republican Era. In October of the 19th year of the Republican Era, paper price soared, and the price of Yuanshu paper, rose from 11 *yuan* to 15 *yuan*. The price of Haifang paper, has risen from 5 *yuan* to 8 *yuan*. With paper price rising, paper households made a fortune."

In addition to Fuyang, Wenzhou paper also reached the highest level in the middle period of the Republican Era. In the early years of the Republican Era, Ping paper in Wenzhou City was known throughout the country, with a large sales volume and a wide sales market, but the paper quality was poor. In 1915, the governor of Zhejiang Province went to Wenzhou City, and in his book *Yongjia County Declaration*: "There is a papermaking trough in the township of Quxi area, which sells paper to Zhenjiang area. Owing to the poor quality, improvement is in demand."

The Annals of Wenzhou City records: In the 25th year of the Republican Era (1936), the paper annual output of the counties in Wenzhou reached the peak, with a total of 362 000 dan. In the 19th year of the Republican Era (1930), Yongjia County (including today's Yongjia County, and Lucheng, Longwan, and Zhenhai areas of Wenzhou City) had 1 333 papermaking troughs, each shared by 3 to 5 papermakers, and 4 244 paper workers; There were 606 papermaking troughs and 1 920 paper workers in Rui'an area; Taishun area had 297 papermaking troughs and 1 945 paper workers. With the development of the paper industry, Wenzhou Paper Industry Association was established. In 1933, Nanping Paper Association was established and Shen Jieqing served as the chairman. In the 25th year of the Republican Era (1936), Yongjia County Paper Association was established.

In addition to considerable output, there was a rich variety of handmade paper products. *Zhejiang Paper Industry* records: "Fuyang ranks first in Zhejiang Province's handmade paper production output, producing 17 kinds of paper, such as Kengbian, Yuandou, Huangshao paper, having more than seven million pieces. Yuanshu paper's output is 100 000 *dan* or so. Xiaoshan area is rich in producing Huang letter paper, with an annual output of 100 000 pieces. Quxian County has produced Nanping paper and Huajian paper, and each has a capacity of 100 000 *dan*. With more than 100 000 dan output, Fanggao of Jiangshan becomes the largest papermaking site at the time. As for other paper products such as Luming paper of Zhuji area and Huajian paper of Taishun area, the annual output is also competitive. Huangshao paper from Yuhang and Lin'an areas reaches 2 000 000 pieces. Doufang paper from Yongjia County was produced 200 000 pieces annually. Qianzhang paper from Huangyan area, was produced with an annual output of more than 300 000 pieces. These are some outstanding counties with amazing outputs. Other counties produce more varieties of paper so the output is scattered. While the output is considerable for places with less paper types."

In addition, because of the political reform in the late Qing Dynasty and the early Republican Era, the government encouraged industrial and commercial development, and held various fairs. Zhejiang's famous paper products frequently wan awards at domestic and international exhibitions. For example, in the fourth year of the Republican Era (1915), Changshan paper, made by Jiang Qinbo, a villager in Shanji Village of Liyuan County in Fuyang County, was awarded by the Ministry of Agriculture and Commerce. In the same year, at the Panama International Commodities Fair, "Changshan Paper" and "Jingfang Paper" as varieties of Bamboo paper in Fuyang County won the second prize; Xindeng County's bast paper wan the "Chinese Paper" honorary award. In the 15th year of the Republican Era (1926), at the China National Goods Exhibition held in Beijing, "Jingfang Paper" and "Changshan Paper" produced in Fuyang County respectively won the second and third prize. In the first Western Lake Expo held in Hangzhou area in the 18th year of the Republican Era (1929), Yuanshu paper and Wujin paper of Fuyang County, Baipi paper of Shanglishan area in Jiangbei were awarded the outstanding prize.

During the period of the Republican Era, Zhejiang handmade paper performed well in terms of output, output value, and variety, but it did not mean that the future would be promising. In fact, at that time, worry and concern about declining paper industry in Zhejiang Province could be seen in Zhejiang Design Committee's investigation over provincial handmade paper industry, investigation of Zhejiang paper industry by individual scholars, and the special investigations on several typical papermaking areas. "While the paper produced in Zhejiang has been famous since the ancient times, it can be described as an important papermaking base in history. Today, because the papermakers do not seek improvement, ancient recipe is lost, and the raw materials such as mulberry barks are also lacking. The price of Zhejiang paper has plummeted. What we have now is only joss paper and rough wrapper to maintain our papermaking status in China. If we fail to reverse the trend immediately, it will be dying very soon. Perhaps one day, Zhejiang paper will extinct."

According to the records of Zhejiang Provincial Mainland Bank in the period of the Republican Era: Around 1930, "Fuyang paper production value is 10 million yuan including that of Yuhang area. Fuyang County alone has 8 million or 7 million. It is predicted that the price and the yield are declining. In terms of the current value, it is estimated to be only 6 or 7 million *yuan* for the low price and output this year. While there are 17 kinds of paper, such as Kengbian paper, Houdou paper, Luming paper, Jingfang paper, Duanfang paper, Huangshaoyuan paper. Among them, only Huajian paper and Yuanshu paper are for writing, while the rest are used for sacrificial rituals or personal uses. Price of paper varies greatly, with different size, brands and quality levels. Papermaking materials consist of bamboo, straw and mulberry bark. The papermakers utilize water power to pound raw materials for papermaking. Papermaking screens have extremely fine workmanship and once wan a 'special award' in Peking International Good Exhibition. However, there is little improvement of papermaking skills. Accordingly, it is difficult to compete with the machine-made paper made by foreigners or meet the needs of the times."

The *Zhejiang Paper Industry* investigation group made analysis of foreign paper: "Imported paper can be divided into two categories, one is paper imported directly, the other is the import of paper raw materials from various trade ports. In the first year of the period of the Republican Era, there was no record on foreign paper imports at Wenzhou Port. As for Ningbo and Hangzhou Ports, they accounted for most of foreign imports, amounting to 60 000 *liang*. From 1916, the import portion increased in Hangzhou Port, even surpassed Ningbo Port. According to statistics, there was paper imported from various trade ports, mostly high-level paper. The medium and low-end Zhejiang paper was self-sufficient. Imports were growing in the past years." Even paper from outside Zhejiang had an exclusive advantage, which had also overwhelmed the local paper. Most of the imported paper was high-quality because the local paper products were enough to meet the local needs. "The foreign paper imported annually increased with value of only 90 000 *liang* in the 10th year of the period of the Republican Era, then increased to 340 000 liang in the 16th year. In recent years, the trend became more dramatic. In the long run, the future of Zhejiang paper is quite vague."

At that time, the researchers reflected: "Most of the paper imported at the port was of high-end paper and few low-end paper. On the contrary, the paper exported from the province was quite small in quantity. Among the exported paper, paper of interior quality was three, four, or even ten times more compared with that of first-class paper. This shows the status of paper production in this province. It is easy to replace the rough paper used for local sacrificial ritual. Therefore, the crisis of the paper industry is inevitable." In the middle period of the Republican Era, although the output value of handmade paper in Zhejiang is amazing, "according to a recent field investigation by the Association, paper with the value of more than 20 million yuan was mostly rough paper. While paper used by artists and printers mainly came from overseas or other provinces as Fujian, Jiangxi, Anhui, etc.

Local newspaper produced in Zhejiang Province during the Republican Era

Of course, even with backward techniques, Zhejiang handmade paper still had some breakthroughs. For example, Wenzhou City provided local paper materials for modern newspapers. According to reports at that time: "Wenzhou will have domestic newsprint. China's newsprint before was mostly imported, totalling 30 million *yuan* which was ignored by most civilians. At present, the maritime merchants—Xu Shiying, Feng Shaoshan, Jin Han and other enthusiastic businessmen have initiated the production of Wenzhou newsprint." Wang Xianchuan founded Daming Zhenji Paper Factory in Xishan area in 1937, introduced technology from Fenghua area, and produced wax paper. Since then, three wax paper factories—Zhongguo, Guanghua and Jianguo—had been successively established, all producing wax paper. In June of the same year, Wang Yunwu, manager of the Shanghai Commercial Press, visited Wenzhou to make preparations for the Wenxi Paper Mill.

With the development of Zhejiang handmade paper in the Republican Era and the increasingly frequent exchanges between China and foreign countries, Zhejiang handmade paper began to go abroad. In the 1940s, Ping paper in Zeya Town entered Shanghai market, and the market further expanded to Shandong, Jiangsu, Fujian, Taiwan Provinces and Southeast Asian countries. In addition to 18 provinces and municipalities directly under the Central Government, Fuyang handmade paper was also exported to Southeast Asian countries such as Japan, the Philippines and Singapore.

During the period of the Republican Era, the marketing channels of the paper industry were further extended. Taking Fuyang County as an example, during and before the period of the Republican Era, the purchase and selling of all kinds of paper products in the county were operated by private businesses (including big papermaking families and paper merchants). At the beginning of the Republican Era, a new type of trading firm emerged, connecting papermakers and paper shops in large and medium-sized cities. Most of the paper shops belonged to large paper merchants. In Fuyang County center, there were two shops, i.e., Guangyuan Paper Shop and Zhenhe Paper Shop. In 1926, the number of Fuyang paper shops increased to 22, with an annual turnover of 3.2 million *yuan* (silver dollar). In the 18th year of the Republican Era (1929), there were still 15 paper shops in Fuyang County, with a total turnover of 435 000 *yuan* (silver dollar). The Xindeng County had a turnover of 40 000 *yuan* (silver dollar) selling mulberry bark paper and straw paper, which was affiliated to Fuyang starting from 1958.

On the eve of the outbreak of Anti-Japanese War in 1937, Zhejiang handmade paper industry began to be strongly influenced by the turbulent political environment and commercial exchanges between China and foreign countries. Frequent exchange of Chinese and foreign business had brought convenience to the export of handmade paper in Zhejiang Province. At the same time, foreign paper products began to be imported into China, which also caused a considerable impact on the Chinese paper market. According to *Zhejiang Paper Industry*, the number of imported paper had been rising since the beginning of the period of the Republican Era.

Table 1.3 Import and export quantity and amount of handmade paper from 1924 to 1927 in Zhejiang Province

Year	Imported paper (unit: *dan*)	Value of imports (unit: *liang*)	Exorted paper (unit: *dan*)	Value of exports (unit: *liang*)
1924	4 105	59 422	86 784	1 208 014
1925	4 044	54 964	87 277	788 670
1926	1 661	58 843	105 191	968 037
1927	1 696	59 245	139 763	146 632

Table 1.3 presents the number and amount value of imports and exports from 1924 to 1927. In 1912, the imports value is 34 194 *liang* against the exports 353 003 liang, ten times larger (from the same source *Zhejiang Paper Industry*). By 1924, the exports value is 1 208 014 *liang*, 21 times larger than imports, amounting to 59 422 *liang*. Then by the end of 1927, the imports value is 59 245 *liang* and the exports value is 1 436 632 *liang*, with the gap enlarged to 24—25 times. Therefore, open ports did bring enhanced imports value while for Zhejiang, a handmade papermaking base, exports benefited more in the early period of the Republican Era.

Another noteworthy change in the business was the high taxes and the exploitation of officials in Zhejiang Province since the Republican Era. A notice issued on June 12 of the 17th year of the Republican Era (1928) gives: Zhejiang Provincial Finance Department ordered, in accordance with Article No.6 of the Interim Regulations promulgated and Article No.13 of the Provincial Supplementary Measures, from July 1st, one-tenth of the price of the paper would be added as tax monthly. Although this is only a record of imposing tax on Anhui paper when entering Zhejiang market, in fact, the local paper was facing a similar high tax burden. The double pressure of the impact of foreign imported paper and taxation had brought crisis to the development of *handmade paper industry in Zhejiang Province*.

In addition to the increase in taxes, the exploitation of the handmade paper industry by local officials and bureaucratic capitalists was also very intense. In the summer of the 27th year of the Republican Era (1938), Huchangji, a paper shop of Quxi area in Wenzhou City tried to collaborate with Chen Daren, a local gentry, to establish "Yongrui Local Paper Cooperative", with the intent to monopolize the sale of toilet paper. Then, Shen Yongnian (the acting secretary of Yongjia County Party Department of the Kuomintang) and Chen Zhuosheng have established Yongjia Paper Agency. In order to control the marketing of toilet paper in Zhishan Mountain area of Zeya Town, he offered "producers' cooperative of Yongrui handmade paper" a subsidy of 1 000 *yuan* per month to control the whole trading in the paper market. As a result, the sales and marketing department cut the price and the interests of the paper farmers were impaired. For example, in the first half

The former site of Huchangji Paper Shop on the Quxi Ancient Street

of the April of the Republican Era (1939), Shanghai toilet paper cost 4.8 to 7.8 *yuan* per piece, while the cost of the paper paid by the sales and purchase office to papermakers was 2 to 3 *yuan* per piece, less than half of that in Shanghai City. Yongjia Paper Agency and Yongrui Paper Cooperative dominated the market, and banned outside paper merchants from offering orders, hence were deeply hated by the local papermakers.

Soon after, due to the influence of the war, the port was blocked and the local paper couldn't find a market. Accordingly, Paper Agency stopped purchase, and the source of papermakers' income was cut off. The conflicts between the two sides became increasingly fierce. The shocking struggle finally erupted, that is, the Zhishan Mountain Uprising. After repeated bloodshed struggles by the papermakers, under the pressure of the resistance, the local Kuomintang authorities finally agreed to the requirements of the papermakers: to heal the wounded and rescue the dying, issue arrears, cancel the Paper Agency and Cooperative, and resume the free trade of toilet paper. The "Zhishan Mountain Uprising" itself was the resistance of the paper farmers who were deeply exploited by unequal monopolized sales to safeguard their own interests.

According to the historical archives of Quzhou area, in 1928, due to the impact of the second Northern Expedition in China, Quzhou paper suffered heavy losses. In 1936, Wang Kaiwei, chairman of the Quzhou Paper Industry Association, said in the petition for the local government's relief fund: "The production output of Jianpin paper is currently about 300 000 *dan* (pieces), less than half of the previous level. In the past, it was sold at 7 *yuan*, but now, it has dropped to about 4 yuan."

In 1931, the "September 18th" Incident broke out, and transportation inside or outside the Shanghai Pass was blocked. Paper sales fell sharply and the price slumped. Around 1934, China's economy was affected by the stagnant world capitalist economy. Thus, industrial and agricultural production also shrank. With the commercial economy declining, paper sales and prices dropped. This situation continued until the anti-Japanese War full-scale broke out.

1.3.2 1937—1945: The War Led to Overall Shrinkage of the Handmade Paper Industry

After the full-scale outbreak of the Anti-Japanese War in 1937, Zhejiang handmade paper industry suffered a huge shock. In the first year of the Republican Era, there were 2 493 factories (including handmade paper workshops) of different types in Zhejiang Province. During the World War I, the economic aggression against China was temporarily relieved because the major capitalist countries were busy with the European War. In addition, the government issued some policy guidelines to encourage industry in the early years of the Republican Era. In a decade from 1927 to 1937, although civil war had continued throughout the country, overall it had little impact on Zhejiang, which in fact benefited the development of its industry. Therefore, after 10 years of construction, Zhejiang's industrialization level had taken a step forward. Before the full-scale outbreak of the Anti-Japanese War, its industrialization process was taking the leading role in China.

During the eight years of the late period of the Anti-Japanese War, Zhejiang suffered great industrial losses. The relatively complete data about the industrial losses during the war period were those collected by the Zhejiang Branch of the National Government's Administrative Relief and Rehabilitation Administration after the war. Before the war broke out, there were 6 815 factories engaged in handmade papermaking industry, with a capital of 1 123 650 *yuan*. It was estimated that there was a loss of 1 123 650 000 *yuan* during the war (only the direct loss was counted here). And the losses were so heavy.

Since most of Zhejiang Province was occupied the by Japanese army, wealthy papermakers and rich commercial businessmen had to flee. Poor farmers were unable to continue production. Moreover, due to internal and external traffic isolation, together with sales disruption, paperworkers were in an extreme difficulty. However, unexpectedly, under the rule of Japanese puppet government, urban and rural superstitious rituals were popular. As a result, local joss paper maintained a developing trend. Therefore, although the total output of handmade paper was reduced by about 30%, the use of joss paper was on the rise. Its output accounted for more than 50% of the total.

The loss of Fuyang County, the leading papermaking county of Zhejiang Province,

A photo of *Fuyang Traditional Handmade Paper*

Tang'ao Village in 2016

was particularly heavy. From invasion of Fuyang County by Japanese army on December 24, 1937, to August 15, 1945 when the Japanese surrendered, the invading Japanese army imposed a blockade of water and land traffic. It was difficult for the paper products to be shipped out of the country, causing serious damage to agriculture and paper production. For example, in October 1939, when the Japanese army went to Dumu Village in Xinyi Town, the important access of the paper products, they burned down the inventory of Fuyang local paper worth more than 6 million *yuan* (silver dollar), accounting for 70% of handmade paper total value in Fuyang County at that time. During the invasion of the Japanese army, prosperous Fuyang paper industry fell into a declining situation, which was followed by collapse for a long period of time. According to the book *Fuyang Traditional Handmade Papermaking*: In the 25th year of the Republican Era (1936), Fuyang bamboo paper production output was 515 500 pieces. In the 35th year of the Republican Era (1946) after victory, Fuyang bamboo paper output was 340 500 pieces, with a decrease of 175 000 pieces, 34% off. The annual output of local Paper in Fuyang County dropped from 13 512 ton in 1936 to 9 480 ton in 1949. The number of papermaking families dropped from 10 069 in 1936 to 3 456 in 1949.

A folk song reflected the dilemma of papermakers at that time, which goes like this:
Umbrella and small bags on shoulder,
in search for papermaking households.
Porriage for 3 meals with little salt,
palm-bark raincoat and hat as my quilt.

Wenzhou City was another papermaking area that had been greatly affected by the war. It had been occupied by the Japanese army three times in the war. The raw materials, finished products, equipment and factory buildings of Wenzhou paper factories had been plundered, burned and destroyed by the Japanese army, causing heavy losses. Thousands of papermakers were affected by this tragedy. In October of 1939, 18 cargo ships were loaded with deckle-edge paper, Ping paper, wood, timber, coal, etc., totaling more than 35 000 *dan*. They were intercepted by the Japanese army in the Oujiang River, goods looted, which led to serious losses. In the 30th year of the Republican Era (1941), from April 19 to May 1, during the first fall of Wenzhou City, most of paper in Wenzhou paper shops was burned by the Japanese army. For example, 1 007 pieces of Siliuping paper and 503 pieces of Tiping paper in Taiyuanxiang local products store located in Doumentou of East Gate were burned; 153 pieces of Nanping paper and 100 pieces of Liujiuping paper in the Dongxing store in front of theXingqian street near East Gate and the Longmao store outside the South Gate were burned by the Japanese army. From the 33rd year of the Republican Era to the 34th year of the Republican Era (1944 to 1945), during the third fall of Wenzhou, economic loss of Ping paper industry reached 11.36 million *yuan*, and the direct economic loss of Wenzhou Daming Industrial Factory which produced wax paper was 5.458 million *yuan*. In the 38th year of the Republican Era (1949), there was only one machine-made paper factory and two wax paper factories in Wenzhou City. Others were small family-based papermaking workshops, indicating a serious decline of the industry.

Also affected by the war was the Ping paper industry in Longyou County. A piece of news on *Zhejiang Economic Information* confirmed that the war affected the sales of Ping paper in Longyou County: Longyou Nanping paper was sold about 140 000 pieces a year, to Shandong, Henan, Tianjin, Beiping, Yingkou and other places. However, in recent years, the sales were not good. The total paper sales this year was less than 400 000 *yuan*.

Ruined papermaking equipment remains

Ping paper in Longyou County was toilet paper and joss paper. In the early years of the Republican Era, the annual output was as high as 300 000 *dan*, which was an important local industry. However, the level of Ping papermaking techniques is relatively primitive, and the paperworkers needed to work very hard to just barely make a living. The war had greatly reduced the demand for Ping paper. Thus, papermakers had nowhere to make a living and no place to live. From 1937 to 1938, due to the fall of Hangzhou City, the traffic to the outside world was interrupted, and the production of Longyou Ping paper was almost completely stopped. In addition to private papermaking mills, large paper factories such as "Huafeng" and "Minfeng" were forcibly occupied by Japanese army. Wenxi Paper Factory, which costed 100 000 *yuan* for construction, could not be operated. Some paper mills that barely managed production were also closed due to the high cost of raw materials.

Although the major handicraft production areas in Zhejiang Province were hit hard during the war, there were still handmade paper production in small mountain villages far from the war fields, supporting the handmade paper market. Tang'ao paper in Fenghua of Ningbo area was a case. Located in the remote area, the locals used to take handmade paper as their business. During the Anti-Japanese War, the handmade paper production workshops in the eastern and southern part of Zhejiang Province, as well as Fuyang County were impacted by the war; however, the villages of Fenghua area such as Tang'ao were less affected by the war, so they undertook the handmade paper production, such as paper for government notices, school textbooks, etc. At the same time, due to the disruption of traffic caused by the war, the import of foreign paper was also affected, which gave Zhejiang handmade paper a chance to survive and revive. After the shock of Zhejiang handmade paper industry in the early days of the war, the production of local paper also recovered temporarily, but still in a difficult situation.

1.3.3 From 1945 to 1949: Handmade Paper Industry Struggling for "Recovery"

After the victory of the Anti-Japanese War,

handmade paper industry recovered slightly. In the 37th year of the Republican Era (1948), *Zhejiang Economic Yearbook* records: "Fuyang, Yuhang, Lin'an areas benefited from local resources, made abundant paper. In the east part of Zhejiang Province, Shaoxing and Quzhou were the most prosperous papermaking sites. Located at the other side of the river facing Fuyang County, Xiaoshan became the center of paper industry of the province. Other areas, such as Jiangshan, Changshan and Longyou all enjoyed a special status in papermaking industry. Recently, Wenzhou area mass produces various types of paper and paper is sold to Shanghai, almost overwhelming Fuyang."

Fuyang handmade papermaking production recovered slightly after the Japanese surrender. Due to the destruction of infrastructure and the subsequent inflation, the price of paper plummeted, causing the paper-based farmers who resumed production to suffer. The production of handmade paper was still in depression. According to *The Annals of Fuyang County* (1993 edition), "the annual output of handmade paper in Fuyang (excluding Xindeng County) decreased from 13 512 tons in 1936 to 9 480 tons in 1949; the number of papermaking families dropped from 10 069 in 1936 to 1 546 in 1949. As a result of the decline in production year after year, papermakers' life became very difficult".

The development of Quzhou paper in the short period of time after the war was not promising either. According to the analysis in the article *Papermaking in Quxian County Before the Founding of the People's Republic of China*: "Quzhou papermakers used bamboo pulp in papermaking, which started in the late Ming and early Qing Dynasties, thrived at the end of the 19th century. Later, paper-based households came into being, and flourished in the late 1920s and early 1930s, developed for more than 30 years. In the meantime, after the Anti-Japanese War, Quzhou paper industry, the papermakers, and the vast working people in the mountainous areas, had assumed that they had passed the hard times in the war, and now were expecting a prosperity. So the paper merchants in the city were busy collecting money and reinvesting. Meanwhile, the papermakers were also busy reusing the papermaking troughs, and the production was resumed. Soon, the Kuomintang authorities set off domestic war. They issued Jinyuan currency, implemented inflation policy, and plundered the people's wealth. The investment of paper merchants one year ago dropped 90% when the paper was produced, shipped to Hangzhou, and sold out. So almost all the papermaking mills were thereafter closed. This is an unprecedented catastrophe. On the eve of the founding of the People's Republic of China, not only in mountainous areas, even merchants in cities couldn't survive. The celebrated paper merchant who was claimed to own half of the city's assets, was also in a struggle."

1.3.4 From 1949 to 1978: Promotion of the Development of Zhejiang Paper Industry in the First 30 Years After the Founding of the People's Republic of China

After the founding of the People's Republic of China in 1949, the government took over the management of the production of handmade paper. In brief, "Before 1956, handmade paper was under the jurisdiction of the Papermaking Office of Central Light Industry Ministry. After 1959, it was transferred to the Second Light Industry Ministry. In 1972, it was administered by the Ministry of Commerce. In 1978, it was supervised by the National Supply and Marketing Cooperative. In 1981, the specific leader of the practice was the Department of Grocery, Waste and Old Materials under the Ministry of Agriculture, Forestry, Animal Husbandry and Fishery." In terms of specific policies issued by the government, it can be divided into two phases.

The first stage was from 1949 to 1958:
After 1949, due to the predicament of the paper industry, the government was quite concerned about the production of handmade paper. Thus, the government offered help, from production to sales, and supported the papermaking workshops and households to resume production. For example, in Fuyang County, the papermaking center, at the beginning of the People's Republic of China, the handmade paper was still operated by private paper merchants. In June 1950, the Fuyang County Government expanded the sales channels, formulated and implemented handmade paper production and marketing policies to help the local papermakers, especially the weak papermakers. These measures contributed to the rapid improvement of production capacity of papermakers and rapid recovery of papermaking willingness.

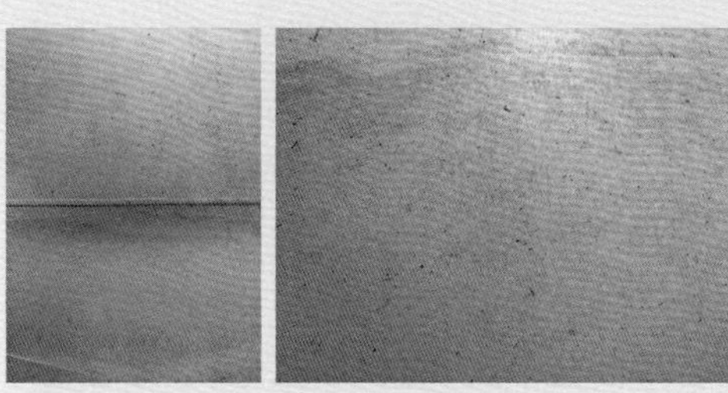

Changshan paper made in the early Republican Era (I)
Changshan paper made in the early Republican Era (II)

Country road leading to Zhishan Mountain of Zeya Town in 2016

Local cloth coupon in Zhejiang Province used in old days

From 1953 to 1956, China began to carry out socialist transformation of the handicraft industry. Zhejiang Province also carried out corresponding transformation measures on the local paper industry. In 1953, private paper industry transformed the private-private partnership model into a public-private partnership model. In December 1956, in accordance with the instruction of the relevant documents released by the National Council, Zhejiang Provincial Government issued the Instructions on *Effectively Strengthening Management of Local Paper Market*, stipulating that "local paper" (traditional handmade paper) was a state-owned materials. None should be involved in its sales, hence opened up the period of more than 20 years of "regulated paper purchase and marketing".

Take Fuyang County as an example. In 1953, there were three large paper shops, i.e. Yongyang, Tongli and Xieji in the Fuyang County, all privately-owned. In the autumn of 1954, there was 1 public-private jointly held paper shop respectively in Fuyang County and Dayuan Town. In June 1954, Fuyang Office and Xindeng Sales Group of Yuhang Branch, State-owned Local Products Company were cancelled together, and its subordinate was transferred to the county collective ownership system and managed by the county Supply and Marketing Cooperative. By June 1956, all private paper shops in the county had been transformed into county or township owned commercial enterprises, and they were under the supervision of local Supply and Marketing Cooperatives. At the same time, the original four handmade paper trading markets were also cancelled. Since then, all kinds of handmade paper in the county were collectively purchased by the local Supply and Marketing Cooperatives.

In Zhishan Mountain area of Zeya Town in Wenzhou City, because of short of rations, papermakers' food was supplied by the government. The toilet paper was the main source of income to exchange foods. In the early 1950s, the local governments were very concerned about the production and life of papermakers. Instead of earning money on toilet paper, the local government would provide money to subsidize the production of toilet paper. For example, the local administration office of Zhishan Mountain area and the People's Commune both had a deputy official in charge of production and sales of toilet paper. After development of nearly a decade, by the 1960s, the production of bamboo paper in Zhishan Mountain area had increased year by year. Meanwhile, the government had also supplied the necessary materials for the expansion of paper production, such as wood and coal, to the Zhishan Mountain area. In order to facilitate the sale of paper, the collection spots would be set anywhere the road traffic is available.

The second stage is from 1959 to 1978:
Between 1959 and 1961, people suffered a harsh life. Prior to it, the "Great Leap Forward" campaign requested rural areas to put emphasis on food production. Papermakers in mountainous areas were forced to abandon the production of paper and go up the mountain to plant grain. Policies of this period had impacted the production of local paper in Zhejiang Province, resulting in a serious decline in the production. The area of bamboo forests destroyed in Fuyang County alone reached more than 1 600 hectares, and the area of deserted bamboo forests reached 4 600 hectares. Fuyang's plain grain-producing areas exploited straw as fertilizer, so the raw materials of Kengbian paper were greatly reduced. The papermakers generally ceased production due to lack of raw materials, and the output of straw paper was severely reduced.

In the spring of 1962, Fuyang County Party Committee timely adjusted the rural work policy, proposing that the mountainous area should rely on mountains for development; the papermaking area should plant bamboo; the grain-producing plain area should focus on agriculture for development, while making Kengbian paper as sideline production. On March 15 of the same year, Fuyang County Local Paper Union was established, undertaking the tasks of technical exchanges and providing market information. In Fuyang a deputy secretary of the County Party Committee and a deputy county mayor both took charge of the production of handmade paper. All districts and townships had also established paper production organizations and assigned specific personnel in charge.
Around 1969, influenced by the "Cultural Revolution", many places in Fuyang took paper production of the team under unified management, the farmers of the team took over the purchase of the raw materials, and the finished products were sold to the Supply and Marketing Cooperative by the production team. The payment of one tiao (*dan*) paper was only 2 *yuan* and 5 *jiao*. Thus, enthusiasm for papermakings was declining. In the early 1970s, stealthy production and sales appeared, which indicated the beginning of aversion from the unified arrangement of production, purchase and sales system.

In conclusion, from the founding of the People's Republic of China in 1949 to the eve of the reform and opening up in 1978, policies of Zhejiang local government concerning papermakers and merchants mainly include the following five points:

(1) To issue a pre-order deposit for handmade paper. Every year when cutting bamboo for raw material, paper farmers needed to pay a lot of money in advance. In the past, most of the papermakers could only take private loans, even from loan sharks. In the new period, local government issued pre-purchase interest-free deposits to papermakers through the Supply and Marketing Cooperative every year to support the development of handmade paper production. According to statistics, during the 32 years between 1951 and 1982 in Fuyang County, a total of 15.64 million *yuan* of pre-purchase deposits for local paper was issued, 489 000 *yuan* per year (considering the currency and purchasing power at that time, the support is substantial.

(2) To provide food subsidy for bamboo cutting workers and rice paste for papermakers. (From the 1960s to the middle and late 1970s, the national government implemented the unified purchase and sale of grain, cotton and oil, and supplied it with designated amount. However, in view of the fact that the bamboo cutting was labor-intensive and exhausting, so the workers would need a lot of food to keep energy. Rice paste for pasting and baking paper was also a necessity. Therefore, the Food Department of Zhejiang government agreed that from 1964 to 1969, Fuyang County would allocate a total of 214 000 kg rice as a subsidy for bamboo cutting laborers (subsidy standard: 5 kg grain for cutting every 5 000 kg green bamboo). From 1972 to 1979, 244 500 kg rice was set aside as rice paste for papermakers (subsidy standard: 0.5 kg for making printing and writing paper per *dan* and 0.25 kg for toilet paper per dan).

(3) Quantitative supply of bag cloth. In the 1960s and 1970s, cotton cloth was strictly supplied with cloth coupon, and the production of Kengbian paper required cloth bags for washing the materials. Each cloth bag needed 14 to 18 *chi* (one *chi* is one third of a meter) cloth. Approved by the local government after discussion and investigation, each piece of Kengbian paper (12 *dao*, 1 080 sheets, size: 21 cm×23 cm) would be assigned 0.1 *chi* cloth, which should be listed in the national production plan, and Fuyang Cotton Textile Factory was designated for its production. Therefore, even in the difficult period when the supply of cotton cloth was tight in the 1960s, the cloth paper bags of Kengbian paper was supplied as usual and the production was guaranteed. According to the statistics, the government had supplied 200 000 *chi* cloth for making bags per year.

(4) Subsidy for raw material price gap. After 1965, raw materials of local green bamboo rose. Meanwhile, price of raw materials from other places was also high due to high transportation fee, which affected the cost of local paper production. However, due to the unified purchase and marketing system, it was not possible to raise the price of paper at will. In order to support the purchase of raw materials and increase the production of local paper, Fuyang County provided certain subsidies for the purchase of raw materials from outside. For example, the fixed price of green bamboo by the government was 260 *yuan* per 5 000 kg while the market price was 320 *yuan*. So the gap of 60 *yuan* would be bridged by the national government.

(5) Material awards. After 1961, the local departments of cities and counties in Zhejiang Province collected scarce resources and implemented temporary or one-off sales in exchange of local paper. From the second half of 1962 to the first half of 1963, the Provincial Cooperative and Commercial Departments decided to give the papermakers cotton and fertilizer in exchange of printing and writing paper, toilet paper and Kengbian paper. Specifically, for printing and writing paper, 0.5 *chi* cotton cloth would be given; for double-layer toilet paper, 0.5 chi cotton cloth would be provided; while for Kengbian paper, 0.5 *chi* of cotton cloth and 0.5 kg of fertilizer would be provided if the papermaker sell 10 *yuan* worth of paper.

2 Current Status of Handmade Paper in Zhejiang Province

From July 2016 to April 2019, the investigation team from the Institute of Handmade Paper which located in the University of Science and Technology of China conducted a systematic and in-depth field investigation and literature research on the contemporary *handmade paper industry in Zhejiang Province*. There are 11 prefecture-level cities in Zhejiang Province, among which Jiaxing, Jinhua, Taizhou and Zhoushan Cities have not found the existence of live hand-made paper production. According to the principles of collecting information on contemporary papermaking sites, 12 of the 89 county-level divisions (including 37 municipal districts, 19 county-level cities, 32 counties, and 1 autonomous county) in 7 prefecture-level cities (Hangzhou, Huzhou, Shaoxing, Ningbo, Wenzhou, Lishui, Quzhou Cities which were still involved in papermaking) had been investigated in Zhejiang Province. All 15 factories in 11 counties except Fuyang District were surveyed and recorded into the database (the survey team had discovered the concentrated distribution of contemporary papermaking sites). For Fuyang District, where papermaking industry was densely distributed so far, 25 representative handmade papermaking workshops and factories and 3 papermaking tool workshops were selected to be included in our field investigation for a complete recording of papermaking information.

Information collected includes: cultural and geographic environment, raw materials, papermaking techniques and tools, history and curren status, business channels, market and sales data, paper varieties and uses, customs and brand culture, challenges and developing plans. At the same time, the standardized physical properties and fiber characteristics of representative paper products from all workshops and manufacturers were analyzed in the Handmade Paper Laboratory of the University of Science and Technology of China. The survey and research analysis have obtained comprehensive information on the current state of Zhejiang Province's contemporary papermaking industry, its product and papermakers, as well as its distribution, structure, and ups and downs, which is shown in Table 1.4.

Table 1.4 Distribution of handmade paper in Zhejiang Province from July 2016 to April 2019 (based on the field investigation and experimental analysis)

Location	Papermaking companies and mills (or paper brand)	Types	Field sample collection / experimental analysis	Paper type / brand name	Raw materials	Status
Longyou County centre	Zhejiang Chengang Xuan Paper Co., Ltd.	bast paper	sample collection / experimental analysis	Chengang *Edgeworthia chrysantha* Lindl. paper in Longyou County	*Edgeworthia chrysantha* Lindl. bark (100%)	in production
Xidong Village in Huabu Town of Kaihua County	Kaihua Paper Research Centre	bast paper	sample collection / experimental analysis	Huang Hongjian's newly-made Kaihua paper	*Wikstroemia monnula* Hance, bamboo pulp	in experiment
Tangzhai Village in Zeya Town of Ouhai District	Pan Xiangyu Bamboo Paper Mill	bamboo paper	sample collection / experimental analysis	Zeya Ping paper	sympodial bamboo (100%)	in production
Ao'wai Village in Zeya Town of Ouhai District	Lin Xinde Bamboo Paper Mill	bamboo paper	sample collection / experimental analysis	Zeya Ping paper	sympodial bamboo and tender bamboo (100%)	production ceased
Wenqiao Village in Xiaocun Town of Taishun County	Weng Shige Bamboo Paper Mill	bamboo paper	sample collection / experimental analysis	Dajiang paper made with foot pestle	tender bamboo (100%)	in production
The third floor of Zhou'aoshang Village Committee in Zeya County	Yazexuan	bast paper	sample collection / experimental analysis	"Luanyunfei", Wenzhou bast paper	mulberry bark and *Edgeworthia chrysantha* Lindl. bark	resuming production
				Lingfeng Huangbojian		
			field sample collection	Wenzhou local bast paper	mulberry bark (100%)	
Songjiadian Village in Pingshui Town of Shaoxing County	Shaoxing Luming Paper	bamboo paper		Luming paper	tender bamboo	production ceased
No. 388 Punan Avenue in Shengzhou City	Shanteng Paper Research Institute in Shengzhou City	Teng paper	sample collection / experimental analysis	Geteng paper (made from mixed materials)	Geteng, cotton pulp, *Eulaliopsis binata*, etc.	in production

(Continued)

Location	Papermaking companies and mills (or paper brand)	Types	Field sample collection / experimental analysis	Paper type / brand name	Raw materials	Status
Longwang Village in Shangshu Town of Anji County	Handmade bamboo paper in Longyu Village of Anji County	bamboo paper	sample collection / experimental analysis	joss paaper	old bamboo	production ceased
				tinfoil paper	old bamboo	in production
				Yuewang paper	pure bamboo	
No. 13 Xixia Anling Dunxia area in Tang'ao Village of Xiaowangmiao Community in Fenghua District	Yuan Hengtong Paper Mill in Tang'ao Village of Fenghua City	bamboo paper	sample collection / experimental analysis	Tang'ao paper	*Pleioblastus amarus* (100%)	in production
				Tang'ao paper	*Pleioblastus amarus*, mulberry bark	
Likeng Village in Anmin Town of Songyang County	Likeng Paper Mill in Likeng Village of Songyang County	bast paper	sample collection / experimental analysis	Likeng Mian paper	*Edgeworthia chrysantha Lindl* bark, (40%), wood pulp (60%)	in production
Fengling Village in Yuqian Town of Lin'an District	Hangzhou Lin'an Fuyutang Paper Co., Ltd.	bast paper	sample collection / experimental analysis	"Yunlong paper" in original color (made from mulberry bark)	mulberry bark	in production
				calligraphy and painting "Xuan paper"(made from mixed materials)	*Edgeworthia chrysantha Lindl* bark, bamboo pulp, *Eulaliopsis binata* pulp, wood pulp	
Pingdu Natural Village in Qianmao Administrative Village of Yuqian Town in Lin'an District	Hangzhou Qianfo Paper Co., Ltd.	bast paper	sample collection / experimental analysis	window paper made from mulberry bark	wood pulp, mulberry bark	in production
Xiapingdu Villagers Group in Qianmao Village of Yuqian Town in Lin'an District	Hangzhou Lin'an Calligraphy and Painting Xuan Paper Factory	bast paper	sample collection / experimental analysis	mulberry paper	mulberry bark	in production
				calligraphy and painting paper	*Edgeworthia chrysantha Lindl bark*, *Eulaliopsis binata* pulp, wood pulp, bamboo pulp	
Guanxingta Villagers Group in Xinsan Administrative Village of Huyuan Town in Fuyang District	XinsanYuanshu Paper Factory	bamboo paper	sample collection / experimental analysis	Yuanshu paper with ancient methods	tender bamboo	in production
Fangjiadi Villagers Group in Datong Village of Dayuan Town in Fuyang District	Hangzhou Fuchunjiang Xuan Paper Co., Ltd.	bamboo paper	sample collection / experimental analysis	bamboo Xuan paper	tender bamboo	in production
Caijiawu Village in Lingqiao Town of Fuyang District	Hangzhou Fuyang Cais Culture and Creativity Co., Ltd.	bamboo paper	sample collection / experimental analysis	super Yuanshu paper	pure bamboo pulp (100%)	in production
Zhujiamen Natural Village in Datong Administrative Village of Dayuan Town in Fuyang District	Hangzhou Fuyang Yiguzhai Yuanshu Paper Co., Ltd.	bamboo paper	sample collection / experimental analysis	"Yiguzhai' *Phyllostachys edulis* Yuanshu paper	tender bamboo	in production
				"Yiguzhai" *Pleioblastus amarus* Yuanshu paper	*Pleioblastus amarus*	
				"Yiguzhai"dyed *Phyllostachys edulis* Yuanshu paper	tender bamboo	

Location	Papermaking companies and mills (or paper brand)	Types	Field sample collection / experimental analysis	Paper type / brand name	Raw materials	Status
The First Villagers' Group in Zhaoji Natural Village of Datong Administrative Village in Dayuan Town of Fuyang District	Hangzhou Fuyang Xuan Paper Lu Factory	bamboo paper	sample collection / experimental analysis	Yuzhu Xuan paper	bamboo pulp (75%), *Edgeworthia chrysantha Lindl* bark (25%)	in production
				paper for printing ancient book	*Eulaliopsis binata* pulp (90%), wood pulp (10%)	
Yanjiaqiao Natural Village in Xinsan Administrative Village of Huyuan Town in Fuyang District	Fuyang Fuge Paper Sales Co., Ltd.	bamboo paper	sample collection / experimental analysis	Yuanshu paper	bamboo, bamboo pulp, *Eulaliopsis binata* pulp	in production
Fangjiadi Villagers Group in Zhaoji Natural Village of Datong Administrative Village in Dayuan Town of Fuyang District	Hangzhou Fuyang Shuangxi Calligraphy and Painting Paper Factory	bamboo paper	sample collection / experimental analysis	Shuangxi superb Yuanshu paper	tender bamboo (97%), *Pteroceltis tatarinowii* Maxim. bark (3%)	in production
Yuanshu Papermaking Park in Xin'er Village of Huyuan Town in Fuyang District	Fuyang Dazhuyuan Xuan Paper Co., Ltd.	bamboo paper	sample collection / experimental analysis a	Baitang paper	tender bamboo	in production
Zhujiamen Natural Village in Datong Administrative Village of Dayuan Town in Fuyang District	Zhu Jinhao Paper Mill	bamboo paper	sample collection / experimental analysis	Yuanshu paper	yellow paper edge (12%), wood pulp (8%), tender bamboo (80%)	in production
No. 46 Zhongta Natural Village in Xin'er Administrative Village of Huyuan Town in Fuyang District	Sheng Jianqiao Paper Mill	bamboo paper	sample collection / experimental analysis	Yuanshu paper	tender bamboo (100%)	in production
No. 241 Qinluo Village Committee in Luocun Village of Dayuan Town in Fuyang District	Xinxiang Xuan Paper Mill	bamboo paper	sample collection / experimental analysis	Yuanshu paper	*Phyllostachys edulis*	in production
Yuanshu Paper Making Park in Xin'er Village of Huyuan Town in Fuyang District	Fuyang Zhuxinzhai Yuanshu Paper Co., Ltd.	bamboo paper	sample collection / experimental analysis	Zhuxinzhai handmade Yuanshu paper applying with ancient methods	pure bamboo pulp (100%)	in production
No.71 Siqian Natural Village in Huangdan Administrative Village of Changlv Town in Fuyang District	Zhang Xiaoping Paper Mill	bamboo paper	sample collection / experimental analysis	bamboo paper for sacrificial purposes	tender bamboo	in production
Zhuangjia Natural Village in Datong Administrative Village of Dayuan Town in Fuyang District	Zhuang Chaojun Paper Mill	bamboo paper	sample collection / experimental analysis	Yuanshu paper	bamboo bark (70%), *Eulaliopsis binata* pulp (30%)	in production
No. 21 Sanzhi Natural Village in Sanling Administrative Village of Dayuan Town in Fuyang District	Jiang Weifa Paper Mill	bamboo paper	sample collection / experimental analysis	deckle-edged paper	cotton (1/3), waste paper deckle and bamboo (2/3)	in production
Xinhua Village in Lingqiao Town of Fuyang District	Li Cairong Paper Mill	bamboo paper	sample collection / experimental analysis	yellow paper for sacrificial purposes	old bamboo (60%), cement bags (20%), cotton (20%)	in production
No. 32 Datian Village in Chang'an Town of Fuyang District	Li Shenyan Joss Paper Mill	bamboo paper	sample collection / experimental analysis	Joss paper	*Pleioblastus amarus*	production ceased temporarily
No. 105 Datian Village in Chang'an Town of Fuyang District	Li Xueyu Ping Paper Mill	bamboo paper	sample collection / experimental analysis	Ping paper	*Pleioblastus amarus*	production ceased temporarily

(Continued)

Location	Papermaking companies and mills (or paper brand)	Types	Field sample collection / experimental analysis	Paper type / brand name	Raw materials	Status
Shanji Village in Lingqiao Town of Fuyang District	Jiang Mingsheng Paper Mill	bamboo paper	sample collection / experimental analysis	joss paper	tender bamboo, paper deckle	in production
Dage Village in Yushan Town of Fuyang District	Qi Wuqiao Paper Mill	bamboo paper	sample collection / experimental analysis	bamboo paper for sacrificial purposes	bamboo (50%), waste paper deckle (50%)	in production
Xinsan Village in Huyuan Town of Fuyang District	Zhang Genshui Paper Mill	bamboo paper	sample collection / experimental analysis	joss paper	bamboo (50%), wasted paper deckle (50%)	in production
Shanji Village in Lingqiao Town of Fuyang District	Zhu Nanshu Paper Mill	bamboo paper	sample collection / experimental analysis	joss paper	bamboo (85%), waste paper deckle (10%), cotton (5%)	in production
Yuanjia Village in Xindeng Town of Fuyang District	Hangzhou Shanyuan Cultural and Creative Co., Ltd.	bamboo paper	sample collection / experimental analysis	Shanyuan *Phyllostachys edulis* Yuanshu paper	tender bamboo (90%), *Pteroceltis tatarinowii* Maxim. bark (10%)	in production
Wusi Village in Lushan Community of Fuyang District	Taohua Paper Mill in Wusi Village	bast paper	sample collection / experimental analysis	newly made Taohua paper in Wusi Village	wild paper mulberry bark	resuming production
Dashan Village in Xindeng Town of Fuyang District	Mulberry Paper Rcoery Site in Dashan Village	bast paper	sample collection / experimental analysis	staw paper in Dashan Village	mulberry bark (60%), straw (40%)	resuming production

2.1 The Existing Resource Characteristics of Handmade Paper in Zhejiang Province

From 1949 to 1958, the People's Republic of China, emphasizing independence and autonomy, interrupted the free dumping of foreign machine-made paper so as to make the market price stable to provide good environmental support for the development of local Zhejiang handmade paper industry. To revive the weather-beaten handmade papermaking industry, a series of measures were undertaken by the local goverenments to promote production, such as issuing loans, setting up workshops to guide production, and establishing channels and platforms to help promote products. Land reform has also freed up productivity. The aggregation of various factors promoted the boom of Zhejiang handmade paper industry and the optimization of its structure. According to statistics, the output of handmade paper in Zhejiang Province was 25 000 tons in 1952, and increased to 72 000 tons in 1956. While the production increased, the product categories also changed significantly. The low-end joss paper decreased from 50% of the total production in the 1940s to 19% in 1957.

However, this fine industrial rejuvenation did not continue, followed by a series of political movements and then the wave of urbanization and industrialization after 1978. By the second decade of the 21st century when the survey was conducted, the form of handmade paper industry had changed dramatically.

2.1.1 Sharp Decrease of Handmade Paper Production

In December 1930, *Zhejiang Paper Industry* compiled by the Zhejiang Provincial Government Design Committee recorded that there were 53 papermaking counties in Zhejiang Province before 1930.

In 1940, *Zhejiang Paper Industry* was revised and reprinted by the Zhejiang Provincial Government Design Committee. It was recorded that the origins of Zhejiang handmade paper industry included 44 counties such as Fuyang, Yuhang, Lin'an, Xindeng, Anji and Fenghua, which basically covered every corner of Zhejiang Province.

In the article *Research on the Protection of Traditional Papermaking Techniques in Zhejiang Province* (hereinafter referred to as Research on Protection), the map drawn by the author based on *Paper Industry in Zhejiang Province* can more intuitively reflect the distribution of handmade papermaking areas in Zhejiang Province in the early 20th century. From the map, it can be seen that around 1940, the *handmade paper industry in Zhejiang Province* was widely distributed, almost all over the province.

Another handmade paper distribution map in the article Research on Protection directly reflects the contemporary distribution of handmade paper in Zhejiang Province. The article, written in 2014, based on his field surveys over the past few years, the author confirmed that Zhejiang Provincewas still doing handmade papermaking. It can be seen from the figure that in 2014, 12 places i.e. Hangzhou City, Anji County, Lin'an County,

Fuyang District, Zhuji City, Shengzhou City, Taizhou City, Ouhai District, Rui'an City, Quzhou City, Kaihua County and Fenghua City were still making paper by hand. From the two figures, it can be clearly seen that the sites still involved in papermaking practice in Zhejiang has dropped sharply.

When the investigation team went to Zhejiang Province for the first round of on-site investigation from July 28, 2016 to November 11, 2016, they also witnessed the sharp decline of the handmade papermaking sites. In many areas, the handmade papermaking practice was once prosperous, yet through years only "single seedling" survived.

From July 2016 to April 2019, the investigation team systematically searched all available areas in Zhejiang, and found that there were still some papermaking-centered areas in Fuyang District of Hangzhou City. Zeya Town of Ouhai District in Wenzhou City had a number of Ping papermaking workshops distributed in many villages, and there were 3 in Lin'an County of Hangzhou City. There are still 1 or 2 handmade paper workshops or factories in Tang'ao Village of Fenghua District, Longyou County center, Longwang Village of Anji City, Shengzhou Science and Technology Park, Likeng Village of Anmin Town, Songyang County, Yunqiao Village of Taishun County, Xidong Village of Kaihua County. In Pingshui Town of Shaoxing City and Cailin Village of Shangyu District, there are still papermaking practice, equipment and skill exhibition halls of historical famous paper Luming paper and Wujin paper.

Zhuji City (county-level), Taizhou City, Rui'an City (county-level) and Hangzhou City have been removed from the 12-region list mentioned in *Research on Protection* (before 2014) in only five or six years. It is certain that some papermaking workshops documented in the previous article *Research on Protection* were not found in our survey. such as in Songyang County, and some workshops are indeed newly restored, for example, Weng Shige Bamboo Paper Workshop in Yunqiao Village of Taishun County was rebuilt in 2015 after the papermaker Weng Shige returned from Northern Fujian Province. Of course, due to the limitation of information obtained, the team may completely miss the information of the papermaking sites in the remote mountainous areas of Zhejiang Province. As a result, this book failed to include them in the book.

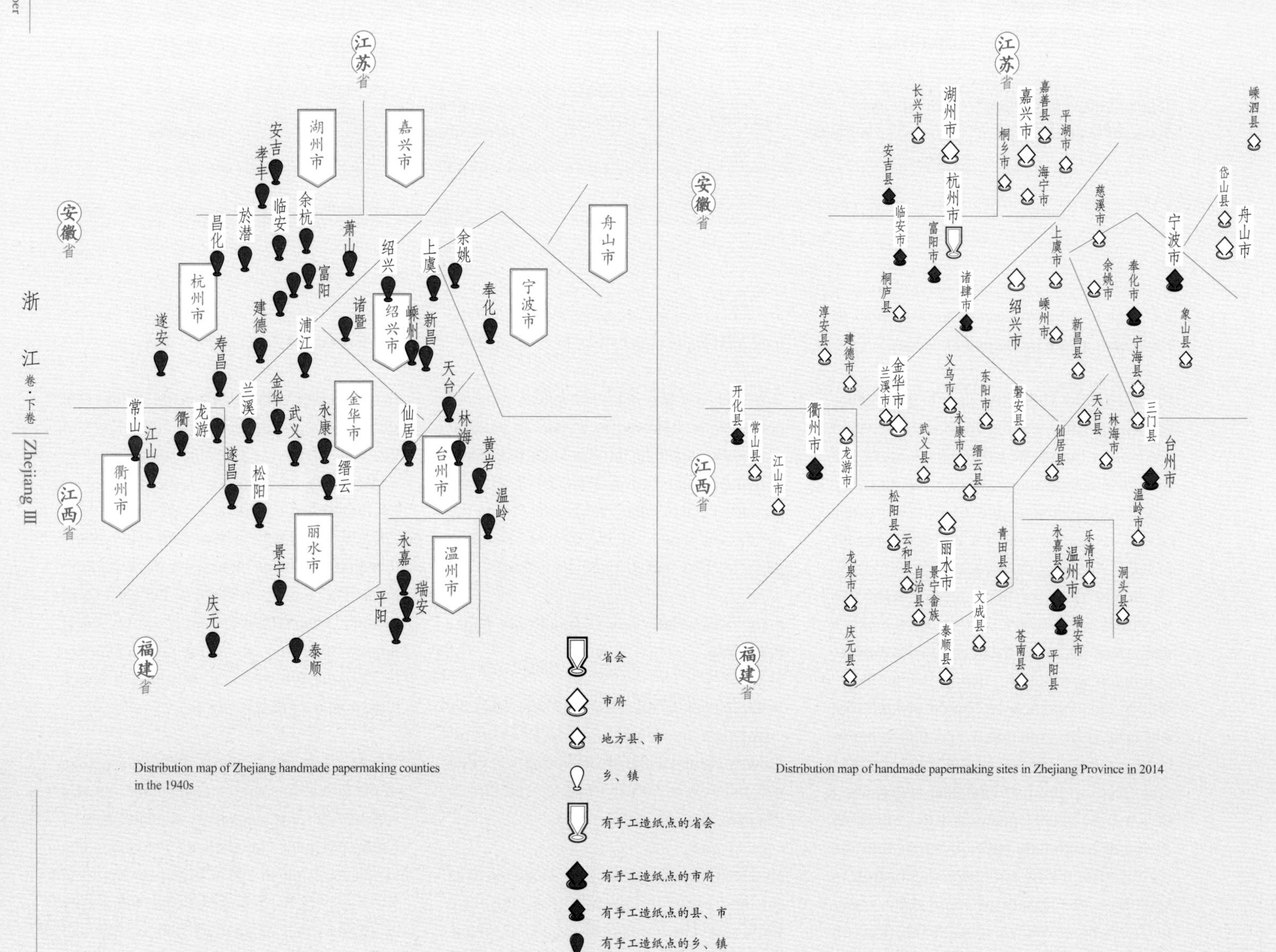

Distribution map of Zhejiang handmade papermaking counties in the 1940s

Distribution map of handmade papermaking sites in Zhejiang Province in 2014

Table 1.5 Papermaking sites and tool worshops discovered by the investigation team from July 2016 to April 2019

Name	Location	Management form	Family inheritors
Zhu Jinhao Paper Mill	Dayuan Town in Fuyang District	family mill	no
Sheng Jianqiao Paper Mill	Huyuan Town in Fuyang District	family mill	no
Fuyang Fuge Paper Sales Co., Ltd.	Huyuan Village in Fuyang District	registered company	no
Xinxiang Xuan Paper Mill	Dayuan Town in Fuyang District	registered company	son
Fuyang Zhuxinzhai Yuanshu Paper Co., Ltd.	Huyuan Village in Fuyang District	registered company	son (making paper, in charge of production management and marketing)
Hangzhou Fuyang Cais Cultural and Creative Co., Ltd.	Lingqiao Town in Fuyang District	registered company developed from a family mill	son-in-law (making paper and production management) and daughter (marketing)
Zhang Xiaoping Paper Mill	Changlv Town in Fuyang District	family mill	no
Fuyang Dazhuyuan Xuan Paper Co., Ltd.	Huyuan Town in Fuyang District	registered company	no
Yuan Hengtong Paper Mill	Fenghua District in Ningbo City	registered company developed from a family mill	children inheriting and taking charge of the company
Xinsan Yuanshu Paper Factory	Huyuan Village in Fuyang District	registered company developed from a family mill	children running the company
Hangzhou Fuyang Shuangxi Calligraphy and Painting Paper Factory	Dayuan Town in Fuyang District	registratered company developed from a handmade paper purchasing station	no
Hangzhou Fuchunjiang Xuan Paper Co., Ltd.	Dayuan Town in Fuyang District	Paper business mill transferred to registered company	daughter running the company and the son attending school in Japan
Zhuang Chaojun Paper Mill	Dayuan Town in Fuyang District	family mill	no
Hangzhou Fuyang Yiguzhai Yuanshu Paper Co., Ltd.	Dayuan Town in Fuyang District	Registered company developed from a family mill	son and nephew making paper
Hangzhou Lin'an Calligraphy and Painting Xuan Paper Factory	Lin'an District in Hangzhou City	Registered company developed from a family mill	no
Hangzhou Lin'an Fuyutang Paper Co., Ltd.	Lin'an District in Hangzhou City	registered company	no
Hangzhou Qianfo Paper Co., Ltd.	Lin'an District in Hangzhou City	registered company	no
Jiang Weifa Paper Mill	Dayuan Town in Fuyang District	family mill	no
Li Cairong Paper Mill	Lingqiao Town in Fuyang District	family mill	no
Li Shenyan Joss Paper Mill	Chang'an Town in Fuyang District	family mill	no
Li Xueyu Ping Paper Mill	Chang'an Town in Fuyang District	family mill	no
Jiang Mingsheng Paper Mill	Lingqiao Town in Fuyang District	family mill	no
Zhu Nanshu Paper Mill	Lingqiao Town in Fuyang District	family mill	no
Weng Shige Bamboo Paper Mill	Taishun County in Wenzhou City	family mill	no

Name	Location	Management form	Family inheritors
Hangzhou Fuyang Xuan Paper Lu Factory	Dayuan Town in Fuyang District	registered company	children in charge of business, son learning to make paper
Qi Wuqiao Paper Mill	Yushan Village in Fuyang District	family mill	no
Pan Xiangyu Bamboo Paper Mill in Tangzhai Village of Zeya Town	Zeya Town in Ouhai District of Wenzhou City	family mill	no
Lin Xinde Bamboo Paper Mill in Aowai Village of Zeya Town	Zeya Town in Ouhai District of Wenzhou City	family mill	no
Zhang Genshui Paper Mill	Huyuan Village in Fuyang District	family mill	no
Zhejiang Chengang Xuan Paper Co., Ltd.	Longyou County in Quzhou City	registered company	daughter and son-in-law involved in production, management and marketing
Bao Jinmiao Paper Mill	Anji County in Huzhou City	family mill	no
Kaihua Paper Research Center	Kaihua County in Quzhou City	Research center	papermaker himself being still young
Yazexuan Brand	Zeya Town of Ouhai District in Wenzhou City	family mill	no
Luming Paper in Shaoxing County	Keqiao District in Shaoxing City	no	no
Shanteng Paper Research Institute	Shaoxing City, Shengzhou City (county level)	Institute & registered company	no
Likeng Paper Mill	Songyang County in Lishui City	family mill	son-in-law involved in papermaking and marketing
Hangzhou Shanyuan Culture and Creative Co. Ltd.	Dongqiao Town in Fuyang District	registered company	The operator being still young
Taohua Paper Mill in Wusi Village	Lushan Street in Fuyang District	family mill	no
Mulberry Paper Recovery Site in Dashan Village	Xindeng Town in Fuyang District	family mill	no
Yongqing Screen-making Mill	Dayuan Town in Fuyang District	family mill	no
Guangming Screen-making Mill	Lingqiao Town in Fuyang District	registered company	workers
Lang Shixun Scrape Knife-making Mill	Dayuan Town in Fuyang District	family mill	no

It took only a few decades for a number of papermaking villages to turn from intensive production to isolated islands of paper mills, or just disappear. Luo Xinxiang Xuan Paper Mill is located in Luocun Village of Dayuan Town in Fuyang District. When our team revisited the village in January, 2019, the mill owner, Luo Xinxiang related that most local farmers traditionally made a living by papermaking, because the mountainous terrain hinders farming. Papermaking practice peaked around 1984, when almost all the villagers, from 50 to 60 local families, were involved. From 1980s, market and machine-made paper challenged the local papermaking practice. When the investigation team visited the village in 2016, only one paper mill, i.e. Luo Xinxiang's family mill, survived, yet failed to find an inheritor. Amount to 85% families are now producing and selling roller shutter.

Another example is that the 1930s was the heyday of Tang'ao paper production in Xiaowangmiao Town of Fenghua County. Three natural villages of Dongjiang, Xijiang and Xixia villages in Tang'ao Village harbored more than 300 papermaking troughs, with more than 1 000 employees. Bamboo paper they made were sold through Ningbo City to the whole country by ship. In 1950s and 1960s, Tang'ao also provided bamboo paper to the local media such as *Daily News and Ningbo Public*. By the end of the 1980s, challenged by the machine-made paper, handmade paper mills mostly failed to sustain. By the time of the survey in 2016, there was only a paper mill run by Yuan Hengtong's family. So the handmade papermaking was basically crashed.

Main door of Yuan Hengtong's family paper mill

From the data in Table 1.5 it can be seen that in the existing production units of handmade papermaking in Zhejiang Province, handmade papermaking family mills are dominant. According to the statistics, of the 42 households officially included in *Chinese Handmade Paper library: Zhejiang*, 22 are run by family mills, more than half of the total. 5 companies were converted from original family mills to registered companies; 2 companies were originally engaged in the purchase and sale of handmade paper, and then they registered as companies and produced handmade paper as companies. The remaining 8 were incorporated in the first place.

2.1.2 Lost of Various Historical Papermaking Techniques and Obvious Decline of Paper Use from the High-end to Low-end

According to the 1930 edition of *Zhejiang Paper Industry*, before the middle period of the Republican Era, Zhejiang Province had a wide variety of handmade paper, which were divided into four categories: yellow and white paper, Huangshao paper, straw paper, and bast paper. Among them, yellow and white paper includes Nanping, Huajian, Changshan, Baijian, Yuanshu, Maobian, Jingfang, Fanggao, Luming, Huangjian, Changhuang, Duanfang, Haifang, Jingfang, Huangyuan, Erxi, Lian paper and other paper products; Huangshao paper includes Huangshao paper, Qianzhang, Cugao, Bianhuang, Zhebian, Maojiaolian; straw paper is divided into Kengbian, Dou paper, Cu paper, Mingcao, Fang paper, straw paper, Sanding, Xiaozhipeng etc.; and bast paper products include Mian paper, bast paper, mulberry paper, Shenpi paper, Taohua paper, Canzhong paper, Yangpi paper and umbrella paper, etc. (see Table 1.6 for details).

Table 1.6 Paper types and names of handmade paper in Zhejiang Province during the middle Republican Era

Yellow and white paper	Nanping, Huajian, Changshan, Baijian, Yuanshu, Maobian, Jingfang, Fanggao, Luming, Huangjian, Changhuang, Duanfang, Haifang, Jingbian, Huangyuan, Erxi, Lian paper, etc.
Huangshao paper	Huangshao paper, Qianzhang, Cugao, Bianhuang, Zhebian, Maojiaolian, etc.
Straw paper	Kengbian, Dou paper, Cu paper, Mingcao, Fang paper, straw paper, Sanding, Xiaozhipeng, etc.
Bast paper	Mian paper, bast paper, mulberry paper, Shenpi paper, taohua paper, Canzhong paper, Yangpi paper and umbrella paper, etc.

According to the statistics of *Zhejiang Paper Industry* (1940 edition), 96 types of paper coexisted before 1940. Among them, bast paper, mulberry paper, straw paper and Yuanshu paper and other 18 types of paper were prodeud locally and are still in production. There are 68 kinds of paper, including Zhushao paper, Chiting paper and Xiegongjian paper, which were produced previously but ceased production now. Paper that was not made in history but produced nowadays include 74 types, such as Kengbian, Houdou, Jingfang and Duanfang (see Table 1.7 for details).

Table 1.7 Paper types made from ancient times and the ones made currently in different counties in Zhejiang Province

	Traditional paper type	Paper type currently in production
Hangxian (former county name, alias Qiantang County)	bast paper, bamboo paper, mulberry paper, Zhushao paper, straw paper, oil paper, Yibo paper	no
Fuyang	bamboo paper, straw paper, mulberry paper, Xiaojing paper, bast paper, Tang paper, Chiting paper, Xiegongjian paper, Lu Ming paper, Huajian paper, Yuanshu paper, Wuqianyuan paper, Cugao paper, Biaoxin paper, Liuqianyuan paper	Kengbian paper, Houdou paper, Lu ming paper, Liuqianyuan paper, Jingfang paper, Duanfang paper, Haifang paper, Wuqianyuan paper, Jingbian paper, Changbian paper, Cugao paper, mulberry bark, Huangshao paper, Changshan paper, Changhuang paper, Kuaitou paper, Dahuangjian paper
Yuhang	Teng paper, Zhushao paper, Pichao paper	Xiaolian paper, Huangshao paper, Haifang paper, Dalian paper, mulberry paper, Huangyuan paper, Yuanshu paper, Jingfang paper, straw paper
Lin'an	Huangshao paper, Chabai paper, Dapi paper, Fanggao paper, Cansheng paper, mulberry paper, Yuanshu paper	Huangshao paper, Biaoxin paper, Wuqianyuanshu paper, Dachangshan paper, Haifang paper, Hongbiao paper, Liuqianyuanshu paper, straw paper
Yuqian	Taohua paper, Qiupi paper, silver paper, mulberry paper	Taohua paper
Xindeng	Mianbai paper, Fanggao paper, Cansheng paper, silver paper, Zhutie paper, Yuanshu paper, mulberry paper, Huangshao paper	Kengbian paper, Ping paper, Umbrella paper, Canzhong paper, bast paper, Dadou paper, Haifang paper
Changhua	Babaizhang paper, Jingbian paper, oil paper	Liuyinsheng paper, Taohua paper, Maochang paper, Kuoshenpichaxiang paper, Qiyinshen paper, Dashenpi paper, Maocao paper, Sanlianmao paper
Jiaxing	Youquan paper, Meilijian paper	It is not included due to low output and quality
Wuxing	mulberry paper, Changqian paper, yellow paper, straw paper, yellow white paper, paper veil	no
Anji	mulberry paper, paper veil	Straw paper
Xiaofeng	Yellow and white paper, mulberry paper, straw paper	Zhebain paper, Fanggao paper, Cugao paper, Huangyuan paper, Banzhe paper, Changshan paper, Hongjian paper, Haifang paper, Huangjian paper, Banjian paper, Nanping paper, Wuqianyuan paper, Liuqianyuan paper, Maojiaolian paper
Yinxian	straw paper, Jianchao paper, bast paper, paper veil, paper quilt, Maijuan paper	no

County name	Traditional paper type	Paper type currently in production
Cixi	Huangliri paper, Bailiri paper	no
Fenghua	Jianchao paper, bamboo paper, bast paper	Luming paper, Zhenliao paper, Xuqing paper, bast paper, Houdou paper, Zhenpi paper, Maqing paper, letter paper
Zhenhai	straw paper	It is not included due to the low output
Dinghai	Chu paper	no
Xiangshan	Jianchao paper	no
Shaoxing	bamboo paper, straw paper, Celi paper, yellow paper, Mian paper	Luming paper, straw paper
Xiaoshan	yellow paper, white paper	Huangjian paper, Huangyuan paper, Luming paper, Huangjingfang paper, Changshan paper, Baijian paper, Lianshi paper, Yuanshu paper, Dajingfang paper
Zhuji	Teng paper, Chabai paper, Wulian paper, Huangpi paper, straw paper, mulberry paper, Qilian paper, yellow paper	Luming paper, Huang paper, Jingfang paper, Dacao paper, Nanping paper, Lianshi paper, Huangchangbian paper, Straw paper, Lianqi paper, Yangpi paper, Yuanshu Paper
Yuyao	bamboo paper	Xiping paper
Shangyu	Dajian paper	Xiping paper
Shengxian	Shanteng paper, Yuban paper, Chuijian paper, Bamboo paper, Qiaobing paper, Xiaozhu paper, Taijian/Huajian paper, Chengxintang paper, Moon-shade and pine pattern paper, Xiaolian paper, Fenyunluo paper, Nanping paper	Nanping paper, bast paper, Lianwu paper, straw paper, yellow paper
Xingchang	Bing paper, Maijuan paper, Moon-shade and pine pattern paper, bamboo paper, straw paper	mulberry paper, Huangjian paper, Straw paper, Xiyuan paper
Linhai	Huangtan paper, Dongchen paper	Qianzhang paper, Zhongqing paper
Huangyan	Teng paper, Yuban paper	Qianzhang paper, Zhongqing paper, Fang paper
Ninghai	Huanggong paper	It is not included due to the low output
Wenling	no	Xiaofang paper
Tiantai	Dadan paper	Huangian paper, Nanping paper, Changlian paper, Pu paper, bast paper, Xiaobai paper
Xianju	Shao paper, bast paper	Shao paper, Pibai paper
Jinhua	Qi paper	Daliancu paper, Xiaocu paper, Xicu paper
Lanxi	paper	no
Dongyan	Yuluan paper	Cu (rough) paper
Yongkang	no	Jinping paper, Zhongfang paper
Wuyi	no	Nanping paper, Erhaoping paper
Pujian	bast paper	Hongding paper, Liuju paper
Tangxi	no	Nanping paper, Huangjian paper
Quxian	paper	Nanping paper, Huajian paper
Longyou	Shao paper	Shao paper, Nanping paper, Huangjian paper, Yuanshu paper
Jiangshan	Mian paper, Teng paper, bamboo paper	Fanggao paper, Huajian paper
Changshan	Bang paper, Dong paper	Huajian paper, Mian paper
Kaihua	Teng paper	no

County name	Traditional paper type	Paper type currently in production
Jiande	mulberry paper	Liuyingshen paper
Chun'an	Gailian paper, bast paper	no
Tonglu	Momei paper, straw paper, Liri paper	Mingcao paper, Sanding paper, Hengda paper, Changdou paper, Changdou paper, Kengbian paper
Sui'an	paper	Huajian paper, Jiaobai paper, Nanping paper
Shouchang	no	Huajian paper
Fenshui	Xuqing paper, silver paper, shao paper	no
Yongjia	Gongliao paper, Zhijuan paper	Nanping paper, Doufang paper, Mian paper
Rui'an	no	Nanping paper, Erxi paper, Yatou paper, Sunke paper, Xiaoxipeng paper
Pingyang	no	Doufang paper, Sihaobo paper
Taishun	no	Huajian paper, Maobian paper
Lishui	Shanli paper	no
Qingtian	Changliao bamboo paper, Duanliao bamboo paper, bast paper	no
Jinyun	Fang paper, Nanping paper, straw paper	Nanping paper, staw paper
Songyang	Baila paper	pine paper, oil paper, Mian paper
Suichang	Dalian paper, Xiaolian paper, Huangbiao paper	no
Longquan	paper	no
Qingyuan	deckle-edged paper	deckle-edged paper
Yunhe	Maotou paper	no
Xuanping	bamboo paper, Fang paper	No
Jingning	Mian paper, wrapped deckle-edged paper	Huajian paper, deckle-edged paper

Handmade Paper Industry in Zhejiang Province also records that before 1930, there were 89 kinds of handmade paper in Zhejiang Province, 96 kinds of handmade paper names in 1940, and the number amounted to 168 in 1956.

The authors assumed the reason for the increase of the handmade paper name is that the name of handmade paper originates from the use of raw materials, production areas, specifications, quality, uses, and some historical reasons, and the historical reason is the main cause for the complicated paper names. "Before liberation, the sales of handmade paper in Zhejiang was almost 100% in the hands of the merchants, and these people always kept the purchase price down to the minimum to increase the profits as much as possible, and sometimes the value of the goods was lower than the actual value." Accordingly, the papermakers had to change the specification and quality of the handmade paper or change the papermaking techniques, hence a new name was used, or cease production to resist the "oppression" of the merchants.

However, it is worth noting that although the types of handmade paper increased, the increase was mostly low-end paper products, such as paper for sacrificial offering, packaging, or daily necessities.

As mentioned above, there are 168 kinds of handmade paper names in Zhejiang Province in 1956, which can be divided into four categories: paper for writing and painting, joss paper, practical paper (including for industrial purpose) and toilet paper. Among them, paper for writing and painting is the high-end paper with only 16 kinds, while the rest are toilet paper, practical paper and joss paper.

The statistics in the book *Traditional Handmade Paper in Fuyang* can also confirm the change of Zhejiang handmade paper products and the trend of product specifications from high-end to low-end.

This book provides the variation of handmade paper varieties in Fuyang County (including the former Xindeng County) in 1930, 1950, 1960, 1980, 1990, 2000 and 2005. For example, there were 24 species in 1930, 46 in 1950, 52 in 1960, 15 in 1980, 14 in 1990, 9 in 2000 and 9 in 2005.

The data of handmade paper varieties in 7 years show that the trend of traditional handmade paper varieties in Fuyang does change from the high end to the low end. Taking the phase from 1930 to 1960 as an example, among the 24 kinds of handmade paper produced in Fuyang (Xindeng) County in 1930, there were 4 kinds of paper for writing and painting, 20 kinds of toilet paper, paper for daily use and joss paper. In 1950, there were 6 kinds of paper for writing and painting, 40 kinds of toilet paper, paper for daily use and sacrificial paper. In 1960, there were 5 kinds of paper for writing and painting, 47 kinds of toilet paper, paper for daily use and joss paper.

Information also mentioned in the report of *Fuyang Daily* on May 8, 2006: From 2001 to 2005, the main types of local paper in Fuyang were Yuanshu paper, calligraphy and painting paper, joss paper, Siliuping paper (a kind of toilet paper) , machine-made paper, and Wujin paper (completely ceased production in 2002). Among Fuyang's three famous paper types: Yuanshu paper, Jing paper and Chiting paper (strictly speaking, the latter two have ceased production), superb Yuanshu paper and first class Yuanshu paper also have not been produced for many years; production of one of the most representative of paper types, mulberry paper in Fuyang has been discontinued and the craft is on the verge of extinction; Wujin paper from Fuyang's famous Chouxi Village gradually stopped production after the late 1980s, and the papermaking technique is on the verge of loss; Once famous superb Yuanshu paper in Fuyang area was no longer produced due to the cancellation of the corresponding paper evaluation agency.

The only surviving paper mill in Longwang Village during the investigation period

Chen Xiaodong, one of the interviewees and general manager of Hangzhou Qianfo Paper Co., Ltd. in Lin'an District, recalled that the Chens family had just started to make paper from their grandfather's generation, and had made a type of paper called Taohua paper, which was widely used to write genealogy and copy Scriptures because of its softness and durability. However, after 1980, due to the decrease in demand and some other reasons, the paper was no longer made locally. Chen Xiaodong said that because he was very young at that time, he failed to identify what materials the paper used or how the paper was made.

The high-end painting and calligraphy paper produced in Longwang Village of Anji County mainly include three types "Jingfang" "Yuanshu" and "Liuping". These three types of quality paper products usually need 5—6 people to cooperate with each other in terms of material selection and workmanship, and the output is limited. Before 1949, about hundreds of villagers were engaged in papermaking in the village, and the number of paper containers was nearly 50. After 1949, due to the loss of traditional techniques, machine-made paper gradually replaced handmade paper as a writing carrier. The local handmade paper industry changed to toilet paper, and the high-end Jingfang paper for painting and calligraphy production techniques were gradually lost.

2.1.3 Loss of Handmade Papermaking Tecniques and Incomplete Crucial Papermaking Steps

Taking the process flow of producing Yuanshu paper in Fuyang District as an example, it is divided into three parts: raw material processing, manufacturing method and subsequent procedures. There are more than 70 procedures from bamboo to making a piece of Yuanshu paper. Dayuan Mountain spreads a proverb-"Making paper is not easy and it requires more than 72 procedures".

During the investigation, it was found that the handmade papermaking process was quietly changing due to the consideration of improving efficiency and reducing costs. Many traditional techniques were neglected or replaced under the pretext of adapting to the market. The traditional "artisan heart" gradually disappeared. On May 8, 2006, *Fuyang Daily* reported that some local paper manufacturers in Huyuan Town had added many modern automated technologies into the traditional process. Li Weijun, who runs a paper factory, said that modern papermaking techniques were incorporated into the process of making Yuanshu paper, such as beating, grinding and stirring, in order to improve the efficiency of the production of local paper. In fact, automatic techniques or modern papermaking techniques spoiled the handmade papermaking techniques, which impact was frequently ignored by the public, while they could save energy and time.

Cai Yuhua, a senior papermaker in charge of The Cais Cultural and Creative Co., Ltd. in Lingqiao Town, said: "the traditional bamboo used in making high-end Yuanshu paper needs to go through the process of removing the outer green bark first, and then take the clean inner bamboo part for further processing. However, scraping demands strong physical labor, while the wage is too high (250 to 300 *yuan* RMB per day). Besides, it is not easy to find skilled hands who are strong enough to do the job. Because the skilled hands are mostly local elderly people, they could barely be competent for the task now while young people were mostly unwilling to undertake such hard task. In order to save the cost and solve the problem of shortage of manpower, most of the paper manufacturers and mills in Fuyang no longer scrape off the outer layer of bamboo. Consequently, no high-quality paper could be made now.

An old man scraping the bamboo (photo provided by Xu Jianhua)

In addition to the change of the raw materials, according to Zhu Youshan, a collector of bamboo paper relics in Fuyang District, Fuyang high grade "Yuanshu paper" was originally made from local bamboo. In the late Qing Dynasty and the Republican Era, due to the reduction of local bamboo output, the raw materials changed to the mixture of *Dianthus chinensis* L. and *Phyllostachys edulis*. Later, with the decrease of local bamboo, the paper was then made from tender bamboo. So in modern times, when referring to Yuanshu paper in Fuyang District, it is alleged is that bamboo is used as raw material for papermaking. However, in fact, there are obvious differences in natural whiteness and durability properties between bamboo paper made from local *Dianthus chinensis* L. and

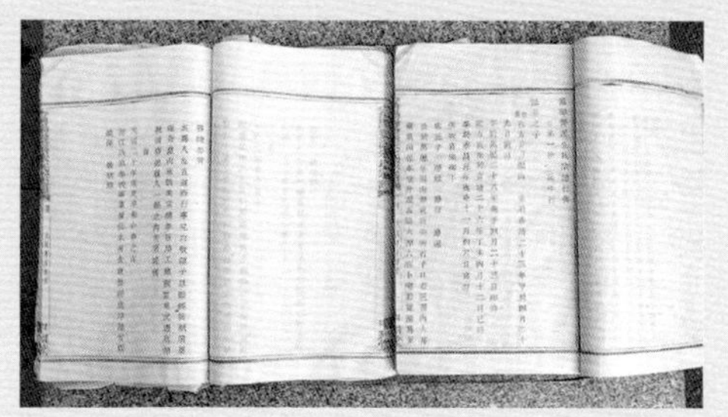
Comparison of printing the same genealogy on *Dianthus chinensis* L. paper (left) and *Phyllostachys edulis* paper (right) in the late Qing Dynasty

Phyllostachys edulis.

2.2 Current Running Status of Handmade Paper in Zhejiang Province

Through the field investigation and literature research of Zhejiang Province from July 2016 to April 2019, the current characteristics of Zhejiang handmade papermaking practice can be summarized as the following four points.

2.2.1 Coexistence of "Inheritance" and "Reform"

Of course, the "inheritance" mentioned here is not a follow-up to the ancient methods, but in the new era maintaining the key process and raw material processing principles, and having a clear concept of making good paper. "Reform" does not mean abandoning the traditional papermaking process, but means adapting to the time, adopting new and appropriate techniques, equipment and materials. Similarly, it is still a prerequisite to have a clear goal of making good paper.

While based on the advantages of natural resources, inheriting traditional techniques and adhering to protection, the handmade paper industry in Zhejiang Province is constantly carrying out paper product improvement and exploration on market consumption innovation, and is seeking development on the basis of inheritance and innovation. With the advent of industrialization, urbanization and the new information age, paper consumption demand, people's living habits and papermaking techniques are undergoing big changes. Zhejiang handmade paper reflects the following characteristics in the contemporary environment in terms of raw materials, product forms, and process and equipment changes.

1. Adherence and flexibility in the raw materials selection and processing step

Take Fuyang District in Hangzhou City, the Zhejiang provincial paper center as well as one of the three major papermaking centers in China, as an example, its mainstream product bamboo paper is inherited and carried forward from the processing techniques of the production of Yuanshu paper using tender bamboo as raw materials in the Song Dynasty. Although the varieties of Fuyang handmade bamboo paper have been rich since the Song Dynasty, they are consistently in the use of the single raw material—bamboo material. In history, there had been three major categories including *Dianthus chinensis* L. paper, *Phyllostachys edulis* paper and *Pleioblastus amarus* paper. From the literature and remains of the late Qing Dynasty or during the period of the Republican Era concerning Fuyang bamboo paper raw materials, various kinds of mixed raw materials such as mulberry bark, paper mulberry bark, and linen were continuously used during the paper development process, which had effectively changed some of the performance defects of bamboo paper. However, the change was never the mainstream. In the 1970s, affected by the trend of making Xuan paper all over the nation, people started to add the raw materials such as *Eulaliopsis binata* and *Pteroceltis tatarinowii* Maxim. bark to the papermaking, and made "Fuyang Xuan" which was specially used for calligraphy and painting. This has not only enriched the types of paper, but improved the traits of paper for special use.

Tender *Phyllostachys edulis* for papermaking (photo provided by Xu Jianhua)

In addition to the main raw materials, auxiliary materials of Fuyang traditional handmade high-end bamboo paper are mainly high-quality lime, human urine (children's urine), soy milk (mainly used for fermenting *Pleioblastus amarus*) etc., while most modern factories have added bleaching powder now. What we have to mention is that, different from other areas that use regular papermaking methods, local Fuyang handmade bamboo paper is supported by unique craftsmanship.

Unique urine fermenting procedure

Plant mucilage not being added to the traditional ways of papermaking is the unique tradition in Fuyang which makes Fuyang bamboo paper special and has strict requirements in the processing of raw materials.

2. The relatively colorful and self-contained product system

In the history of Zhejiang Province, the types of handmade paper products are relatively complete. In the six major categories of handmade paper, bamboo paper, Teng paper, mulberry paper and straw paper were once the best-selling products in China, which formed peak of paper products in one or more dynasties. And a number of Chinese famous paper brands have been harbored, which is still rare in other Chinese provinces.

Take the extended products of paper for an example, a number of paper-producing areas such as Fuyang, Longyou, Fenghua, Lin'an, Wenzhou, etc. not only produce handmade paper in original color, bleached paper, paper for repairing ancient works, but produce many types of hand-dyed paper and art processing paper, thus having the influence on the innovation of types, quality and applications. In terms of Fuyang area, with the development of handmade paper industry, the use of the original paper as the carrier has promoted the rise and development of dyed paper, woodcut and engraving printing, decoration and repair, sacrificial paper products, and other industries, forming a relatively complete traditional paper-related industrial chain.

Hangzhou Fuyang Yiguzhai Yuanshu Paper Co., Ltd. is a small handmade paper mill in Datong Village of Dayuan Town. The mill has only five or six employees, but the family papermaking history is long. From the beginning of the Wanli period in the Ming Dynasty, to the beginning of 2019, the inheritance of the papermaking process of 14th generation papermakers was clearly recorded in the *Genealogy of the Zhus in Fuchun*. It can be said that the clan calling for creating good paper is strong. As a mill for making bamboo paper, its status from 2016 to 2019 is that "Yiguzhai" pursues the restoration of the traditional bamboo paper in original color produced by the ancestors, and also trials and manufactures other paper types with bamboo as the raw materials, such as the *Pleioblastus amarus* restoration paper, *Pleioblastus amarus* Wujin paper and *Dianthus chinensis* L. Yuanshu paper. On

Engraved ancient works printed on "Yiguzhai" Yuanshu paper in Dongzhai Cultural Workshop of Fuyang District

the basis of Yuanshu paper of original color, "Yiguzhai" tentatively made multicolor dyed Yuanshu paper, bamboo paper for woodcut printing, bamboo paper for buddhist scripture printing, bamboo paper for engraved ancient works printing and so on. It is unique that it extends from high-quality paper and maintains a clear development context and system.

3. Papermaking techniques and equipments

Fuyang is the only central area for contemporary handmade papermaking in Zhejiang Province. Therefore, the path of process and equipment improvement and evolution is relatively clear and representative. In 2006, Fuyang bamboo papermaking techniques were all listed in the first batch of national intangible cultural heritage protection list by the Ministry of Culture in China. Fuyang handmade papermaking procedures are diverse and the techniques are complex. The popular saying is that its production process entails 72 different steps. Therefore, there is a proverb among Fuyang papermakers, "Every piece of paper is not easy to make because it has to go through 72 steps of processing." Among the steps, the core ones include: chopping the bamboo, trimming the branches, cutting the bamboo into sections, stripping the bamboo, beating the white bark, soaking the bamboo, chopping the materials, fermenting materials, piling the bamboo for fermenting, putting the bamboo in the utensil, boiling the materials, picking the materials out of the utensil, cleaning the materials, piling them up after splashing urine, waiting for papermaking, making the pulp, papermaking, drying the paper, packing and sealing the paper.

Of course, the papermaking process and tool requirements of Zhejiang handmade papermaking were relatively stable in the traditional period, but in different periods accompanied by the production of different paper products, the process and equipment were constantly changing and improving. Therefore, research expert of Fuyang bamboo papermaking, Li Shaojun held that while inheriting the traditional papermaking techniques, the industry was constantly adjusting and improving in the process and tool.

Then, what are the main manifestations of this adjustment and progress (and of course, possible regression)?

Firstly, traditional papermaking process continues to improve. The efforts to improve and upgrade the handmade papermaking process have run through the entire process of papermaking, and are concentrated in the processing of raw materials. For example, in order to ensure the quality of the bamboo materials, Fuyang handmade bamboo paper directly uses the quicklime purchased from the market to replace the lime obtained by the villagers themselves by burning grass and wood in fermenting. From the division of labor, on the one hand, a lot of time, manpower and material resources can be saved, because the lack of manpower is the bottleneck of the current development of the handmade papermaking industry; on the other hand, the finished paper fiber will not be contaminated by impurities caused by poor calcination of the lime, which will affect the quality and price of the paper.

Secondly, there are some changes and innovations in Zhejiang handmade paper production tools. Zhejiang Province is a developed area with modern wood pulp system in the production of machine-made paper. With the development and technological advancement in papermaking industry, some papermaking mills have adopted much simpler and more efficient power machinery equipment in the processes of cooking, pulping and drying. For instance, use a high-pressure steamer instead of a wooden barrel and the time for cooking bamboo materials will be shortened from the original 7 days or more to about 1 day. Electricity equipment such as electric stone roller, electric beater have been employed to save much human costs. These improvements to equipment inspired by the mechine-made paper industry have a significant effect on improving production efficiency, shortening production cycles, increasing output, saving labor, reducing labor intensity and so on.

In the meanwhile, negative effects have appeared when the traditional stone roller, hydraulic pestle were replaced by electric equipment, which might do harm to bamboo fiber. At present, it has a significant impact on the production of superb bamboo paper, thus it is necessary to have a comprehensive understanding of tool improvement.

2.2.2 Coexistence of Regional Commonness and Diversities

Zhejiang handmade papermaking has distinct regional characteristics. Based on the distribution of natural resources and material living conditions, the distribution of handmade paper, the inheritance of techniques and the development of the industry all have typically regional culture characteristics.

First of all, with moderate climate and abundant mountains and water, Zhejiang Province is conducive to bamboo forests' growth, so it is one of the main producing areas of bamboo resources in China. The distribution of these natural resources is beneficial to the development of the handmade paper industry, especially the bamboo paper industry. From the perspective of topography and natural resources in Zhejiang Province, it can be divided into three areas: mountainous areas, semi-mountainous areas and plain areas. Due to many bamboo forests and trees in mountainous areas, the residents living there, are mostly engaged in the production of handmade bamboo paper. In addition to trees, in the semi-mountainous areas, rice also can grow there, so the residents mainly use bamboo and straw as raw materials for handmade paper. In the plain areas, the terrain is suitable for planting rice, and bamboo

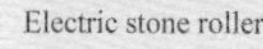
Electric stone roller

Paper mill in the remote mountain area of Wenqiao Village in Taishun County

forests and water resource are not that dense as mountainous or semi-mountainous areas. Therefore, the main production areas of handmade paper in Zhejiang Province lie in mountainous and semi-mountainous areas with abundant water and bamboo.

In addition, Zhejiang handmade paper is closely related to the local economy, cultural development and papermakers' livelihood. With the transfer of Chinese economic and cultural center to Jiangnan area since the Song Dynasty, especially since the Southern Song Dynasty, Hangzhou area becoming the capital and central city of China, and Mingzhou (ancient name for Ningbo) area becoming a big international trade port city since the Tang Dynasty, the economy and culture in Zhejiang region has witnessed its boom. Paper has not only become the necessity of cultural life and daily life in Jiangnan area, but also was sold to a wider consumer market. The increased demand for paper and the development of the market have shed a positive impact on Zhejiang handmade paper industry, which has superior papermaking conditions and profound craftsmanship. Handmade paper production has become the main source of income for papermakers and their livelihood.

The natural environment, historical inheritance and living conditions have formed the regional commonness of Zhejiang handmade papermaking, but the papermaking industry in Zhejiang Province also presents the diversified characteristics formed by different regions. Till the contemporary era, the industry has shrunk seriously as a whole, but it still maintains the coexistence of regional commonness and diversity. The diversity of handmade papermaking in Zhejiang is mainly shown in the following four points.

1. Diverse distribution of papermaking sites

According to the results of a field survey conducted by the investigation team from July 2016 to April 2019, under the background of general threat of shrinkage and extinction, the handmade papermaking sites are still distributed in many cities, districts, counties all over the province. Even in one region, there are still many papermaking sites distributed in different places. This distribution feature is not available even in Anhui Province, which has the largest number of handmade paper mills. According to the results of the field survey conducted by the investigation team, the papermaking sites in Zhejiang Province are mainly located in Fuyang District of Hangzhou, Lin'an City

Ping Paper Mill on Zhishan Mountain in Zeya Town of Wenzhou City

of Hangzhou, Fenghua District of Ningbo City, Longyou County of Quzhou City, and Zeya Town in Ouhai District of Wenzhou City. At the beginning of 2019, according to the statistics by Fuyang Bamboo Paper Industry Association, about 201 workshops or enterprises were distributed in diverse towns and communities in Fuyang District.

2. Diversity and richness of papermaking raw materials

According to the field survey conducted by the investigation team from July 2016 to April 2019, the raw materials for large quantities of handmade paper in Zhejiang Province are all kinds of bamboo. The most used ones are *Phyllostachys edulis*, *Phyllostachys*

Edgeworthia chrysantha Lindl. for papermaking in Likeng Village of Songyang Mountain area

heteroclada Oliver, *Pleioblastus amarus*, sympodial bamboo, and *Dianthus chinensis* L.. Bark is the second largely used raw material used in papermaking, including *Edgeworthia chrysantha* Lindl. bark, *Wikstroemia monnula* Hance bark, mulberry bark, paper mulberry bark, *Wikstroemia pilosa* Cheng bark and *Pteroceltis tatarinowii* Maxim. bark. Moreover, some raw materials such as straw, *Eulaliopsis binata*, hemp, and rattan are also used for papermaking. At the same time, mixed raw materials are used to make paper, for instance "calligraphy and painting" paper made from *Eulaliopsis binata* and other materials.

3. Diverse papermaking techniques as regional traditions

Zhejiang handmade papermaking process has distinct features of diversity and richness. For example, in terms of the pulping process, Zhejiang has employed different ways such as stone roller drawn by manpower, electric stone grinding and machine beating. In terms of papermaking methods, there are papermaking undertaken by one person, or two-person cooperation with a movable or fixed screen, with papermaking method employing pulp-shooting technique, or similar to Tibetan ways. It can be said that all contemporary papermaking methods in China are available in the region. Even for the way the local papermakers using the papermaking screen, the area employs

Scooping pulp four times to make original *Pleioblastus amarus* Wujin paper in Dayuan Town of Fuyang District

different methods: scooping and getting out of water for once as for Yuanshu paper, scooping for four times as for *Pleioblastus amarus* Wujin paper, or even scooping for a dozen times for making Taohua paper which made of mulberry bark, which is quite a special procedure even nationwide. In the fermentation process, unrine and soy bean milk are used in Fuyang District, which is special and demanding. It is also noteworthy that the process of papermaking in Fuyang District is different from that of other handmade papermaking areas in that plant mucilage is not added, which is a unique process in China.

4. Diverse paper types and uses

As one of the most important representative provinces of Chinese handmade papermaking, Zhejiang has a wide range of handmade varieties and uses, covering almost all the mainstream uses of ancient and modern handmade paper: for production or daily life, for calligraphy and painting, for printing or repairing, for folk or official contract, for government documents, for sacrificial purposes, or for rituals and festivals. Even in the early 1920s, local paper was widely used for newspaper printing, which was called the "local newspapers". In addition to paper with original color, a variety of plants, minerals and chemicals are used to dye paper of different colors, hence processed for art creation.

The use of handmade paper in Zhejiang covers writing and printing central or local government documents (Teng paper and Yuanshu paper), copying ancient works, calligraphy and painting (bamboo paper), making tissue paper, wax paper, gold and silver tin-foil (original Wujin paper and Luming paper) etc. In addition, paper is also used to copy family genealogy or anecdotes, or to make joss paper, or to copy scriptures, or for pasting windows, or for wrapping and so on. In history, handmade paper has penetrated widely and deeply into all aspects of production, daily life and cultural activities of the Zhejiang people and the influence could still be traced to today.

Bamboo paper in Taishun County for sacrificial purposes

2.2.3 Traditional Papermaking Mode Coexists with Modern Transformation Exploration

Through the field survey and local literature research conducted by the survey team from July 2016 to April 2019, it is found that in 42 handmade paper (tools) mills in Zhejiang Province, except for Luming paper in Shaoxing City which completely ceased production, there are 22 family mills engaged in production, and 19 engaged in production as enterprises. The handmade papermaking practice presents a coexistence of traditional agricultural economic production mode and modern commercial production mode, with traditional family-based production mode slightly dominant.

Zhejiang handmade papermaking has the typical characteristics of the combination of agricultural economy and modern management, which is mainly manifested as follows:

(1) Till the early 2019, one of the characteristics of Zhejiang hadmade papermaking industry is that individual household or small-size family mills dominated in the villages, mostly located in the mountainous areas. With only one or a couple of family members making paper accounts for a larger part. The paper mills are small in scale and low in output, mainly as low-end paper producers and suppliers of handmade paper. Most handmade papermakers are the owners of the mills or the key family members.

(2) Since 2005, the Ministry of Culture and Tourism (former the Ministry of Culture) and

Raw materials processing with the traditional papermaking techniques in a small paper mill—scraping the bamboo

Zhejiang Province have jointly promoted the comprehensive inheritance and protection of the intangible cultural heritage, and the folk and cultural circles have formed the "fashion" to restore ancient papermaking methods, especially for the Fuyang ancient Yuanshu paper. In order to control the technical process to ensure the quality of the products, the papermakers sticked to pure handmaking in every single step, hence increased the cost and energy invested, and decreased the output. Among them, Yiguzhai, the representative papermaking mill of Yuanshu paper in Fuyang District is one typical example pursuing the ancient method.

An attempt on the mode of "Industry-University-Research" of handmade paper in Fuyang District

(3) The paper products made by the mills have diverse sales channels. Currently, paper mills are scattered, and paper industry cluster could hardly be found in villages. The industry chain is often broken, especially for the local marketing section. Hence, the small-scaled paper mills have to adapt to the market and quickly correspond to the demands so as

An attempt on the mode of "Industry-University-Research" of handmade paper in Fuyang District

to survive.

Due to the geographical, cultural and economic advantages of Zhejiang Province, Zhejiang papermakers are quite flexible, and have formed the business-oriented traditions including seeking export for profit, small-scale production of diverse products, customized products for presents, production of auxiliary products, etc. If one section of the handmade paper market is challenged, other paper production sales will not be blocked totally, and will still enjoy a relatively stable market.

Compared with other provinces in China, the modern operation ideas of Zhejiang handmade papermaking have better compensated for the defects of agricultural handicraft workshops, and formed a benchmark in terms of production scale, product selection, innovation, sales system and operation management. The operation system is mutually connected and complementary with the traditional production of handmade paper.

2.2.4 Coexistence of Paper Factories Targeting Domestic Market and Export-oriented Paper Factories

The characteristic of Zhejiang handmade paper sales is that it has wide sales channels in history, especially the paper producing regions and large manufacturers have relatively stable markets.

1. The broad domestic market recorded in literature

In ancient times, there was no detailed documentary record. In terms of the modern literature, the data in the 1940 edition of *Zhejiang Paper Industry* stated that in the middle period of the Republican Era, Zhejiang handmade paper was widely sold. Handmade paper produced in Hangzhou City area, Jianggan District and Hushu District was sold to Tianjin City, Pukou area and Qingdao Ciity of Shandong Province, Jiangsu Province and other places; handmade paper produced in Fuyang area was sold to the counties in Zhejiang and Jiangsu Provinces; handmade paper produced in Yuhang area was sold to Jiangsu and Zhejiang ports; handmade paper produced in Ningbo City was sold to Hebei Province, Shandong Province, Dalian City, Jiangsu Province, Yanshan County of Jiangxi Province, Zhenhai area and other places; handmade paper produced in Haimen area of Linhai County was sold to the local north bank, and handmade paper produced in Yongjia County was sold to Xiannv Town,

Taizhou City and other places.

Fuyang is the most representative papermaking area. From the records of *Traditional Handmade Paper in Fuyang*, Fuyang handmade paper has always played an important role in the domestic paper market because of its large output, diverse varieties and wide sales. Before 1951, Fuyang handmade paper was purchased by the local paper merchants (paper shops), and then sold to paper merchants in Wuxi, Suzhou City, Shanghai City and other places, and then sold to Jiangsu Province, Shandong Province, Tianjin City, Beijing City, etc. Fuyang handmade paper was purchased by the local supply and marketing cooperatives. Almost all the paper was sold outside the county. It still enjoyed a broad market in China, involving 22 provinces, municipalities and autonomous regions such as Jiangsu, Anhui, Jiangxi, Guangdong, Shanxi, Shaanxi and Hubei, as well as local cities in Zhejiang, e.g. Hangzhou, Jiaxing, Huzhou, Ningbo and Shaoxing. Among them, bamboo toilet paper and joss paper for sacrificial purposes were mainly sold to Jiangsu, Shandong, Tibet, Heilongjiang and other provinces and autonomous regions; while the straw Kengbian paper was mostly sold to Shanghai City and Southern Jiangsu Province and Songjiang area of Jiangsu Province, and cities in Zhejiang Province, e.g. Hangzhou, Jiaxing, Huzhou, Ningbo, Shaoxing.

2. The export market reflected in the literature and field investigation

Still in the case of the former Fuyang County, Fuyang handmade paper began to have records of being exported to Japan and other countries since the period of the Republican Era, but the number was small. After the founding of the People's Republic of China, the export sales volume increased year by year. After 1960, Zhejiang government issued an annual export quota of about 3 000 tons to Fuyang handmade paper. According to *Traditional Handmade Paper in Fuyang* records: during the 13 years from 1965 to 1983, though the foreign trade (export) of handmade paper in Fuyang County was not stable, it had an overall growth trend. For example, in 1965, Fuyang County's handmade paper exports amounted to 550 tons, in 1983 it was 2 785 tons, and the highest year (1982) exported nearly 5 000 tons (4 990 tons). The target countries mainly include Japan, Singapore, Malaysia and other countries.

In the return visit survey in January 2019, the investigation team found that there are several handmade paper mills in Zhejiang Province that are completely export-oriented or mainly export-oriented.

Located in Lin'an City, Hangzhou Qianfo Paper Co., Ltd. takes window paper and calligraphy and painting paper as its main product, 95% of which are exported to Japan and Korea. The painting and calligraphy paper produced by the factory are all under the requirements of Japanese and Korean customers, so not sold in the domestic market.

Located in Dayuan Town of Fuyang District, Hangzhou Fuchunjiang Xuan Paper Co., Ltd. produces more than 50 000 dao handmade paper annually, with 80% of its products sold to Japan and South Korea.

In August 2016, Zhejiang Chengang Xuan Paper Co., Ltd., located in Longyou County, was also found by our investigation that the main body of the product was exported to Japan and South Korea.

Samples in the exhibition hall of bast paper in Longyou County

2.3 Challenges to the Inheritance and Development of Zhejiang Handmade Paper Industry

Since the reform and opening up, China has undergone tremendous changes, mainly from the agriculture-centered production form to the industrial production form; consumption habit turned from the rural life to the urbanized pattern. The channel of traditional retail sales has transformed into an e-commerce sales network. Today's handmade paper industry is indeed facing unprecedented shocks and challenges.

As a representative province of urbanization and industrialization in China, Zhejiang Province has built the most developed e-commerce system, which has indeed brought direct pressure on the survival and development of the very "classical" handmade paper industry.

2.3.1 Lack of Papermaking Inheritors, Decrease and Aging of the Papermakers

The lack of successors of traditional papermaking techniques is a serious problem in the inheritance of Zhejiang handmade paper.

Of 42 handmade papermaking mills and factories, the survey team found that, 29 papermaking mills and factories do not have inheritors, and only 9 have their children as inheritors.

Meanwhile, inheritance here includes participation in the operation of the company, not just the inheritance of papermaking techniques.

It is worth noting that the nine papermaking families in which children are involved are mostly registered companies. Of the 9 paper mills, 5 have inheritors who can really make paper.

On the other hand, in the 21st century, in the economically developed Zhejiang Province, the number of young people willing to engage in papermaking has fallen sharply, and the aging of papermakers has become a normal state.

For example, Datian Village in Chang'an Town of Fuyang District, has a long history of papermaking. In the heyday, more than 700 people were involved in papermaking. However, with the growing number of migrant workers to big cities, papermakers decreased, and fewer young people are willing to learn papermaking.

When the investigation team visited the village in 2016, it was found that only 8 to10 papermaking troughs were left in Datian Village, and the papermakers were basically in their 60s or 70s. The trend of aging was

A papermaker from Guizhou Province making paper in"Zhuxinzhai"paper mill

Weng Shige making paper at home all by himself

very serious, and papermaking has reached a point where it is difficult to continue.

Li Shengyu, in charge of Zhuxinzhai Yuanshu Paper Co., Ltd., in Huyuan Town of Fuyang District, also encountered a similar situation. Because of the low salary, long working hours and high labor intensity, it is difficult to find local young people who are willing to engage in handmade papermaking. Consequently, the workers in Li Shengyu's paper factory are all from other places, mainly from Jinping County of Qiandongnan Miao and Dong Autonomous Prefecture in Guizhou Province, and are all over 45 years old.

2.3.2 Due to the Impact of Machine-made Paper, the Phenomenon of "Cutting Corners" is Very Common in Order to Survive the Crisis

Of course, due to the strong impact of high efficiency and low cost of machine-made papermaking, as well as the changes of modern living style and working conditions, Zhejiang, like other provinces in China, has obviously reduced the sales of handmade paper and its cost advantage continuously declines, except for paper for specific purposes, such as bamboo paper in Fuyang and Fenghua Districts, which are specially used for the restoration of ancient works.
In order to maintain the operation of paper mills and reduce losses, it is common for paper mills to save labour and materials, and to reduce costs by using the shoddy raw materials and auxiliary materials.

In January 2019, the person in charge of "Zhuxinzhai" in Fuyang said: the minimum wage of paper workers is 200 yuan per day. There are six workers in the paper mill, including papermaking, paper-pressing and pulping workers. So the wage expenditure is 1200 yuan per day. While the cost of making a bundle of paper in the regular way is 450 yuan, and a bundle of paper can only be sold at 430 *yuan* in the current market, "So the business is at a loss".

Therefore, if papermakers do not want to lose money, they can only "cut corners", by, for example, mixing waste paper (0.8—1.0 *yuan* per kg) into bamboo pulp, unless they want to upgrade the paper quality or seek other specific uses.

In some family-centered handicraft workshops, in order to reduce wage cost, the couple would not hire workers, but strived to make paper all by themselves or even all by the husband, so as to maintain the paper mill. Weng Shige Paper Mill in Taishun County, Paper mill in Likeng Village of Songyang County, and Zhang Genshui Paper Mill in Fuyang District are some of the examples.

2.3.3 Handmade Papermaking is in Lack of Industry Standards

There is no uniform standard in the industry, which is also one of the difficulties encountered in the development of Zhejiang handmade paper.

Luo Xinxiang, a papermaker in Luocun Village of Fuyang District, said that at present, there is no uniform standard in many areas of the handmade paper industry, so the chaos has brought many problems.

Luo Xinxiang gave one example: because the government does not set a unified standard for the paper used for the printing and restoration of ancient works, many paper mills sell the paper processed by caustic soda to the public and private institutions for the printing and repair of ancient works. On the one hand, this practice hurts the paper mills making quality paper sticking to the ancient methods. On the other hand, using such flawed paper for preservation would do fatal damage to the ancient works.

Another example is the toilet paper produced in Zhishan Mountain area of Wenzhou City, which was once livelihood commodity in large demand. Then the purchase price from other places for a piece of paper was 10 *yuan*, while the local price was 8.5 *yuan*.

Thus, local supply and marketing cooperatives in Ouhai area set up toilet paper checkpoints at every intersection of Zhishan Mountain area in Zeya Town to prevent outside companies from purchasing toilet paper from Zeya. Zeya's toilet paper then formed a special sales way: first transported to local grocery shops or supply and marketing cooperatives, and then wholesaled to the regional villages and communities for sale.

After 1987, Wenzhou private economy and commercial circulation industry rose. The Zeya local government saw that the sales of toilet paper was disorderly, the paper specifications were not uniform, the quality was not guaranteed, and the tax revenue was lost. Therefore, the government united the supply and marketing cooperatives, toilet paper companies and private papermakers association to purchase toilet paper and then sold to private merchants.

From the second half of 1987, papermakers

Ping paper of poor quality made with cement bag paper added in the raw materials in Zeya Town

from villages first delivered goods by car to the local headquarters; from 1988, they directly delivered goods to the supply and marketing cooperative stores in Nantong City of Jiangsu Province and other places; and from 1989, they began to sell directly to big merchants in towns and markets.

As the cost of toilet paper increases along the hierarchy of sales chain, the size of the toilet paper becomes narrower and narrower, the thickness of the paper also becoming increasingly thinner, and even adulteration is unscrupulously practiced in the field. Consequently, the machine-made toilet paper replaced the handmade paper in the market relentlessly.

2.4 Representative Folk Customs and Cultural Memories of Handmade Paper in Zhejiang Province

The paper industry in Zhejiang Province has a long history and can be traced back to the Eastern Jin Dynasty according to the existing historical records. From the Eastern Jin Dynasty, about 1 700 years ago, to the Tang Dynasty, more than 1 300 years ago, famous paper represented by Teng paper in Shanxi area and Teng paper in Quancun area had been popular throughout the country, and was famous nationwide. In the development of papermaking techniques in the past 1 700 years, there are abundant folk stories and papermaking culture with regional characteristics in Zhejiang Province. Based on the typical characteristics and value of legends, the important cases can be listed as follows.

2.4.1 A Distinctive Paper Cultural Zone: Zhishan Mountain in Wenzhou City

1. Source of Zhishan Mountain in Wenzhou City

As the name implies, Zhishan Mountain (paper mountain) refers to the specific mountainous area in which generations and thousands of families are mainly engaged in papermaking practice. "Zhishan Mountain in Wenzhou City"is not a historical concept, but a concept of cultural geography with modern connotations, specifically referring to the entire territory of Zeya Town of Ouhai District and Huling Town of Rui'an City (a county-level city under the jurisdiction of Wenzhou City) as well as the traditional handmade papermaking area located in Quxi Town of Ouhai District and in Tengqiao Town of Lucheng District.

"Zhishan Mountain in Wenzhou City" is not named by the local papermakers to their own production and living environment, nor a name given by the literati, but a jargon used by guerrilla groups belonging to the Communist Party of China in the southern Zhejiang during the Republican Era. The "mountain" is located in Zeya Town of Ouhai District, which refers to the papermaking mountainous area enclosed by the two small mountains of Qiyun Mountain and Lingyun Mountain in Zeya Town. It does not include the traditional papermaking mountainous areas located in the southern Yandang Mountain of Pingyang County, Taishun County and Wencheng County in Wenzhou City. In 1939, the "Zhishan Mountain Uprising" incident, which was launched by Wenzhou papermakers, shocked the country, and the term "Zhishan Mountain" was widely spread ever since.

Paper Mill on Zhishan Mountain in Zeya Town

2. The Papermaking Case

In the late period of the Anti-Japanese War from 1937 to 1945, the shipping roads from Wenzhou City to Shanghai City were interrupted, and the paper transportation and sales agency in Yongjia County and the Yongrui Handmade Paper Production Cooperatives monopolized the market. Because the sales channel was damaged, the paper farmers could not get their full payment for a long time, and the purchase of handmade paper was even once stopped. In July 1939, the papermakers went to the Quxi Paper Shop, a famous paper trading center in Quxi Ancient Street in Wenzhou City, asking for payment in arrear and resuming paper purchase. There was a fierce conflict between the two sides in which a papermaker was killed. The person in charge of the paper transportation and sales agency of Yongjia County was brought into the Zhishan Mountain area as a hostage by the angry papermakers. After the multi-party negotiation, the cooperative paid off and resumed the free trade of handmade paper. This movement was called the "Papermaking Case" and the media called it the "Zhishan Mountain Uprising". The dominant fighting force of the "papermaking case" was the ordinary papermakers in the core areas of Zhishan Mountain in Zeya Town of Ouhai District and Huling Town of Rui'an City. Since then, "Zhishan Mountain" and "Papermaking Case" have been linked together.

3. *Zhishan Mountain Defence War*

Zhishan Mountain Defence War is a film written, directed and acted by Wenzhou people, reflecting the traditional papermaking techniques and protection of Wenzhou culture. *Zhishan Mountain Defence War* was jointly produced by Wenzhou Tianyun Zongheng Media Co., Ltd. and Wenzhou Ouhua Media Co., Ltd. During the Anti-Japanese War, a young Japanese scholar named Watanabe served in the Japanese army. After the occupation of the area, he ordered a military officer named Okamura taking advantage of the military to get "ancient technique of papermaking employing hydraulic pestle" in Zeya Zhishan mountain area and the "wooden movable type printing techniques " in Dongyuan Village of Rui'an County, but the Japanese army was met with the stubborn resistance of the Chinese local guerrillas, and finally was defeated. In this case, Wenzhou's traditional papermaking process was well protected. The film was shot in Fanshan area of Wenzhou City in the second half of 2016.

Wood Movable Type Heritage Museum in Dongyuan Village of Rui'an City

2.4.2 Anecdotes Related to Papermaking

1. Ragged Landlore with 48 Papermaking Steamers—Shi Yaocheng

During the investigation, Zhu Zhonghua from Fuyang Yiguzhai Yuanshu Paper Co., Ltd. related a story: In the history of papermaking in Fuyang County, the construction of a papermaking steamer (also known as Zengguo) was a big event in a village or even needed cooperation of several villages. Considerable investment and manpower were required. Generally, only rich families would afford one. Poor papermakers usually had to jointly build one steamer, or rent one to steam bamboo. About in the Qing Dynasty (unclear which year it was), a papermaker named Shi Yaocheng in Fuyang County unexpectedly had 48 papermaking steamers. One day, when the county magistrate of Fuyang was inspecting in the countryside, he saw the mark of Shi's family in the tens of kilometers of bamboo in the county (it's a custom of Fuyang County that in order to avoid disputes caused by chopping others' bamboo, the owners would inscribe the name the owner or other marks on each bamboo of the family). He was so surprised that he decided to go to Shi's house to find out what the rich papermaker was like.

Bamboo forest marked with the name of its household in Caijiawu Village of Fuyang District during the investigation

When the county magistrate was led to the residence of Shi Yaocheng, he saw rows of steamers. A ragged old man was tending the fire by a steamer. The county magistrate went forward and asked him where Shi Yaocheng was. The old man answered he was the person whom the magistrate was looking for. The county magistrate was furious, and summoned Shi Yaocheng to Yamen (government office in ancient China). Shi Yaocheng was puzzled and said: "I did not break the law!" The county magistrate said: "You are already so rich, but you were still tending the fire by yourself. You have everything and how can others make a living? You should be punished!" So the famous landlord was beaten and embarrassed.

2. Transport and Sale Landlord—Smart Millionaire Li

Now all the bamboo paper made in Datian Village of Chang'an Town in Fuyang District is sold to Jiangsu Province, but a few years ago, the paper made in the village was sold to Huzhou area (based on Li Shenyan's relation, who is a villager of Datian Village of Fuyang). In the early period of the Republican Era, a merchant in Huzhou, specialized in handmade paper trade. Finding the high cost performance of the bamboo paper made in Datian Village, he opened a store there to purchase paper from the villagers. All the paper made in the village was transported to Huzhou, and then to Jiangsu through the waterway by this businessman. It solved the problem of sale of the local paper, and the businessman also made a fortune. Because the businessman's surname was Li, and the villagers thought he was very rich, they called him "the Millionaire Li". As time went by, he became famous in paper trade.

3. "Papermaking Champion" She Fujin

Papermaking in Zhejiang has a history of about 1 700 years with countless famous paper. Naturally, there is no shortage of "paper champions" with the leading papermaking technology. This story is about one of the champions named She Fujin from Fuyang in the Republican Era. According to the story told by Yu Renshui, a papermaker from Zhaoji Village in Dayuan Town, there was a papermaking master named She Fujin in Zhaoji Village. In 1945, a papermaker named Tang Baoshan cut a batch of tender bamboo at Grain Full. After processing, it was transported back to the workshop to finish the papermaking. There was always water left on the wet paper piles, forming bubbles. Dozens of papermakers tried to solve the problem but failed, which greatly frustrated Tang Baoshan.

One day, Tang Baoshan, who was sighing all the time, met the 31-year-old "papermaking champion" She Fujin at the Dayuan Street. He rushed to ask She Fujin for advice. She Fujin said that he could have a try. Tang Baoshan, who was overjoyed, said quickly that as long as She Fujin was willing to have a try, he would pick up She Fujin with a palanquin. In the end, he did what he promised to pick up She Fujin. On the first day, She Fujin went to the mill and did nothing. He just looked at the two papermaking boards and told Tang Baoshan to do another piece as soon as possible. He changed the original 500 sheets of wet paper a pile into 250 sheets a pile. He also asked the papermakers scoop slowly to make the wet paper evener. Sure enough, after the seemingly effortless instruction of She Fujin, the problem of wet paper bubbles was solved, and the paper made turned out to be better. She Fujin was known as the "papermaking champion" in Fuyang. Later, he was invited to Lin'an County as a master papermaker.

2.4.3 Legends of Papermaking in Caijiawu Village

Caijiawu Village is located in Lingqiao Town of Fuyang District, in which there are several legends related to papermaking.

1. The Country Legend of Inventing the "Hanging Screen"

Formerly when making paper, a worker hade to support and lift the papermaking screen by two hands. Since the process needed to be repeated continuously, it was laborious for workers. After this process was done, the screen frame must be placed in the corner in order to pick up the papermaking screen to separate and pile the wet paper. Later, it saved a lot of effort with the hanging screen, but who invented the hanging screen and when was it invented?

According to the papermaking households in Caijiawu Village, the hanging screen was invented by a highly skilled papermaker in a village of Lingqiao Town during the Republican Era. From then on, the papermakers saved the procedure to place the screen frame.

Wooden frame hanging above the papermaking trough in Fuyang District

2. Local Legend of "Wooden Hammer" in Fuyang District

It is said that in the room for drying paper, since the pressed paper sheets were not easy to separate, there was a cloth bag with grass ash on the top of the paper block. Workers hit the bag and sprinkling a little ash on the paper after separating one piece. Hence, the pieces of paper could be separated smoothly.

One day, a papermaker chatted with a stranger while drying paper. He forgot to hit the bag, and failed to separate paper. Seeing this, the stranger said: "You go to find a sheep's hoof." Then he left. The papermaker was puzzled: "Do I have to find a real sheep's hoof?" He thought it was not feasible, so he made a wooden hoe in the shape of a sheep's hoof. Then the paper layers can be easily separated with a hoe scraping on the paper. Because the shape of the hoe is a bit like a goose head, the local papermakers call the wooden hammer "goose hoe".

3. "Heating Process Should be Done in Summer, and Cooling Process Done in Winter"

There is an old saying in Fuyang that "Heating process should be done in summer, and cooling process done in winter". In order to make best bamboo paper, the processes, such as fermenting and boiling, involving fire or sunshine should be done in summer, in which way bamboo material would be processed easily in hot days; the processes, such as cleaning and papermaking, involving water should be carried out in winter, in which way bamboo paper would be more compact and smooth in the the low water temperature. So what do papermakers do between cold winters and hot summers? Stacking the materials is

Using the wooden hammer to separate the paper layers

Bamboo soaking pools in Xin'er Village

mainly completed in the half year between the fermenting and the cleaning steps of making bamboo paper in Fuyang District.

2.4.4 Sacrificial Ceremonies by Papermakers in Different Villages Of Fuyang District

Sacrificial ceremony is an important part of the papermaking culture. The tradition of papermakers in Zhejiang region is to hold different forms of rituals related to papermaking at different time points each year:

(1) According to Li Wende of Fuyang Dazhuyuan Xuan Paper Co., Ltd., in the 1960s and 1970s, members of the production team cut bamboo all together in the mountain in Xin'er Village of Huyuan Town in Fuyang District. There was a mountain opening ceremony on the day by offering sacrifices such as pig head and fish to the ancestors, accompanying with gongs and drums. In this way, they expressed their gratitude to their ancestors for bringing good craftsmanship to the younger generations. They also worshiped the mountain god, hoping everything went well and that business became prosperous. At the beginning of the 21st Century, the custom has been changed to treating papermakers with a feast before departing for the mountain.

(2) Zhu Zhonghua of Fuyang Yiguzhai Yuanshu Paper Co., Ltd. told a different story: the custom of sacrificing is popular in the history of Dayuan Town in Fuyang District. Dayuan Town pays attention to people in the rirual, and the person who presides over the mountain opening ceremony must be a man or owner of the mill recognized by the village as having good fortune and charisma. However, it is very strange that the scripture must be chanted by women. When the members of the investigation team wanted to get further details, Zhu Zhonghua did not know the reason either.

(3) The opening ceremony of Caijiawu Village in Lingqiao Town of Fuyang District is not the same: generally, three days before Grain Full (usually one day between May 20 and 22), the bamboo is cut. In the 1960s and 1970s, there was a tradition that before cutting the bamboo, all the villagers should dine together. Before the meal, they would pray to Guanyin Bodhisattva for protecting workers in the mountain. The rituals and banquets were held at the big house. Usually a dozen tables were set, serving more than 100 people in the village. And the production team paid for the treatment.

2.4.5 Special Taboos in Papermaking

In fact, in the over one-thousand-year history of papermaking in Zhejiang, due to technical secrecy and beliefs, taboos are diverse, and almost every papermaking area has its own taboos. Of course, there are many taboos in common because of similar technical process. According to the taboos collected by the investigation team, 5 cases will be introduced in the following.

1. Women Should Not Visit the Place Where Paper is Made

According to Li Wende of Fuyang Dazhuyuan Xuan Paper Co., Ltd., the traditional custom of Huyuan Town was that women were not allowed to visit papermaking sites before the mountains and papermaking troughs were distributed to different families in the early 1980s. Regarding the origin of this taboo, Li Wende speculated that because people were generally poor at that time, and would not afford new clothes. So when men were cleaning the materials, being afraid to damage clothes, men would take off their clothes as much as possible. Or when transporting

Wokers fermenting the materials by a lime pool

fermented bamboo bundles for steaming, men wore less clothes; drying paper must be done in high temperature, and men basically didn't wear tops; when pressing wet paper, men generally didn't wear tops and sat on the press bar, for the task was very labor-intensive. Due to these reasons, women were not allowed to visit papermaking sites.

2. Women Should Not Go to the Lime Pool

The emphasis of the custom of Datian Village is different from that of Xin'er Village in Huyuan Town. According to Li Shenyan, a paper mill owner, the tradition was that women cannot go to the lime pool during the fermenting process. The villagers believed that once this tradition was violated, the paper would not be made well. Li Shenyan believed that there were two reasons: first, when men were doing the task, they must do their best and work together to complete the work. A person by the lime pool would interfere with their work, especially a woman, who would distract workers' attention. Second, the temperature near the lime pool was very high, and it's a laborious work, so most of the men only wore underpants. Women could not help with the work, while the scene would be embarrassing to both men and women.

3. Women were Prohibited from Passing Through the Soaking Pool

Zhuang Fuquan from Hangzhou Fuchunjiang Xuan Paper Co., Ltd., located in the Zhuangjia Natural Village of Datong Administrative Village in Dayuan Town of Fuyang District in Hangzhou City, said that the ancient people were superstitious so they thought that since women got monthly period, they were not clean. If a woman who was having a period walked around the soaking pool, the paper would be produced with worse quality. Therefore, before abandoning superstition (in the mid 1950s), the pool would be required to be closed for 24 hours in advance to avoid women walking through when the paper would be made in the village.

4. Three Generations of Papermaking Families Who used to Steal Raw Materials were Not Allowed to Build Papermaking Troughs

According to Zhu Zhonghua, there were taboos in the history of Fuyang area that three generations of households who stealing raw materials should not build papermaking troughs, and that the family cutting others' bamboo secretly could not build papermaking troughs for one generation either. This rule

Legendary small papermaking village of the tribute paper in Kaihua County

had rarely been violated in the past century. About 150 years ago, a person whose family name is Xu went to Songjia Village to steal bamboo, but was discovered by the villagers. The villagers in Songjia Village asked a group of prestigious people, such as the Baozhang (officer of the village), to demand the raw materials back. Moreover, the person who had stolen and caught on the spot wrote a guarantee that three generations of his family would not build papermaking troughs. It is said that until the grandson of Mr. Xu married a girl from Songjia Village, the guarantee written by Mr. Xu was returned to the Xu family.

5. Secrets of Making Tribute Paper Should be Kept Confidential

Kaihua paper of the Ming and Qing Dynasties is very famous, which was the most famous tribute paper in the late Ming Dynasty to the middle of the Qing Dynasty. Others could not have the access to the papermaking tips for its top quality and the highly classified information on its materials and making techniques. According to the description in Guo Chu written by Sun Hongqi, the techniques of making paper must be kept as a secret. In the village producing the tribute paper, it was strictly required that only sons could inherit the techniques. If there was only a daughter in the family, it was possible to choose grandsons as the successors, rather than the son-in-law, even if he was taken in to bear his wife's family name. Kaihua paper did such a great job in preventing the techniques from leaking outside that even the raw materials of the tribute paper could not be ascertained when Kaihua County was trying to resume Kaihua papermaking.

2.4.6 A Long Papermaking Song Named *Zhu San and Liu Erjie*

The long narrative song *Zhu San and Liu Erjie*, which was widely circulated in Fuyang area, was created and sung among papermakers in the handmade paper industry, showing the ups and downs of Fuyang handmade papermaking practice.

The slang "papermaking condition is harsh, and papermakers enjoy rough talk" is popular in Fuyang papermaking area. People in the town worked hard all day long with repeated and boring papermaking procedures. In order to eliminate the fatigue and make the boring work more interesting, paper makers liked to use jingles, witticisms, and ironies, while local scholars created jokes, stories and songs based on those funny words.

It is said that the famous song *Zhu San and Liu Erjie* was created collectively by papermakers in Fuyang area, whose rhythm is very consistent with the process of making paper. The song was originally inspired by the story that Liu Erjie of Yuhang area eloped with Zhu San of Suzhou area during the Hongzhi Reign of the Ming Dynasty. At first, the story was spread among papermakers, and later it was adapted into local dramas, such as Hu opera and Yue opera.

No later than the end of the Ming Dynasty, the folk song *Zhu San and Liu Erjie* was sung in the Fuyang area. In the Qing Dynasty, more printed versions of *Zhu San and Liu Erjie* appeared. Hulu Bridge in Yuhang County, which is bordered by Fuyang area, is the starting point for the Fuyang handmade paper to be transported to Jiangsu region, Shanghai and other places. According to the local old residents, Fuyang papermakers often shouldered handmade paper to Yuhang area in groups and skillful papermakers from Fuyang also helped to make paper in Yuhang at that time. As a result, the romantic stories between Fuyang men and Yuhang women happened all the time, which was the basis of long songs. With the marketing of Fuyang handmade paper, long songs were also spreaded to the northern Zhejiang Province and southern Jiangsu Province, following the footsteps of paper merchants, becoming the popular narrative song created by workers in the paper industry in Jiangnan area.

According to the old people, singing *Zhu San and Liu Erjie* became fashionable in Fuyang in the 1920s and 1930s. There are still Fuyang folk songs related to *Zhu San and Liu Erjie*, describing and confirming how popular the long song was at that time. For example, "Do not sing *Zhusan Song* during papermaking, because the more you sing, the more you feel toilsome; do not sing *Zhusan Song* while working in the field, because the seeds will not germinate." It shows that people sang *Zhusan Song* in the past in order to relieve fatigue and make physical labor less boring. Another example is "if a man sings *Zhusan Song* well, he can marry a women without bride-price". It shows that young men expressed their love to the beloved girls by singing *Zhusan Song*. The girls were so deeply moved by Zhu San wooing Liu Erjie that they would just pursue love even without bride-price.

A photo of collecting information of the folk song *Zhu San and Liu Erjie* in the 1980s

3 Current Protection, Inheritance and Researches of Handmade Paper in Zhejiang Province

3.1 Inheritance and Protection of Handmade Papermaking in Zhejiang Province

3.1.1 Policy-driven Practices, Intentions and Effects

The handmade papermaking practice in Zhejiang Province enjoys a history of 1,700 years. Doubtlessly there have been various supportive and protective forces contributing to its inheritance and protection, from the government to the public, from the royal family to the clan. For example, the famous tribute paper system in royal family and government, the routine revision of genealogy among the clan, the official school and the large-scale book printing of booksellers are all the supportive factors in developing the handmade paper.

Genealogy in the late Qing Dynasty printed on *Dianthus chinensis* L. paper in Pingshui Town of Shaoxing City

However, the inheritance and protection of papermaking techniques are made top priority due to the large-scale industrialization, urbanization, and information age. Because the new era has spurred a strong impact on the traditional agricultural, local, and writing civilization, and the traditional papermaking techniques and consumption customs have been replaced by new ones largely, facing the crisis of extinction. The vigorous skills and industry in the past have become an "intangible cultural heritage", so protection and inheritance have been quickly put on the agenda.

The change initially originated from the systematic input of the western science and technology in the late Qing Dynasty and the Republican Era, and the dramatic revolutions came from the intense industrialization and urbanization that began in the 1980s and 1990s. After the first 20 years of the beginning of the 21st Century, combined with information and network, the intensity of changes is further enhanced. Therefore, the inheritance and protection work carried out in Zhejiang Province mentioned in this section mainly refers to the protection and promotion measures that began in the 1990s.

After the founding of the People's Republic of China in 1949, a series of measures were adopted to restore the people's handicraft industry and revitalize the country's handmade papermaking industry.

In the spring of 1950, the Industrial and Mining Department of Zhejiang Government (later reformed to the Industrial Department) established the Rural Handicraft Reconstruction Center (reformed to the Industrial Improvement Institute in July 1950), taking handmade paper recovery and development as one of the main tasks, and setting up offices in Fuyang, Lin'an, Quzhou, etc. The provincial government required that the handmade papermaking industry should be improved, focusing on the following tasks: to provide loans for bamboo cutting and fermenting, to help the troubled handmade paper factories to resume production; to help the joss paper mills in stagnation to tranfer to bamboo pulp production, and support the supply of raw materials for machine-made paper industry in Shanghai, Jiaxing and other places. The above measures effectively promoted the industrial recovery in the handmade paper central area in Zhejiang Province.

Since then, the government has organized scattered and weak papermakers and households to jointly establish papermaking cooperative, use the supply and marketing cooperative system in all counties, districts and towns, establish a unified purchase and marketing system, and adopt the collective papermaking mode by the production team and so on. Until the early 1980s, the tide of marketization began, so the forest resources and papermaking facilities were allocated to the farmers, and the unified purchase and marketing system of supply and marketing cooperatives withdrew from the historical stage. At the same time, the upsurge of industrialization has been launched, and continues ever since.

The protection of folk crafts and intangible cultural heritage in Zhejiang Province is at the forefront of China, as evidenced by the promulgation and implementation of relevant regulations and policies.

In 2004, the government in Zhejiang Province promulgated the Notice on *Strengthening the Protection of Ethnic Folk Art* and began to focus on the protection of intangible cultural heritage. At that time, the inheritance and protection of the intangible cultural heritage (the name at the national level was national folk culture, and the competent authority was the National and Folk Cultural Center of the Ministry of Culture) led by the Ministry of Culture on behalf of the state were also at the tentative stage.

On June 10, 2006, Zhejiang government released the *Opinions on Further Strengthening the Protection of Cultural Heritage*, requiring that the issue of cultural heritage protection and ecological environment should be fully considered in the reconstruction of countryside.

On March 15, 2007, the Department of Culture in Zhejiang Province formulated and promulgated the *Evaluation Standards of the Intangible Cultural Heritage of Zhejiang Province*. On December 4, 2007, the Department of Culture in Zhejiang Province issued the *Notice from Zhejiang Department of Culture on the First Batch of Inheritors of the Intangible Cultural Heritage in Zhejiang Province*. The first batch of 270 intangible cultural heritage representative inheritors in Zhejiang Province had been acertained, among which handmade paper craftsman Zhuang Fuquan and Li Fa'er in Fuyang area were selected as the representative inheritors of bamboo papermaking techniques.

浙江省文化厅关于印发《浙江省非物质文化遗产名录评审工作规则》（试行）的通知

发布时间：2015-07-29　　点击数：1837

各市、县（市、区）文化广电新闻出版局：

为建立和完善评审工作机制，进一步规范我省非物质文化遗产名录评审工作，根据《国务院办公厅关于加强我国非物质文化遗产保护工作的意见》（国办发[2005]18号）、《浙江省政府关于进一步加强文化遗产保护的意见》精神，特制订《浙江省非物质文化遗产名录评审工作规则》（试行），现印发给你们，请参照执行。

二〇〇七年三月十五日

附件

浙江省非物质文化遗产名录
评审工作规则（试行）

第一条　为建立和完善评审工作机制，进一步规范我省非物质文化遗产名录评审工作，根据《国务院办公厅关于加强我国非物质文化遗产保护工作的意见》（国办发[2005]18号）精神，特制定本规则。

第二条　省文化厅设立浙江省非物质文化遗产保护工作专家库，评审将根据每次工作需要，从专家库中分类抽选专家，组成省非物质文化遗产名录评审委员会（以下简称"评审委员会"），负责对申报浙江省非物质文化遗产保护名录项目进行评审，提出浙江省非物质文化遗产名录推荐名单。

第三条　评审委员会设主任委员一名，副主任委员、委员若干，秘书长一名，实行主任委员负责制，副主任委员协助主任委员。主任委员由文化厅主要领导担任，副主任委员由省文化厅分管领导担任，委员及办公室主任由省文化厅有关职能部门负责同志和相关领域的专家组成，经随机抽选并报经主任委员同意后聘任，聘期自该次评审工作结束时中止。

评审委员会下设省民族民间艺术保护工程办公室（以下简称"评审办公室"），具体负责评审委员会日常事务。

第四条　入选评审委员会委员应具备以下条件：

（一）从事文化遗产保护工作，具有相当职务和经历，实践经验丰富；

（二）从事文化遗产保护研究，一般应具有高级专业技术职务；

（三）热爱文化遗产保护事业，遵纪守法，具有良好的职业道德；

（四）专家须以独立身份参加浙江省非物质文化遗产名录项目评审工作，并愿意接受省文化厅的监督与管理。

Work plan on *Evaluation Standards of Intangible Cultural Heritage List in Zhejiang Province* (screenshot of a web page)

On April 1, 2008, the official website of the Department of Culture in Zhejiang Province issued an announcement of the representative inheritors of Zhejiang national intangible cultural heritage projects. According to the document, the Ministry of Culture announced the first and second batch of representative inheritors of the national intangible cultural heritage projects in June 2007 and January 2008 (the first batch includes five categories: folk literature, acrobatics and competition, folk art, traditional craftsmanship, and traditional medicine; the second batch also includes five categories: folk music, folk dance, traditional drama, Quyi, and folklore). There are 777 representative inheritors of the two batches of national intangible cultural heritage projects, and 43 are from Zhejiang Province. Among them, Zhuang Fuquan in Fuyang County was listed as a representative inheritor of the national intangible cultural heritage project of bamboo papermaking techniques.

At the end of 2007 and the beginning of 2008, Zhejiang government built Research and Protection Base for the Intangible Cultural Heritage in six universities, i.e. Zhejiang University, China Academy of Art, Zhejiang Normal University, Zhejiang Media College, Hangzhou Normal University and Zhejiang Art Vocational College. The purpose is to take advantages of the universities to protect and work on the intangible cultural heritage.

On June 16, 2008, in order to scientifically and effectively preserve the intangible cultural heritage in Zhejiang Province and make rational use of it, the government decided to carry out pilot work on the construction of intangible cultural heritage database and distribution map, and explore the path to maximizc the experiences.

On June 2, 2008, in order to thoroughly implement the protection principles of "protection-oriented, rescue first, rational utilization, inheritance and development" of intangible cultural heritage, and implement the regulation in article 24 of the *Regulations on the Protection of Intangible Cultural Heritage of Zhejiang Province* so as to "encourage and support educational institutions to incorporate intangible cultural heritage in their syllabus, establish heritage teaching bases, and cultivate intangible cultural heritage talents", and give full play to the positive role of various schools in the education and inheritance of intangible cultural heritage, explore new mechanisms of inheritance, and promote the protection and inheritance of intangible cultural heritage in Zhejiang Province. The government encouraged the application and nomination of the intangible

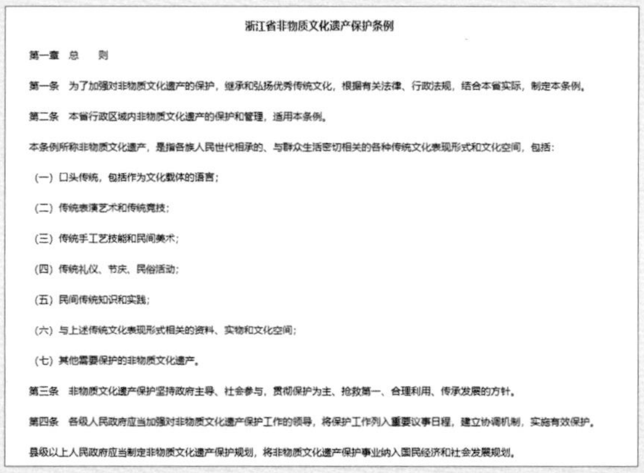

浙江省非物质文化遗产保护条例

第一章 总 则

第一条 为了加强对非物质文化遗产的保护，继承和弘扬优秀传统文化，根据有关法律、行政法规，结合本省实际，制定本条例。

第二条 本省行政区域内非物质文化遗产的保护和管理，适用本条例。

本条例所称非物质文化遗产，是指各族人民世代相承的、与群众生活密切相关的各种传统文化表现形式和文化空间，包括：

（一）口头传统，包括作为文化载体的语言；

（二）传统表演艺术和传统竞技；

（三）传统手工艺技能和民间美术；

（四）传统礼仪、节庆、民俗活动；

（五）民间传统知识和实践；

（六）与上述传统文化表现形式相关的资料、实物和文化空间；

（七）其他需要保护的非物质文化遗产。

第三条 非物质文化遗产保护坚持政府主导、社会参与，贯彻保护为主、抢救第一、合理利用、传承发展的方针。

第四条 各级人民政府应当加强对非物质文化遗产保护工作的领导，将保护工作列入重要议事日程，建立协调机制，实施有效保护。

县级以上人民政府应当制定非物质文化遗产保护规划，将非物质文化遗产保护事业纳入国民经济和社会发展规划。

Work plan on *Evaluation Standards of Intangible Cultural Heritage List in Zhejiang Province* (screenshot of a web page)

cultural heritage teaching base. Eligible higher education institutions, secondary vocational technical schools and primary and secondary schools in Zhejiang Province were all involved.

On September 17, 2009, the Department of Culture in Zhejiang Province launched the "First Chinese (Zhejiang) Intangible Cultural Heritage Expo and the 6th China Time-honored Quality Goods Expo", and held the street performance of intangible cultural heritage programs in Zhejiang Province, the activities harboring representative inheritors of intangible cultural heritage, masters of arts and crafts, etc. All these activities held in Hangzhou Oriental Cultural Park.

In April 2009, the Department of Culture in Zhejiang Province, together with *Zhejiang Daily, Qianjiang Evening News*, *Today's Zhejiang and Zhejiang Voice*, jointly organized 10 new discoverys activity to survey the intangible cultural heritage in Zhejiang Province. Different places in Zhejiang actively participated in the competition. On June 7, 2009, 10 items such as "foot-binding bitterness" were selected as ten new discoveries of intangible cultural heritage in Zhejiang Province, and 10 items such as "Indigo Print" were believed to be competitive. Among them, "Shangyu Wujin Papermaking Techniques", mainly distributed in Cailin Village of Shangyu City, was listed as the ten new discoveries in the survey of intangible cultural heritage in Zhejiang Province and was included in the final evaluation project.

On September 16, 2014, the Department of Culture in Zhejiang Province issued the *Guidelines from the Department of Culture on Strengthening the Productive Protection of Intangible Cultural Heritage in Zhejiang Province*. The file pointed out that it would focus on five promotions, one is to promote the inheritance mode reform of traditional handicrafts, training more inheritors, and laying a sustainable talent base for the protection of intangible cultural heritage. The second is to promote the integration and development of the protection of intangible cultural heritage with tourism, cultural creativity, design services and other related industries, and exploit the practical value of intangible cultural heritage. The third is to promote the in-depth exploration of resources of intangible cultural heritage, to shape the regional cultural image, to highlight the urban cultural characteristics and the charm of beautiful countryside. The fourth is to promote intangible cultural heritage to more closely integrate into the production and life of the people and to meet the moral and aesthetic needs of the people. The fifth is to promote the combination of intangible cultural heritage and people's daily life, promoting cultural consumption, expand employment opportunities, contributing to economic restructuring and industrial transformation and upgrading. The protection level of intangible cultural heritage in Zhejiang Province should be impoved.

In 2015, the "Bamboo Papermaking Techniques (Zeya Ping Papermaking Techniques)" submitted by the Ouhai District Cultural Center in Wenzhou City of Zhejiang Province, was listed as the fourth batch of national intangible cultural heritage representative projects and the District Cultural Center served as the supervisor.

In 2011, Longyou Bast Papermaking Technique was admitted in the third batch of national intangible cultural heritage protection list. In December 2012, Li Fa'er from Fuyang District of Hangzhou City and Wan Aizhu from Quzhou City, were selected as the representative successors of the fourth batch of national intangible cultural heritage for Fuyang Bamboo Papermaking Techniques and Longyou Bast Papermaking Techniques. In May 2018, Lin Zhiwen from Ouhai District of Wenzhou City was selected as the representative inheritor of Zeya Ping Papermaking Techniques in the fifth batch of national intangible cultural heritage.

In addition to the representative inheritors of national intangible cultural heritage, in October 2013, Cai Yuhua, the inheritor of Fuyang Bamboo Papermaking Techniques from Hangzhou City, and Yin Shoulian, the inheritor of Nanping Papermaking

Techniques from Rui'an City, were selected in the fourth batch of representative inheritor of Zhejiang Intangible Cultural Heritage Project. In November 2017, Chen Xudong, the inheritor of Qianhong Taohua and Xuan Papermaking Techniques in Lin'an District of Hangzhou City, were selected as the representative inheritor in the fifth batch of Zhejiang Intangible Cultural Heritage Project.

3.1.2 Symbolic Achievements of the Inheritance and Protection of Handmade Papermaking in Zhejiang Province

Zhejiang Province has made rich achievements in the inheritance and protection of contemporary handmade papermaking techniques. According to the characteristics of the major papermaking areas in the province, the investigation team selected its landmark achievements as follows:

1. Cultural Display Center of Zeya Ping Papermaking Techniques in Wenzhou City

After the completion of the Zeya Reservoir in 1998, considering the quality of the water source, the government strictly controlled the development of the paper industry in the upstream of the reservoir, and did not allow papermakers to switch to machine-made papermaking. However, at that time, Zeya Ping paper has been employing machines to improve efficiency of production. The new regulations had a strong impact on the papermaking practice. A number of papermakers left home to seek a living out in the city, and papermakers who stayed for a variety of reasons could not find other livelihood at that moment, so they could only continue traditional handmade papermaking to make a living using the traditional tools.

Since about 2000, Ouhai local government launched the "Zhishan Mountain Culture" brand building, starting the systematic protecting, exploiting and research of the cultural resources of Zeya Zhishan Mountain. According to the introduction of Lin Zhiwen and Zhou Yinchai, the landmark achievements include:

In 2001, Siliandui Paper Mill was announced by the State Council as the fifth batch of national key cultural relics protection units.

In 2005, Huangkeng Village and Shuiduikeng Village, both with a long history of papermaking culture, were announced by the Zhejiang Provincial Government as provincial-level historical and cultural villages.

In 2007, Zeya Papermaking Techniques was selected as the intangible cultural heritage of Zhejiang Province.

In 2007, the Ouhai District Political Consultative Conference published the book of "Zhishan Mountain Culture".

In 2008, the Bureau of Culture, Radio and Television of Ouhai District cooperated with Zhejiang Normal University to establish the research group of "Zeya Paper Protection and Inheritance Research".

In 2009, the National "Compass Plan" project—"Construction of Chinese Traditional Papermaking Techniques Inheritance and Demonstration Base" settled in Zeya Zhishan Mountain and was listed as a key construction project in Ouhai District in 2010. The scope of the project includes the Siliandui Papermaking Mill of the state-level cultural protection unit and the representative papermaking villages such as Yangkeng Village, Tangzhai Village, Hengyang Village and Shuiduikeng Village, etc. covering 120 hectares, four zones, i.e. the paper culture display zone, papermaking exhibition area, local culture display area, papermaking techniques recovery site and Siliandui exhibition area.

The interior view of Zeya Traditional Papermaking Exhibition Hall

In December 2010, the Zeya Traditional Papermaking Exhibition Hall was built near Tangzhai Village.

In October 2010, Lin Zhiwen and Zhou Yinchai wrote the book *Zeya Papermaking*, which was published by China Drama Press, and this was the first monograph on papermaking techniques and culture in Zeya Zhishan Mountain.

From September to October in 2012, the first China (Wenzhou) Zeya Zhishan Mountain Culture Festival was held with the theme of "Millennium Zhishan Mountain, Charm of Ouhai".

In 2015, Zeya Ping paper production techniques were included in the fourth batch of national "intangible cultural heritage" protection list.

In 2018, Lin Zhiwen was selected as the representative inheritor of Zeya Ping Papermaking Techniques in the fifth batch of National Intangible Cultural Heritage.

2. Beautiful Rural Papermaking Culture Experience Zone Built in Longwang Village of Anji County in Huzhou City

In 2014, during the construction of beautiful villages and mountain village farmhouses, the village committee of Longwang Village in Shangshu Town of Anji County raised and invested 2.5 million *yuan* to create the handmade papermaking exhibition hall and the cultural corridor of paper mill, and launched the traditional cultural experience tour projects like the rural cultural experience tour about "Handmade Papermaking Techniques" and "King Dragon New Year Painting", the family tour, etc. The village committee introduced the handmade papermaking process with a long history and culture of Longwang Village as a characteristic theme tourism resource into the beautiful rural construction. When the investigation team entered the village in December 2016, the paper culture of Longwang Village was closely integrated with beautiful landscapes and Tourist Farm. There were 52 farmhouses and 5 homestays, and more than 300 villagers were engaged in tourism. In 2015, Longwang Village attracted 80 000 tourists and the tourism income was about 4.8 million *yuan*.

According to the investigation conducted

A section of Papermaking Culture Corridor in Longwang Village

Opening ceremony of Handmade Paper Research Institute and Yiguzhai "Industry-University-Research" Base

by the investigation team, the inheritors of the bamboo paper production technique of Longwang Village still make handmade paper, preserve the complete handmade papermaking process, and direct the handmade papermaking experience project regularly.

3. The Experimental Mode of In-depth Industry-University-Research Cooperation of "Yiguzhai" in Datong Village of Dayuan Town in Fuyang District

From June to July 2016, Zhu Zhonghua, the papermaking master of Hangzhou Fuyang Yiguzhai Yuanshu Paper Co., Ltd., was recommended by the "Intangible Cultural Heritage Center" in Fuyang District, and participated in the 30-day national intangible cultural heritage research and training program, sponsoreded by the Ministry of Culture and hosted by the University of Science and Technology of China, with the handmade papermaking master Yu Renshui from "Fuyang Xuan Paper Factory". After graduation, they were hired as practical teachers by the Institute of Handmade Paper of the University of Science and Technology of China and *Fuyang Daily* made a special edition report. In 2016, Zhu Zhonghua received more than 1000 schools and institutions, and volunteered to introduce and display Fuyang bamboo papermaking techniques.
The Institute of Handmade Paper of the University of Science and Technology of China and Beijing Decheng Tribute Paper Mill jointly established the Industry-University-Research base with Hangzhou Fuyang Yiguzhai Co., Ltd. on December 13, 2016, and jointly worked on the restoration research of Fuyang high-end bamboo paper.

For this cooperation, *Hangzhou Daily* reported as follows: The Institute of Handmade Paper of the University of Science and Technology of China is a representative institution of Chinese handmade papermaking research and technical analysis. At present, it is the only handmade paper research institute and professional handmade paper test and analysis laboratory in China. And it is the host unit of the national major research project "Library of Chinese Handmade Paper". Based on the need of in-depth Industry-University-Research collaboration, the Institute of Handmade Paper of the University of Science and Technology of China established the handmade papermaking techniques cooperation mechanism with Yiguzhai Company in Datong Village to help Fuyang District create a cultural golden business card—"Hometown of Bamboo Paper in China".

4. Fudan University and Kaihua County jointly researched and restored the production techniques of Kaihua paper

Since the end of 2008, the cultural department of Kaihua County has begun to explore Kaihua papermaking techniques. According to Sun Hongqi's description in the article *The Magic of the Kaihua Paper System*: they searched for the inheritors of Kaihua paper, and explored the papermaking troughs and tools left in the production area of Kaihua paper. Then they recorded the production processes of collecting papermaking materials, boiling bark, fermenting bark, rubbing bark, beating, cleaning, adding mucilage, picking paper out and drying, etc. based on field investigation. The local government attached great importance to this traditional culture. Under the direct promotion of governments at all levels, the Zhejiang Provincial Department of Culture listed Kaihua paper in the third batch of intangible cultural heritage in Zhejiang Province. The Kaihua County Government provided about 1300 m^2 of land free of charge and invested about 2 million *yuan* (excluding tools), and jointly established an academician workstation to restore Kaihua paper with Fudan University.

On March 24th, 2017, the academician workstation jointly established by Kaihua County Government and Fudan University—"Kaihua Paper Research Laboratory: Yang Yuliang Academician Workstation" was officially opened. This station was the first cultural academician workstation in Quzhou City.

Opening ceremony of Yang Yuliang Kaihua Paper Academician Workstation (photo provided by Huang Hongjian)

Yang Yuliang, academician of the Chinese Academy of Sciences and dean of the China Ancient Books Protection Research Institute of Fudan University, said at the opening ceremony that his team would use modern science and technology requirements to work with the head of the "Kaihua Paper Restoration Techniques and Protection Development Project" to restore the Kaihua papermaking techniques and produce high-end paper that can be used to repair ancient works at the international level.

3.2 Current Overview of Handmade Paper Researches in Zhejiang Province

Zhejiang handmade paper production is mainly distributed in Fenghua, Longyou, Wenzhou, Shaoxing, and Fuyang areas. The research on Zhejiang handmade paper mainly focuses on Fuyang and Wenzhou areas. Contemporary Zhejiang handmade paper research started in the 1980s and has gradually increased in recent years. The research problems mainly focus on the combing and discussion of the historical origin of handmade paper, as well as the research on the status quo of handmade paper, including the production techniques, management, inheritance and protection, to strive to provide constructive opinions for further development and protection of handmade paper. Due to the late start, the related research on contemporary Zhejiang handmade paper is not rich enough in terms of history, current situation and development directions.

The following part is a summary of the researches on Zhejiang handmade paper (classified by area):

3.2.1 Fenghua District of Ningbo City

(1) In 2011, Li Dadong's research report *Survey on Traditional Papermaking Techniques in Tangyun Village of Fenghua District*, briefly introduced the origin and development of Tangyun Village's papermaking and introduced original handmade papermaking techniques maintained by Yuan Hengtong, including ingredients, tools, etc. in detail.
Since the ancient times, Tangyun Village

has had two big families, Jiang and Yuan. Although Jiang was interested in the papermaking industry, his descendants believed that would be very toilsome, and gradually they gave up this ancestral business. With the constant competition, the prosperity of the past (three hundred papermaking troughs were used then) in Tangyun Village has returned to the original era with only one papermaking trough owned by one family.

With the changes of the times, the rise and fall of the paper industry and the increase and decrease of papermakers, Master Yuan still maintains the original handmade mill and retains the ancient papermaking memories for us, which is a rare "living papermaking cultural relic". With the rapid development of high technology, papermaking finds it more difficult to survive in the original handmade mill. While from the point of view of "living papermaking cultural relic", it is a great wealth for future generations to demonstrate the general situation and actual situation of ancient papermaking, and it is a living first-hand teaching material.

(2) In 2016, Chen Weiquan's article *The Legend of Tangyun Bamboo Paper* cuts in from the contribution of bamboo paper in Tangyun County to Tianyi Pavilion's book restoration, and tells the story of the field investigation. Through an interview of Yuan Hengtong, the only papermaker remaining paper mill in Tang'ao Village, this paper introduces the papermaking techniques. In addition, the evaluation and protection of Tangyun Papermaking by UNESCO Paper Protection Project is also introduced.

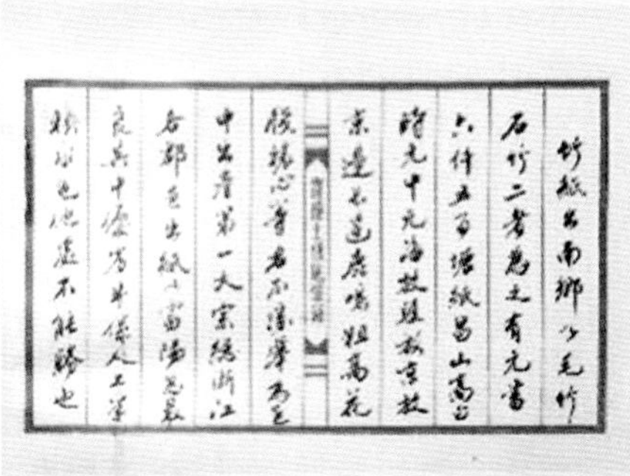

A photo of *A Record of Handmade Paper in Fuyang District*

3.2.2 Fuyang District of Hangzhou City

(1) In November 1991, the Publicity Department of the Fuyang County Party Committee of China, the Fuyang County Cultural Bureau and the Fuyang County Literary Federation co-compiled the *A Record of Fuyang Local Paper*, which became one of the few books devoted to the introduction of Fuyang local paper.

(2) In 1992, Qian Yunxiang's research paper *Fuyang Local Paper in Archival Historical Records*, from the advent of Fuyang bamboo paper in the Eastern Jin and Southern and Northern Dynasties, sorted out the Fuyang local paper information recorded in the historical materials in chronological order, and analyzed the reasons for the prosperity of Fuyang local paper from the aspects of geography, history, economy and society, etc.

(3) In 1993, Zhu Xuelin and Xu Gengfa introduced the basic information of Fuyang local paper in the article *Some Experiences of Writing 'Fuyang Paper Industry'*, and then introduced the writing process and summarized the writing experience of *Fuyang Paper Industry* from the materials to the screening and writing.

(4) In 1995, the origin of Fuyang paper was discussed in the research paper *Discussion on the Origin of Fuyang Paper*, and the handmade papermaking and machine-made papermaking were described respectively.

(5) In 1999, Zhou Yongquan's article *Can Fuyang Maintain the Title of 'Hometown of Papermaking'* analyzed the advantages of Fuyang papermaking industry, including historical, geographical and environmental advantages, focusing on the advantages like technique advantages, price advantages and investment advantages, operating mechanism advantages, and pointed out the problems in the development of Fuyang papermaking industry, including environmental pollution, scattered distribution, single variety of products, brand awareness and management level, as well as safety issues in production.

史话

□ 富阳县档案局 钱云祥

档案史料记载中的富阳土纸

Fuyang Local Paper in Archival Historical Records (screenshot)

纸坛概貌

浙江富阳纸业的基本情况

● 缪大经

Basic Information of Fuyang Paper Industry in Zhejiang Province (screenshot)

(6) In 2005, Miao Dajing's research paper *Basic Information of Fuyang Paper Industry in Zhejiang Province* introduced the development and current situation of Fuyang Paper, including the distribution of Fuyang Paper, analysis of major economic indicators and construction of papermaking industrial zone, etc.

(7) In 2008, Wang Huan and Hu Zhan's article *Unique Treasure of Millennium*: *Fuyang Paper* told the glory and decline of Fuyang handmade papermaking. On the investigation of the technology and present situation of Fuyang handmade papermaking, the inheritance and protection of Fuyang handmade papermaking were considered.

(8) In 2009, Hong An's paper *Fuyang Bamboo Papermaking Techniques* described the techniques of bamboo papermaking.

(9) In 2010, Li Shaojun's article *Fuyang Bamboo Paper*: *Interpretation of Bamboo Paper and Exploration of Bamboo Paper Culture* was published.

(10) Zhou Anping's master's thesis released in 2013, *Handmade Paper Industry in Fuyang County of Zhejiang Province in the 1950s and 1960s* examined the handmade paper industry in Fuyang County of Zhejiang Province, which had the reputation of the "hometown of papermaking" from a macro perspective, and briefly introduced the historical evolution and contemporary profile of Fuyang handmade paper industry. The contemporary profile includes the relationship between Fuyang's handmade paper industry and people's lives, the natural environment in which it was harbored, the classification and manufacturing procedures, and production costs.

This paper analyzed the cost-benefit of handmade paper among Fuyang papermaker families. From the perspective of division of labor and professionalization, it studied the role of Fuyang handmade paper industry,

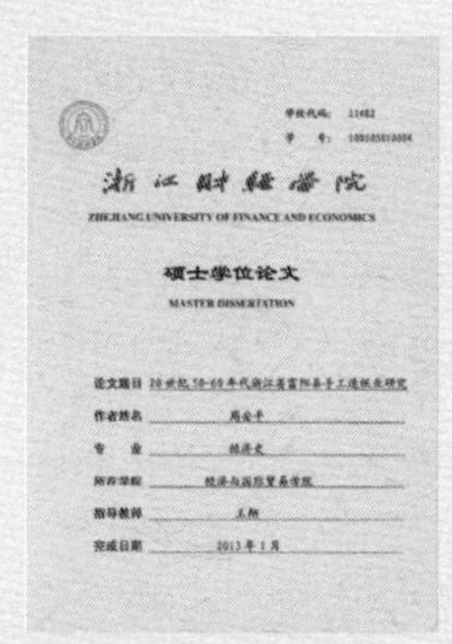

Dissertation cover of *Handmade Paper Industry in Fuyang County of Zhejiang Province in the 1950s and 1960s*

A photo of *Watching Bamboo Paper*

discussed the influence of government organizations and local social communities on the professionalization of Fuyang handmade paper industry, and examined the transformation of Fuyang handmade paper industry into machine-made paper industry and the causes for the change.

(11) In 2013, Zhang Zaiqing's article Bamboo Paper Fragrance in Yuanzhu Mill introduced the family paper mill of Li Fa'er in Guanxingta Natural Village of Xinsan Administrative Village in Huyuan Town of Fuyang City, including the history of Fuyang paper, the development of paper mill and the introduction to Yuanshu paper, the main paper product of Yuanzhu Mill.

(12) In 2015, Zhang Yaru's paper *Inheritance and Development of Fuyang Bamboo Paper* studied the production conditions of Fuyang bamboo paper, including the natural environment and social environment; and summarized the production process, development and innovation of Fuyang bamboo paper.

(13) In 2016, Ye Qin's research paper *The Gentleman in Paper: Fuyang Bamboo Paper* introduced the characteristics, honors and marketing of Fuyang bamboo paper, and introduced the types and techniques of Fuyang bamboo paper.

(14) In 2019, Tang Shukun and Tang Yumei's research paper *Research on Wujing Paper* systematically examined the historical famous paper: the origin, place of production, materials, crafts and uses of Wujin paper, and the technical problems encountered in the process of recovering Wujin paper with the papermakers of Fuyang bamboo paper.

(15) In 2005, the People's Political Consultative Conference of Literature and History Committee of Fuyang County compiled the book *Fuyang Paper Industry in China*, which was divided into Fuyang Paper Industry Overview, Traditional Industry, Modern Industry, Enterprises, Selected Literature, and Appendix. This book was a collection of 53 research articles and literatures, covering almost every aspect of Fuyang paper. It was the first collection of contemporary articles discussing Fuyang paper. Although handmade paper only accounted for about 50%, it still had a great reference significance. The book was accompanied by a wealth of pictures.

(16) In 2010, Zhou Guanxiang's book *Fuyang Traditional Handmade Papermaking* gave a detailed description of the historical production of Fuyang traditional handmade paper in modern times, providing rich industry data at different times, especially the modern papermaking data information has a good reference value.

(17) In 2010, Li Shaojun wrote *Fuyang Bamboo Paper*, which was characterized by the raw materials and complete process of Fuyang bamboo papermaking. This book introduced the specific techniques and key requirements of each step in a very in-depth and meticulous manner with the coorperation of pictures and texts. It could be called the guideline for making Fuyang bamboo paper.

(18) In 2016, Chen Gang's *Bamboo Paper: Proceedings of the 2015 China Bamboo Paper Protection and Development Symposium* was published. This was a collection of papers from the National Bamboo Paper Symposium held in Fuyang County in 2015. A total of 22 paper research articles, 1 review and 3 speeches were collected, which explored the issue of the inheritance and development of bamboo paper. Also, there were 7 papers devoted to Fuyang bamboo paper.

3.2.3 Kaihua County in Quzhou City

(1) In 1985, Cheng Guang's paper *Kaihua Paper and Kaihua Announcement Paper: On Ancient Paper* introduced the characteristics and differences of Kaihua paper and Kaihua announcement paper.

(2) In 2007, Yang Jurang's research paper *Inquiry of the Collection of Good Books: Paper Used by the Collection of Ancient and Modern Books* pointed out that *The Collection of Ancient and Modern Books* has a version written on Kaihua paper, which was "white as jade".

(3) In 2013, Sun Hongqi's research article *The Magic of Kaihua Paper* introduced the origin and prosperity of Kaihua paper, and studied the decline and inheritance of Kaihua paper.

(4) In 2015, Wang Chuanlong's research paper *Research of Kaihua Paper* studied the origin, naming, usage, and original place of the Kaihua paper, and believed that it was wrong for the scholars to generally accept that the Kaihua paper appeared at the end of Ming Dynasty or could also be called Kaihua (blossom) paper or Taohua (peach blossom) paper.

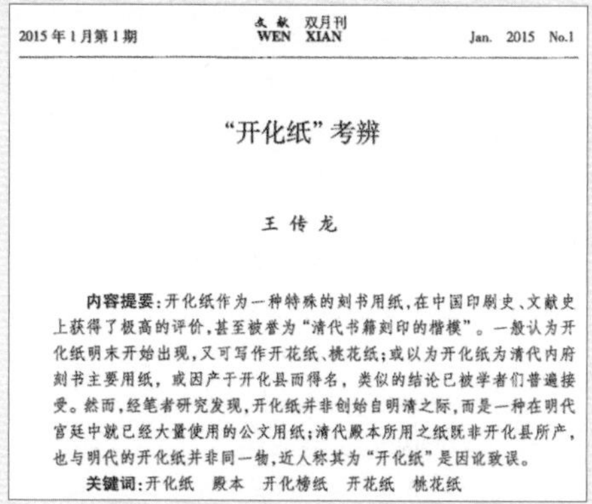
2015年1月第1期　　文献 双月刊 WEN XIAN　　Jan. 2015 No.1

“开化纸”考辨

王传龙

内容提要：开化纸作为一种特殊的刻书用纸，在中国印刷史、文献史上获得了极高的评价，甚至被誉为“清代书籍刻印的楷模”。一般认为开化纸明末开始出现，又可写作开花纸、桃花纸；或以为开化纸为清代内府刻书主要用纸，或因产于开化县而得名，类似的结论已被学者们普遍接受。然而，经笔者研究发现，开化纸并非创始自明清之际，而是一种在明代宫廷中就已经大量使用的公文用纸；清代殿本所用之纸既非开化县所产，也与明代的开化纸并非同一物，近人称其为“开化纸”是因讹致误。

关键词：开化纸　殿本　开化榜纸　开花纸　桃花纸

Research of Kaihua Paper (screenshot)

(5) In 2015, Sun Hongqi's novel Guo Chu told the story of the twists and turns of a generation of famous papermaking families and their brilliant achievements.

(6) On November 23, 2017, an international academic seminar on the production of Kaihua paper and documents on Kaihua paper was held in Kaihua County. At the seminar, experts and scholars from the University of California at Berkeley, Stanford University, Fudan University in China, and well-known collectors such as Wei Li gathered together to share the research results of Kaihua paper.

(7) In 2018, Yi Xiaohui put forward his own unique views on Kaihua paper in the article *Investigation of the Source of Kaihua Paper in the Qing Dynasty*. The "Kaihua Paper" and "Kaihua Announcement Paper", which were widely used in the Qing Dynasty, had been considered to be produced in Kaihua County of Zhejiang Province. However, after research, it was found that the "Kaihua Paper" used in the Qing Royal Palace was a kind of "Liansi Paper", and the "Kaihua Announcement Paper" was actually "the announcement paper made in Jingxian County", both of which were produced in the area of Jingxian County in Anhui

Province. It was found that the fiber content of the relevant paper samples was 100% *Pteroceltis tatarinowii* Maxim. bark, and that raw material was unique to Jingxian County in Anhui Province. The conclusions of the document research were consistent with the results of the paper test analysis, which indicated that the "Kaihua Paper" and "Kaihua Announcement Paper" used in the Qing Dynasty government should belong to the system of Xuan paper in Jingxian County.

(8) In December 2018, the papermaking techniques of Kaihua tribute paper were selected in the first batch of traditional craft revitalization catalogue in Zhejiang Province.

(9) In the Stockholm International Stamp Exhibition held on May 23, 2019, the stamp "Sailboat" was an engraved gravure print designed by the international stamp engraving master Martin Merck. The paper used for the stamp was the "sample" paper printed on the newly restored Kaihua paper.

3.2.4 Longyou County in Quzhou City

(1) Li Zhongkai's *History of Mulberry Bark Paper (I)* in 1990 introduced the properties of mulberry tree, studied the source of mulberry planting in China, and summarized the situation of mulberry papermaking in the past, especially in the Wei, Jin, Southern and Northern Dynasties and the mulberry papermaking of the Sui, Tang and the Five Dynasties. Li Zhongkai's *History of Mulberry Bark Paper (II)*, which was published in 1991, continued to talk about the history of the mulberry papermaking in the Song and Yuan Dynasties, the Ming and Qing Dynasties, and the modern days. This book also looked forward to the development of contemporary mulberry paper, in which there was a series of discussion on mulberry papermaking in Zhejiang Province.

(2) In 2013, Wu Xinghui's research paper *Art Characteristics of Longyou Bast Paper* analyzed the human history, raw materials, paper characteristics and production techniques of Longyou bast paper, and introduced its production process and artistic characteristics as well as the current protection measures.

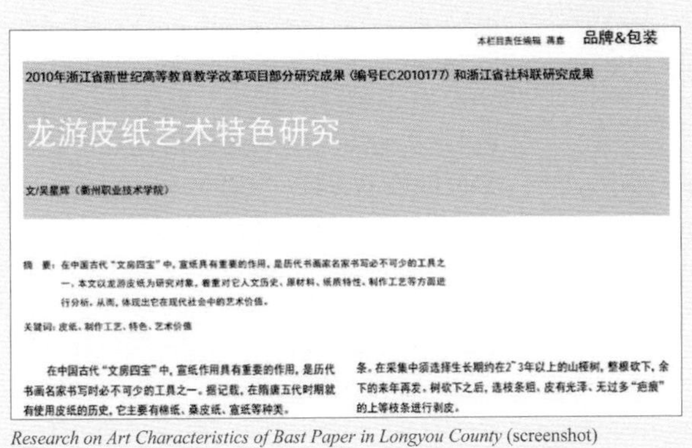
本栏目责任编辑 冯鑫 品牌&包装

2010年浙江省新世纪高等教育教学改革项目部分研究成果（编号EC2010177）和浙江省社科联研究成果

龙游皮纸艺术特色研究

文/吴星辉（衢州职业技术学院）

摘 要：在中国古代"文房四宝"中，宣纸具有重要的作用，是历代书画家名家书写必不可少的工具之一。本文以龙游皮纸为研究对象，着重对它人文历史、原材料、纸质特性、制作工艺等方面进行分析，从而，体现出它在现代社会中的艺术价值。

关键词：皮纸、制作工艺、特色、艺术价值

在中国古代"文房四宝"中，宣纸作用具有重要的作用，是历代书画名家书写时必不可少的工具之一。据记载，在隋唐五代时期就有使用皮纸的历史，它主要有棉纸、桑皮纸、宣纸等种类。

条，在采集中须选择生长期约在2~3年以上的山桠树，整根砍下，余下的来年再发。树砍下之后，选枝条粗、皮有光泽、无过多"疤痕"的上等枝条进行剥皮。

Research on Art Characteristics of Bast Paper in Longyou County (screenshot)

(3) In 2018, Wu Jianguo made a detailed explanation of the history and techniques of the traditional bast paper in Longyou County and the current development in the article *Techniques of Bast Paper in Longyou County*.

3.2.5 Pingshui Town in Keqiao District of Shaoxing City, Shengzhou City, and Shangyu City

(1) In 1987, the book *Shaoxing Practical Encyclopedia* written by Tao Renkun described the characteristics, making techniques and legends of Luming paper in Shaoxing City.

(2) In 2008, Peng Yan's paper *On Ancient Papermaking and Engraving Printing in Shaoxing City* reviewed the rise and development of ancient papermaking and engraving printing in Shaoxing City from three aspects. This paper described the emergence of Teng paper and the rise of block printing in Shaoxing City, the flourishing of bamboo paper and block printing in Shaoxing City, the great development of papermaking in the Yuan, Ming and Qing Dynasties and the rise and fall of the rise of block printing in Shaoxing City, and explored its historical origin in the light of social, historical and cultural changes.

(3) In 2016, in the article *On Shanteng Paper*, through the literature study on the famous paper made in ancient Shanxian County (now Shengzhou City) —Shanteng paper, Wang Chuntian combed the historical development of Shanteng paper, pointed out that its peak period was in Tang Dynasty, and listed its other special functions besides for writing, such as storing tea, making quilts, making paper bed curtain, pharmaceuticals and so on.

(4) In the 1960s, *Ningbo Public* magazine published the article *The Spring and Autumn of Wujin Paper* written by Shen Jiwang, proposing that "Wujin Paper was created by Wei Liangzai in the Jin Dynasty, and it had a history of 1500 years". However, the source of this view was not stated.

(5) In 2016, Liu Renqing mentioned that Wujin paper was originally produced at Shangyu City and Shaoxing City in Zhejiang Province in the research article of *Ancient Paper Names (Part 4): Paper Names in the Tang Dynasty (I)*. The paper was so named because the surface of this paper was black and shiny, like gold shining. Wujin paper was one of the key materials for the production of thin golden foil.

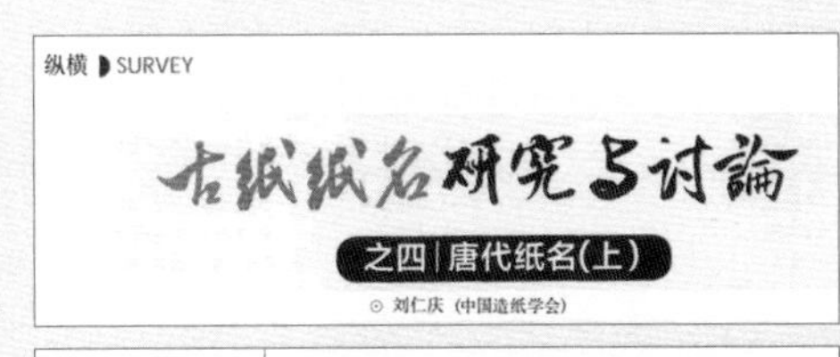
纵横 SURVEY

古纸纸名研究与讨论

之四｜唐代纸名(上)

⊙ 刘仁庆（中国造纸学会）

◆ 乌金纸

唐代纸名。原产于浙江上虞、绍兴等地。此纸的表面乌黑发亮，似金子一样，故而得名。乌金纸是生产薄如蝉翼金箔的关键用品之一。金箔是从古至今进行"贴金"工序不可缺少的材料。所谓贴金，就是用涂胶把金箔黏贴到织物、皮革、纸张、器具及建筑物等表面，以起到装饰、美化的作用。据说，乌金纸始创于晋代。明·宋应星（约1587-1655年）在《天工开物》中介绍说："凡乌金纸由苏杭造成，其纸用东海巨竹膜为质，用豆油点灯，闭塞周围，只留针孔通气，熏染烟光，而成此纸"。上虞县"赵开堂"店制造的乌金纸远近闻名，其生产精工细作，首先要选用当年山阴竹笋所长之嫩竹，在"大暑"时砍下，流水中浸泡五个伏天（故此纸又称五伏纸），制成竹浆。加入黄豆，制成乌金纸坯，刷以梓油或麻油熏的烟子（炭黑），待其慢慢阴干即成乌金纸。乌金纸的机械强度、韧性和耐磨性都很好，多次捶打亦不损破。它用于碑文拓本，人称乌金拓。图27。

Research and Disscussion on Ancient Paper Names (Part 4): Paper Names in the Tang Dynasty (I) (screenshot)

(6) In 2017, Ding Ting mentioned in her master's thesis Research on the Protection and Inheritance of Traditional Forging Techniques of Nanjing Gold Foil: it can be seen from *Exploitation of the Works of Nature* that Wujin paper can only be made in Zhejiang Province and Jiangsu Province because the bamboo used to make Wujin paper is suitable for growing in Jiangsu and Zhejiang Provinces only. In China, the quality of Wujin paper produced in Shangyu District of Shaoxing City in Zhejiang Province was very good. In May 2009, Wujin paper's production technqiues were included in the third batch of Zhejiang Province's intangible cultural heritage list, and it was incorporated into the Ten New Discoveries Project of the Intangible Cultural Heritage Survey in Zhejiang Province. The production techniques of Wujin paper had regained people's concern.

3.2.6 Zhishan Mountain in Zeya Town of Ouhai District in Wenzhou City

(1) In 2005, Huang Zhousong's research report *Siliandui Paper Mill Relics in Zeya Town of Wenzhou City* was the preliminary research on the protection of the national key cultural relics protection unit, i.e. Siliandui Paper Mill in Wenzhou City, including three aspects: historical research of Wenzhou papermaking, value evaluation and protection. The basic point of view is: Wenzhou papermaking began in the Tang Dynasty, flourished in the Song and Ming Dynasties, and bloomed in the 1930s and 1940s; Wenzhou paper had a very complete

and reasonable production and sales model; Siliandui Paper Mill in Wenzhou City is the "living fossil of Chinese papermaking". It not only has a high historical value, but also has many scientific and artistic value; its protection should include two aspects: equipment and process flows, two environments points: nature and humanities, and two centers: production and sales.

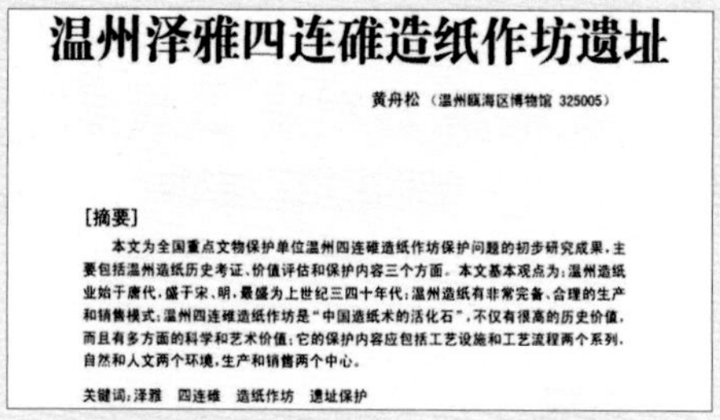

温州泽雅四连碓造纸作坊遗址

Siliandui Paper Mill Relics in Zeya Town of Wenzhou City (screenshot)

(2) In 2009, Fan Jialu's research paper Survey on Ping Papermaking *Techniques in Wenzhou City of Zhejiang Province* briefly described the history of Zhejiang Ping paper, and introduced the production process of Wenzhou Ping paper in detail.

(3) In 2010, Wang Yajun's research article *The Connotation, Inheritance and Development of Zhishan Mountain Culture in Hengyang Village of Zhejiang Province* proposed that "Zhishan Mountain Culture" was a unique regional culture in Zeya Town of Wenzhou City, which had rich connotations. The article introduced the history and current situation of Pan's clan in Hengyang Village through a case study of the clan of Hengyang Village in Zeya Town. Then it explored the connotation of the Zhishan Mountain Culture, interpreted the inheritance and development of the Zhishan Mountain Culture, and hoped to enrich and deepen the research of the intangible cultural heritage protection.

(4) In 2011, Lin Zhiwen and Zhou Yinchai's research article *Papermaking in Wenzhou and Zeya Zhishan Mountain* was published in *Wenzhou Daily*. Lin Zhiwen and Zhou Yinchai spent 3 years walking into 81 villages in Zeya Town. They visited hundreds of families and traced the papermaking techniques in Wenzhou history. The article introduced Wenzhou bast paper, wax paper, straw paper and their uses, as well as the usage, processing method and history of Juan paper. It also depicted the geographical environment of Zeya Zhishan Mountain. In addition, it described the distribution and production profile of Wenzhou handmaking paper for a thousand years.

(5) In 2011, Li Linlin adopted a multidisciplinary approach in her master's thesis *Research on the Change of Traditional Cultural Areas: Taking the "Zeya Zhishan Mountain Cultural District" in Wenzhou as an Example*. This thesis explored Zeya which represented the traditional cultural district, and investigated the unique historical stage and the corresponding human-land relationship at that time. The research method combining data analysis and field investigation reflected the change of Zeya Zhishan Mountain cultural district. Her thesis explored the institutional change of traditional cultural districts, the influence of institutional changes and the role of technology in cultural districts, and proposed that institutional changes and technological progress were the internal driving force for cultural district development.

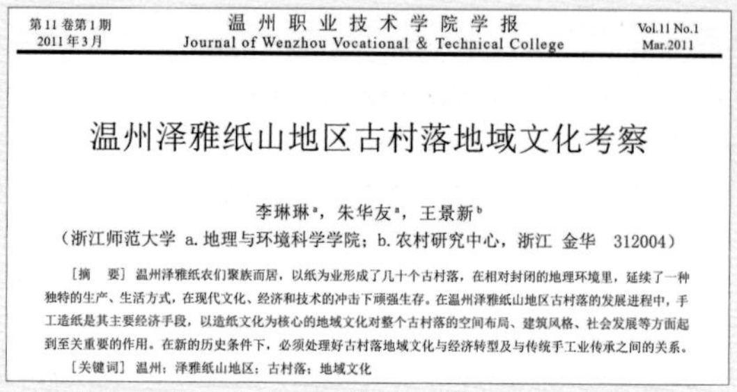

In the third chapter of the book, the development of "Zeya Zhishan Mountain Cultural Zone" were summarized. The techniques, processes and uses of handmade paper were sorted out and the connotation of the Zhishan Mountain cultural district was defined. That is to say, handmade paper was the core cultural factor, including more than 20 existing papermaking processes, papermaking tools, paper mills and traditional cultural districts in the form of market, street and traditional houses formed by the sale of handmade paper.

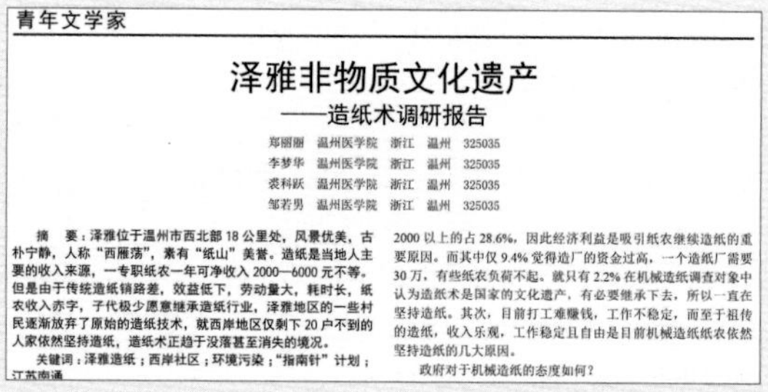

(6) In 2011, Li Linlin, Zhu Huayou, and Wang Jingxin's research article *Investigation on the Ancient Villages in the Zhishan Mountain Area of Zeya in Wenzhou* studied the ancient villages involving ancient papermaking in the Zhishan Mountain area of Zeya Town in Wenzhou City, including spatial distribution, major ancient villages, introduction to ancient architecture and folk customs. The article also introduced the historical origins and processes of handmade paper in Zeya Zhishan Mountain area in Wenzhou City, and put forward some thoughts on the culture of the ancient villages in the Zhishan Mountain area of Zeya Town in Wenzhou City.

(7) In 2012, *Fujian Paper Information* published the "Six One Program" for creating a Zhishan mountain cultural brand in Ouhai District of Wenzhou City in Zhejiang Province.

温州泽雅纸山地区古村落地域文化考察

Cultural Investigation on the Ancient Villages in Zhishan Mountain Area of Zeya Town in Wenzhou City (screenshot)

泽雅非物质文化遗产
——造纸术调研报告

Intangible Cultural Heritage in Zeya Town: Papermaking Research Report (screenshot)

(8) In 2012, Han Yuxi's research paper *Analysis of the Spatial Form of Zeya Zhishan Mountain Ancient Village in Wenzhou* studied Zeya Zhishan Mountain Ancient Village in Wenzhou City, in terms of the street network, water system, paper production facilities, and village boundary. The layout factors were used to analyze the shape and characteristics of the spatial layout of the ancient village of Zeya Zhishan. The study found that the site selection of the ancient village of Zeya Zhishan Mountain was greatly affected by paper production.

(9) In 2012, Meng Zhaoqing's article *Zhishan Mountain Culture: Ancient Papermaking* showed the scene and papermaking on the west bank of Zeya in Ouhai District of Wenzhou City with text and pictures.

(10) In 2012, Zheng Lili, Li Menghua, Qiu Keyue, and Zou Ruonan's research report *Zeya Intangible Cultural Heritage: Papermaking Research Report* was conducted by a survey and random sampling methods in five areas including Xiaoyuan Village to research on machine-made and handmade paper. It analyzed why the number of papermakers was decreasing, why some papermakers still insisted on papermaking, the government's attitude towards machine-made paper and other crucial problems in the industry, and put forward targeted suggestions on the sales difficulties faced by handmade paper.

(11) In 2014, Zhou Jimin and Zheng Gaohua's article *The Charm of Zeya Ancient Papermaking* described the history and geographical environment of ancient papermaking in Zhejiang Province,

introduced the daily life of papermakers and the process of papermaking. It also mentioned the status quo of Zhishan Mountain's paper industry and the protection of related techniques.

(12) In 2014, Zhou Jimin and Ni Zhijian's article *Papermaking Techniques of Zeya Ping Paper* outlined the history and current situation of Zeya Ping paper and gave a detailed introduction to the production process of it.

(13) The book *Zeya Papermaking*, edited by Lin Zhiwen and Zhou Yinchai, published by China Drama Publishing House, firstly combining words with pictures, introduced the traditional papermaking technology and the manufacturing technology of tools such as hydraulic pestle, papermaking screen, and papermaking trough of Zeya Town in a systematic and detailed way. This was the first time that the traditional papermaking process of Zeya was completely described. The "living fossil of Chinese papermaking", which was left in the Zhishan Mountain area of Wenzhou, had been systematically studied for the first time.

Zeya Papermaking comprehensively introduced the whole process and technical essentials of Zeya paper from the aspects of operation procedures, technical requirements, other issues, working experience, which preserved detailed historical data for the inheritance of the millennium traditional handmade papermaking process. The book introduced the manufacturing technology of hydraulic pestle, papermaking screen and papermaking trough, and described the production status and sales of toilet paper. The book was sponsored by Ouhai District government in the name of "Compass Plan" special project entitled "Chinese Traditional Paper Technology Inheritance and Demonstration Base Construction". And this book was published with the support of Ouhai District Government.

A photo of *Zeya Papermaking*

3.2.7 Other Researches

(1) In 1986, Liu Yizhu's research article *On the Raw Materials of Handmade Paper in China* introduced the raw materials of handmade paper in China. The Zhejiang Maobian (deckle-edged) paper mentioned in the article belongs to the category of raw materials.

(2) In 2002, Guan Chuanyou's research article *Exploration of Chinese Bamboo Paper History* examined and discussed the origin and history of ancient Chinese bamboo paper. It was believed that Chinese bamboo paper originated in the Jin Dynasty, and the bamboo paper industry developed rapidly during the Tang and Song Dynasties. The development of the bamboo paper industry in the Ming Dynasty flourished and reached its peak in the Qing Dynasty, and the development of the modern bamboo paper industry was still prosperous. This article explored the process and techniques of ancient bamboo paper production and its impact on ancient social economy and culture. Part of the introduction to the bamboo paper in the Qing Dynasty refered to the Zhejiang papermaking areas, and introduced the main production sites and the characteristics of the paper products produced in the area.

(3) In 2009, Miao Dajing's article *Retrospecting Zhejiang Handmade Paper and Features of Machine-made Paper* started with Zhejiang handmade paper, briefly introduced the development of Zhejiang handmade paper, focusing on the origin and development of Zhongri bast paper, and the influence of the introduction of long fibre paper machine on the development of handmade paper after the mechanization of Zhejiang handmade paper.

(4) In 2014, Gu Yu's master thesis *Research on the Protection of Traditional Papermaking Process in Zhejiang Province*, summed up the status of five traditional papermaking processes through field investigation. After preliminary analysis of their respective characteristics, Gu selected several models with reference significance, conducted SWOT analysis, found problems in the traditional papermaking process protection practice in Zhejiang Province, and proposed protection strategies for specific problems, hoping to promote the inheritance of traditional papermaking processes and development.

(5) In 2015, Yan Jieyu's *The Status Quo and Outlook of Traditional Handmade Paper in Western Zhejiang Province* introduced the development and production status of traditional handmade paper in Lin'an City of Zhejiang Province. Through the investigation of Lin'an Fuyutang Paper Co., Ltd. in Hangzhou City, the paper introduced the making techniques of handmade paper, and discussed how to protect and inherit the handmade paper crafts of Western Zhejiang Province from five aspects: new rural construction, eco-tourism, education, cultural and creative industries, and local government.

(6) In 2015, Liu Renqing's research article *On the Status and Problems of Traditional Handmade Paper in China* introduced the distribution and production techniques of traditional Chinese handmade paper, as well as the existing problems, and put forward constructive opinions on future development. It refered to the Zhejiang region, and introduced the climate and environment of the area and several paper products produced in Zhejiang.

(7) In 2016, Liu Renqing's research article *Identification of Handmade Paper Names* introduced the raw materials, techniques and characteristics of Yuanshu paper produced in Fuyang area of Zhejiang region in the Song Dynasty.

Although in terms of research standard and research depth, the research results of contemporary Zhejiang handmade papermaking process, economic and cultural achievements are not rich compared with those on Anhui Xuan paper and Jiajiang bamboo paper, Zhejiang still ranks high among Chinese handmade papermaking provinces.

Raw materials of *Pleioblastus amarus* Wujin paper exposed to the sun and rain

图目
Figures

章节	图中文名称	图英文名称
	正在使用的焙壁	Dryibg wall in use
	焙壁屋外部	Exterior of the drying wall
	袁恒通造纸坊纸品展示	Paper displayed in Yuan Hengtong Paper Mill
	展示馆摆放的乌金纸（左）和乌金原纸（右）	Wujin paper (left) and the original Wujin paper (right) in the exhibition hall
	袁恒通与天一阁图书馆来往的信件	Correspondences between Yuan Hengtong and Tianyige Library
	朱信灿1991年从袁恒通纸坊购买的乌金原纸	Zhu Xincan purchased original Wujin paper from Yuan Hengtong Paper Mill in 1991
	朱信灿用袁恒通的原纸所造的乌金纸	Zhu Xincan made Wujin paper with Yuan Hengtong's original paper
	国家图书馆张平给袁恒通的来信	Letter from Zhang Ping of the National Library to Yuan Hengtong
	各地图书馆下订单的信件	Orders from various libraries
	调查时正在搭建的技艺展示间	Craft showroom in construction during the investigation
	袁恒通（右）和袁建增（左）父子合影	Yuan Hengtong (right) and his son Yuan Jianzeng (left)
	纯苦竹修复用纸透光摄影图	A photo of pure *Pleioblastus amarus* paper for repairing seen through the light
	"苦竹+毛竹+桑皮"书写用纸透光摄影图	A photo of "*Pleioblastus amarus* + *Phyllostachys edulis* + paper mulberry bark" writing paper seen through the light
	"苦竹+桑皮"修复用纸透光摄影图	A photo of "*Pleioblastus amarus* and mulberry bark" paper for repairing seen through the light
	"苦竹+桑皮+麻"古籍印刷用纸透光摄影图	A photo of "*pleioblastus amarus* + mulberry bark + hemp" paper for repairing ancient workes seen through the light

章节	图中文名称	图英文名称
第七章	丽水市	Chapter VII Lishui City
	松阳县李坑村李坑造纸工坊	Likeng Paper Mill in Likeng Village of Songyang County
	箬寮原始森林的一条小山沟	A small valley in Ruoliao Primeval Forest
	李坑村现存的《张氏宗谱》（修于1936年）	*Genealogy of the Zhangs* kept in Likeng Village (revised in 1936)
	李坑村村头的村名景观石	Landscape stone craved the name of the village in Likeng Village at the entrance of the village
	李坑村的导览图牌	Guide map of Likeng Village
	李坑造纸工坊的主要生产场地	Main production site of Likeng Paper Mill
	用李坑绵纸制成的油纸伞	Oil-paper umbrella made of Mian paper in Likeng Village
	书写在李坑绵纸上的旧日契约	Old contract written on cotton paper in Likeng Village
	榨完纸的张祖献与傅珠莲	Zhang Zuxian and Fu Zhulian after pressing the paper
	准备"流水袋料"的潘黎明	Pan Liming preparing to "clean the materials with a bag"
	"李坑牌"绵纸成品	Mian paper product of "Likeng Brand"
	"李坑牌"绵纸纤维形态图（10×）	Fibers of Mian paper of "Likeng Brand" (10× objective)
	"李坑牌"绵纸纤维形态图（20×）	Fibers of Mian paper of "Likeng Brand" (20× objective)
	"李坑牌"绵纸润墨性效果	Writing performance of Mian paper of "Likeng Brand"
	李坑造纸工坊附近的山桠皮树	*Edgeworthia chrysantha* Lindl. surrounding Likeng Paper Mill
	李坑造纸工坊所用的猕猴桃枝	Chinese gooseberry branches used in Likeng Paper Mill
	调查组成员在测水的pH	Researchers measuring the pH value of the water
	李坑造纸工坊门前小溪	Stream by Likeng Paper Mill
	砍山桠皮树枝	Cutting the branches of *Edgeworthia chrysantha* Lindl.
	蒸煮皮料的锅灶	Stove for steaming and boiling the bark materials
	剥皮	Peeling the bark
	刮皮	Scraping the bark
	浸泡好的山桠皮	Soaked *Edgeworthia chrysantha* Lindl. bark
	沤皮	Fermenting the bark
	刚沤制好的山桠树皮	Prepared *Edgeworthia chrysantha* Lindl. bark
	蒸煮	Steaming and boiling
	漂洗山桠皮料	Cleaning *Edgeworthia chrysantha* Lindl. bark materials
	拣皮	Picking out the impurities
	打浆	Beating the papermaking materials
	流水袋料	Cleaning the materials with a bag
	潘黎明正在抄纸桶里抄纸、扣纸	Pan Liming scooping and lifting the papermaking screen out of water and turning it upside down on the board
	榨纸	Pressing the paper
	分纸	Separating the paper
	晒纸	Drying the paper
	打捆	Binding the paper
	柴刀	Sickle
	蒸料桶	Wooden barrel for steaming the materials
	大凳	Wooden bench
	皮刀	Bamboo tool for beating the bark
	料袋	Material bag
	纸捅	Tool for pounding the bark
	抄纸桶	Wooden barrel for papermaking
	纸桶架	Shelf for holding the wooden barrel
	纸药桶	Wooden barrel for holding papermaking mucilage
	搅料棍	Bamboo stick for stirring the papermaking materials
	大号纸帘	Large-size papermaking screen
	小号纸帘	Small-size papermaking screen
	体验纸帘	Papermaking screen for visitors to experience papermaking
	大号帘架	Frame for supporting the large-size papermaking screen
	小号帘架	Frame for supporting the small-size papermaking screen
	体验帘架	Frame for supporting the visitors' Papermaking screen
	抄纸底板	Wooden board
	木榨	Wooden pressing device
	民宿"隐舍李坑"正门	Door of the homestay "Likeng House"
	李坑村唯一遗留下的旧日造纸场所	The only old papermaking place left in Likeng Village
	潘黎明的大伯潘昌平	Pan Liming's uncle, Pan Changping
	用李坑绵纸制成的台灯	Table lamp made of Mian paper in Likeng Village
	"李坑牌"染色纸	Dyed paper of "Likeng Brand"
	"李坑牌"山桠皮绵纸透光摄影图	A photo of "Likeng Brand"*Edgeworthia chrysantha* Lindl. Mian paper seen through the light

章节	图中文名称	图英文名称
第八章	杭州市	Chapter VIII Hangzhou City
第一节	杭州临安浮玉堂纸业有限公司	Section 1 Hangzhou Lin'an Fuyutang Paper Co., Ltd.
	浙江省级"非遗"生产性保护基地标牌	Plaque of Zhejiang Provincial Intangible Cultural Heritage Protection Base
	杭州市级"非遗"生产性保护基地标牌	Plaque of Hangzhou Municipal Intangible Cultural Heritage Protection Base
	调查组成员访谈陈旭东（右）	Researchers interviewing Chen Xudong (right)

	陈旭东现在的纸厂厂区外景	External view of Chen Xudong's factory
	浮玉堂书画用途的"宣纸"	"Xuan paper" of Fuyutang Brand used for calligraphy and painting
	浮玉堂包装纸	Wrapping paper of Fuyutang Brand
	浮玉堂"本色云龙纸"纤维形态图（4×）	Fibers of "Yunlong paper in original color" of Fuyutang Brand (4× objective)
	浮玉堂"本色云龙纸"纤维形态图（10×）	Fibers of "Yunlong paper in original color" of Fuyutang Brand (10× objective)
	浮玉堂"本色云龙纸"纤维形态图（20×）	Fibers of "Yunlong paper in original color" of Fuyutang Brand (20× objective)
	浮玉堂"本色云龙纸"润墨性效果	Writing performance of "Yunlong paper in original color" of Fuyutang Brand
	浮玉堂混料书画"宣纸"纤维形态图（10×）	Fibers of calligraphy and painting "Xuan paper" with mixed materials of Fuyutang Brand (10× objective)
	浮玉堂混料书画"宣纸"纤维形态图（20×）	Fibers of calligraphy and painting "Xuan paper" with mixed materials of Fuyutang Brand (20× objective)
	浮玉堂混料书画"宣纸"润墨性效果	Writing performance of calligraphy and painting "Xuan paper" with mixed materials of Fuyutang Brand
	购自龙游县的山桠皮粗加工原料	*Edgeworthia chrysantha* Lindl. bark materials bought from Longyou County
	正在浸泡的檀皮浆料	Soaking *Pteroceltis tatarinowii* bark
	正在浸泡的龙须草浆板	Soaking *Eulaliopsis binata* pulp board
	化学纸药及存放池	Chemical papermaking mucilage and the container
	厂区门口流经的河水	River flowing through the factory
	正在浸泡的不同浆料	Soaking different pulp materials
	打浆机	Machine for beating materials
	喷浆法捞纸	Papermaking by pulp shooting technique
	将纸帘揭下	Peeling the papermaking screen down
	便于后续揭纸的尼龙线	Nylon threads for peeling the paper down
	捞纸当晚的压纸	Pressing the paper at the night of papermaking
	千斤顶榨纸	Pressing the paper with a hydraulic lifting jack
	放置在纸架上的纸沓	Paper pile on papermaking frame
	将纸一张张分离并揭下	Separating the paper piece by piece
	刷纸上墙	Pasting the paper on a drying wall
	火墙烘干	Drying the paper on a drying wall
	质量检验	Quality checking
	使用切纸机裁切	Cuting the paper with a paper cutting machine
	纸帘	Papermaking screen
	纸帘架	Frame for supporting the papermaking screen
	千斤顶	Hydraulic lifting jack
	榨纸架	Frame for pressing the paper
	放浆池	Pulp pool
	半自动喷浆机	Semi-automatic pulp shooting machine
	捞纸槽	Papermaking trough
	小烘墙	Small drying wall
	大烘墙	Big drying wall
	纸架	Papermaking frame
	生物燃料	Biofuel
	刷把	Brush
	切纸机	Paper cutting machine
	尼龙线	Nylon threads
	正在捞纸的工人	Worker scooping and lifting the papermaking screen out of water
	浮玉堂制作的手工纸扇子	Handmade paper fans made in Fuyutang Paper Co., Ltd.
	浮玉堂制作的印有《兰亭序》的加工纸	Processed paper printed with *the Orchid Pavilion* made in Fuyutang Paper Co., Ltd.
	浮玉堂"宣纸"品牌的厂牌	Plaque of Fuyutang "Xuan Paper" Brand
	浮玉堂成品纸仓库	Paper products in the warehouse of Fuyutang Paper Co., Ltd.
	浮玉堂手工纸生产车间	Handmade paper workshop of Fuyutang Paper Co., Ltd.
	陈旭东参展时与客户交流	Chen Xudong communicating with clients at the exhibition
	"本色云龙纸"透光摄影图	A photo of "Yunlong Paper in Original Color" seen through the light
	混料书画"宣纸"透光摄影图	A photo of calligraphy and painting "Xuan paper" with mixed materials seen through the light

章节	图中文名称	图英文名称
第 二 节	杭州千佛纸业有限公司	Section 2 Hangzhou Qianfo Paper Co., Ltd.
	千佛纸业有限公司厂区雨后内景	Internal view of Qianfo Paper Co., Ltd. after raining
	生产车间里的陈晓东	Chen Xiaodong at the workshop
	原千洪纸厂老厂房	Old factory of former Qianhong Paper Factory
	千佛纸业有限公司生产的窗户纸	Window paper produced in Qianfo Paper Co., Ltd.
	千佛纸业有限公司生产的皮纸	Bast paper produced in Qianfo Paper Co., Ltd.
	千佛纸业有限公司生产的书画纸	Calligraphy and painting paper produced in Qianfo Paper Co., Ltd.
	千佛纸业窗户纸纤维形态图（10×）	Fibers of window paper in Qianfo Paper Co., Ltd. (10× objective)
	千佛纸业窗户纸纤维形态图（20×）	Fibers of window paper in Qianfo Paper Co., Ltd. (20× objective)
	千佛纸业窗户纸润墨性效果	Writing performance of window paper in Qianfo Paper Co., Ltd.
	工人正在拣选和裁切山桠皮料	A worker selecting and cutting *Edgeworthia chrysantha* Lindl. bark
	千佛纸业购买的楮皮湿浆料	Paper mulberry bark materials bought by Qianfo Paper Co., Ltd.
	正在打浆的楮皮浆料	Beating paper mulberry bark materials
	裁纸过程中产生的纸边	Deckle edges made during the cutting process
	正在纸药缸中过滤的纸药	Filtering the papermaking mucilage in a vat
	厂区附近的河流	River nearby the factory
	正在浸泡的浆料	Soaking the pulp materials
	荷兰式打浆机加工浆料	Dutch beating machine processing the pulp materials
	捞纸	Scooping and lifting the papermaking screen out of water
	捞纸工人正在放置尼龙线	Papermakers fixing the nylon threads
	将纸帘揭下	Peeling the papermaking screen down
	尼龙线	Nylon threads
	纸帖下方的腈纶网	Acrylic net under the paper pile
	初压	Pressing the paper for the first time
	榨纸器械	Machine for pressing the paper
	揭纸	Peeling the paper down
	刷纸上墙	Pasting the paper on the drying wall
	质量检验	Quality checking
	使用切纸机裁切	Cutting the paper with a paper cutting machine
	包装好的纸	Packed paper products
	生产窗户纸的纸帘	Papermaking screen for making window paper
	帘床	Frame for supporting the papermaking screen
	千斤顶和榨纸架	Hydraulic lifting jack and pressing frame

朱有善收集的老纸帘与纸架（97 cm×38 cm）	Old papermaking screen and its supporting frame collected by Zhu Youshan (97cm×38 cm)
老切纸刀 (68 cm×34.5 cm)	Old paper cutting knife (68 cm×34.5 cm)
老磨纸刀	Old rubbing knife
老刮青刀	Old Stripping knife
老断料刀	Old cutting knife
清代磨纸石	Stone for rubbing the paper in the Qing Dynasty
生产队和合作社时期的印章	Stamps during the period from 1958 to 1984
染色一级元书纸	Dyed Yuanshu paper of first class
"一级元书"章	Stamp of "Yuanshu Paper of First Class"
"行楷小品"章	Stamp of "Xingkai Xiaopin"
行楷、小楷用途纸样	Paper sample for Xingkai and Xiaoka
毛竹山桠混料纸纸样	Paper sample made from mixed materials
"高级书画纸"章	Stamp of "Super Calligraphy and Painting Paper"
逸古斋混料"富阳宣"纸样	Paper sample of "Fuyang Xuan" paper in Yiguzhai
"古籍修复"章	Stamp of "Paper for Mending Ancient Books"
为浙江大学图书馆做的修复纸纸样（苦竹30%，毛竹70%）	Paper sample for mending ancient books for the Library of Zhejiang University (*Pleioblastus amarus* 30% and *Phyllostachys edulis* 70%)
不同色调的纯毛竹修复纸	Different hues of paper for mending ancient books made from *Phyllostachys edulis*
古籍印刷用纸	Paper for printing ancient books
"木刻水印"章	Stamp of "Muke Shuiyin"(traditional Chinese craft for duplication)
苦竹纸刷印（良才墨业油烟墨）清代朱师辙著《黄山樵唱》	*Woodman Antiphonal Singing in Mount Huangshan* written by Zhu Shizhe in the Qing Dynasty printed on *Pleioblastus amarus* paper (lampblack provided by Liancai Ink Company)
新刷印的古版画作品	Newly printed ancient woodcut
石墨打印经卷竹纸	Bamboo paper for buddhist scripture printed with graphite
逸古斋毛竹元书纸纤维形态图（10×）	Fibers of *Phyllostachys edulis* Yuanshu paper in Yiguzhai (10× objective)
逸古斋毛竹元书纸纤维形态图（20×）	Fibers of *Phyllostachys edulis* Yuanshu paper in Yiguzhai (20× objective)
逸古斋毛竹元书纸润墨性效果	Writing performance of *Phyllostachys edulis* Yuanshu paper in Yiguzhai
逸古斋苦竹元书纸纤维形态图（10×）	Fibers of *Pleioblastus amarus* Yuanshu paper in Yiguzhai (10× objective)
逸古斋苦竹元书纸纤维形态图（20×）	Fibers of *Pleioblastus amarus* Yuanshu paper in Yiguzhai (20× objective)
逸古斋苦竹元书纸润墨性效果	Writing performance of *Pleioblastus amarus* Yuanshu paper in Yiguzhai
逸古斋毛竹染色元书纸纤维形态图（10×）	Fibers of dyed *Phyllostachys edulis* Yuanshu paper in Yiguzhai (10× objective)
逸古斋毛竹染色元书纸纤维形态图（20×）	Fibers of dyed *Phyllostachys edulis* Yuanshu paper in Yiguzhai (20× objective)
逸古斋毛竹染色元书纸润墨性效果	Writing performance of dyed *Phyllostachys edulis* Yuanshu paper in Yiguzhai
大同村附近的毛竹林	*Phyllostachys edulis* forest near Datong Village
苦竹林	*Pleioblastus amarus* forest
山溪水源头与流经纸坊附近的山溪	Source of stream and the stream flowing by the papermaking mill
用小推车运到料塘边的石灰	Lime carried by cart beside the soaking pool
豆浆发酵苦竹原料工序	Procedures of fermenting raw material of *Pleioblastus amarus* with soybean milk
砍竹	Cutting the bamboos
山间就地而建的刮皮工具（削竹马）	Tools for stripping the bark in the mountain area
朱中华演示刮皮	Zhu Zhonghua showing how to strip the bark
刮青皮	Stripping the bark
拷白	Beating the bamboo into pieces
落塘浸泡	Soaking
洗坯	Cleaning the bamboo pieces
断料	Cutting the bamboo sections
捆好的料	Bundles of bamboo sections
浆石灰	Liming the bamboo materials
堆蓬	Stacking the bamboo sections up
皮镬	Stone utensil
落镬	Putting the bamboo sections into the utensil
蒸煮	Steaming and boiling the bamboo materials
蒸煮后朱中华观察竹纤维分丝帚化程度	Zhu Zhonghua observing the fibrillation of bamboo
出镬	Taking the bamboo materials out of the stone utensil
山溪水清洗浸泡	Cleaning and soaking the bamboo materials in a stream
最后一次清洗	Cleaning the bamboo materials for the last time
捆扎	Tie the bamboo materials
淋尿工序	Pouring urine on the bamboo materials
落塘20～30天后发青黑色的池水	Dark green water after 20 to 30 days of soaking
甘草覆盖	Covering the soaking pool with hay
压榨竹页	Pressing the bamboo materials
木碓碓打	Beating the materials with a wooden pestle
电动打浆机的回形浆料池	Oval pulp pool applying with electric beating machine
吊帘抄纸的主要工序动作	Major procedures of papermaking with a hanged papermaking screen
朱起扬在用千斤顶压榨湿纸垛	Zhu Qiyang pressing the wet paper pile with a lifting jack
晒纸工序	Procedures of drying the paper
裁切纸边	Trimming the paper edges
压平纸	Flattening the paper
捆扎	Binding the paper
磨边	Trimming the paper dege
盖边	Stamping the paper
断料刀	Bamboo cutting knife
浆料二齿耙	Stirring rake with two teeth
磨刮青刀	Sharpening the stripping knife
拷白榔头	Hammer for beating the stripped bamboo
和单槽棍	Stirring stick
沉淀浆（等待抄纸）	Precipitating the pulp (waiting for papermaking)
纸帘	Papermaking screen
帘架	Frame for supporting the papermaking screen
鹅榔头	Wooden hammer for separating the paper Layers
松毛刷	Brush made of pine needles
2017年时正在新砌传统的土焙墙	Building the traditional drying wall in 2017
浙江省图书馆原馆长徐晓军（中）深夜在煮料炉膛旁	Xu Xiaojun, former curator of Zhejiang Provincial Library standing beside the steamer at midnight
朱中华在仔细判断发酵的程度	Zhu Zhonghua checking the fermentation degree of bamboo carefully
调查组访谈傅尚公	Researchers interviewing Fu Shanggong

章节	图中文名称	图英文名称
	浆腌	Fermenting the materials
	堆放发酵	Piled materials for fermenting
	煮料	Boiling the materials
	翻滩漂洗	Repeat cleaning the materials
	堆蓬	Piled materials
	落塘	Soaking the materials in the pool
	打好的浆料	Pulp materials after beating
	捞纸	Papermaking
	压榨用的榨床	Pressing device used in the pressing procedure
	压榨用的千斤顶	Lifting jack used in the pressing procedure
	牵纸	Peeling the paper down
	牵出待晒的湿纸	Wet paper to be dried
	晒纸	Drying the paper
	晒纸竹架	Bamboo frame for drying the paper
	数纸检纸	Counting and checking the paper
	包扎成捆的成品纸	Bundles of paper
	完成包装入库存放的成品竹纸	Packaged bamboo paper in the storehouse
	一隔三的纸帘	Papermaking screen that can make three piece of paper simultaneously
	帘床	Frame for supporting the papermaking screen
	电动打浆机	Electric beating machine
	灰耙	Grey rake
	两齿耙	Two tine rake
	料耙	Stirring rake
	榨床	Pressing device
	纸槽	Papermaking trough
	断料凳	Bench for cutting the materials
	砍竹刀	Knife for cutting the bamboo
	仓库中待出售的纸	Paper to be sold in the storehouse
	姜仁来纸坊外观	External view of Jiang Renlai
	敞开或虚掩上门的纸坊	Paper mill with unclosed doors
	山基村村名碑石	Village name monument stone of Shanji Village
	村里仍在造纸的老人	Old man who still making paper in the village
	山基村盖着毛竹的料塘	Soaking pool covered with *Phyllostachys edulis* in Shanji Village
	山基村登山休闲步道介绍	Introduction to the hiking pathway of Shanji Village
	祭祀纸（毛竹+纸边+废纸袋）透光摄影图	A photo of joss paper(*Phyllostachys edulis*+paper edges+waste paper bag) seen through the light

章节	图中文名称	图英文名称
第七节	戚吾樵纸坊	Section 7 Qi Wuqiao Paper Mill
	大西庵遗址	Relics of Daxi Nunnery
	石牛山	Shiniu Mountain
	戚吾樵纸坊	Qi Wuqiao Paper Mill
	纸坊中的戚吾樵	Qi Wuqiao in the paper mill
	葛夏芹正在揭纸	Ge Xiaqin peeling the paper down
	捆扎好的祭祀竹纸	A bundle of bamboo paper for sacrificial purposes
	戚吾樵纸坊祭祀竹纸纤维形态图（10×）	Fibers of bamboo paper for sacrificial purposes in Qi Wuqiao Paper Mill (10× objective)
	戚吾樵纸坊祭祀竹纸纤维形态图（20×）	Fibers of bamboo paper for sacrificial purposes in Qi Wuqiao Paper Mill (20× objective)
	老竹碎末	Scrap of old bamboo
	废纸边	Abandoned paper edge
	山溪源头水	Source of the mountain stream
	水源pH测试	Testing the water source pH
	纸坊前堆放的毛竹	*Phyllostachys edulis* piled before the paper mill
	砍竹刀	Knife for cutting the bamboo
	落塘	Soaking the materials in the pool
	榨干的竹料	Dried bamboo materials
	石磨磨料	Grinding the materials with a stone roller
	抄纸	Papermaking
	压榨后的湿纸块	Wet paper pile after pressing
	分开纸张	Separating the paper
	院子里晒的纸	Drying paper in the yard
	数纸	Counting the paper
	压纸	Pressing the paper
	捆扎纸品	Binding the paper
	石磨	Stone roller
	打浆槽	Beating though
	抄纸槽	Papermaking trough
	纸帘	Papermaking screen
	帘架	Frame for supporting the papermaking screen
	鹅榔头	Goose hammer
	浆瓢	Ladle for scooping the pulp
	和单槽棍	Stirring stick
	戚吾樵家中阴干的祭祀竹纸	Paper for sacrificial purposes drying in Qi Wuqiao's house
	叠元宝的老人	The elderly folding the paper ingot
	小屋前燃烧的香和红烛	Joss sticks and red candles burned in front of the hut
	捆扎好的纸品	A bundle of paper
	废弃的蒸锅	Abandoned steamer
	废弃的料塘	Abandoned soaking pool
	祭祀纸（毛竹+废纸边）透光摄影图	A photo of joss paper(*Phyllostachys edulis*+waste paper edges) seen through the light

章节	图中文名称	图英文名称
第八节	张根水纸坊	Section 8 Zhang Genshui Paper Mill
	纸坊外的乡村小景	View of the village outside the paper mill
	张根水纸坊的准确位置标识	The exact location of Zhang Genshui Paper Mill
	纸坊附近的生态环境	Ecological environment near the paper mill
	纸坊外观	The external view of the paper mill
	张根水在料塘边介绍原料制作工艺	Zhang Genshui introducing the raw material production procedure by the soaking pool
	张茶花在晒纸间晒纸	Zhang Chahua drying the paper in the drying room
	纸坊主要纸品——祭祀纸“白纸”	The main paper type of the paper mill – joss paper named “white paper”
	堆放的祭祀纸	Piled joss paper
	张根水纸坊祭祀纸纤维形态图（10×）	Fibers of joss paper in Zhang Genshui Paper Mill (10× objective)

章节	图中文名称	图英文名称
	五四村新制桃花纸纤维形态图（10×）	Fibers of newly made Taohua paper in Wusi Village (10× objective)
	五四村新制桃花纸纤维形态图（20×）	Fibers of newly made Taohua paper in Wusi Village (20× objective)
	五四村新制桃花纸润墨性效果	Writing performance of newly made Taohua paper in Wusi Village
	上里的桑树	Mulberry tree in Shangli Area
	富阳本地的构树	Native mulberry tree in Fuyang District
	做“油水”用的梧桐树枝	Chinese parasol parasol tree branches used for papermaking mucilage
	野生猕猴桃枝	Wild Chinese geoseberry branches
	敲皮	Beating the bark
	叶汉山在切皮料	Ye Hanshan cutting the back materials
	切成15 mm长度的树皮小块	Bark cutting into small pieces 15 mm in length
	老技工吴金坤在野外煮料	Skilled mechanic Wu Jinkun boiling the materials outside the mill
	叶汉山冲洗黑液	Ye Hanshan cleaning the bark materials
	水椎碓舂料	Beating the materials with a hydraulic wooden pestle
	舂碎的细料	Fine beaten material
	老技工吴庆荣在河里洗黑液	Skilled mechanic Wu Qingrong cleaning the bork materials in the River
	水缸漂料	Bleaching the materials
	摇白	Shaking the materials in a bucket
	打槽	Stirring the materials
	打前浪	Scooping the papermaking screen from the front side
	抄后浪	Scooping the papermaking screen from the back side
	揭纸帘	Uncovering the papermaking screen
	把湿纸覆盖到纸桩上	Piling the wet paper on a board
	压榨	Pressing the paper
	从湿纸块上揭下湿纸	Peeling wet paper down from wet paper stack
	将湿纸贴到烘纸墙上	Pasting wet paper on the drying Wall
	收纸	Peeling the paper down
	叶汉山示范检验桃花纸	Ye Hanshan demonstrating how to check Taohua paper
	新制桃花纸成品	Newly made Taohua paper
	蒸煮原料的不锈钢锅	Stainless steel cooker for steaming and boiling the raw materials
	皮笪	Bamboo basket
	水椎碓（中间的木制工具）	Hydraulic wooden pestle (the middle one)
	旧的草纸料袋（重新制作皮纸料袋仿制样本）	Old straw paper material bags (remaking imitation samples of bast paper bags)
	退休教师夏荣芳正在缝制料袋	Retired teacher Xia Rongfang sewing the material bags
	料耙（右起第一个工具）	Stirring rake (the first tool from the right)
	料缸	Matenrial vat for storing and making the materials
	吴金坤在制作摇白桶	Wu Jinkun making the shaking bucket
	摇白桶	Shaking bucket
	新制的纸槽与帘床	Newly made papermaking trough and the screen frane
	汪美英新制作的皮纸纸帘	Newly made bark papermaking screen made by Wang Meiying
	细密的纸帘局部	A part of fine papermaking screen
	搁置纸帘的帘床	Frame for supporting the papermaking screen
	叶汉山手持槽棒打槽	Ye Hanshan stirring the materials with a stirring stick
	压榨床（摇白桶与纸槽中间的设施）	Pressing device (between the shaking bucket and papermaking trough)
	叶汉山在压榨床上加放小木条	Ye Hanshan adding small wooden strips to the pressing device
	湖源乡颜小平正在制作松毛刷	Yan Xiaoping from Huyuan Town making brushes made of pine needles
	竹笪、敲皮木榔头、料耙、铁锹	Bamboo basket, wooden hammer for beating the bark, material rake and spade
	吴明村保存的新华造纸厂请帖	Invitation card of Xinhua Papermaking Factory stored by Wu Mingcun
	吴明富收藏的“爱国皮纸”方印	Seal of “Patriot Bast Paper” collected by Wu Mingfu
	曾任五四造纸厂厂长的何关元（左一）和夏明汉（右一）	He Guanyuan (first one from the left) and Xia Minghan (first one from the right) - former heads of Wusi Papermaking Factory
	叶汉山（站立者）讲述皮纸生产中的伤人事故	Ye Hanshan (standing one) relating the accidents happened in bast paper production procedure
	调研桃花纸制作工艺	Researching production procedure of Taohua paper
	展示叶汉山新制桃花纸用墨效果	Demonstrating the writing performance of newly made Taohua paper by Ye Hanshan
	新制桃花纸上的书写效果	Writing effect on newly made Taohua paper
	新制杨皮桃花纸透光摄影图	A photo of newly made paper mulberry bark Taohua paper seen through the light

章节	图中文名称	图英文名称
第 二 节	大山村桑皮纸恢复点	Section 2 Mulberry Paper Recovery Site in Dashan Village
	流经村边的大山溪（左下方）	Dashan Stream running through the village (bottom left) `
	大山村村口的路标	Road sign at the entrance of Dashan Village
	俯瞰大山村	Overlooking Dashan Village
	清道光年间《新登县志》中的文字记载	Written records in *The Annals of Xindeng County* during Daoguang Reign of the Qing Dynasty
	浇纸纸帘（当地人称“戗”）	Papermaking screen (the locals calling it “Qiang”)
	浇纸托架（当地人称“戗架”）	Frame for supporting the papermaking screen (the locals calling it “Qiangjia”)
	黄洪渭家附近的山地	Mountain field near Huang Hongwei's house
	大山村纸农保存的20世纪80年代初手工造的桑皮纸	Handmade mulberry paper produced in the early 1980s,preserving by in the Dashan Village
	访谈年老卧床的黄玉铨老人	Interviewing Huang Yuquan, the old man who stayed in bed
	大山村旧日造的草皮纸	Straw paper produced in the old days in Dashan Village
	大山村草皮纸纤维形态图（10×）	Fibers of straw paper in Dashan Village (10× objective)
	大山村草皮纸纤维形态图（20×）	Fibers of straw paper in Dashan Village (20× objective)
	刚刚砍下的新鲜桑树枝条	Fresh cut mulberry branches
	沙田早灿米稻草	Early rice straw in sandy field
	大山村附近山间生长的滑涅树	*Litsea cubeba* tree growing on the mountains near Dashan village
	流经村里的山溪水	Mountain stream flowing through the village
	老纸工沈健根在敲打剥取桑皮	Shen Jiangen, an old papermaker beating and stripping the mulberry bark
	刚剥出的桑皮	Freshly peeled mulberry bark
	沈健根在化石灰	Shen Jiangen melting the lime
	黄洪渭（左）和沈健根（右）在腌料	Huang Hongwei (left) and Shen Jiangen (right) fermenting the materials
	缪红潮在蒸煮桑皮和稻草	Miao Hongchao steaming and boiling the mulberry back and straw
	缪红潮、沈健根在活水漾塘里漾料漂洗	Miao Hongchao and Shen Jiangen cleaning the materials in the cleaning pool
	沈健根在踩细稻草料	Shen Jiangen stamping fine straw materials
	用小竹条或小木条多次搅拌	Stirring the pulp multiple times with bamboo strips or wooden strips
	捡掉粗、长的纸筋	Picking out the thick, long paper residues
	罗忠火用棒槌敲打桑皮料	Luo Zhonghuo beating mulberry bark materials with a wooden mallet
	踩蹚滑涅叶	Stamping the leaves of *Litsea Cubeba*
	过滤滑涅汁	Filtering *Litsea cubeba* sap
	缪柏堂在调制皮水（纸浆）	Miao Baitang making paper pulp
	倒浆浇纸	Pouring the pulp to make paper
	用羽毛刷平纸浆	Flattening the paper pulp with feather brushes

	光明制帘厂厂区	Guangming Screen-making Factory
	汪美英正在织帘	Wang Meiying making the screen
	光明村周边环境	Surrounding environment of Guangming Village
	灵桥中学内遗存的灵岩寺旧水池	The old pool of Lingyan Temple in Lingqiao Middle School
	抽好的苦竹丝	*Pleioblastus amarus* filament after threading
	涤纶线	Polyester thread
	生漆	Raw paint for painting the screen
	竹片切簧、去芯	Cutting the bamboo and removing the core
	汪美英演示抽丝	Wang Meiying demonstrating the threading procedure
	帘丝密度特写	Close-up of density of the screen filament
	编帘	Making the screen
	汪美英在装箸竹	Wang Meiying packing the bamboo for fixing
	绷架	Stretcher
	方如堂在漆帘	Fang Rutang painting the screen
	劈竹刀	Iron tool for cutting the bamboo
	织帘木机	Wooden room for making the Screen
	木钻	Wooden drill
	绷架	Stretcher
	作坊内刷漆间场景	Scene of the paint room in the workshop
	光明制帘厂生产的茶帘	Screen for tea produced by Guangming Screen-making Factory
	工作中的方如堂和汪美英	Fang Rutang and Wang Meiying at work
	汪美英参加富阳市竹文化艺术展的感谢状	Wang Meiying's certificate of appreciation for participating in Fuyang Bamboo Culture Art Exhibition
	汪美英被评为第三批杭州市非物质文化遗产项目代表性传承人证书	The certificate of Wang Meiying as the third representative inheritor of Hangzhou Intangible Cultural Heritage Project

章节	图中文名称	图英文名称
第三节	郎仕训刮青刀制作坊	Section 3 Lang Shixun Scraping Knife-making Mill
	郎仕训刮青刀制作坊	Lang Shixun Scraping Knife-making Mill
	在打铁铺与郎仕训交流	Interviewing Lang Shixun in the iron shop
	在打铁铺与庄美华交流	Interviewing Zhuang Meihua in the iron shop
	刮青刀	Scraping knife
	演示刮青刀削竹动作	Demonstrating scraping the bamboo with a scraping knife
	A3钢	A3 steel
	45号钢（夹在中间的为45号钢）	No. 45 steel (middle)
	裁料	Cutting the materials
	炉灶生火	Making fire in a stove
	鼓风机鼓风	Blowing with a blast furnace
	打钢时夹住钢片的钳子	Pliers holding the steel sheets while hitting the steel
	烧红铁块	Burning iron block
	空气锤第一轮锤打	Air hammer first round hammering
	铁锤第一轮锤打	Iron hammer first round hammering
	铁锤第二轮锤打	Iron hammer second round hammering
	空气锤第二轮锤打	Air hammer second round hammering
	空气锤第三轮锤打	Air hammer third round hammering
	铁锤第三轮锤打	Iron hammer third round hammering
	空气锤第四轮锤打	Air hammer fourth round hammering
	铁锤第四轮锤打	Iron hammer fourth round hammering
	初具刀型	Original knife form
	砂轮	Grinding wheel
	砂轮打磨	Grinding the blade with a grinding wheel
	铲口造刃	Sharpening the blade with a shovel
	砂轮打光	Polishing the blade with a grinding wheel
	锉刀打磨	Grinding the blade with a file
	砂轮打磨	Grinding the blade with a grinding wheel
	上销	Fitting the knife into a pine handle
	冷却	Cooling
	空气锤	Air hammer
	铁锤（上边为大锤，下边为小锤）	Iron hammer (sledgehammer on the top and small hammer on the bottom)
	各种用途的钳子	Pliers for various purposes
	锉刀	File
	铲子	Shovel
	锉刀容器	File Container
	砂轮机	Grinding machine
	墩头	A support tool when hammering the knife with a hammer
	刮青刀	Scraping knife
	作坊里谈到市场显得无奈的郎仕训	Helpless Lang Shixun when talking about the marketing status
	郎家湾公园	Langjiawan Park
	锄头上的勾刀	Hook on the hoe
	锤子	Hammer
	郎仕训为调研组制作的刮青刀	A scraping knife made by Lang Shixun for the researchers
	富阳区大源镇郎仕训刮青刀制作坊周围环境	Surrounding environment of Lang Shixun Scraping Knife-making Mill in Dayuan Town of Fuyang District

表目
Tables

术语
Terminology

地　　理　　名 Places

汉语术语 Term in Chinese	英语术语 Term in English
安岱后村	An'aohou Village
安吉县	Anji County
安民乡	Anmin County
岙底村	Ao'di Village
岙外村	Ao'wai Village
宝山村	Baoshan Village
报福镇	Baofu Town
碧子坞村	Biziwu Village
蔡家坞村	Caijiawu Village
苍南县	Cangnan County
昌东镇	Changdong Town
昌化镇	Changhua Town
常安镇	Chang'an Town
常绿镇	Changlv Town
常山县	Changshan County
常熟市	Changshu City
朝阳南路	Chaoyangnan Road
城关镇	Chengguan Town
城阳乡	Chengyang Town
崇安县	Chong'an County
稠溪村	Chouxi Village
淳安县	Chun'an County
慈溪市	Cixi City
村头镇	Cuntou Town
大葛村	Dage Village
大潘坑村	Dapankeng Village
大山村	Dashan Village
大山坞	Dashanwu Village
大田村	Datian Village
大同村	Datong Village
大同坞村	Datongwu Village
大源镇	Dayuan Town
点口村	Diankou Village
定海区	Dinghai District
东边大山坞	Dongbiandashanwu Village
东坞村	DongwuVillage
东阳市	Dongyang City
渡头村	Dutou Village
墩头村	Duntou Village
方家地村	Fangjiadi Village
分水镇	Fenshui Town
枫凌村	Fengling Village
奉化区	Fenghua District
福安县	Fu'an County
福鼎县	Fuding County
富阳县	Fuyang County
富阳区	Fuyang District
高桥镇	Gaoqiao Town
葛村	Gecun Village
观音塔村	Guanyinta Village
冠形塔村	Guanxingta Village
光明村	Guangming Village
杭县	Hangxian County
后溪村	Houxi Village
湖岭镇	Huling Town
湖源乡	Huyuan County
华埠镇	Huabu Town
华家村	Huajia Village
华堂村	Huatang Village
黄弹村	Huangdan Village
黄家村	Huangjia Village
黄岩区	Huangyan District
嘉兴市	Jiaxing City
建德市	Jiande City
江家村	Jiangjia Village
江山市	Jiangshan City
蒋家村	Jiangjia Village
金华市	Jinhua City
缙云县	Jinyun County
景宁县	Jingning County

开化县	Kaihua County
柯桥区	Keqiao District
坑口村	Kengkou Village
葵溪	Kuixi Village
李坑村	Likeng Village
丽水市	Lishui City
连城县	Liancheng County
林岙村	Lin'ao Village
临安市	Lin'an City
临海市	Linhai City
灵桥镇	Lingqiao Town
岭北乡	Lingbei County
刘家村	Liujia Village
柳峰乡	Liufeng County
柳溪村	Liuxi Village
六石村	Liushi Village
六渚乡	Liuzhu County
龙王村	Longwang Village
龙王殿村	Longwangdian Village
龙游县	Longyou County
渌渚镇	Luzhu Town
骆村	Luocun Village
沐尘乡	Muchen County
南安乡	Nan'an County
南平市	Nanping City
倪家滩村	Nijiatan Village
宁波市	Ningbo City
宁海县	Ninghai County
瓯海区	Ou'hai District
潘家村	Panjia Village
泮家坞村	Panjiawu Village
平渡村	Pingdu Village
平水镇	Pingshui Town
平阳县	Pingyang County
浦江县	Pujiang County
千洪乡	Qianhong County
千茂村	Qianmao Village
潜阳镇	QianyangTown
桥上山村	Qiaoshangshan Village
秦骆村	Qinluo Village
秦坞村	Qinwu Village
勤建村	Qinjian Village
青田县	Qingtian County
庆元县	Qingyuan County
瞿溪镇	Quxi Town
瑞安市	Rui'an City
三合村	Sanhe Village
三岭村	Sanling Village
三桥乡	Sanqiao County
三山乡	Sanshan County
三支村	Sanzhi Village
山基村	Shanji Village
山坑里村	Shankengli Village
山坑坞里村	Shankengwuli Village
山口镇	Shankou Town
山毛坞口	Shanmaowukou Village
上官乡	Shangguan County
上里片	Shanglipian Village
上墅乡	Shangshu County
上旺村	Shangwang Village
上叶村	Shangye Village
上虞区	Shangyu District
绍兴市	Shaoxing City
绍兴县	Shaoxing County
胜总村	Shengzong Village
嵊县	Shengxian County
嵊州市	Shengzhou City
施阮村	Shiruan Village
寿昌镇	Shouchang Town
寿宁县	Shouning County
双联村	Shuanglian Village
双桥村	Shuangqiao Village
水碓坑村	Shuiduikeng Village
寺前村	Siqian Village
寺前村	Siqian Village
泗州村	Sizhou Village
松溪县	Songxi County
松溪乡	Songxi Town
松阳县	Songyang County
宋家店村	Songjiadian Village
遂安县	Sui'an County
遂昌县	Suichang County
泰顺县	Taishun County
汤溪县	Tangxi County
唐宅村	Tangzhai Village
棠岙村	Tang'ao Village
桃坑村	Taokeng Village
藤桥镇	Tengqiao Town
天台县	Tiantai County
桐庐县	Tonglu County
外水良村	Waishuiliang Village
温州市	Wenzhou City
榅桥村	Wenqiao Village
文成县	Wencheng County
乌浪村	Wulang Village
吴兴区	Wuxing District
五四村	Wusi Village
五玉岭村	Wuyuling Village
坞口村	Wukou Village
西坞村	Xiwu Village
溪口镇	Xikou Town
溪下村	Xixia Village
下叶村	Xiaye Village
夏履乡	Xialv County
仙居县	Xianju County
象山县	Xiangshan County
萧山区	Xiaoshan District
小源镇	Xiaoyuan Town
小章村	Xiaozhang Village
筱村镇	Xiaocun Town
孝丰镇	Xiaofeng Town
新昌县	Xinchang County

新登镇	Xindeng Town
新店村	Xindian Village
新二村	Xin'er Village
新华村	Xinhua Village
新浦村	Xinpu Village
新浦片	Xinpupian Village
新三村	Xinsan Village
新桐乡	Xintong County
新祥村	Xinxiang Village
新新村	Xinxin Village
新一村	Xinyi Village
形边村	Xingbian Village
旭光村	Xuguang Village
宣平县	Xuanping County
牙阳	Yayang
颜家村	Yanjia Village
颜家桥村	Yanjiaqiao Village
窑口村	Yaokou Village
鄞县	Yinxian County
永嘉县	Yongjia County
於潜镇	YuqianTown
於潛镇	Yuqian Town
余杭区	Yuhang District
余姚市	Yuyao City
渔山乡	Yushan County
虞山镇	Yushan Town
玉岩镇	Yuyan Town
袁家村	Yuanjia Village
月台村	Yuetai Village
越州	Yuezhou City
云和县	Yunhe County
泽雅镇	Zeya Town
张村坞村	Zhangcunwu Village
张村乡	Zhangcun County
兆吉村	Zhaoji Village
柘荣县	Zherong County
桢祥村	Zhenxiang Village
镇海区	Zhenhai District
钟塔村	Zhongta Village
周岙上村	Zhou'aoshang Village
周八坞	Zhoubawu Village
朱家门村	Zhujiamen Village
诸暨市	Zhuji City
庄家村	Zhuangjia Village
庄家坞村	Zhuangjiawu Village

纸　　品　　名 Paper names

汉语术语 Term in Chinese	英语术语 Term in English
“白泉”牌温州皮纸	Wenzhou Baiquan bast paper
蔡侯纸	Caihou paper
常山纸	Changshan paper
“进三”纸	Jinsan paper
“李坑”牌绵纸	Likeng Mian paper
“山山居”元书纸	Shanshanju Yuanshu paper
上虞纸	Shangyu paper
十色笺	Shise (Ten-color) paper
“寿牌宣纸”	Shoupai Xuan paper
“曙光”牌打字蜡纸	Shuguang waxed paper for type-writing
“宣纸”	Xuan paper
左伯纸	Zuobo paper
案纸	An paper
八百张	Babaizhang paper
白构皮“本色云龙纸”	Yunlong unbleached white mulberry bark paper
白蜡纸	White waxed paper
白历日纸	Bailiri paper
白钱纸	Baiqian paper
白唐纸	Baitang paper
白纸	White paper
榜纸	Bang paper (used in the imperial examination and official notices)
包扎毛边纸	Writing Paper Made from Bamboo for packing
本色仿古元书纸	Unbleached bamboo Paper made by Traditional Method
本色元书纸	Unbleached bamboo paper
边黄	Bianhuang paper
裱心	Biaoxin paper
冰纸	Bing paper (moistened with salt water)
泊纸	Bo paper
彩色粉笺	Delicate colored paper
蚕生纸	Cansheng paper
蚕种纸	Canzhong paper
草皮纸	Caopi paper
草纸	Straw paper
侧理纸	Celi paper
茶白纸	Chabai (tea-leave-color paper)
昌山纸	Changshan paper
厂黄	Changhuang paper
澄心堂笺	Chengxintang paper
赤亭纸	Chiting paper
楮皮纸	Mulberry bark paper
楮纸	Mulberry bark paper
处牌行移伞纸	Chupai Xingyi Umbrella paper
窗户纸	Window paper shades
纯皮纸	Pure bark paper
瓷青纸	Porcelain green paper
次纸	Ci paper
粗高	Cugao paper
粗纸	Cu paper
大淡	Dadan paper
大笺纸	Dajian paper
大簾粗纸	Dalian coarse paper
大簾纸	Dalian paper
大皮纸	Dapi paper
稻秆纸	Rice straw paper
东陈	Dongchen paper
东纸	Dong paper
斗方纸	Doufang paper
斗坊	Doufang paper
斗纸	Dou paper
短聊竹纸	Duanliao Bamboo paper
段放	Duanfang Paper

二细纸	Erxi paper
矾红纸	Fanhong paper
方高纸	Fanggao paper
方稿纸	Fanggao paper
方纸	Fang paper
粉云罗笺	Fenyunlong letter paper
富阳油纸	Fuyang oiled paper
改连	Gailian paper
工业滤油纸	Industrial oil filter paper
贡聊	Gongliao paper
古法纯手工元书纸	Handmade paper made with ancient methods
古籍修复用纸	Paper for ancient books restoration
古籍印刷纸	Paper for printing ancient books
谷纸	Gu paper
海放	Haifang paper
红鼎	Hongding paper
花笺纸	Huaxian letter paper
画仙纸	Huaxian paper
黄白纸	Huangbai paper
黄标纸	Huangbiao paper
黄表纸	Huangbiao paper
黄公纸	Huanggong paper
黄笺	Yellow letter Paper
黄历日纸	Huangliri paper
黄皮纸	Huangpi Paper
黄烧纸	Huangshao paper (paper burned as sacrificial offerings)
黄檀	Huangtan paper
黄纸	Yellow Paper
会稽竹纸	Kuaiji bamboo paper
混合原料书画纸	Mixed material calligraphy and painting paper
火烧纸	Huoshao paper
极品蚕丝纸	Super Cansi paper
祭祀用纸	Sacrificial paper
监钞纸	Jianchao paper
交白	Jiaobai paper
金钱纸	Jinqian paper
金粟山藏经纸	Jinsushan scripture paper
京边纸	Jingbian paper
京放纸	Jingfang paper
蠲纸	Juan paper
军工绵纸	Mian paper for military purposes
开化纸	Kaihua paper
坑边	Kengbian paper
苦竹屏纸	*Pleioblastus amarus* Ping paper
历日纸	Liri paper
六尺条屏	6-chi Tiaoping paper
六局纸	Liuju paper
六千元书	Liuqianyuan paper
龙游屏纸	Longyou Ping paper
龙游宣纸	Longyou Xuan paper
鹿鸣纸	Luming paper
麦纸	Mai paper
毛边纸	Deckle-edged paper
毛角连	Maojiaolian paper
毛头纸	Maotou paper
梅里笺	Meili letter paper
迷信纸	Mixin paper
绵白纸	Mianbai paper
绵纸	Mian paper
名槽	Mingcao paper
墨煤	Momei paper
南屏纸	Nanping paper
皮抄纸	Pichao paper
皮纸	Bark paper
屏纸	Ping paper
七连纸	Qilian paper
漆纸	Qi paper
千张纸	Qianzhang paper
敲冰纸	Qiaobing paper
秋皮纸	Qiupi paper
衢州纸	Quzhou paper
日历纸	Rili paper
如厕草纸	Toilet paper
三顶	Sanding paper
桑皮桃花纸	Mulberry bark Taohua paper
桑皮纸	Mulberry bark paper
山里纸	Shanli paper
山桠皮纸	Shanyapi paper
烧纸	Shao paper
绍兴竹纸	Shaoxing bamboo paper
书画纸	Painting and calligraphy paper
四尺本色元书纸	4-chi unbleached Yuanshu paper
四尺生宣纸	4-chi uncooked Xuan paper
苦竹乌金原纸	*Pleioblastus amarus* Wujin paper
四号薄	Sihaobu paper
四六屏纸	Siliuping paper
松纸	Song paper
笋壳纸	Bamboo shoot shell paper
苔笺	Tai letter paper
棠岙纸	Tang'ao paper used for restoration of ancient books
塘纸	Tang Paper
桃花纸	Taohua paper
特级元书纸	Super Yuanshu paper
藤纸	Rattan paper
铁笔蜡纸原纸	Tiebi wax paper
铁笔蜡纸	Tiebi wax paper
乌金原纸	Wujin raw paper
乌金纸	Wujin paper
五连纸	Wulian Paper
五千元书	Wuqianyuan paper
锡箔纸	Tin foil paper
细粗纸	Xicu paper
细黄状纸	Xihuangzhuang Paper
小井纸	Xiaojing Paper
小粗纸	Xiaocu Paper
小簾笺	Xiaolian letter Paper
小细蓬	Xiaoxipeng paper
小元书	Xiaoyuanshu Paper
小纸蓬	Xiaozhipeng Paper
小竹纸	Xiaozhu bamboo paper
谢公笺	Xiegong letter paper

徐青纸	Xuqing paper
押头纸	Yatou paper
剡藤纸	Shanteng paper
剡溪藤纸	Shanxi rattan paper
雁皮纸	*Wikstroemia pilosa Cheng* bark paper
一道水元书纸	Yuanshu paper (with single scooping process)
一级元书纸	Super Yuanshu paper
银皮纸	Yinpi paper
银色纸	Yinse paper
印刷纸	Printing paper
由拳藤纸	Youquan Rattan paper
由拳纸	Youquan paper
油纸	Oiled paper
鱼卵纸	Yuanluan paper
雨伞纸	Oiled paper used to make umbrella
玉版纸	Yuban paper (high quality Xuan paper)
元书纸	Yuanshu Paper
月面松纹纸	Yuemian Songwen paper
越王纸	Yuewang paper
越纸	Yue Paper
泽雅屏纸	Zeya Ping paper
长帘纸	Changlian Paper
长聊竹纸	Changliao Bamboo Paper
长钱纸	Changqian Paper
折边	Zhebian Paper
正号绵纸	Zhenghao Mian Paper
纸被	Zhibei Paper
纸躅纸	Juan Paper
纸帐	Zhizhang Paper
竹浆纸	Zhujiang paper
竹皮纸	Zhupi paper
竹烧纸	Zhushao paper
竹帖纸	Zhutie paper
竹纸	bamboo paper
硾笺	Zhuijian paper
奏本纸	Zouben paper

原料与相关植物名 Raw materials and plants

汉语术语 Term in Chinese	英语术语 Term in English
5号钢	No. 45 steel
A3 钢	A3 steel
白竹肉	Bai Bamboo shoots
楮皮	Mulberry Bark
楮树皮料	Mulberry Bark
次氯酸钠	Sodium hypochlorite
丛生竹	Group of bamboo
稻草	Straw
稻草浆	Straw pulp
涤纶线	Polyester yarn
多毛荛花茎皮	Stem bark of *Wikstroemia pilosa*
废纸边	Deckle Edge
干桑皮	Dry mulberry bark
干旱稻草	Dry early straw
葛藤	Kudzu vine
构皮	Mulberry bark
构树皮	Paper mulberry bark
滑涅叶	Litsea cubeba leaf
姜黄粉	Turmeric powder
浆板	Pulp board
腈纶线	Acrylic yarn
聚丙烯酰胺	Polyacrylamide
苦竹	*Pleioblastus amarus*
苦竹丝	*Pleioblastus amarus* strip
矿泉水	Mineral water
老毛竹	Mature bamboo
燎草	Processed straw
硫酸	Sulphuric acid
龙须草	*Eulaliopsis binata*
龙须草浆	Pulp of *Eulaliopsis binata*
龙须草浆板	Board of *Eulaliopsis binata*
罗汉松根	*Momordica grosvenori* root
麻	Rattan
毛竹	*Phyllostachys edulis*
毛竹浆板	Pulp board of *Phyllostachys edulis*
毛竹料	*Phyllostachys edulis* material
毛竹肉	*Phyllostachys edulis* shoots
煤	Coal
猕猴桃藤	Branches of *kiwi fruit*
绵浆	Pulp
棉花	Cotton
木浆	Wood pulp
嫩黄染料	Bright yellow pigment
嫩毛竹	Tender *Phyllostachys edulis*
嫩竹	Tender bamboo
尿素	Urea
尿液	Urine
漂白粉	bleaching powder
青檀皮	*Pteroceltis tatarinowii Maxim*. bark
山桠皮	*Edgeworthia chrysantha* Lindl.
桑皮	Mulberry bark
山棉皮	*Wikstroemia monnula* Hance
山泉水	Mountain spring
烧碱	Caustic soda
生漆	Raw lacquer
石灰	Lime
石竹	*Dianthus chinensis* L.
水	Water
水泥袋	Cement bag
水竹	*Phyllostachys heteroclada* Oliver
檀皮	*Wingceltis* Bark
檀树皮	*Wingceltis* Bark
童子尿	Juvenile urine
土漆	Local lacquer
香椿叶	Chinese toon leaves
鲜毛竹	Fresh bamboo
颜料	Pigment
雁皮	*Wikstroemia pilosa Cheng* bark
野生黄柏	Wild golden cypress
野生猕猴桃汁	Juice of kiwi fruit
长梗结香	*Edgeworthia longipes Lace*

纸边	Deckle Edge
竹	Bamboo
竹浆	Bamboo Pulp
竹浆板	Bamboo board
竹青	Bamboo bark

工艺技术和工具设备 Techniques and tools

汉语术语 Term in Chinese	英语术语 Term in English
黑房	Dyehouse
烟房	Room for collecting tobacco tar
爊刷	Steaming bamboo
扒	Beating materials
掰料	Separating bamboo materials
半自动喷浆机	Semi-automatic pulp spraying machine
包装	Packaging
包装入库	Stocking in warehouse
焙壁	Baking wall
焙垅	Baking wall
焙墙	Baking wall
焙纸	Baking paper
绷架	Screen frame
壁弄	Papermaking Wall
扁刷	Flat brush
玻璃片	Glass pieces
剥皮	Peeling tree bark
裁料	Cutting materials
裁纸	Cutting
裁纸刀	Paper-cutting scissors
裁纸机	Paper-cutting machine
采皮	Picking bark
踩料	Treading materials
槽棒	Stick
槽耙	Rake
草耙	Brush
拆纸	Separating paper
柴刀	Chopping knife
铲口造刃	Polishing knife
铲子	Shovel
抄纸	Papermaking
抄纸槽	Papermaking container
抄纸底板	Papermaking board
抄纸桶	Papermaking barrel
朝天	Chaotian (stick for picking out papermaking materials)
沉沙	Settling sand
沉沙池	Pool for settling sand
成品包装	Product packaging
成品捆扎	Product bundling
舂捣	Beating
舂料	Beating materials
舂竹	Beating bamboo
抽丝	Drawing bamboo wire
抽丝工具	Tools for drawing bamboo wire
出镬	Getting out of wok
出镬翻洗	Turning over and washing
出镬清洗	Washing
储料槽	Material container
敲纸药榔头	Hammer for beating paper mucilage
捶打	Beating
锤子	Hammer
淬火定性	Quenching
存放	Storing
锉刀	File
锉刀打磨	Polishing with a file
锉刀容器	File container
打包	Packaging
打包成捆	Bundling
打槽	Beating materials
打槽棍	Beating stick
打浆	Beating
打浆槽	Beating container
打浆池	Beating pool
打浆机	Beating machine
打浆拷白	Beating and crashing bamboo to make pulp
打结头	Fixing the bamboo screen
打捆	Binding Paper
打料	Beating materials
打料碓	Pestle for beating materials
打料耙	Rake for beating materials
大凳	Big stool
掸清	Wiping away impurities while painting the screen
捣烂	Beating
捣料	Beating
捣刷	Beating and washing 打浆，方言
滴管	Dipping
第一次洗料	Washing for the first time
第二次洗料	Washing for the second time
第三次洗料	Washing for the third time
电动脚碓	Electric foot pestle
吊帘	Hanging curtain
跌打拆松	Patting to separate paper
断料	Cutting materials
断料刀	Knife for cutting
断切刀	Knife for cutting
断青	Beating bamboo
断竹	Beating bamboo
断竹个	Beating a bundle of bamboo sections
堆积蒸晒	Stack for drying
堆料场	Stockyard
堆蓬	Covering bamboo with plastic cloth
堆蓬落镬	Putting bamboo into a container for boiling
碓打	Beating
墩头	Board supporting the hammer
鹅榔头	Goose-head hammer used to separate paper
额沿	Trimming the deckle edge
发酵	Fermenting
翻滩漂洗	Washing the materials
翻滩	Turning over materials
翻滩凳	Tool for turning over materials
反复烧热锤打	Beating the boiled material repeatedly

放浆池	Pulp pool
放烧碱水	Adding Alkaline water
分段	Cutting into segments
分劈	Splitting
分纸	Separating paper
敷料	Processing materials
腐竹个	Fermenting bamboo sections
缚料	Binding material
缚纸	Binding paper
盖印	Sealing
盖纸	Covering paper
缸	Cylinder
购毛竹	Purchasing bamboo
购竹	Purchasing bamboo
刮刀	Scraper
刮皮	Scraping bark
刮青刀	Scraper
和单槽棍	Stirring pulp
和料	Mixing materials
烘墙	Baking wall
烘纸	Baking paper
烘纸壁	Baking wall
滑板	Stick for stirring paper pulp
化验袋	Bag holding materials for testing
灰耙	Rake for stirring lime water
回浆池	Pulp pool
夹子	Clip
尖门针	Pointy needle for picking out impurities on the screen
拣料	Picking materials
拣皮	Picking bark
拣选	Sorting and picking
检料	Checking materials
检皮	Checking bark
检验	Testing and checking
检纸	Checking paper
剪纸	Cutting paper
浆板浸泡	Soaking pulp board
浆池	Pulp Pool
浆灰	Soaking in lime water
浆料	Soaking materials
浆料耙	Rake for stirring materials
浆尿	Soaking materials in urine
浆瓢	Pulp ladle
浆石灰	Lime water
浆竹个	Soaking bamboo in lime water
浇纸	Papermaking by ladling pulp into a fixed screen
浇注工场	Papermaking field
绞结刀	Knife for scraping bamboo
搅拌	Stirring
搅棍	Stirring stick
搅料棍	Sticks for stirring materials
揭纸	Peeling paper down
截断	Cutting bamboo into sections
截竹	Cutting bamboo into sections
浸泡	Soaking
浸坯	Soaking raw material
浸皮	Soaking bark
锯竹	Sawing bamboo
砍刀	Chopper
砍断	Cutting
砍料	Cutting
砍料凳	Stool for cutting
砍毛竹	Cutting bamboo
砍青	Cutting bamboo
砍树	Cutting down trees
砍竹	Cutting bamboo
拷白	Beating
拷白榔头	Beating with a hammer
靠贴	Wet piles of paper for drying
空气锤	Air hammer
捆件	Binding materials
捆料	Binding
捆扎	Binding
捆扎、盖印	Binding and sealing
捆纸	Binding paper
榔头	Hammer
捞纸	Papermaking
捞纸槽	Papermaking container
老式压榨机	Traditional machines for pressing and squeezing paper
冷却	Cooling down
冷却、调整	Cooling down and adjusting
理纸	Collecting paper
沥水	Drying by airing
帘棒竹烧直	Bamboo sticks
帘床	Screen frame
帘隔	Screen partition
帘架	Papermaking screen frame
帘滤	Screen filter
帘丝上漆	Painting screen
帘衣	papermaking screen
两齿耙	Two teeth rake
量角测正及号字	Checking the screen painting
料池	Material container
料袋	Material bag
料缸	Material vat
料耙	Rake
淋尿	Soaking material in urine
淋尿板	Board for spraying urine
淋尿堆蓬	Piling material for spraying urine
淋刷	Filtering and washing
流水袋料	Bag holding materials for washing
龙刨	Tool for flattening paper
落塘	Soaking materials in a pool
落塘储藏	Storing materials in a pool
篾索	Rope made of bamboo strips
磨料	Grinding
磨料机	Grinding machine
磨碎	Milling
磨纸	Trimming deckle edge
磨纸刀	Tool for trimming deckle edge

磨纸工具	Tool for trimming deckle edge
磨纸石	Stone for trimming deckle edge
磨纸刷	Brush for trimming deckle edge
木锤	Wooden hammer
木耙	Wooden rake
木榨	Wooden presser
木纸槽	Wooden papermaking container
木制千斤顶	Wooden hydraulic lifting jack
木钻	Wooden drilling
尼龙线	Nylon thread
碾料	Grinding materials
碾磨	Grinding
碾碎	Grinding
碾压	Grinding and pressing
碾竹	Grinding bamboo
沤皮	Fermenting bark
耙子	Rake
排稀密	Arranging according to density
刨竹	Cutting bamboo
泡料池	Pool for soaking
配浆	Making pulp
烹槽	Stirring to make pulp
皮篚	Bamboo basket
皮刀	Knife for cutting bark
皮镬	Pot for cooking bark
皮镬灶	Stove for cooking bark
皮镬蒸煮	Boiling in a pot
漂白	Bleaching
漂料	Bleaching
漂洗	Bleaching and washing
漂洗漾滩	Washing
剖竹刀	Knife for cutting bamboo
漆帘	Painting curtain
漆刷	Painting brush
千斤顶	Hydraulic lifting jack
牵纸	Pliers for separating paper
钳开口	Pliers for separating paper
钳子	Pliers
敲皮	Beating bark
敲碎	Beating and smashing
敲竹	Beating bamboo
切割机	Cutting machine
切皮	Slicing
切条	Slicing
切小口	Making a small cut
切纸	Cutting paper
切纸刀	Knife for cutting
切纸机	Machine for cutting paper
清水浸泡	Soaking in water
清洗	Washing
去节、刮青	Getting rid of exterior stains
燃烧炉	Burning furnace
溶解分散剂	Dissolving and dispersing agent
入槽	Putting materials into papermaking vat
砂轮打光	Polishing with a grinding-wheel

砂轮打磨	Grinding wheel
砂轮机	Grinding machine
煞帘丝	Aligning curtain
晒干	Drying in the sun
晒皮	Drying bark
晒纸	Drying paper
晒纸焙壁	Baking wall
晒帚	Broom
上绷架	Frame supporting the papermaking screen
上大榨	Pressing paper in the modern way
上镬	Putting materials in wok for boiling
上山娘竹	Preserved high-quality bamboo
上销	Installing knife handle
上小榨	Pressing paper in the traditional way
烧煮	Boiling and steaming
生物燃料	Biofuels
石碓碾碎	Grinding with stone pestle
石灰池	Limestone pool
石灰浸泡	Soaking in lime water
石灰腌制	Fermenting in lime water
石灰煮烂	Boiling in lime water
石磨	Stone grinder
石磨磨碎	Grinding
石碾	Stone rollers
石台	Stone table
收纸	Collecting paper
手工抄纸槽	Handmade papermaking container
数纸	Counting paper
刷把	Brush handle
刷子	Brush
水棚碓	Hydraulic pestle
水洗	Rinsing and washing
撕纸板	Board for separating paper layers
松毛刷	Brush made of pine branch
松刷	Brush
踏料	Treading materials
踏刷	Treading and washing
踏洗	Washing by treading
摊晾	Drying
添加纸药	Adding papermaking mucilage
挑料	Picking out materials
挑料耙	Rake for picking out materials
挑漆渣	Picking out waste residue during painting process
挑选	Checking and selecting
帖架	Shelf holding paper
铁锤	Iron hammer
桶	Barrel
筒席	Bamboo mat
筒牙头	Tool for peeling paper down
囤纸	Storing paper
温手锅	Pot for warming hands
温水锅	Hot water pot
洗白料	Washing bleached materials
洗黑液	Washing unbleached materials
洗料	Washing

洗料袋	Washing bag
洗皮	Washing bark
洗刷	Washing and brushing
选皮	Choosing bark
削皮	Peeling bark
削青	Scraping bamboo bark
削青刀	Knife for scraping bamboo bark
削青皮	Scraping bamboo bark
削竹	Cutting bamboo
熏染	Dyeing by smoking
压板	Pressing board
压晒	Pressing and drying
压榨	Pressing
压榨床	Pressing board
压榨机	Pressing machine
压榨去水	Pressing to dry
压榨纸架	Paper frame for pressing
压纸	Squeezing paper
腌草塘	Pond for soaking materials
腌料	Soaking materials
腌刷	Soaking and brushing
摇臼	Shaking materials to make pulp
摇臼桶	Barrel for shaking materials
页杆	Tool for measuring bamboo length
阴料	Fermenting materials in a shady place
造纸用帘	Papermaking screen
扎捆	Bundling
斫竹	Cutting bamboo
斫竹工具	Tools for cutting bamboo
榨床	Pressing board
榨干	Squeezing
榨料	Pressing
榨水	Pressing
榨纸	Pressing paper
榨纸架	Pressing shelf
斩藤	Cutting rattan
长帘网	Long screen
蒸	Steaming
蒸锅	Wok for steaming
蒸料	Steaming
蒸料桶	Steaming barrel
蒸煮	Steaming
整纸	Sorting paper
织机	Weaving
织帘	Weaving screen
织帘木机	Wooden weaving machine
纸架	Screen frame
纸浆槽	Paper pulp container
纸帘	Papermaking screen
纸戗	Papermaking site
纸捅	Papermaking barrel
纸桶架	Shelf supporting papermaking barrel
纸线	Joss paper
纸砑	Paper calendering
纸药桶	Papermaking mucilage barrel
制浆	Making pulp
治印	Sealing
竹帘捞起	Lifting bamboo screen
竹帘上漆	Painting bamboo screen
竹镊子	Bamboo tweezers
竹耙	Bamboo rake
煮	Boiling
煮料	Boiling
注水蒸煮	Boiling with water
装箬竹	Bamboo rod used to fix the papermaking screen
斫竹	Cutting bamboo
自舂	Beating with a hydraulic beater
做料	Making materials
做摞	Making materials

历　史　文　化 History and culture

汉语术语 Term in Chinese	英语术语 Term in English
朝拜王羲之故居	Visiting the former residence of Wang Xizhi
吊九楼	Diao Jiu Lou (stacking up nine old-fashioned tables for eight people, symbolizing nine layers of heaven)
“方茂林”品牌	Fangmaolin
“浮玉堂”商标	Fuyutang
“福利岗”	Position set up for philanthropic purposes
“冠形塔”	Guanxingta (pyramid tower, name of a place)
祭蔡伦	Ceremony in memory of Cai Lun
“贱骨金口”	A folklore relating a local man foretelling the flourishing of bamboo
“姜芹波生记”	Jiang Qinbo Shengji
“姜芹波纸号”	Jiang Qinbo Paper Store
“姜芹波忠记”	Jiang Qinbo Zhongji
“京都状元富阳纸，十件元书考进士”	Fuyang paper helps cultivate the top scholars, and one can become a successful candidate in the highest imperial examinations as long as he practices with ten bundles of Yuanshu paper.
开槽请菩萨	Ceremony held before making paper
“开山”祭祀	Ceremony held before cutting bamboo for papermaking
开山祭	Ceremony held before cutting bamboo for papermaking
开山请菩萨	Inviting bodhi-sattva on the opening ceremony for blessings
开山宴	Feast before making paper
判山	Purchasing bamboo from other places
平安经	Scriptures for Ping'an (Scriptures for safety)
“山山居”	Shanshanju
“烧焙人”	People help baking and drying paper
“十二烧佛”	A local tradition, with 12 old women (each with different Chinese zodiac title) burning scriptures in memory of the deceased
太平经	Scriptures for Taiping (Scriptures for peace and tranquility)
天一阁	Tianyi Pavilion
“仙人指路”	God's guidance (leaving a cut on the screen so that the location will be marked and help align paper)
“向阳牌”	Xiangyang
“小红旗”	Xiaohongqi (small red flag)
心经	Heart Sutra
“剡纸光如月”	Shanxi rattan paper is as smooth as moon

元书	Yuanshu paper
“造户董(东)家”	A papermaking family with steady output
“长木匠短铁匠”	Carpenters prefer longer wood while blacksmiths prefer shorter iron, both to avoid waste of materials
“纸山”	Zhishan (Paper mountain)
“朱禄年的谷，史尧臣的竹”	Zhu Lunian's grain and Shi Yaochen's bamboo
“48 口蒸锅”	48 papermaking steamers
“爱国皮纸”	Aiguo bast paper (paper of patriotism)
拜菩萨	Praying to bodhi-sattva
拜山祭祀	Praying to local gods for blessings before cutting bamboo
抄纸名师	Papermaking master
抄纸之术不外传	The crafts of making paper must be kept as a secret.
传媳不传女	Daughters-in-law rather than daughters could inherit the skills
吹气揭纸	Separate the layers of paper by blowing
发明吊帘	Inventing the hanging screen
富阳钟塔山	Zhongta Moutain in Fuyang
膏药用纸	Paper for making plaster
公共石碾房排队	Lining up in front of public stone mill
贡纸	Tribute paper
旱地竹戗浇纸法	A papermaking method by pouring pulp onto papermaking screen frame
红烛	Red candle
花笺纸传说	Legend of Huajian paper
画家试纸	Artists testing paper
皇帝赐名纸帘	Papermaking screen named by the emperor
祭拜山神	Worshipping the mountain god
祭祀槽桶菩萨	Worshipping Caotong Bodhisattva for good paper production
祭祖烧“真经”	Burning Taoist Scriptures to worship ancestors
姜氏宗祠	Jiang's ancestral hall
“洁癖制纸”	Observing cleanliness in the process of making paper
禁女性过料塘	Women are prohibited from passing through the pond
开工择日祭拜	Starting papermaking and offering sacrifice on an auspicious day
李煜	Li Yu
亮篾	Thin bamboo strips
骆宾王后裔	Descendants of Luo Binwang
媒头纸	Meitou Paper
男漆女织	Men lacquering and women weaving
女人不造纸	Women are not allowed to make paper
奇妙鹅榔头	Wooden Hammer for Separating Layers of Paper (also nicknamed “goose hoe”)
山棉皮采摘难(山棉皮)	Shanmian (*Wikstroemia monnula* Hance) is hard to get
上香	Offering incense
烧纸祭祖	Burning joss paper to worship ancestors
手工画心纸	Handmade Huaxin (Xuan) Paper
水碓与电碓共建共享	Employing both hydraulic pestle and electric pestle
诵经	Chanting scriptures
偷料人家三代不开槽	Three generations of papermaking families who stealed raw materials are not allowed to practice papermaking
舞龙灯	Dragon lantern dance
香烟纸上的水滴	Water drops on the cigarette paper
写圣旨	Writing the imperial edict
宜家家具用纸	Paper for making Yijia furniture
造纸师傅多为男	Most of the papermakers are male
纸灰能“飞”	Joss paper ash can fly (a good message according to the tradition, meaning the ancestors having taken the money)
纸农菜文化	The culture of papermakers' dishes
纸农同书法名家结缘	Papermakers attaching to calligraphers
纸商合作十八载	Paper merchants cooperate for 18 years
纸寿千年	Paper is durable for a thousand years
纸元宝	Paper golden ingot
中国造纸术活化石	Chinese papermaking, a living fossil
钟塔石凉亭	Zhongtashi pavilion

后 记

《中国手工纸文库·浙江卷》的田野调查起始于2016年7月下旬到8月上旬，先后到富阳区（原富阳县）大源镇大同村调查了杭州富阳逸古斋元书纸有限公司和杭州富阳宣纸陆厂的手工纸车间。说起来也特别有缘分，在这一年的6月至7月，文化部在全国推动的第一批8所高校中国非物质文化遗产传承人群驻校研修研习培训计划中，中国科学技术大学承办的手工造纸“非遗”传承人第一届研修班到富阳访学，而富阳竹纸制作技艺选送的研修学员正是富阳逸古斋与富阳宣纸陆厂的两位造纸技艺传承人，因缘际会之下，浙江手工造纸的调查工作就与访学计划同步开展了。

从2016年盛夏大同村的开端到2019年季春对丽水市松阳县李坑村最后一个皮纸作坊的调查，田野调查研究历经了近3年的时间。其间，深入浙江省各手工纸造纸点的调查、采样按照既定规划持续不懈地进行，而根据需要随时走乡串户一次又一次的补充调查以及文献求证则几乎贯穿始终。其中仅仅“概述”部分引文注释一手文献的核对，一个负责文献研究的小组就先后在浙江省图书馆、杭州市图书馆蹲点查核了近20天（2019年6月25～29日、7月1～14日）。

Epilogue

Field investigation of *Library of Chinese Handmade Paper: Zhejiang* started around late July and early August of 2016, when the researchers visited papermaking workshops of Hangzhou Fuyang Yiguzhai Yuanshu Paper Co., Ltd. and Hangzhou Fuyang Lu (meaning six in Chinese, connoting that everything goes smoothly) Xuan Paper Factory located in Datong Village of Dayuan Town in Fuyang District (former Fuyang County). At almost the same time, i.e.,around June and July in 2016, Ministry of Culture initiated an Intangible Cultural Heritage Protection Training Program of China, funding the inheritors to study in campus (8 universities for the 1st session) for their production and protection explorations. University of Science and Technology of China hosted the handmade papermaking inheritors, among whom were two bamboo papermakers from Yiguzhai Yuanshu Paper Co., Ltd. and Fuyang Lu Xuan Paper Factory in Fuyang District. So with their recommendation, all the grantees visited Fuyang District, and our researchers took advantage of the occasion and started their initial investigations.

Our field investigation lasted for almost three years, from the first visit to Datong Village in the summer of 2016, to the spring of 2019, when we finished our investigation of a bast paper mill in Likeng Village of Songyang County in Lishui City.

浙江是中国历史上非常著名的手工纸产区，剡溪藤纸、温州皮纸、越州与富阳竹纸等，早在唐宋时期就享誉中国、畅销四海。但到《浙江卷》田野调查时段，前三类名纸基本上已经处于中断后尝试恢复、一丝苟存的衰微状态，只有富阳竹纸虽然比起高峰时期有明显收缩，但中高端用纸依然富有生机并拥有较好的市场空间。也正是由于杭州市富阳区当代以竹纸为主业态的手工纸的丰富多样，田野调查获得的信息较为充足，因此《浙江卷》分为上、中、下三卷，其中中卷和下卷都是富阳手工纸，这是浙江当代手工纸业态现状的如实呈现。

具体到每一章节，田野调查及文献研究通常由多位成员合作完成，前后多轮补充修订多数也不是由一人从头至尾独立承担，因而事前制定的作为指导性工作规范的田野调查标准、撰稿标准、示范样稿实际执行起来依然具有差异，田野信息采集格式和初稿表达存在诸多不统一、不规范处，在初稿基础上的统稿工作因而显得相当重要。

初稿合成后，统稿与补充调查工作由汤书昆、朱赟、朱中华、沈佳斐主持，

During the period, the researchers studied on the papermaking sites in Zhejiang Province repeatedly and sedulously for sample collection, planned or spontaneous investigations and verification. For instance, a group working on literature review in the Introduction part, stayed in Zhejiang Library and Hangzhou Library for literature study for almost twenty days (June 25-29, and July 1-14, 2019).

Zhejiang Province is a historically famous papermaking area, with Teng paper in Shanxi Area, bast paper in Wenzhou City, bamboo paper in Yuezhou Area and bamboo paper in Fuyang District as its representative famous paper types enjoying a national reputation since the Tang and Song Dynasties. However, when we started our field investigation, the former three paper types were experiencing a declining status, managing to recover from ceased production. Among them, bamboo paper in Fuyang District, high and middle-end paper, was well developed and enjoyed a flourishing market, though not comparable to its historical boom. Therefore, due to abundant data we obtained in our investigation in Fuyang District of Hangzhou City, which is a dominant area harboring bamboo paper production, Zhejiang volume actually consists of three sub-volumes. The second and third volumes of Zhejiang series are both focusing on handmade paper in Fuyang District, which vividly shows its dominance in current status of handmade paper industry in Zhejiang Province.

Field investigation and literature studies of each section and chapter are accomplished by the cooperative efforts of multiple

从2018年12月开始，以几乎马不停蹄的节奏和驻点补稿补图的方式，共进行了3轮集中补稿修订，最终形成定稿。虽然我们觉得浙江手工纸调查与研究有待进一步挖掘与完善之处仍有不少，但《浙江卷》从2016年7月启动，纸样测试、英文翻译、示意图绘制、编辑与设计等团队的成员尽心尽力，所呈现内容的品质一天天得到改善，书稿的阅读价值和图文魅力也确实获得了显著提升，可以作为目前这个工作阶段的调查与研究成果出版和接受读者的检验。

《浙江卷》书稿的完成和完善有赖于团队成员全心全意的投入与持续不懈的努力，在即将付梓之际，特在后记中对各位同仁的工作做如实的记述。

researchers, and even the modification was undertaken by different people. Therefore, investigation rules, writing norms and format set beforehand may still fail to make amends for the possible deviation in our first manuscript, and modification is of vital importance in our work.

Modification and supplementary investigation were headed by Tang Shukun, Zhu Yun, Zhu Zhonghua, and Shen Jiafei after the completion of the first manuscript. Since December 2018, the team members have put into three rounds of sedulous efforts to modify the manuscript, and revisit the papermaking sites for more information and photos. Of course, we admit that the volume cannot claim perfection, yet finally, through meticulous works in sampling testing, translation, map drawing, editing and designing since we started our handmade paper odyssey in July 2016, the book actually has been increasingly polished day by day. And we can be positive that the book, with fluent writing and intriguing pictures, is worth reading, and ready for publication.

On the verge of publication, we acknowledge the consistent efforts and wholehearted dedication of the following researchers:

第一章 浙江省手工造纸概述

撰稿	初稿主执笔：汤书昆、朱赟、陈敬宇 修订补稿：汤书昆、朱赟、沈佳斐 参与撰稿：王圣融、潘巧、王怡青、姚的卢、陈欣冉、廖莹文、孔利君、郭延龙、叶珍珍

第二章 衢州市

第一节	浙江辰港宣纸有限公司（地点：龙游县城区灵山江畔）
田野调查	汤书昆、朱中华、朱赟、陈彪、刘伟、何瑗、程曦、郑斌、潘巧、江顺超、钱霜霜
撰稿	初稿主执笔：汤书昆、汪竹欣 修订补稿：汤书昆、江顺超 参与撰稿：朱赟
第二节	开化县开化纸（地点：开化县华埠镇溪东村、村头镇形边村）
田野调查	汤书昆、朱中华、朱赟、姚的卢、陈欣冉、沈佳斐、潘巧、江顺超、钱霜霜
撰稿	初稿主执笔：汤书昆、朱赟 修订补稿：汤书昆、江顺超 参与撰稿：王怡青

第三章 温州市

第一节	泽雅镇唐宅村潘香玉竹纸坊（地点：瓯海区泽雅镇唐宅村）
田野调查	林志文、汤书昆、朱赟、朱中华、陈彪、刘伟、何瑗、程曦、沈佳斐、潘巧、江顺超、钱霜霜
撰稿	初稿主执笔：朱赟 修订补稿：汤书昆、钱霜霜
第二节	泽雅镇岙外村林新德竹纸坊（地点：瓯海区泽雅镇岙外村）
田野调查	林志文、朱赟、姚的卢、陈欣冉、沈佳斐、潘巧、江顺超、钱霜霜
撰稿	初稿主执笔：姚的卢 修订补稿：汤书昆、林志文、江顺超 参与撰稿：沈佳斐
第三节	泰顺县榅桥村翁士格竹纸坊（地点：泰顺县筱村镇榅桥村）
田野调查	汤书昆、朱中华、朱赟、黄飞松、姚的卢、陈欣冉、沈佳斐、潘巧、江顺超、钱霜霜
撰稿	初稿主执笔：汤书昆、潘巧 修订补稿：汤书昆、朱中华 参与撰稿：姚的卢
第四节	温州皮纸（地点：泽雅镇周岙上村）
田野调查	汤书昆、林志文、朱赟、朱中华、潘巧、黄飞松、沈佳斐、江顺超、钱霜霜
撰稿	初稿主执笔：潘巧、汤书昆 修订补稿：汤书昆、朱赟 参与撰稿：沈佳斐

第四章 绍兴市

第一节	绍兴鹿鸣纸（地点：柯桥区平水镇宋家店村）
田野调查	汤书昆、朱中华、王圣融、郑斌、朱有善、王黎明、潘巧、江顺超、钱霜霜
撰稿	初稿主执笔：汤书昆、王圣融 修订补稿：朱中华、汤书昆 参与撰稿：郑斌
第二节	嵊州市剡藤纸研究院（地点：嵊州市浦南大道 388 号）
田野调查	汤书昆、朱中华、沈佳斐、朱起杨
撰稿	初稿主执笔：沈佳斐、汤书昆 修订补稿：汤书昆、沈佳斐

第五章 湖州市

	安吉县龙王村手工竹纸（地点：安吉县上墅乡龙王村）
田野调查	朱中华、朱赟、姚的卢、陈欣冉、郑久良、潘巧、江顺超、钱霜霜
撰稿	初稿主执笔：陈欣冉 修订补稿：汤书昆、朱赟 参与撰稿：潘巧

第六章　宁波市

	奉化区棠岙村袁恒通纸坊（地点：奉化区萧王庙街道棠岙村溪下庵岭墩下13号）
田野调查	朱中华、朱赟、何瑗、王圣融、尹航、郑久良、潘巧、江顺超、钱霜霜
撰稿	初稿主执笔：何瑗 修订补稿：汤书昆、朱中华 参与撰稿：钱霜霜

第七章　丽水市

	松阳县李坑村李坑造纸工坊（地点：松阳县安民乡李坑村）
田野调查	汤书昆、朱中华、沈佳斐、石永宁
撰稿	初稿主执笔：沈佳斐、石永宁 修订补稿：汤书昆

第八章　杭州市

第一节	杭州临安浮玉堂纸业有限公司（地点：临安区於潜镇枫凌村）
田野调查	朱中华、朱赟、何瑗、王圣融、叶婷婷、尹航、沈佳斐、潘巧、江顺超、钱霜霜
撰稿	初稿主执笔：王圣融 修订补稿：汤书昆、沈佳斐
第二节	杭州千佛纸业有限公司（地点：临安区於潜镇千茂行政村平渡自然村）
田野调查	朱中华、朱赟、何瑗、王圣融、尹航、沈佳斐、潘巧、江顺超、钱霜霜
撰稿	初稿主执笔：王圣融 修订补稿：汤书昆、潘巧、朱中华
第三节	杭州临安书画宣纸厂（地点：临安区於潜镇千茂行政村下平渡村民组）
田野调查	朱中华、朱赟、何瑗、王圣融、尹航、沈佳斐、潘巧、江顺超、钱霜霜
撰稿	初稿主执笔：何瑗 修订补稿：汤书昆、钱霜霜、朱中华

第九章　富阳区元书纸

第一节	新三元书纸品厂（地点：富阳区湖源乡新三村冠形塔村民组）
田野调查	朱中华、朱赟、汤书昆、刘伟、何瑗、程曦、沈佳斐、潘巧、江顺超、钱霜霜
撰稿	初稿主执笔：王圣融 修订补稿：汤书昆、江顺超
第二节	杭州富春江宣纸有限公司（地点：富阳区大源镇大同村方家地村民组）
田野调查	朱中华、刘伟、何瑗、沈佳斐、潘巧、江顺超、钱霜霜
撰稿	初稿主执笔：汪竹欣、王圣融 修订补稿：汤书昆、钱霜霜
第三节	杭州富阳蔡氏文化创意有限公司（地点：富阳区灵桥镇蔡家坞村）
田野调查	朱中华、汤书昆、朱赟、何瑗、程曦、叶婷婷、沈佳斐、潘巧、江顺超、钱霜霜
撰稿	初稿主执笔：何瑗 修订补稿：汤书昆、朱赟 参与撰稿：潘巧
第四节	杭州富阳逸古斋元书纸有限公司（地点：富阳区大源镇大同行政村朱家门自然村）
田野调查	汤书昆、朱赟、汤雨眉、刘伟、何瑗、程曦、姚的卢、陈欣冉、沈佳斐、陈彪、潘巧、江顺超、钱霜霜
撰稿	初稿主执笔：汤雨眉、朱赟 修订补稿：汤书昆 参与撰稿：王圣融、王怡青
第五节	杭州富阳宣纸陆厂（地点：富阳区大源镇大同行政村兆吉自然村第一村民组）
田野调查	汤书昆、朱赟、朱中华、刘伟、王圣融、尹航、沈佳斐、潘巧、江顺超、钱霜霜、汤雨眉、陈彪
撰稿	初稿主执笔：朱赟 修订补稿：汤书昆 参与撰稿：江顺超
第六节	富阳福阁纸张销售有限公司（地点：富阳区湖源乡新三行政村颜家桥自然村）
田野调查	朱赟、朱中华、刘伟、何瑗、程曦、沈佳斐、桂子璇、江顺超、钱霜霜

撰稿	初稿主执笔：程曦 修订补稿：汤书昆、沈佳斐
第七节	杭州富阳双溪书画纸厂（地点：富阳区大源镇大同行政村兆吉自然村方家地村民组）
田野调查	朱中华、朱赟、刘伟、何瑗、沈佳斐、桂子璇、江顺超、钱霜霜
撰稿	初稿主执笔：何瑗 修订补稿：汤书昆、朱中华 参与撰稿：桂子璇
第八节	富阳大竹元宣纸有限公司（地点：富阳区湖源乡新二村元书纸制作园区）
田野调查	汤书昆、朱中华、朱赟、何瑗、程曦、沈佳斐、朱起杨、潘巧、桂子璇、江顺超、钱霜霜
撰稿	初稿主执笔：汤书昆、朱赟 修订补稿：汤书昆、沈佳斐 参与撰稿：潘巧
第九节	朱金浩纸坊（地点：富阳区大源镇大同行政村朱家门自然村 20 号）
田野调查	朱赟、朱中华、程曦、沈佳斐、桂子璇、江顺超、钱霜霜
撰稿	初稿主执笔：程曦 修订补稿：汤书昆、钱霜霜
第十节	盛建桥纸坊（地点：富阳区湖源乡新二行政村钟塔自然村 46 号）
田野调查	朱中华、朱赟、刘伟、何瑗、程曦、沈佳斐、潘巧、江顺超、钱霜霜
撰稿	初稿主执笔：程曦 修订补稿：汤书昆、江顺超
第十一节	鑫祥宣纸作坊（地点：富阳区大源镇骆村（行政村）秦骆自然村 241 号）
田野调查	朱中华、朱赟、刘伟、何瑗、程曦、沈佳斐、潘巧、江顺超、钱霜霜
撰稿	初稿主执笔：程曦 修订补稿：汤书昆、钱霜霜
第十二节	富阳竹馨斋元书纸有限公司（地点：富阳区湖源乡新二村元书纸制作园区）
田野调查	汤书昆、朱赟、朱中华、刘伟、何瑗、程曦、潘巧、沈佳斐、桂子璇、江顺超、钱霜霜
撰稿	初稿主执笔：潘巧、汤书昆 修订补稿：汤书昆、沈佳斐 参与撰稿：刘伟
第十三节	庄潮均作坊（富阳区大源镇红霞书画纸经营部）（地点：富阳区大源镇大同行政村庄家自然村）
田野调查	朱中华、朱赟、刘伟、何瑗、程曦、汪竹欣、沈佳斐、潘巧、江顺超、钱霜霜
撰稿	初稿主执笔：何瑗 修订补稿：汤书昆、钱霜霜 参与撰稿：汪竹欣
第十四节	杭州山元文化创意有限公司（地点：富阳区新登镇袁家村）
田野调查	汤书昆、朱中华、朱赟、刘伟、何瑗、王圣融、王怡青、沈佳斐、潘巧、江顺超、钱霜霜
撰稿	初稿主执笔：王圣融、朱赟 修订补稿：朱中华、汤书昆 参与撰稿：汤书昆、朱赟、王怡青

第十章　富阳区祭祀竹纸

第一节	章校平纸坊（地点：富阳区常绿镇黄弹行政村寺前自然村 71 号）
田野调查	朱中华、朱赟、刘伟、何瑗、程曦、汪竹欣、沈佳斐、潘巧、江顺超、钱霜霜
撰稿	初稿主执笔：汪竹欣 修订补稿：汤书昆、沈佳斐
第二节	蒋位法作坊（地点：富阳区大源镇三岭行政村三支自然村 21 号）
田野调查	朱赟、朱中华、何瑗、刘伟、沈佳斐、江顺超、钱霜霜、潘巧
撰稿	初稿主执笔：刘伟 修订补稿：汤书昆、江顺超
第三节	李财荣纸坊（地点：富阳区灵桥镇新华村）
田野调查	朱中华、朱赟、刘伟、何瑗、汪竹欣、沈佳斐、潘巧、江顺超、钱霜霜
撰稿	初稿主执笔：汪竹欣 修订补稿：汤书昆、潘巧
第四节	李申言金钱纸作坊（地点：富阳区常安镇大田村 32 号）
田野调查	汤书昆、朱赟、朱中华、刘伟、何瑗、程曦、沈佳斐、潘巧、江顺超、钱霜霜

撰稿	初稿主执笔：何瑗 修订补稿：汤书昆、沈佳斐
第五节	李雪余屏纸作坊（地点：富阳区常安镇大田村105号）
田野调查	朱中华、汤书昆、朱赟、刘伟、何瑗、程曦、沈佳斐、潘巧、江顺超、钱霜霜
撰稿	初稿主执笔：何瑗 修订补稿：汤书昆、沈佳斐
第六节	姜明生纸坊（地点：富阳区灵桥镇山基村）
田野调查	朱赟、朱中华、叶婷婷、尹航、沈佳斐、潘巧、江顺超、钱霜霜
撰稿	初稿主执笔：叶婷婷 修订补稿：汤书昆、钱霜霜 参与撰稿：江顺超
第七节	戚吾樵纸坊（地点：富阳区渔山乡大葛村）
田野调查	朱中华、朱赟、刘伟、何瑗、程曦、汪竹欣、沈佳斐、桂子璇、江顺超、钱霜霜
撰稿	初稿主执笔：汪竹欣 修订补稿：汤书昆、江顺超
第八节	张根水纸坊（地点：富阳区湖源乡新三村）
田野调查	朱中华、朱赟、何瑗、叶婷婷、沈佳斐、桂子璇、江顺超、钱霜霜
撰稿	初稿主执笔：叶婷婷 修订补稿：汤书昆、桂子璇
第九节	祝南书纸坊（地点：富阳区灵桥镇山基村）
田野调查	朱赟、朱中华、叶婷婷、尹航、沈佳斐、桂子璇、江顺超、钱霜霜
撰稿	初稿主执笔：叶婷婷 修订补稿：汤书昆、江顺超

第十一章　富阳区皮纸

第一节	五四村桃花纸作坊（地点：富阳区鹿山街道五四村）
田野调查	方仁英、陈彪、李少军、朱中华、汤书昆
撰稿	初稿主执笔：方仁英 修订补稿：汤书昆 参与撰稿：陈彪
第二节	大山村桑皮纸恢复点（地点：富阳区新登镇大山村）
田野调查	方仁英、李少军、朱中华
撰稿	初稿主执笔：方仁英 修订补稿：汤书昆、李少军

第十二章　工具

第一节	永庆制帘工坊（地点：富阳区大源镇永庆村）
田野调查	朱中华、朱赟、何瑗、叶婷婷、尹航、沈佳斐、桂子璇、江顺超、钱霜霜
撰稿	初稿主执笔：尹航 修订补稿：汤书昆、钱霜霜
第二节	光明制帘厂（地点：富阳区灵桥镇光明村）
田野调查	朱赟、刘伟、朱中华、何瑗、程曦、沈佳斐、桂子璇、江顺超、钱霜霜
撰稿	初稿主执笔：刘伟 修订补稿：汤书昆、沈佳斐
第三节	郎仕训刮青刀制作坊（地点：富阳区大源镇朝阳南路二弄）
田野调查	朱中华、朱赟、王圣融、叶婷婷、尹航、王怡青、沈佳斐、桂子璇、江顺超、钱霜霜
撰稿	初稿主执笔：朱赟 修订补稿：汤书昆

二、技术与辅助工作

实物纸样测试分析	主持：朱赟、陈龑 成员：朱赟、陈龑、王圣融、刘伟、何瑗、汪竹欣、王怡青、姚的卢、叶珍珍、尹航、孙燕、廖莹文、郭延龙
手工纸分布示意图绘制	郭延龙

实物纸样纤维图及透光图制作	朱赟、王圣融、刘伟、何瑗、汪竹欣、王怡青、姚的卢、廖莹文、陈龑、郭延龙
实物纸样拍摄	黄晓飞
实物纸样整理	朱赟、汤书昆、刘伟、何瑗、汪竹欣、王圣融、倪盈盈、沈佳斐、郑斌、付成云、蔡婷婷、潘巧、王怡青、姚的卢、尹航、陈欣冉、廖莹文、孔利君、郭延龙、叶珍珍
附录及参考文献整理	汤书昆、朱赟、沈佳斐

三、 总序、编撰说明、附录与后记

	总序
撰稿	汤书昆
	编撰说明
撰稿	汤书昆、朱赟
	附录
名词术语整理	朱赟、沈佳斐、倪盈盈、唐玉璟、蔡婷婷
	后记
撰稿	汤书昆

四、 统稿与翻译

统稿主持	汤书昆、朱中华
统稿规划	朱赟、沈佳斐
翻译主持	方媛媛
统稿阶段其他参与人员	陈敬宇、林志文、李少军、徐建华、潘巧、桂子璇、江顺超、钱霜霜

Chapter I Introduction to Handmade Paper in Zhejiang Province

Writer	First manuscript written by: Tang Shukun, Zhu Yun, Chen Jingyu Modified by: Tang Shukun, Zhu Yun, Shen Jiafei Wang Shengrong, Pan Qiao, Wang Yiqing, Yao Dilu, Chen Xinran, Liao Yingwen, Kong Lijun, Guo Yanlong and Ye Zhenzhen have also contributed to the writing

Chapter II Quzhou City

Section 1	Zhejiang Chengang Xuan Paper Co., Ltd. (location: Lingshan Riverside of Longyou County)
Investigators	Tang Shukun, Zhu Zhonghua, Zhu Yun, Chen Biao, Liu Wei, He Ai, Cheng Xi, Zheng Bin, Pan Qiao, Jiang Shunchao, Qian Shuangshuang
Writers	First manuscript written by: Tang Shukun, Wang Zhuxin Modified by: Tang Shukun, Jiang Shunchao Zhu Yun has also contributed to the writing
Section 2	Kaihua Paper in Kaihua County (location: Xidong Village of Huabu Town and Xingbian Village of Cuntou Town in Kaihua County)
Investigators	Tang Shukun, Zhu Zhonghua, Zhu Yun, Yao Dilu, Chen Xinran, Shen Jiafei, Pan Qiao, Jiang Shunchao, Qian Shuangshuang
Writers	First manuscript written by: Tang Shukun, Zhu Yun Modified by: Tang Shukun, Jiang Shunchao Wang Yiqing has also contributed to the writing

Chapter III Wenzhou City

Section 1	Pan Xiangyu Bamboo Paper Mill in Tangzhai Village of Zeya Town (location: Tangzhai Village of Zeya Town in Ouhai District)
Investigators	Lin Zhiwen, Tang Shukun, Zhu Yun, Zhu Zhonghua, Chen Biao, Liu Wei, He Ai, Cheng Xi, Shen Jiafei, Pan Qiao, Jiang Shunchao, Qian Shuangshuang
Writers	First manuscript written by: Zhu Yun Modified by: Tang Shukun, Qian Shuangshuang
Section 2	Lin Xinde Bamboo Paper Mill in Aowai Village of Zeya Town (location: Aowai Village of Zeya Town in Ouhai District)
Investigators	Lin Zhiwen, Zhu Yun, Yao Dilu, Chen Xinran, Shen Jiafei, Pan Qiao, Jiang Shunchao, Qian Shuangshuang
Writers	First manuscript written by: Yao Dilu Modified by: Tang Shukun, Lin Zhiwen, Jiang Shunchao Shen Jiafei has also contributed to the writing
Section 3	Weng Shige Bamboo Paper Mill in Wenqiao Village of Taishun County (location: Wenqiao Village of Xiaocun Town in Taishun County)
Investigators	Tang Shukun, Zhu Zhonghua, Zhu Yun, Huang Feisong, Yao Dilu, Chen Xinran, Shen Jiafei, Pan Qiao, Jiang Shunchao, Qian Shuangshuang
Writers	First manuscript written by: Tang Shukun, Pan Qiao Modified by: Tang Shukun, Zhu Zhonghua Yao Dilu has also contributed to the writing
Section 4	Bast paper in Wenzhou City (location: Zhouaoshang Village of Zeya Town)
Investigators	Tang Shukun, Lin Zhiwen, Zhu Yun, Zhu Zhonghua, Pan Qiao, Huang Feisong, Shen Jiafei, Jiang Shunchao, Qian Shuangshuang
Writers	First manuscript written by: Pan Qiao, Tang Shukun Modified by: Tang Shukun, Zhu Yun Shen Jiafei has also contributed to the writing

Chapter IV Shaoxing City

Section 1	Luming Paper in Shaoxing County (location: Songjiadian Village of Pingshui Town in Keqiao District)
Investigators	Tang Shukun, Zhu Zhonghua, Wang Shengrong, Zheng Bin, Zhu Youshan, Wang Liming, Pan Qiao, Jiang Shunchao, Qian Shuangshuang
Writers	First manuscript written by: Tang Shukun, Wang Shengrong Modified by: Zhu Zhonghua, Tang Shukun Zheng Bin has also contributed to the writing
Section 2	Shanteng Paper Research Institute in Shengzhou City (location: No.388 Punan Ave., Shengzhou City)
Investigators	Tang Shukun, Zhu Zhonghua, Shen Jiafei, Zhu Qiyang
Writers	Fist manuscript written by: Shen Jiafei, Tang Shukun Modified by: Tang Shukun, Shen Jiafei

Chapter V Huzhou City

	Handmade Banboo Paper in Longwang Village of Anji County (location: Longwang Village of Shangshu Town in Anji County)
Investigators	Zhu Zhonghua, Zhu Yun, Yao Dilu, Chen Xinran, Zheng Jiuliang, Pan Qiao, Jiang Shunchao, Qian Shuangshuang
Writers	The first manuscript written by: Chen Xinran Modified by: Tang Shukun, Zhu Yun Pan Qiao has also contributed to the writing

Chapter VI Ningbo City	
	Yuan Hengtong Paper Mill in Tang'ao Village of Fenghua District (location: No.13 Xixia Anling Dunxia of Tang'ao Village in Xiaowangmiao Residential District of Fenghua District)
Investigators	Zhu Zhonghua, Zhu Yun, He Ai, Wang Shengrong, Yin Hang, Zheng Jiuliang, Pan Qiao, Jiang Shunchao, Qian Shuangshuang
Writers	First manuscript written by: He Ai Modified by Tang Shukun, Zhu Zhonghua Qian Shuangshuang has also contributed to the writing

Chapter VII Lishui City	
	Likeng Paper Mill in Likeng Village of Songyang County (location: Likeng Village of Anmin Town in Songyang County)
Investigators	Tang Shukun, Zhu Zhonghua, Shen Jiafei, Shi Yongning
Writers	First manuscript written by: Shen Jiafei, Shi Yongning Modified by: Tang Shukun

Chapter VIII Hangzhou City	
Section 1	Hangzhou Lin'an Fuyutang Paper Co., Ltd. (location: Fengling Village of Yuqian Town in Lin'an District)
Investigators	Zhu Zhonghua, Zhu Yun, He Ai, Wang Shengrong, Ye Tingting, Yin Hang, Shen Jiafei, Pan Qiao, Jiang Shunchao, Qian Shuangshuang
Writers	First manuscript written by: Wang Shengrong Modified by: Tang Shukun, Shen Jiafei
Section 2	Hangzhou Qianfo Paper Co., Ltd. (location: Pingdu Natural Village of Qianmao Administrative Village of Yuqian Town in Lin'an District)
Investigators	Zhu Zhonghua, Zhu Yun, He Ai, Wang Shengrong, Yin Hang, Shen Jiafei, Pan Qiao, Jiang Shunchao, Qian Shuangshuang
Writers	First manuscript written by: Wang Shengrong Modified by: Tang Shukun, Pan Qiao, Zhu Zhonghua
Section 3	Hangzhou Lin'an Calligraphy and Painting Xuan Paper Factory (location: Xiapingdu Villagers' Group of Qianmao Administrative Village in Yuqian Town of Lin'an District)
Investigators	Zhu Zhonghua, Zhu Yun, He Ai, Wang Shengrong, Yin Hang, Shen Jiafei, Pan Qiao, Jiang Shunchao, Qian Shuangshuang
Writers	First manuscript written by: He Ai Modified by: Tang Shukun, Qian Shuangshuang, Zhu Zhonghua

Chapter IX Yuanshu Paper in Fuyang District	
Section 1	Xinsan Yuanshu Paper Factory (location: Guanxingta Villagers' Group of Xinsan Village in Huyuan Town of Fuyang District)
Investigators	Zhu Zhonghua, Zhu Yun, Tang Shukun, Liu Wei, He Ai, Cheng Xi, Shen Jiafei, Pan Qiao, Jiang Shunchao, Qian Shuangshuang
Writers	First manuscript written by: Wang Shengrong Modified by: Tang Shukun, Jiang Shunchao
Section 2	Hangzhou Fuchunjiang Xuan Paper Co., Ltd. (location: Fangjiadi Villagers' Group of Datong Village in Dayuan Town of Fuyang District)
Investigators	Zhu Zhonghua, Liu Wei, He Ai, Shen Jiafei, Pan Qiao, Jiang Shunchao, Qian Shuangshuang
Writers	First manuscript written by: Wang Zhuxin, Wang Shengrong Modified by: Tang Shukun, Qian Shuangshuang
Section 3	Hangzhou Fuyang Caishi Cultural and Creative Co. Ltd. (location: Caijiawu Village of Lingqiao Town in Fuyang District)
Investigators	Zhu Zhonghua, Tang Shukun, Zhu Yun, He Ai, Cheng Xi, Ye Tingting, Shen Jiafei, Pan Qiao, Jiang Shunchao, Qian Shuangshuang
Writers	First manuscript written by: He Ai Modified by: Tang Shukun, Zhu Yun Pan Qiao has also contributed to the writing
Section 4	Hangzhou Fuyang Yiguzhai Yuanshu Paper Co. Ltd. (location: Zhujiamen Natural Village of Datong Administrative Village in Dayuan Town of Fuyang District)
Investigators	Tang Shukun, Zhu Yun, Tang Yumei, Liu Wei, He Ai, Cheng Xi, Yao Dilu, Chen Xinran, Shen Jiafei, Chen Biao, Pan Qiao, Jiang Shunchao, Qian Shuangshuang
Writers	First manuscript written by: Tang Yumei, Zhu Yun Modified by: Tang Shukun Wang Shengrong and Wang Yiqing have also contributed to the writing
Section 5	Hangzhou Fuyang Xuan Paper Lu Factory (location: No.1 Villagers's Group of Zhaoji Natural Village in Datong Administrative Village of Dayuan Town in Fuyang District)
Investigators	Tang Shukun, Zhu Yun, Zhu Zhonghua, Liu Wei, Wang Shengrong, Yin Hang, Shen Jiafei, Pan Qiao, Jiang Shunchao, Qian Shuangshuang, Tang Yumei, Chen Biao
Writers	First manuscript written by: Zhu Yun Modified by: Tang Shukun Jiang Shunchao has also contributed to the writing

Section 6	Fuyang Fuge Paper Sales Co., Ltd. (location: Yanjiaqiao Natural Village of Xinsan Administrative Village in Huyuan Town of Fuyang District)
Investigators	Zhu Yun, Zhu Zhonghua, Liu Wei, He Ai, Cheng Xi, Shen Jiafei, Gui Zixuan, Jiang Shunchao, Qian Shuangshuang
Writers	First manuscript written by: Cheng Xi Modified by: Tang Shukun, Shen Jiafei
Section 7	Hangzhou Fuyang Shuangxi Calligraphy and Painting Paper Factory (location: Fangjiadi Villagers' Group of Zhaoji Natural Village of Datong Administrative Village in Dayuan Town of Fuyang District)
Investigators	Zhu Zhonghua, Zhu Yun, Liu Wei, He Ai, Shen Jiafei, Gui Zixuan, Jiang Shunchao, Qian Shuangshuang
Writers	First manuscript written by: He Ai Modified by: Tang Shukun, Zhu Zhonghua Gui Zixuan has also contributed to the writing
Section 8	Fuyang Dazhuyuan Xuan Paper Co., Ltd. (location: Yuanshu Papermaking Park in Xin'er Village of Huyuan Town in Fuyang District)
Investigators	Tang Shukun, Zhu Zhonghua, Zhu Yun, He Ai, Cheng Xi, Shen Jiafei, Zhu Qiyang, Pan Qiao, Gui Zixuan, Jiang Shunchao, Qian Shuangshuang
Writers	First manuscript written by: Tang Shukun, Zhu Yun Modified by: Tang Shukun, Shen Jiafei Pan Qiao has also contributed to the writing
Section 9	Zhu Jinhao Paper Mill (location: No.20 Zhujiamen Natural Village of Datong Adminstrative Village in Dayuan Town of Fuyang District)
Investigators	Zhu Yun, Zhu Zhonghua, Cheng Xi, Shen Jiafei, Gui Zixuan, Jiang Shunchao, Qian Shuangshuang
Writers	First manuscript written by: Cheng Xi Modified by: Tang Shukun, Qian Shuangshuang
Section 10	Sheng Jianqiao Paper Mill (location: No.46 Zhongta Natural Village of Xin'er Adminstrative Village in Huyuan Town of Fuyang District)
Investigators	Zhu Zhonghua, Zhu Yun, Liu Wei, He Ai, Cheng Xi, Shen Jiafei, Pan Qiao, Jiang Shunchao, Qian Shuangshuang
Writers	First manuscript written by: Cheng Xi Modified by: Tang Shukun, Jiang Shunchao
Section 11	Xinxiang Xuan Paper Mill (location: No.241 Qinluo Natural Village of Luocun Village in Dayuan Town of Fuyang District)
Investigators	Zhu Zhonghua, Zhu Yun, Liu Wei, He Ai, Cheng Xi, Shen Jiafei, Pan Qiao, Jiang Shunchao, Qian Shuangshuang
Writers	First manuscript written by: Cheng Xi Modified by: Tang Shukun, Qian Shuangshuang
Section 12	Fuyang Zhuxinzhai Yuanshu Paper Co., Ltd. (location: Yuanshu Papermaking Park in Xin'er Village of Huyuan Town in Fuyang District)
Investigators	Tang Shukun, Zhu Yun, Zhu Zhonghua, Liu Wei, He Ai, Cheng Xi, Pan Qiao, Shen Jiafei, Gui Zixuan, Jiang Shunchao, Qian Shuangshuang
Writers	First manuscript written by: Pan Qiao, Tang Shukun Modified by: Tang Shukun, Shen Jiafei Liu Wei has also contributed to the writing
Section 13	Zhuang Chaojun Paper Mill (Hongxia Calligraphy and Painting Paper Sales Department in Dayuan Town of Fuyang District) (location: Zhuangjia Natural Village of Datong Administrative Village in Dayuan Town of Fuyang District)
Investigators	Zhu Zhonghua, Zhu Yun, Liu Wei, He Ai, Cheng Xi, Wang Zhuxin, Shen Jiafei, Pan Qiao, Jiang Shunchao, Qian Shuangshuang
Writers	First manuscript written by: He Ai Modified by: Tang Shukun, Qian Shuangshuang Wang Zhuxin has also contributed to the writing
Section 14	Hangzhou Shanyuan Cultural and Creative Co., Ltd. (location: Yuanjia Village of Xindeng Town in Fuyang District)
Investigators	Tang Shukun, Zhu Zhonghua, Zhu Yun, Liu Wei, He Ai, Wang Shengrong, Wang Yiqing, Shen Jiafei, Pan Qiao, Jiang Shunchao, Qian Shuangshuang
Writers	First manuscript written by: Wang Shengrong, Zhu Yun Modified by: Zhu Zhonghua, Tang Shukun Tang Shukun, Zhu Yun and Wang Yiqing have also contributed to the writing

Chapter X Bamboo Paper for Sacrificial Purposes in Fuyang District

Section 1	Zhang Xiaoping Paper Mill (location: No.71 Siqian Natural Village of Huangdan Administrative Village in Changlü Town of Fuyang Distict)
Investigators	Zhu Zhonghua, Zhu Yun, Liu Wei, He Ai, Cheng Xi, Wang Zhuxin, Shen Jiafei, Pan Qiao, Jiang Shunchao, Qian Shuangshuang
Writers	First manuscript written by: Wang Zhuxin Modified by: Tang Shukun, Shen Jiafei
Section 2	Jiang Weifa Paper Mill (location: No.21 Sanzhi Natural Village of Sanling Administrative Village in Dayuan Town of Fuyang Distict)
Investigators	Zhu Yun, Zhu Zhonghua, He Ai, Liu Wei, Shen Jiafei, Jiang Shunchao, Qian Shuangshuang, Pan Qiao
Writers	First manuscript written by: Liu Wei Modified by: Tang Shukun, Jiang Shunchao

Section 3	Li Cairong Paper Mill (location: Xinhua Village of Lingqiao Town in Fuyang Distict)
Investigators	Zhu Zhonghua, Zhu Yun, Liu Wei, He Ai, Wang Zhuxin, Shen Jiafei, Pan Qiao, Jiang Shunchao, Qian Shuangshuang
Writers	First manuscript written by: Wang Zhuxin Modified by: Tang Shukun, Pan Qiao
Section 4	Li Shenyan joss Jinqian Paper Mill (location: No.32 Datian Village of Chang'an Town in Fuyang Distict)
Investigators	Tang Shukun, Zhu Yun, Zhu Zhonghua, Liu Wei, He Ai, Cheng Xi, Shen Jiafei, Pan Qiao, Jiang Shunchao, Qian Shuangshuang
Writers	First manuscript written by: He Ai Modified by: Tang Shukun, Shen Jiafei
Section 5	Li Xueyu Ping Paper Mill (location: No.105 Datian Village of Chang'an Town in Fuyang District)
Investigators	Zhu Zhonghua, Tang Shukun, Zhu Yun, Liu Wei, He Ai, Cheng Xi, Shen Jiafei, Pan Qiao, Jiang Shunchao, Qian Shuangshuang
Writers	First manuscript written by: He Ai Modified by: Tang Shukun, Shen Jiafei
Section 6	Jiang Mingsheng Paper Mill (location: Shanji Village of Lingqiao Town in Fuyang District)
Investigators	Zhu Yun, Zhu Zhonghua, Ye Tingting, Yin Hang, Shen Jiafei, Pan Qiao, Jiang Shunchao, Qian Shuangshuang
Writers	Manuscript written by: Ye Tingting Modified by: Tang Shukun, Qian Shuangshuang Jiang Shunchao has also contributed to the writing
Section 7	Qi Wuqiao Paper Mill (location: Dage Village of Yushan Town in Fuyang District)
Investigators	Zhu Zhonghua, Zhu Yun, Liu Wei, He Ai, Cheng Xi, Wang Zhuxin, Shen Jiafei, Gui Zixuan, Jiang Shunchao, Qian Shuangshuang
Writers	First Manuscript written by: Wang Zhuxin Modified by: Tang Shukun, Jiang Shunchao
Section 8	Zhang Genshui Paper Mill (location: Xinsan Village of Huyuan Town in Fuyang District)
Investigators	Zhu Zhonghua, Zhu Yun, He Ai, Ye Tingting, Shen Jiafei, Gui Zixuan, Jiang Shunchao, Qian Shuangshuang
Writers	First manuscript written by: Ye Tingting Modified by: Tang Shukun, Gui Zixuan
Section 9	Zhu Nanshu Paper Mill (location: Shanji Village of Lingqiao Town in Fuyang District)
Investigators	Zhu Yun, Zhu Zhonghua, Ye Tingting, Yin Hang, Shen Jiafei, Gui Zixuan, Jiang Shunchao, Qian Shuangshuang
Writers	First manuscript written by: Ye Tingting Modified by: Tang Shukun, Jiang Shunchao

Chapter XI Bast Paper in Fuyang District

Section 1	Taohua Paper Mill in Wusi Village (location: Wusi Village of Lushan Residential District in Fuyang District)
Investigators	Fang Renying, Chen Biao, Li Shaojun, Zhu Zhonghua, Tang Shukun
Writers	First manuscript written by Fang Renying Modified by: Tang Shukun Chen Biao has also contributed to the writing
Section 2	Mulberry Paper Recovery Site in Dashan Village (location: Dashan Village of Xindeng Town in Fuyang District)
Investigators	Fang Renying, Li Shaojun, Zhu Zhonghua
Writers	First manuscript written by: Fang Renying Modified by: Tang Shukun, Li Shaojun

Chapter XII Tools

Section 1	Yongqing Screen-making Mill (location: Yongqing Village of Dayuan Town in Fuyang District)
Investigators	Zhu Zhonghua, Zhu Yun, He Ai, Ye Tingting, Yin Hang, Shen Jiafei, Gui Zixuan, Jiang Shunchao, Qian Shuangshuang
Writers	First Manuscript written by: Yin Hang Modified by: Tang Shukun, Qian Shuangshuang
Section 2	Guangming Screen-making Factory (location: Guangming Village of Lingqiao Town in Fuyang District)
Investigators	Zhu Yun, Liu Wei, Zhu Zhonghua, He Ai, Cheng Xi, Shen Jiafei, Gui Zixuan, Jiang Shunchao, Qian Shuangshuang
Writers	First manuscript written by: Liu Wei Modified by: Tang Shukun, Shen Jiafei
Section 3	Lang Shixun Scraping Knife-making Mill (location: 2nd lane of Chaoyangnan Rd. in Dayuan Town of Fuyang District)

Investigators	Zhu Zhonghua, Zhu Yun, Wang Shengrong, Ye Tingting, Yin Hang, Wang Yiqing, Shen Jiafei, Gui Zixuan, Jiang Shunchao, Qian Shuangshuang
Writers	First manuscript written by: Zhu Yun Modified by: Tang Shukun

2. Technical Analysis and Other Related Works

Paper sample test and analysis	Headed by: Zhu Yun, Chen Yan Members: Zhu Yun, Chen Yan, Wang Shengrong, Liu Wei, He Ai, Wang Zhuxin, Wang Yiqing, Yao Dilu, Ye Zhenzhen, Yin Hang, Sun Yan, Liao Yingwen, Guo Yanlong
Distribution maps of handmade paper	Drawn by: Guo Yanlong
Fiber pictures and those showing through the light	Made by: Zhu Yun, Wang Shengrong, Liu Wei, He Ai, Wang Zhuxin, Wang Yiqing, Yao Dilu, Liao Yingwen, Chen Yan, Guo Yanlong
Paper sample pictures	Photographed by: Huang Xiaofei
Paper samples	Sorted by: Zhu Yun, Tang Shukun, Liu Wei, He Ai, Wang Zhuxin, Wang Shengrong, Ni Yingying, Shen Jiafei, Zheng Bin, Fu Chengyun, Cai Tingting, Pan Qiao, Wang Yiqing, Yao Dilu, Yin Hang, Chen Xinran, Liao Yingwen, Kong Lijun, Guo Yanlong, Ye Zhenzhen
Appendices and references	Arranged by: Tang Shukun, Zhu Yun, Shen Jiafei

3. Preface, Introduction to the Writing Norms, Appendices and Epilogue

	Preface
Writer	Tang Shukun
	Introduction to the Writing Norms
Writers	Tang Shukun, Zhu Yun
	Appendices
Terminology	Sorted by: Zhu Yun, Shen Jiafei, Ni Yingying, Tang Yujing, Cai Tingting
	Epilogue
Writer	Tang Shukun

4.Modification and Translation

Director of modification and verification	Tang Shukun, Zhu Zhonghua
Planners of modification	Zhu Yun, Shen Jiafei
Chief translator	Fang Yuanyuan
Other members contributed to the modification efforts	Chen Jingyu, Lin Zhiwen, Li Shaojun, Xu Jianhua, Pan Qiao, Gui Zixuan, Jiang Shunchao, Qian Shuangshuang

在历时多个月的集中修订、增补与统稿工作中，汤书昆、朱赟、朱中华、沈佳斐、方媛媛、陈敬宇、郭延龙等作为主持人或重要模块的负责人，在文稿内容、图片与示意图的修订增补，代表性纸样的测试分析，英文翻译，文献注释考订，数据与表述的准确性核实等方面做了大量力求精益求精的工作。另一方面，从2019年8月开始，责任编辑团队、北京敬人工作室设计团队、北京雅昌艺术印刷有限公司印制团队接手书稿后不辞辛劳地反复打磨，使《浙江卷》一天天变得规范和美丽起来。从最初的对田野记录进行提炼整理，到能以今天的面貌和品质问世，上述团队全心全意的工作是不容忽视的基础。

在《浙江卷》的田野调查过程中，先后得到富阳朱家门村竹纸文物收藏与研究者朱有善先生、温州瓯海区非物质文化遗产保护中心潘新新先生、绍兴平水镇传统工艺民宿创办人宋汉校先生、富阳历史名纸桃花纸造纸老师傅叶汉山先生、富阳稠溪村乌金纸造纸老师傅郑吉申先生、富阳大同村元书纸年轻造纸师傅朱起杨先生等多位浙江手工造纸传统技艺和非物质文化遗产研究与保护专家的帮助与指导，在《中国手工纸文库·浙江卷》正式出版之际，我谨代表田野调查和文稿撰写团队，向所有这项工作进程中的支持者与指导者表达真诚的感谢！

汤书昆

于中国科学技术大学

2019年10月

Tang Shukun, Zhu Yun, Zhu Zhonghua, Shen Jiafei, Fang Yuanyuan, Chen Jingyu and Guo Yanlong, et al., who were in charge of the writing, modification and other related works, all contributed their efforts to the completion of this book. Their meticulous efforts in writing, drawing or photographing, mapping, technical analysing, translating, format modifying, noting and proofreading should be recognized and eulogized in the achievement of the high-quality work. Since August 2019, the editors of the book, Beijing Jingren Book Design Studio, Bejing Artron Printing Service Co., Ltd. have been dedicated to the polishing and publication of the book, whose efforts enable a field investigation-based research to be presented in a stylish and quality way.

Many experts from the field of handmade paper production and intangible cultural heritage research and protection have helped in our investigations: Zhu Youshan, a collector and researcher of bamboo paper from Zhu Jiamen Village in Fuyang District; Pan Xinxin from Intangible Cultural Heritage Protection Centre in Ouhai District of Wenzhou City; Song Hanxiao, Traditional Handicraft Homestay Program initiator from Pingshui Town of Shaoxing city; Ye Hanshan, Taohua paper (historically famous paper) maker from Fuyang District; Zheng Jishen, Wujin papermaker from Chouxi Village of Fuyang District; and Zhu Qiyang, a young papermaker of Yuanshu paper from Datong Village in Fuyang District, et al. On the verge of publication, sincere gratitude should go to all those who have supported and recognized our efforts!

Tang Shukun

University of Science and Technology of China

October 2019